Accretion Tectonics in the Circum-Pacific Regions

Advances in Earth and Planetary Sciences

Accretion Tectonics in the Circum-Pacific Regions

Proceedings of the Oji International Seminar
on Accretion Tectonics
September, 1981, Tomakomai, Japan

Edited by

M. Hashimoto
Department of Earth Sciences, Ibaraki University, Japan

and

S. Uyeda
Earthquake Research Institute, University of Tokyo, Japan

Terra Scientific Publishing Company/Tokyo
D. Reidel Publishing Company / Dordrecht, Boston, London

Library of Congress Cataloging in Publication Data

Oji International Seminar on Accretion Tectonics (1981 :
    Tomakomai-shi, Japan)
    Accretion tectonics in the circum-Pacific regions

    (Advances in earth and planetary sciences)
    Bibliography: p.
    Includes index.
    1. Geology, Structural—Congresses.   2. Geology—Pacific Area—Congresses.
I. Hashimoto, Mitsuo, 1925-   . II. Uyeda, Seiya, 1929-   . III. Title. IV. Series.
QE601.048   1981   551.1′36   83-8651
ISBN-13:978-94-009-7104-2   e-ISBN-13:978-94-009-7102-8
DOI: 10.1007/978-94-009-7102-8

Published by Terra Scientific Publishing Company (TERRAPUB),
307 Shibuyadai-haim, 4-17 Sakuragaoka-cho, Shibuya-ku, Tokyo 150, Japan,
in co-publication with D. Reidel Publishing Company, Dordrecht, Holland

Sold and distributed in the U.S.A. and Canada
by Kluwer Boston Inc.,
190 Old Derby Street, Hingham, MA 02043, U.S.A.,
in Japan by Terra Scientific Publishing Company (TERRAPUB),
307 Shibuyadai-haim, 4-17 Sakuragaoka-cho, Shibuya-ku, Tokyo 150, Japan

In all other countries, sold and distributed
by Kluwer Academic Publishers Group,
P. O. Box 322, 3300 AH Dordrecht, Holland

D. Reidel Publishing Company is a member of the Kluwer Academic Publishers Group

Preface

Accretion-collision tectonics in mobile belts is one of the most important new topics in solid earth science. A special seminar on this subject, the Oji International Seminar on Accretion Tectonics, was held from the 10th to 16th September, 1981, in Tomakomai, Hokkaido, Japan. It was sponsored by the Japan Society for the Promotion of Science and the Fujihara Foundation of Science, organizers were S. Uyeda, Earthquake Research Institute, University of Tokyo, Amos Nur, Department of Geophysics, Stanford University, and M. Hashimoto, Department of Geology, National Science Museum, Tokyo. More than fifty geoscientists, thirty from Japan and twenty from abroad, met together to present findings from their recent studies and exchange ideas about accretion-collision phenomena in the circum-Pacific mobile belts. Two field days were also spent in the Horokanai area, central Hokkaido, to examine the Kamuikotan high-pressure metamorphic rocks and the Horokanai ophiolite. The latter is a fragment of ancient ocean floor crust that has been obducted onto the Kamuikotan rocks at the time of collision of the Okhotsk Micro-continent with Asia. The seminar was by no means a large congress. However, to the best of our knowledge, it was the first international conference on accretion-collision tectonics. The meeting was highly successful and we believe that it has opened an important new era in the study of plate tectonics.

This volume constitutes the proceedings of the seminar. Although a few of the papers presented at the seminar have been published elsewhere (listed below*), twenty one papers that were read at the meeting are included along with four additional papers.

We wish to express our sincere gratitude to the Japan Society for the Promotion of Science and the Fujihara Foundation of Science for their sponsorship, and to the staff of the Hotel New Oji for their kind hospitality during the seminar. For the very fine arrangement and management of the meeting, primarily by Dr. Y. Saito, Department of Geology, National Science Museum and Mrs. F. Sato, we are greatly indebted. Our appreciation is extended to all the contributors, and to Mr. K. Oshida for his great help in assembling this volume.

Mitsuo HASHIMOTO
Department of Earth Science, Ibaraki University,
Mito 310, Japan

Seiya UYEDA
Earthquake Research Institute, University of Tokyo,
Tokyo 113, Japan

---

*List of papers presented at Oji Seminar and published elsewhere (continued to next page).

BARBER, A. J., Interpretations of the tectonic evolution of Southwest Japan, *Proc. Geol. Assoc.*, **93** (2), 131–145, 1982.

BEN-AVRAHAM, Z. and S. UYEDA, Entrapment origin of marginal seas, AGU/GSA Geodynamics Series, 1982 (in press).

CHURKIN, M. JR., H. L. FOSTER, R. M. CHAPMAN, and F. R. WEBER, Terranes and suture zones in east central Alaska, *J. Geophys. Res.*, **87**, 3718–3730, 1982.

HILDE, T. W. C. and S. UYEDA, Trench depth: variation and significance, AGU/GSA Geodynamics Series, 1982 (in press).

LEGGETT, J. K., W. S. MCKERROW, and D. W. CASEY, Anatomy of a Lower Palaeozoic accretionary fore-arc; the Southern Uplands of Scotland, *Trench-Fore-Arc Geology.*, Spec. Publ. Geol. Soc. London, edited by J. K. Leggett, pp. 495–520, 1982.

MONGER, J. W. H., R. A. PRICE, and D. J. TEMPELMAN-KLUIT, Tectonic accretion and the origin of the two major metamorphic and plutonic welts in the Canadian Cordillera, *Geology*, **10**, 70–75, 1982.

NAKAMURA, K. and K. SHIMAZAKI, Sagami and Suruga troughs and subduction of the Philippine Sea plate, *Kagaku*, **51**, 490–497, 1981 (in Japanese).

SAITO, Y. and M. HASHIMOTO, South Kitakami region: an allochtonous terrane in Japan, *J. Geophys. Res.*, **87**, 3691–3696, 1982.

SASAJIMA, S., Magnetic properties of red cherts with special references to the associated greenstones, Southwest Japan: A Rock Magnetic Approach, *Mem. Fac. Sci., Kyoto Univ. Ser. Geol. Mineral.*, **48**, 43–61, 1982.

TAIRA, T., Y. SAITO, and M. HASHIMOTO, The role of oblique subduction and strike-slip tectonics in the evolution of Japan, AGU/GSA Geodynamics Series, 1982 (in press).

TORIUMI, M., Strain, stress and uplift, *Tectonics*, **1**, 57–72, 1982.

Contents

Preface . . . . M. HASHIMOTO and S. UYEDA   v

GENERAL

Break-up and Accretion Tectonics . . . . A. NUR and Z. BEN-AVRAHAM   3

NORTH AMERICA TO NORTHEAST ASIA

Recognition, Character, and Analysis of Tectonostratigraphic Terranes in
Western North America . . . .
    D. L. JONES, D. G. HOWELL, P. T. CONEY, and J. W. H. MONGER   21
Tectonostratigraphic Terranes of Alaska and Northeastern USSR—A Record
of Collision and Accretion . . . . M. CHURKIN, Jr.   37
Accretionary Terranes and Tectonic Evolution of Northeast Siberia . . . . .
. . . . : . . . . K. FUJITA and J. T. NEWBERRY   43
Accretional and Collisional Eugeosynclinal Folded Systems of the
Northwestern Pacific Rim . . . . B. A. NATAL'IN and L. M. PARFENOV   59
Space-Time Distribution of Late Mesozoic to Early Cenozoic Magmatism
in East Asia and Its Tectonic Implications . . . . M. TAKAHASHI   69

HOKKAIDO, NORTH JAPAN

Collision Orogenesis and Sedimentation in Hokkaido, Japan . H. OKADA   91
Mesozoic Arc-Trench Systems in Hokkaido, Japan . . . .
. . . . K. KIMINAMI and Y. KONTANI   107
Collision Tectonics in Hokkaido and Sakhalin . . . .
. . . . G. KIMURA, S. MIYASHITA, and S. MIYASAKA   123
From Subduction to Paleosubductions in Northern Japan . . . .
. . . . J.-P. CADET and J. CHARVET   135
Disclosing of a Deepest Section of Continental-Type Crust Up-Thrust as the
Final Event of Collision of Arcs in Hokkaido, North Japan . . . .
M. KOMATSU, S. MIYASHITA, J. MAEDA, Y. OSANAI, and T. TOYOSHIMA   149

SOUTHWEST JAPAN

Hida and Mino: Tectonostratigraphic Terranes in Central Japan . . . .
. . . . S. MIZUTANI and I. HATTORI   169
Accretion Tectonics Inferred from Paleomagnetic Measurements of Paleozoic
and Mesozoic Rocks in Central Japan . . . . K. HIROOKA,
T. NAKAJIMA, H. SAKAI, T. DATE, K. NITTAMACHI, and I. HATTORI   179
Accreted Oceanic Reef Complex in Southwest Japan . . . .
. . . . K. KANMERA and H. NISHI   195
Tectonic Environments and Crustal Section of the Outer Zone of Southwest
Japan . . . . S. HADA and T. SUZUKI   207

Accretionary Melange of Cretaceous Age in the Shimanto Belt in Japan
.... T. Suzuki and S. Hada   219
Paleomagnetism of the Shimanto Belt in Shikoku, Southwest Japan ....
.... K. Kodama, A. Taira, M. Okamura, and Y. Saito   231

JAPAN CENOZOIC OPHIOLITE

Mineoka Ophiolite Belt in the Izu Forearc Area—Neogene Accretion of
    Oceanic and Island Arc Assemblages on the Northeastern Corner of the
    Philippine Sea Plate .... Y. Ogawa   245
Paleomagnetic and Geotectonic Investigation of Ophiolite Suites and Sur-
    rounding Rocks in South-Central Honshu, Japan ....
.... S. Tonouchi and K. Kobayashi   261

SOUTHWEST TO SOUTH PACIFIC

A Possible Mechanism of Episodic Spreading of the Philippine Sea .....
.... S. Uyeda and R. McCabe   291
Tertiary Accretion of Ophiolite Seamounts, North Island, New Zealand .
.... R. N. Brothers   307
Roles of Seamount, Rise, and Ridge in Lithospheric Subduction ....
.... Y. Tomoda and H. Fujimoto   319

GEOPHYSICS AND METALLOGENESIS

Forces Acting on the Subducted Lithosphere Revealed by the Geometry of
    Deep-Seismic Zones .... Y. Ida   335
Overthrust and Underthrust as a Fundamental Process of Continental
    Accretion .... W. Zhang and J. Zhong   345
Accretion Tectonics and Metallogenesis .... C. Nishiwaki and S. Uyeda   349

Index   357

General

Accretion Tectonics in the Circum-Pacific Regions, edited by M. Hashimoto and S. Uyeda, 3–18.

Break-up and Accretion Tectonics

Amos Nur and Zvi Ben-Avraham

Department of Geophysics, Stanford University,
Stanford, California 94305, U.S.A.

Numerous oceanic rises, several of which are submerged continental fragments, are embedded in the oceanic plates and are fated to be consumed at active margins in the future. This consumption causes gaps in active volcanic chains, disorders the normal seismic pattern associated with subduction of oceanic crusts, and emplaces ophiolites. Some past oceanic rises have become accreted terranes, now found in ancient active margins. We suggest that some of these accreted terranes resulted from the break-up of Gondwana. From Permian fragments in southern Europe to present Arabia and Somalia, a host of continental slivers, microcontinents, and related rises have migrated from Gondwana to be accreted into Europe and Asia. Orogenic deformation has resulted, well before full continent-continent collision took place.

Numerous accreted continental and noncontinental rises, have been identified also in the northern circum Pacific also involving extensive orogenic deformation. A possible source for these continental terranes is Pacifica, an extension of Gondwana beyond New Zealand and Australia. Just as fragments of northern Gondwana and Africa are found today in the Alpine-Himalaya Chain, so are fragments of eastern Gondwana—possibly Pacifica—found in the northern and western circum Pacific.

1. Introduction

Although the tectonic complexities in some large mountain chains of the world appear to be the results of major plate collisions, other chains are related instead to numerous juxtaposed slivers, often termed "allochthonous terranes," which have arrived from distant origins. For example, large parts of western North America are composed of such allochthonous terranes.

Modern analogues of many allochthonous terranes may be the plateaus which comprise about 10% of the floor of today's oceans (Fig. 1). These plateaus, embedded in and migrating with their oceanic plates, are fated to accrete to continents at subduction zones. This accretion involves volcanic gaps at active margins, irregular seismicity patterns, the emplacement of ophiolites, orogenic deformation, continental growth by accretion, and the entrapment of marginal basins.

Some oceanic plateaus (Fig. 2) rise thousands of meters from the sea floor to above sea level (e.g. the Seychelles Bank), or to 2,000 m below sea level (e.g. the Ontong Java Plateau). Many plateaus have crustal thicknesses between 20 km and 40 km (Fig. 2) with an upper 10–15 km crust of compressional wave velocities of 6.0–6.3 km/sec. These values are typical of granitic rocks in the continental crust. The Ontong Java Plateau and the Seychelles Bank could thus be submerged continental fragments.

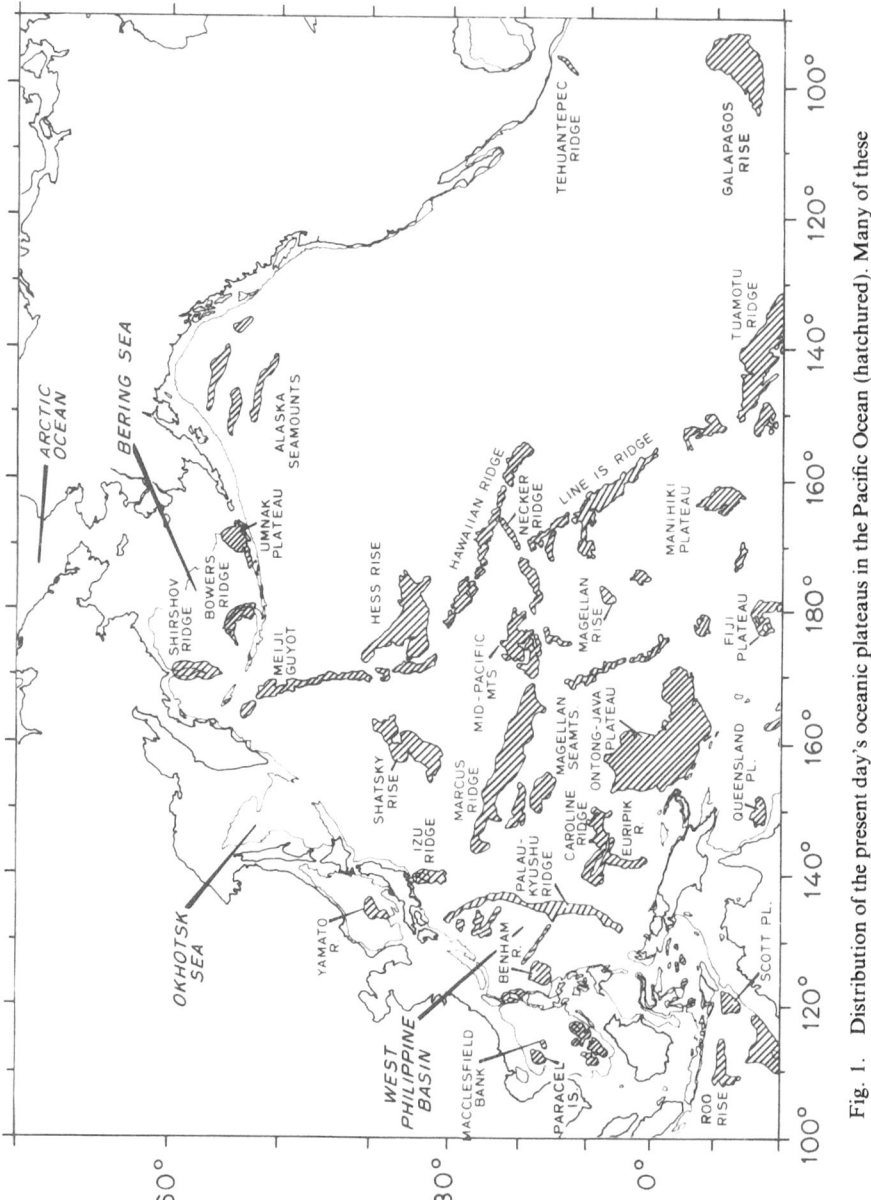

Fig. 1. Distribution of the present day's oceanic plateaus in the Pacific Ocean (hatchured). Many of these plateaus may be continental fragments in spite of their situation well within oceanic plates. Others are of different origins, such as extinct arcs and hot spot traces (after Ben-Avraham *et al.*, 1981).

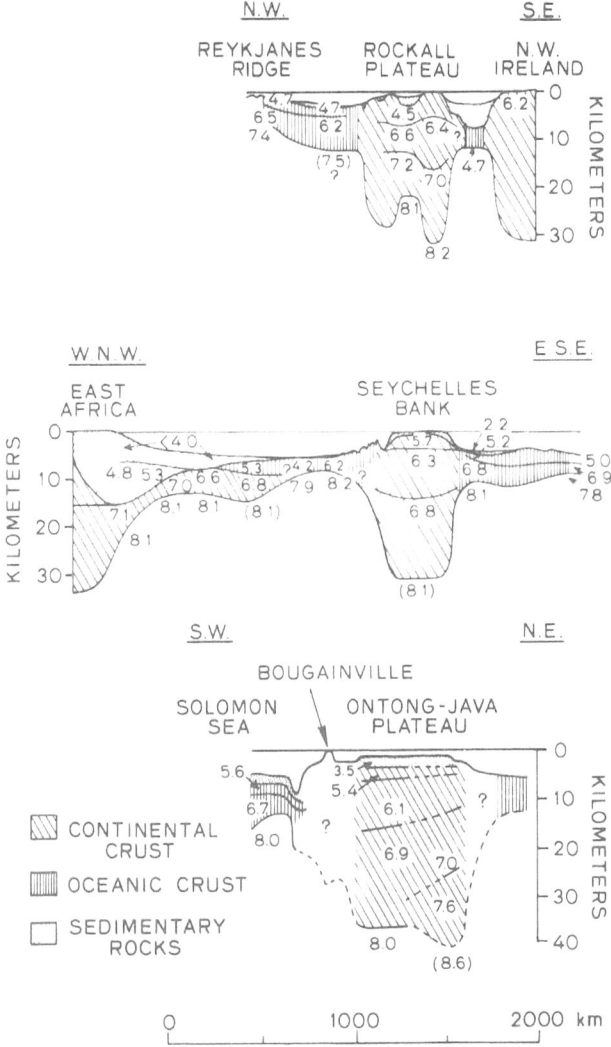

Fig. 2. Cross sections of Rockall Plateau in the Atlantic Ocean, Seychelles Bank in the Indian Ocean (both after SCRUTTON, 1972), and Ontong Java Plateau in the Pacific Ocean. The numbers are compressional wave velocities in km/sec. All three plateaus may be composed in part of continental crust (after BEN-AVRAHAM et al., 1981).

Continental slivers must have originated from continental edges, transported either along strike slip faults (e.g. Baja California), or as separated fragments involving complex rifting and ridge jumping such as for the Seychelles or Madagascar (McKENZIE and SCLATER, 1973).

Several plateaus are presently being consumed at subduction zones, causing reduction of seismicity and gaps in volcanic activity (NUR and BEN-AVRAHAM, 1981), such as associated with collisions of the Juan Fernandez and Nazca Ridges with the western margins of South America. The collisions have (Fig. 3), resulted in a remarkable variety of combinations of seismicity, volcanism, and morphology (Vogt et al., 1976; Barazangi and Isacks, 1976, 1979; Isacks and Barazangi, 1977). The Marcus Necker Ridge and the Magellan Seamounts in the western Pacific, now colliding with their respective trenches (Fig. 3), show similar patterns.

Disrupted slabs are associated with gaps in deep seismic activity (Fig. 4): These deep seismic gaps, possibly the remnant effect of accreted rises, have been observed not only in South America, but also in other complex zones such as in the Hellenic Arc (Rotstein and Kefka, 1981).

Plateaus arriving at active margins may create marginal seas by entrapment. A hypothetical future example—the arrival of the Emperor Seamount chain, the Shatski rise and finally the Hess rise at east Asia—is shown in Fig. 5. Possible past examples include the entrapment of the *Arctic Ocean Basin* (Churkin and Trexler, 1980), the *Bering Sea* (Cooper et al., 1976), by the encroachment of the Umnak Plateau (Ben-Avraham and Cooper, 1981), or the *Sea of Okhotsk*, probably a micro-continental block of unknown origin (Den and Hotta, 1973; Dickinson, 1978) which collided with the eastern margin of Eurasia during early Cenozoic time.

The most extensive record of consumption however is that of accretion in the two Mesozoic-Cenozoic mountain belts on earth: the Alpine-Himalaya belt and the circum Pacific belt.

2. Accreted Fragments Around the Pacific Rim

The record of the consumption of former rises or plateaus which have migrated with the Pacific Ocean plates comes from the northern rim of the Pacific Ocean in Mexico, western North America, Alaska, Siberia, and Japan. Data for several terranes in Alaska and northeast Asia show migrations of several thousand kilometers over periods of tens of millions of years (Hillhouse, 1977; Stone, 1977). Paleomagnetic azimuths and declinations are often anomalous, suggesting that many terranes have also undergone substantial rotations (Cox, 1980).

Some of the allochthonous terranes of western North America such as the Cache Creek Terrane (Monger, 1977), are clear candidates for ancient oceanic plateaus now incorporated into the continental framework. The presence of Tethyan Permian fusulinids led to the early recognition of this terrane as allochthonous (Monger and Ross, 1971). Some allochthonous terranes were separate continental slivers at one time (Coney et al., 1980) such as the Nixon Fork, Ruby and Klamath Mountains. In contrast, the Wrangellia terrane is characterized by an enormous accumulation of Middle to Upper Triassic *subaerial* basalts over an upper Paleozoic volcanic arc assemblage (Jones et al., 1977). These subaerial basalts probably represent rifting of Wrangellia at southern paleolatitudes (Hillhouse, 1977).

In Central America, Gose and Swartz (1977) have suggested that a Honduras Continental terrane was situated far in the Pacific Ocean during Cretaceous times. The Chiapas Massif and the Yucatan platform are also of unknown origins. An accreted terrane or terranes, similar to Salinia, can be identified also in northwestern Mexico (C.

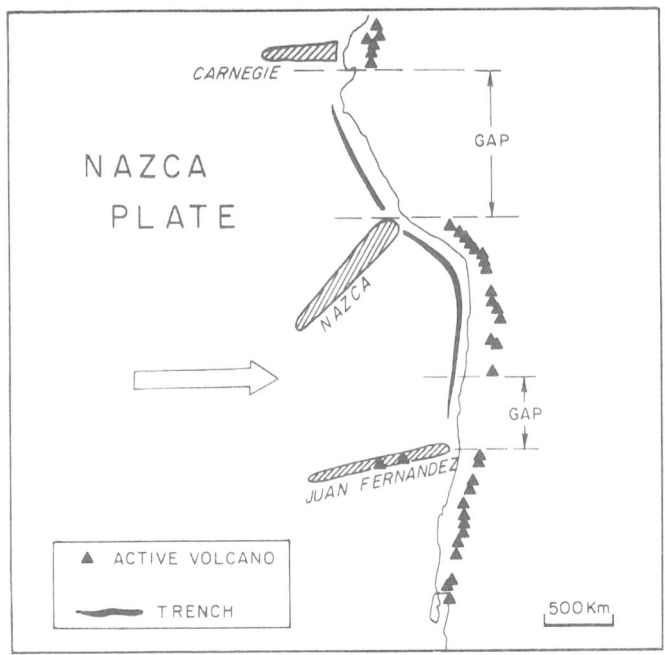

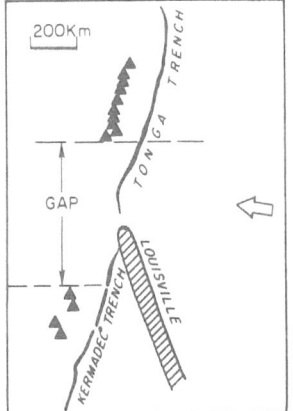
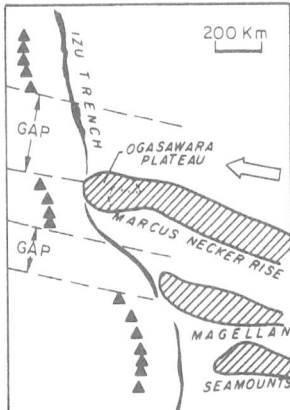

Fig. 3. The oblique consumption of the Nazca and Juan Fernandez Ridges in the eastern South Pacific and the Marcus-Necker and Louisville Rises in the Western Pacific is associated with volcanic gaps, possibly due to disruption of the subduction process. These gaps are temporary trailing behind the moving contract between the ridge and the over-riding plate (NUR and BEN-AVRAHAM, 1981).

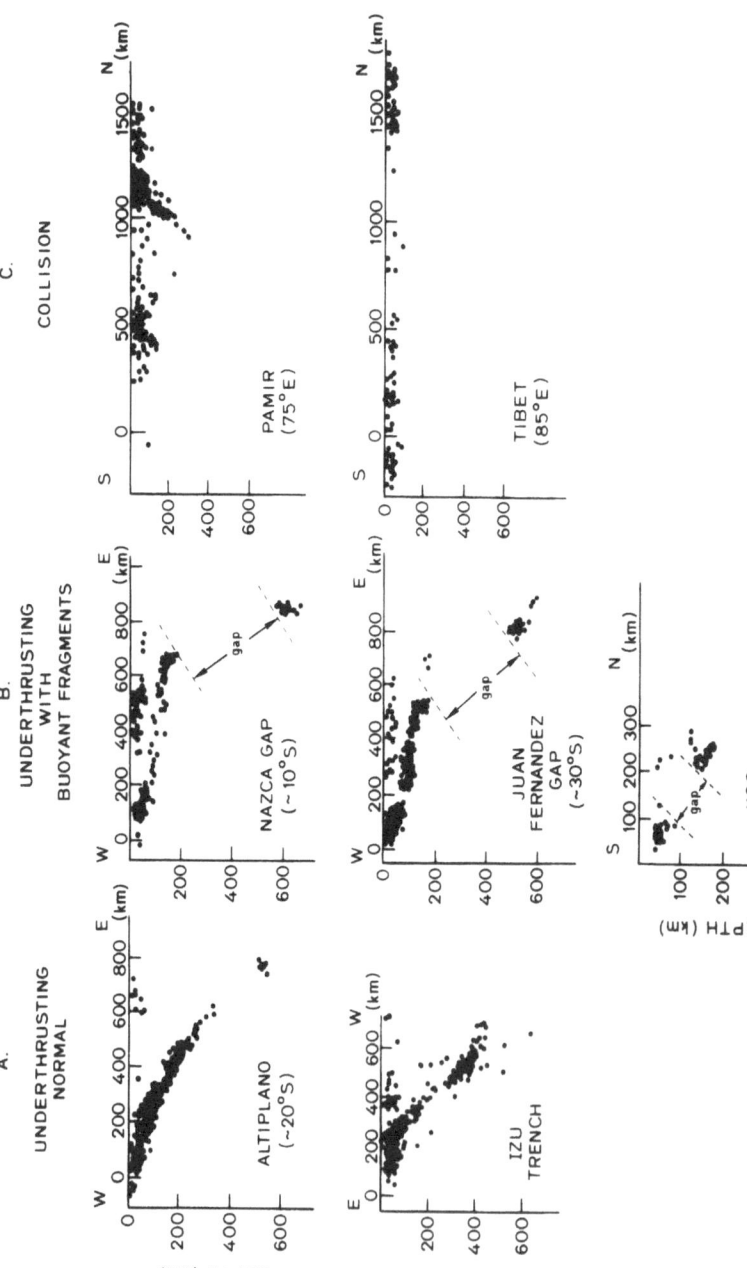

Fig. 4.   Comparison of seismicity patterns in subduction vs. collision tectonics: (A) Subduction related seismicity clearly defines a continuous and coherent slab which contains shallow and deep earthquake, with systematic fault plane solutions. (b) Seismicity in "mixed" cases, where normal subduction is disrupted by a ridge, such as Nazca or Juan Fernandez, or a continental fragment, such as in the Hellenic arc, near the Island of Kos (ROTSTEIN and KAFKA, 1981). (C) Collision related earthquakes are mostly shallow, show mixed fault plane solutions, are widely scattered and typically define relatively short faults.

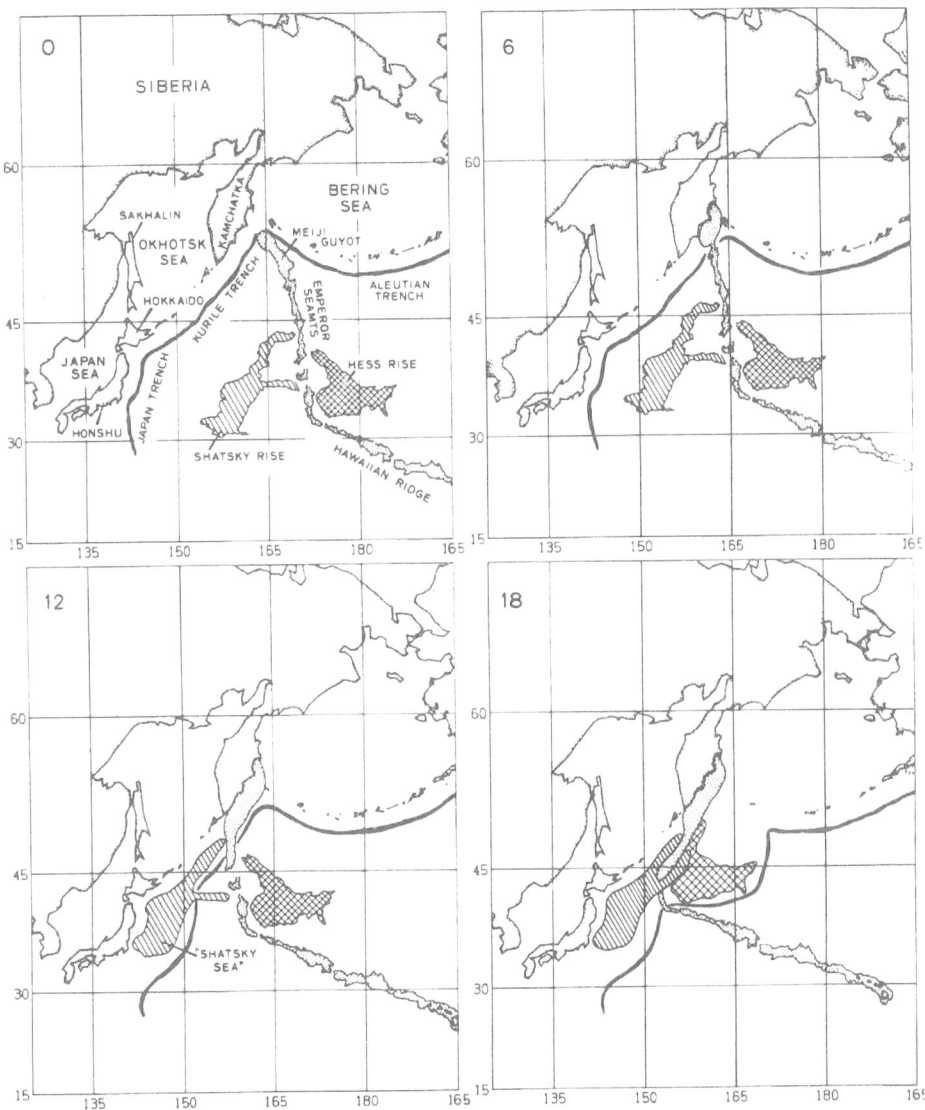

Fig. 5. Sketches of possible future events in the northwest Pacific, based on present-day plate motion parameters. (A) Present-day configuration of Shatsky Rise, Hess Rise, Emperor Seamounts, and Hawaiian Ridge. (B) In 6 million years, all the plateaus will have moved to the northwest, and the Meiji Guyot, after colliding with the subduction zone, will become part of the Kamchatka margin. (C) In 12 million years, the Shatsky Rise will collide with north Honshu, Hokkaido, and the Kuriles. At this stage the trench might move to the oceanic side of the plateau and a new marginal sea, the "Shatsky Sea," could form. (D) In 18 million years, the Shatsky Rise, Hess Rise, and Emperor Seamounts will be part of the Eurasia plate, and new plate boundaries will form in this region.

Rangin, personal communication, 1980).

In northeast Siberia, the Kolyma composite micro-continent (CHURKIN and TREXLER, 1980; FUJITA, 1978) probably arrived with the Kula Plate as indicated by paleomagnetic data (MCELHINNY, 1973). Other microcontinents have also been involved in a series of accretion-type collisions through the end of the Mesozoic (MARKOV et al., 1980).

Paleomagnetic studies show that southern parts of Japan (HATTORI and HIROOKA, 1979) rifted away in the late Paleozoic (HARATA et al., 1978) from an unknown continental source at equatorial latitudes, possibly Pacifica (NUR and BEN-AVRAHAM, 1977). Collisions have been suggested also for the older Maizuru and Tamba belts in Japan, in late Paleozoic or early Mesozoic (SHIMIZU et al., 1978), the younger Miura Belt (OGAWA and HORIUCHI, 1978) and the Quaternary collision of the Izu Peninsula with central Honshu (MATSUDA, 1978; BEN-AVRAHAM and COOPER, 1981).

Late Paleozoic Australian fauna in the Kitakami Mountains, Japan (SAITO and HASHIMOTO, 1982) and the ancient suture zone between west and east Hokkaido (S. Uyeda, personal communication) also suggest collisions. The inferred Oyashio land mass off Honshu (VON HUENE et al., 1980) which served as a sediment source in late cretaceous times has probably moved north with the Kula Plate leaving behind small continental fragments on the Japanese coast.

The Sikhote Alin Complex (MCELHINNY, 1973; HAMILTON, 1970), parts of Korea (KAWAI et al., 1969) and the northern and southern China platforms (e.g. TERMAN, 1973; MCELHINNY, 1973; MCELHINNY et al., 1981) appear also to have accreted from the south or southeast, possibly as fragments of a Pacifica land mass (NUR and BEN-AVRAHAM, 1978).

Micro-continents in southeast Asia include Sundaland, (POWELL and JOHNSON, 1980) the Reed Bank west of Palawan, northern Palawan, Calamian Islands, Cuyo Shelf, the Paracel Islands and Macclesfield Bank (LUDWIG et al., 1979; TAYLOR and HAYES, 1980), the Kontum Massif in eastern Vietnam, Banggai-Sula and the Buton near Celebes (HAMILTON, 1979), the Halmahera fragments, the Seram and Buru regions where Precambrian rocks are exposed, and the Sumba Island continental fragment (HAMILTON, 1979).

The late Paleozoic and Mesozoic tectonic evolution of New Zealand also involves collisions such as with the Torlesse terrane (HOWELL, 1980; KAMP, 1980; and PIRAJNO, 1980), which has been associated with a large unknown continent, possibly Pacifica (KAMP, 1980). This continent could have served as a source for the sediments for the Canterbury Suite.

Many other Precambrian and Paleozoic continental fragments of unknown origins, in part unexposed, probably exist in the circum Pacific margins (e.g. HATTORI and HIROOKA, 1979). The collisions and accretion of all these fragments have probably played a crucial role in the orogenic evolution of these margins.

3. Accreted Fragments in the Alpine-Himalaya Chain

Accreted terranes play a leading role also in the Alpine-Himalaya Chain (Fig. 6). Included are the Calabria, Adria, and Toscana microplates (e.g. KRUEZER et al., 1979),

Fig. 6. Sketch of allochthonous terranes, most of which are fragments of continents, or have clear continental affinities. Shaded area indicates the Mesozoic-Cenozoic orogenic belts. Allochthonous terranes within the belt are most likely parts of Gondwana, which have migrated towards and collided with Europe, S. E. Asia, and the Pacific margins. Oceanic plateaus, some of which are shown, are moving towards consuming boundaries, possibly to become accreted allochthonous terranes. Several fragments of Africa (e.g. Madagascar, Somalia) will probably collide with the Euroasian Plate in the future (after NUR and BEN-AVRAHAM, 1982).

now part of Italy; the Paikon, Palegonian, and Gavrono platforms (e.g. BIJU-DUVAL *et al.*, 1977; GIESE *et al.*, 1979) and the Apulian, Moesian, and Rhodopian fragments (BURCHFIELD, 1980) which may be responsible for the Carpathians and Balkan orogenic belts; the central Anatolian Massif with the Sakaraya micro-continent (SENGOR and YILMAZ, 1981); the transcaucasus median mass (KHAIN and SESLAVINSKY, 1979); the Rezaiyd micro-continent of northern Iran (e.g., KING, 1973); and the Lut Block in central Iran (e.g., GEALY, 1977). Many Gondwana fragments which were accreted over a wide span of time have been recognized further east (BOULIN, 1981) such as the Herat and Panjchir zones in Afghanistan (MATTAUER *et al.*, 1980), the Helmand block (GANSSER, 1980a, b), the Pamir block (GEALY, 1977), the accreted blocks in the Altyn Tagh and Kun Lun sutures in Tibet (e.g., BURKE *et al.*, 1980; GANSSER, 1980a, b) including the Tsaidam block, Chang Thang, north and south Tibet platforms (CHANG and CHENG, 1973). Sundaland (e.g., POWELL, 1979), containing several massifs, e.g., the Kantum Massif in Vietnam, was probably accreted in Cretaceous times (KLOMPÉ, 1957). Finally the South China platform (TERMAN, 1973) and Sikhote Alin originated also as separate terranes. Sikhote Alin has migrated approximately 30 degrees northward since Permian time (MCELHINNY, 1973). Very recent paleomagnetic results (MCELHINNY *et al.*, 1981) suggest that fragments in east China originated in southern latitudes as well, possibly as parts of a Pacifica mass.

Many of these continental fragments are thought to have rifted away from Gondwana, moved across the proto-Tethys and accreted into northern continents, causing extensive deformation and thrusting without actual collisions between two full-sized continents. Indeed, it is questionable whether a full collision between Africa and Europe, presumably responsible for the development of the European Alps, actually occurred. Recent seismic refraction results in the eastern Mediterranean (BEN-AVRAHAM *et al.*, 1979) reveal that the crust is oceanic, clearly separating the central Qnatolian Massif from Africa where it is assumed to have originated. At present, the Arabia sub-plate, now being accreted by collision to Asia, is separated from Africa by oceanic crust in the spreading Red Sea. We suggest that much of the orogenic complexity in the Alpine Chain is due to the consumption of the separate fragments and not necessarily due to massive continent-continent collision.

As shown in Fig. 6, the accretion of terranes is a very general aspect of these orogenic belts and, by inference, older ones as well.

4. The Emplacement of Ophiolites

The accretion of microplates and continental fragments may provide an important mechanism for ophiolite emplacement. This process has been recognized in the Alpine system where emplacement can often be associated with the accretion of relatively small microplates (BURCHFIELD, 1980) well before full continental collision. Ophiolites may, therefore, represent the arrival of fragments at subduction zones rather than the closing of marginal basins or large ocean basins (CHURCH, 1980; BEN-AVRAHAM *et al.*, 1982).

Examples include Cyprus, where new seismic results (BEN-AVRAHAM *et al.*, 1979) reveal oceanic crust in the eastern Mediterranean south of Cyprus, thus ruling out full continental collision as the mechanism for the emplacement of the Troodos ophiolite complex (COLEMAN, 1977; GASS and MASSON-SMITH, 1963; MOORES and VINE, 1971; VINE et al., 1973). Instead, the collision of Cyprus with the prominent Eratosthenes Seamount, south of Cyprus, may be responsible (BEN-AVRAHAM et al., 1982).

The ophiolites in Hispaniola and Cuba could have been emplaced via the collision between the Cuban Arc and the Bahama Platform and Banks (GEALY, 1980). In the southwest Pacific the emplacement of the Papua ophiolite may be the approach of the Trobriand Platform, now just north of eastern Papua. The arrival of the Ontong Java Plateau (PARROT and DUGAS, 1980) could have obducted the Solomon Islands ophiolites, and the arrival of the Benham Rise on the east side of Luzon during the Miocene may have caused the emplacement of ophiolites in the Sierra Madre Range. Similarly, the Sula continental fragment from New Guinea (HAMILTON, 1979) could have emplaced the ophiolite complex on the eastern side of Celebes.

In California the continuous accretion of seamounts and small plateaus into the subduction zones is a possible explanation for emplacement of the coast range ophiolite. The Yolla Bolly Terrane (BLAKE and JONES, 1974) may be the remains of these oceanic rises.

The mechanism for ophiolite emplacement may thus be one and the same in both the Alpine- and Andean-type orogenes: in both cases obduction of oceanic lithosphere onto the continental lithosphere is caused by the convergence of light buoyant bodies in the form of oceanic plateaus, continental slivers, island arcs, or old hot spot traces. Without such light material, obduction may not occur.

5. The Origins of Continental Fragments

Most of the accreted continental blocks in the Alpine chain probably originated from Gondwana, since the Permian and through the present. The sizes of fragments range from small slivers such as in Italy to large ones such as India and Arabia. The continental circum Pacific accreted terranes may have a similar source in a Pacifica which might be also an extended Gondwana domain (Fig. 7). Unlike more westerly parts of the northern rim of Gondwana, Pacifica faced a vast ocean so that its fragments, if they existed, became more widely dispersed over much greater distances.

6. The Andes and the Andean- vs. Collision-Type Orogenies

The orogenic deformation in the northern Pacific rim may indeed have required collisions with accreted fragments and arcs and is not simply due to normal subduction of oceanic lithosphere. Fragment collisions are responsible also for orogenic defor- mation in the Alpine-Himalaya chain, not necessarily requiring two coherent con- tinents. Consequently, both orogenic systems may have a single mechanical origin: collisions with fragments. Whether this is true for all orogenic belts depends on the nature of the Andean chain, which is, after all, the classical example of a noncollision orogenic system.

Some evidence suggests that the orogenic history of the Andes involves not just simple subduction. Several tectonically and stratigraphically distinct geologic assem-

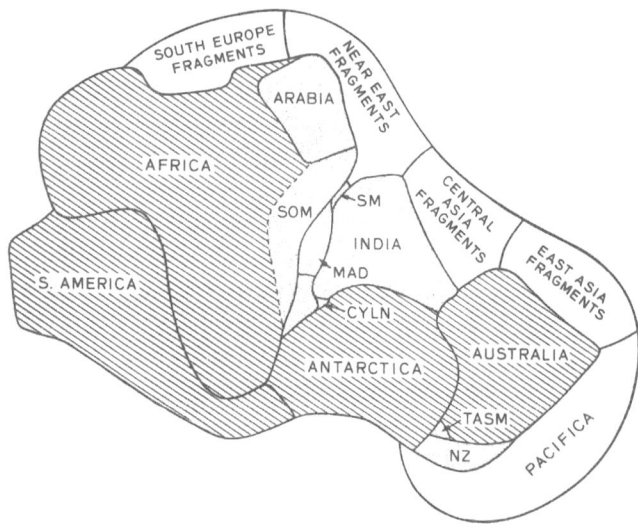

Fig. 7. Speculation about the origin of many of the accreted allochthonous terranes in the Alpine and Pacific Mesozoic-Cenozoic orogenic belts. Possibly, the fragmentation of Gondwana, and particularly its margins gave rise to a host of large and small fragments, which have become embedded in oceanic plates, moving towards subduction zones along the boundaries of other continental masses (after Nur and Ben-Avraham, 1982).

blages, possibly allochthonous terranes, have been welded together over a wide range of geological time (Zeil, 1979). Continental basement rocks are exposed along the western coast from Tierra del Fuego to Peru, between the trench and the young Andean Cordillera, with ages ranging from 1.8 m.y.B.P. to 300 m.y.B.P. Furthermore, continental sources to the west of the Andes fed voluminous late Paleozoic and early Mesozoic conglomerates and sandstones now found in the Andean chain. Arc terranes incorporated from the west were invoked to explain the presence of old continental basement off Peru.

We thus believe that the concept of Andean orogeny, which means orogeny produced by subduction of normal oceanic crust beneath a continent, may be open to question. It may turn out that a single process is responsible for orogenic deformation everywhere on earth: namely, collision.

7. Conclusions

We suggest that the many *continental fragments* now found as allochthonous terranes in the Mesozoic-Cenozoic orogenic belts of the world resulted from the break-up of Gondwana and led to the growth by accretion of other continents. Through the break-up of northern and eastern Gondwana, continuing through the more recent break-up of Africa, a host of continental slivers and related rises have migrated from the south to be accreted into Europe and Asia. Substantial orogenic deformation has resulted from the accretion of these bodies, well before full continent-continent collision took place.

Accreted continental fragments, as well as a variety of noncontinental rises, have been identified also in the Cordillera of the northern circum Pacific. As in the Alpine chain, the accretion of these terranes involves extensive orogenic deformation, also without full continent-continent collisions.

If the deformational role of fragments can also be established for the Andes, then an important generalization may be made: The accretion of continental fragments is the main cause for orogenic deformation. Subduction of normal oceanic crust is insufficient to cause such deformation, whereas full continental collision is necessary.

This study was supported by a grant from the Geophysics Program, Division of Earth Sciences, U.S. National Science Foundation.

REFERENCES

BARAZANGI, M. and B. ISACKS, Spatial distribution of earthquakes and subduction of the Nazca plate beneath South America, *Geology*, **4**, 686–692, 1976.

BARAZANGI, M. and B. ISACKS, Subduction of the Nazca plate beneath Peru: Evidence from spatial distribution of earthquakes, *Roy. Astron. Soc., Geophys. J.*, **57**, 537–555, 1979.

BEN-AVRAHAM, Z., A. BEHLE, J. MAKRIS, A. GINSBURG, P. GEISE, L. STEINMETZ, and R. B. WHITMARCH, Crustal structure of the eastern Mediterranean: A seismic refraction profile from Israel to Cyprus (abs.), *Trans. Am. Geophys. Union*, **60**, 886–887, 1979.

BEN-AVRAHAM, Z. and A. K. COOPER, Early evolution of the Bering Sea by collision of oceanic rises and north Pacific subduction zones, *Geol. Soc. Am., Bull.*, in press, 1981.

BEN-AVRAHAM, Z., A. NUR, D. JONES, and A. COX, Continental accretion and orogeny: From oceanic plateau to allochthonous terranes, *Science*, **213**, 47–54, 1981.

BEN-AVRAHAM, Z., A. NUR, and D. JONES, The emplacement of ophiolites by collision, *J. Geophys. Res.*, in press, 1982.

BIJU-DUVAL, B., J. DERCOURT, and X. LEPICHON, From the Tethys Ocean to the Mediterranean Seas: A plate tectonic model of the evolution of the western alipine system, in *Int. Symposium on the Structural History of the Mediterranean Basins, Split (Yugoslavia), 25–29 Oct. 1976*, edited by B. Biju-Duval and L. Mondadert pp. 143–164, Editons Technip, Paris, 1977.

BLAKE, M. C. JR. and D. J. JONES, Origin of Franciscan melanges in northern California, in *Modern and Ancient Geosynclinal Sedimentation, Society of Economic Paleontologists and Mineralogists, Spec. Pub.*, **19**, 345–357, 1974.

BOULIN, Jean, Afghanistan structure, greater India concept and eastern tethys evolution, *Tectonophysics*, **72**, 261–287, 1981.

BURCHFIELD, B. C., Eastern European alpine system and the Carpathian orocline as an example of collision tectonics, *Tectonophysics*, **63**, 31–61, 1980.

BURKE, K., J. DEWEY, W. S. F. KIDD, and A. M. C. SENGOR, The Alpine-Himalayan zone of continental collision and collision as a stage in continental evolution, in *Proceedings of Symposium on Quinghai-Xizang (Tibet) Plateau, Beijing, China, May–June 1980*, pp. 72–73, Organizing Committee Symposium on Quinghai-Xizang (Tibet) Plateau, Academia Sinica, 1980.

CHANG. C. and H. CHENG, 1973, Some tectonic features of the Mt. Polmo Lungma area, southern Tibet, China, *Sci. Sinica*, **16**, 257–265, 1973.

CHURCH, W. R., 1980, Late Proterzoic ophiolites, in *Orogenic Mafic and Ultramafic Association*, edited by C. Allegre and J. Aubouin, *Coll. Inter. Centre National de la Recherche Scientifique*, Paris, **272**, 105–117, 1980.

CHURKIN, M., JR. and J. H. TREXLER, Circum-artic plate accretion-isolating part of a Pacific plate to form the nucleus of the Arctic Basin, *Earth Planet. Sci. Lett.*, **48**, 356–361, 1980.

COLEMAN, R. G., *Ophiolites, Ancient Oceanic Lithosphere?*, pp. 229, Springer-Verlang, Berlin, 1977.

CONEY, P. J., D. L. JONES, and J. W. H. MONGER, Cordilleran suspect terranes, *Nature*, **288**, 329–333, 1980.

COOPER, A. K., D. W. SCHOLL, and M. S. MARLOW, Mesozoic magnetic lineations in the Bering Sea marginal basin, *J. Geophys. Res.*, **81**, 1916–1934, 1976.

COX, A., Rotation of microplates in western North America, in *The Continental Crust and its Mineral Deposits*, edited by D. W. Strangway *Can., Geol. Surv., Spec. Pap.*, **20**, 305–321, Toronto, Canada, 1980.

DEN, N. and H. HOTTA, Seismic refraction and reflection evidence supporting plate tectonophysics in Hokkaido, *Meteror. Geophys. Pap.* **24**, 31–54, 1973.

DICKINSON, W. R., Plate tectonic evolution of north Pacific rim, *J. Phys. Earth, Suppl.*, **26**, S1–19, 1978.

FUJITA, K., Pre-Cenozoic tectonic evolution of northeast Siberia, *J. Geol.*, **86**, 159–172, 1978.

GANSSER, A., The timing and significance of orogenic events and the Himalaya, in *Proceedings of Symposium on Quinghai-Xizang (Tibet) Plateau, Beijing, China, May–June 1980*, pp. 69–70, Organizing Committee Symposium on Quinghai-Xizang (Tibet) Plateau, Academia Sinica, 1980a.

GANSSER, A., The Peri-Indian suture zone, in *Geology of the Alpine chains born of the tethys, 26th Inter. Geol. Cont., June 7–17, Paris*, pp. 140–148, 1980b.

GASS, I. G. and D. MASSON-SMITH, The geology and gravity anomalies of the Troodos Massif, Cyprus, *Roy. Soc. Lond., Philos. Trans.*, **A-255**, 417–467, 1963.

GEALY, W. K., Ophiolite obduction and geological evolution of the Oman Mountains and adjacent areas, *Geol. Soc. Am., Bull.*, **88**, 1183–1191, 1977.

GEALY, W. K., Ophiolite obduction mechanism, in *Ophiolites*, edited by A. Panayiotu, pp. 228–243, Cyprus Geol. Surv. Dept., Cyprus, 1980.

GEISE, P., K. GORLER, V. JACOBSHAGEN, and K. J. REUTTER, Geodynamic evolution of the Apennies and Hellenides, in *Mobil Earth*, edited by M. Closs Inter. Geodyn. Proj., Final Report of the Federal Republic of Germany, Boppard, pp. 71–87, 1979.

GOSE, W. A. and D. K. SWARTZ, Paleomagnetic results from cretaceous sediments in Honduras: Tectonic implications, *Geology* **5**, 505–508, 1977.

HAMILTON, W., The Uralides and the motion of the Russian and Siberian platform, *Geol. Soc. Am., Bull.*, **81**, 2553–2576, 1970.

HAMILTON, W., Tectonics of the Indonesian region, *U. S. Geol. Surv., Prof. Pap.*, **1078**, 346, 1979.

HARATA, T., K. HISATOMI, F. KUMON, K. NAKAZAWA, M. TATEISHI, H. SUZUKI, and T. TOKUOKA, Shimanto geosyncline and Kuroshio paleoland, *J. Phys. Earth, Suppl.*, **26**, S357–S366, 1978.

HATTORI, I. and K. HIROOKA, Paleomagnetic results from Permian greenstones in central Japan and their geologic significance, *Tectonophysics*, **57**, 211–235, 1979.

HILLHOUSE, J. W., Paleomagnetism of the Triassic Nikolai greenstone, McCarthy quadrangle, Alaska, *Can. J. Earth Sci.*, **14**, 2578–2992, 1977.

HOWELL, D. G., Mesozoic accretion of exotic terranes along the New Zealand segment of Gondwanaland, *Geology* (Boulder), **8**, 487–491, 1980.

ISACKS, B. L. and M. BARAZANGI, Geometry of Benioff zones: Lateral segmentation and downwards bending of the subducted lithosphere, in *Island Arcs, Deep Sea Trenches and Back-arc Basins*, edited by M. Talwani and W. C. Pitmann III, *Am. Geophys. Union Maur. Ewing Ser.*, **1**, 99–114, 1977.

JONES, D. G., N. J. SILBERLING, and J. HILLHOUSE, Wrangellia—a displaced terrane in northwestern North America, *Can. J. Earth Sci.*, **14**, 2565–2577, 1977.

KAMP, P. J. J., Pacifica and New Zealand: Proposed eastern elements in Gondwanaland's history, *Nature*, **288**, 659–664, 1980.

KAWAI, N., K. HIROOKA, and T. NAKAJIMA, 1969, Paleomagnetic and Postassium-Argon age information supporting Cretaceous-Tertiary hypothetic bend of the main island Japan, *Paleogeogr. Paleoclimatol. Paleoecol.*, **6.**, 277–282, 1969.

KHAIN, V. W. and K. B. SESLAVINSKY, Comments and reply on 'the Siberian connection: A case for Precambrian separation of the North American and Siberian Cratons,' *Geology* (Boulder), 7, 466–467, 1979.

KING, L., An improved reconstruction of Gondwanaland, in *Implications of Continental Drift to the Earth Sciences*, edited by D. H. Tarling and S. K. Runcorn, pp. 853–863, Academic Press, London, 1973.

KLOMPÉ, Th. H. F., Pacific and Variscan orogeny in Indonesia: A structural systhesis, *Indones. J. Nat. Sci.*,

113, 43–89, 1957.

KREUZER, H., G. MOREANI, and D. ROEDER, Geodynamic evolution on the eastern Alps along a geotraverse, in *Mobile Earth*, edited by M. Closs pp. 51–64, Int. Geodyn. Proj., Final Report of the Federal Republic of Germany, Boppard, 1979.

LUDWIG, J. W., C. C. WINDISH, R. E. HOUTZ, and J. I. DWING, Structure of Falkland Plateau and offshore Tierra del Fuego, Argentina, in *Geological and Geophysical Investigations of Continental Margins*, edited by J. S. Watkins, L. Montadert, and P. W. Dickerson, Tulsa, Oklahoma, *Am. Assoc. Petrol. Geol., Memoir*, **29**, 125–127, 1979.

MARKOV, M. S., M. YU, M. PUSHCHAROVSKY, and S. M. TILMAN, Active continental margins of the northwest Pacific, *Tectonophysics*, **70**, 1–8, 1980.

MATSUDA, T., Collision of the Izu-Bonin Arc with central Honshu: Cenozoic tectonics of the fossa magna, Japan, *J. Phys. Earth, Suppl.*, **26**, S409–S421, 1978.

MATTAUER, M., P. TAPPONIER, and F. PROUST, Some analogies between the tectonic histories of Afghanistan and Tibet, in *Proceedings of Symposium on Quinghai-Xizang (Tibet) Plateau, Beijing, China, May-June 1980*, p. 62, Organizing Committee Symposium on Quinghai-Xizang (Tibet) Plateau, Academia Sinica, 1980.

McELHINNY, M. W., Paleomagnetism and plate tectonics, Cambridge Univ. Press, pp. 358, 1973.

McELHINNY, M. W., B. J. J. EMBLETON, X. H. Mz, and Z. K. ZHANG, Fragmentation of Asia in the Permian, *Nature*, **293**, 212–216, 1981.

McKENZIE, D. P. and J. G. SCLATER, The evolution of the Indian Ocean, *Sci. Am.*, **288**, 62–72, 1973.

MONGER, J. W. H. and C. A. Ross, Distribution of fusulinarceans in the western Canadian cordillera, *Can. J. Earth Sci.*, **8**, 259–278, 1971.

MONGER, J. W. H. Upper Paleozoic rocks of the western Canadian Cordillera and their bearing on Cordilleran evolution, *Can. J. Earth Sci.* **14**, 1832–1859, 1977.

MOORES, E. M. and F. J. VINE, The Troodos Massif, Cyprus and other ophiolites as oceanic crust: Evaluation and implications, *Roy. Soc. Lond., Philos. Trans.*, **A-268**, 443–460, 1971.

NUR, A. and Z. BEN-AVRAHAM, Lost Pacifica continent, *Nature*, **270**, 41–43, 1977.

NUR, A. and Z. BEN-AVRAHAM, Speculations on mountain building and the lost Pacifica continent, *J. Phys. Earth, Suppl.*, **26**, S21–S37, 1978.

NUR, A. and Z. BEN-AVRAHAM, Volcanic gaps and the consumption of aseismic ridges in South America, *Geol. Soc. Am., Memoir*, in press, 1981.

NUR, A. and Z. BEN-AVRAHAM, Oceanic plateaus, the fragmentation of continents and mountain building, *J. Geophys. Res.*, in press, 1982.

OGAWA, Y. and K. HORIUCHI, 1978, Two types of accretionary fold belts in central Japan, *J. Phys. Earth, Suppl.*, **26**, S321–S336, 1978.

PARROT, J. F. and F. DUGAS, The disrupted ophiolitic belt of the southwest Pacific: Evidence of an Eocene subduction zone, *Tectonophysics*, **66**, 349–372, 1980.

PIRAJNO, F., Origin of the eastern geotectonic domain of New Zealand and a Pacifica continent, *Geol. Soc. New Zealand, Newslett.*, **49**, 19–21, 1980.

POWELL, CMcA., Speculative tectonic history of Pakistan and surroundings; some constraints from the Indian Ocean, in *Geodynamics of Pakistan*, edited by Farah and DeJong, pp. 5–24, Geol. Surv. Pakistan, Quetta, 1979.

POWELL, C. McA. and B. D. JOHNSON, Constraints on the Cenozoic position of Sundaland, *Tectonophysics*, **63**, 91–109, 1980.

ROTSTEIN, Y. and A. L. KAFKA, Seismotectonics of the Levant Basin, eastern Mediterranean, *EOS*, **52**, 404, 1981.

SAITO, Y. and M. HASHIMOTO, South Kitakami region: An allochthonous terrain in Japan, *J. Geophys. Res.*, in press, 1982.

SCRUTTON, R. A., 1972, The crustal structure of Rockall Plateau microcontinent, *Roy. Astron. Soc., Geophys. J.*, **27**, 181–241, 1972.

SENGOR, A. M. C. and Y. YILMAZ, Tethyan evolution of Turkey: A plate tectonic approach, *Tectonophysics*, **75**, 181–241, 1981.

SHIMIZU, D., N. IMOTO, and M. MUSAHINO, Permian and Triassic sedimentary history of the Honshu geosyncline in the Tamba Belt, southwest Japan, *J. Phys. Earth, Suppl.*, **26**, S337–S344, 1978.

Stone, D. B., Plate Tectonics, Paleomagnetism and the Tectonic history of the N. E. Pacific, *Geophys. Surv.*, **3**, 3–37, 1977.

Taylor, B. and D. Hayes, The tectonic evolution of the South China Basin, in *The tectonic and Geologic Evolution of Southeast Asian Seas and Islands*, edited by D. E. Hayes, *Am. Geophys. Union, Washington, D. C., Geophys. Mono.*, **23**, 89–104, 1980.

Terman, M. J., Tectonic map of China and Mongolia, *Geol. Soc. Am., Boulder, Co., Map and Chart Series, MC-4*, 1973.

Vine, F. J., C. K. Poster, and I. G. Gass, Aeromagnetic survey of the Troodos igneous massif, Cyprus, *Nature Phys. Sci.*, **244**, 34–38, 1973.

Vogy, P. R., A. Lowrie, D. R. Bracey, and R. H. Hey, Subduction of aseimic ridges: Effects on shape, seismicity, and other characteristics of consuming plate boundaries, *Geol. Soc. Am., Spec. Pap.*, **172**, 59, 1976.

Von Huene, R., M. Langseth, and N. Nasu, Summary, Japan trench transect, in *Initial Reports of the Deep Sea Drilling Project, leg. 56–57*, edited by M. Lee and L. Strout, pp. 473–488, National Science Foundation and University of California, Scripps Inst. of Oceanography, Washington, D. W., 1980.

Zeil, W., The Andes—a geological review, Gebruder Borntraeger, Berlin, *Beitrage zur Regionalen Geologie der Erde*, **13**, 260, 1979.

North America to Northeast Asia

Accretion Tectonics in the Circum-Pacific Regions, edited by M. Hashimoto and S. Uyeda, 21–35.
Copyright © 1983 by Terra Scientific Publishing Company (TERRAPUB), Tokyo.

Recognition, Character, and Analysis of Tectonostratigraphic Terranes in Western North America

D. L. JONES,* D. G. HOWELL,*

P. J. CONEY,** and J. W. H. MONGER***

*U.S. Geological Survey, 345 Middlefield Rd., Menlo Park,
California 94025, U.S.A.
**University of Arizona, Tucson, Arizona, U.S.A.
***Geological Survey of Canada, Vancouver, Canada

Continental growth involves accretion of allochthonous terranes driven by plate tectonic processes. The Pacific region is rimmed by a tectonic collage composed of accreted terranes. Individual tectonostratigraphic terranes are fault-bounded geologic entities of regional extent, each characterized by a geologic history which is distinct from that of neighboring terranes. Terranes comprise one of four types: (1) stratigraphic terranes composed of coherent stratigraphic sequences which reflect geologic environments of continental fragments, ocean basins and/or volcanic arcs; (2) disrupted terranes characterized by blocks of heterogeneous lithology and age set in a matrix of sheared graywacke or serpentinite, (3) metamorphic terranes represented by areas with a regional penetrative metamorphic fabric that obscures and is more distinctive than original petrogenetic aspects; and (4) composite terranes represented by two or more terranes that amalgamated prior to accretion onto a continental margin. Terranes vary enormously in size from some that require mapping at 1:25,000 to others that are easily depicted at 1:20,000,000. Post-accretionary tectonic processes cause terrane dispersion further complicating the geometry of terrane distribution.

1. Introduction

Most of the Cordillera of western North America is a tectonic collage composed of accreted crustal fragments or terranes. The relations of the terranes to one another during the time of formation of their component rocks is unknown or uncertain, (CONEY et al., 1980). These lithospheric fragments can be discriminated through a process known as terrane analysis. In this brief report, we explain the methodology of terrane analysis, including how terranes are identified and how their boundaries are established. A critical feature of terrane analysis is that geologic and geophysical data must be clearly separated from plate tectonic interpretations concerning the genetic relations among terranes. Because of the large amount of differential movement that various terranes have undergone, it is no longer safe or reasonable to assume that nearby terranes are genetically related—such relations now must be vigorously proven before meaningful paleogeographical and paleotectonic reconstructions should be attempted.

2. Terrane Definition

Terranes are fault-bounded geologic entities of regional extent, each characterized by a geologic history that is different from the histories of contiguous terranes. Ideally, such histories are determined from the stratigraphic succession preserved in a terrane, but in some cases such histories are largely or completely destroyed by tectonic or sedimentological disruptions or by metamorphic overprinting. In the latter cases the disruptive or metamorphic event itself may characterize the terrane. In addition, some terranes are composite entities produced by amalgamation of two or more terranes into a single terrane. In cases where juxtaposed terranes possess coeval strata, one must demonstrate different and unrelated geologic histories as well as the absences of intermediate lithofacies that might link the two terranes. The basic question that must be asked while analyzing stratigraphic sequences of possibly distinct terranes is whether or not the inferred geologic histories are compatible with the present spatial relations. This decision is not always easy to make, and is heavily dependent on the quality and quantity of geologic controls that are available to the analyzer. The degree of differences noted between terranes is thus variable, and classifications will differ according to the judgement, experience, and competency of the analyzer. New data always require reexamination of existing terrane classifications, and it is expected that new combinations or subdivisions will result from additional paleontologic, geologic, and geophysical research. In this regard, terrane nomenclature is similar to stratigraphic nomenclature, and is subject to continuous revision as data accumulate and concepts evolve. Terranes are conveniently categorized into four general types (1) stratified, (2) disrupted, (3) metamorphic, and (4) composite; examples of each are discussed below.

2.1 Stratigraphic terranes

These terranes are characterized by coherent stratigraphic sequences in which depositional relations between successive lithologic units can be demonstrated. Basement rocks may or may not be preserved. Rock sequences within stratified terranes may be subdivided into three broad categories (some terranes have passed through successive tectonic phases encompassing two or three of these categories):

    a. Fragments of continents: These terranes are characterized by the presence of a Precambrian basement with an overlying sequence of shallow-water sedimentary rocks of Paleozoic and Mesozoic ages. Included are sedimentary rocks of continental derivation that are detached from their basement substratum. Examples include the Nixon Fork terrane of Alaska and the Tujunga terrane of southern California.

    b. Fragments of oceanic basins: These terranes are characterized by sequences of mafic and ultramafic rocks characteristic of oceanic crust with overlying deep-sea sedimentary deposits, e.g. the Del Puerto terrane of California and the Chulitna terrane of Alaska. Both have ophiolitic basements and deep-sea sediments, yet younger strata in both indicate continental margin depositional environments. These progradational sequences reflect translational mobility characteristic of many terranes. Also included are deep-sea deposits that are detached from their basement substratum, e.g. the Pingston terrane of Alaska.

c. Fragments of volcanic arcs: These terranes are composed dominantly of volcanic rocks, or the plutonic roots of arcs, and sedimentary debris derived from volcanos that are similar in composition to rocks of presently active volcanic arcs such as the Aleutians. Examples include the Stikine terrane of British Columbia, the Peninsular terrane of Alaska, and the Salinia terrane of California.

2.2 Disrupted terranes

These terranes are characterized by blocks of heterogeneous lithology and age, usually set in a matrix of sheared shale, flysch, or serpentinite. Most of these terranes contain fragments of ophiolitic rocks, blocks of shallow water limestone, deep-water chert, and packages of graywacke with lenses of conglomerate; in addition, many disrupted terranes contain blue schists both as exotic blocks or as a regional metamorphic overprint. Some disrupted terranes have been interpreted as subduction complexes. Examples include parts of the Chugach terrane of Alaska, the Franciscan terrane of California, and parts of the Cache Creek terrane of British Columbia.

2.3 Composite terranes

These terranes are composed of two or more distinct terranes that became amalgamated and subsequently shared or common geologic history prior to their accretion to North America (see discussion below). Examples of amalgamated terranes include: arc-arc amalgams (Alexander-Wrangellia); arc-continental-oceanic-disrupted amalgam (Salinia-Tujunga-Stanley Mountain-San Simeon terrane amalgam that equals the Santa Lucia-Orocopia allochthon of southern California; see Fig. 4). The Tujunga terrane by itself is a composite with basement rocks comprising at least three distinct Precambrian terranes and the Alexander terrane is a composite composed of the pre-Triassic Craig, Admiralty and Annette terranes (see Fig. 3).

2.4 Metamorphic terranes

These terranes are characterized by a regional, terrane-wide penetrative metamorphic fabric and development of metamorphic minerals to such a degree that original stratigraphic features and relations are obscured. In addition to metamorphic differences, protolithic contrasts with adjoining terranes also must be demonstrable. Examples include the Yukon-Tanana terrane of Alaska and the Baldy terrane of California composed of the Pelona, Orocopia and Rand Schists.

3. Size of Terranes

The terranes of the western Cordillera vary enormously in size. Some are of subcontinental dimensions whereas others cover only a few hundred square kilometers or less. A few terranes are not now continuous bodies, but consist of separate, disjunct patches that can be unequivocally correlated. The best example of a terrane that is disjunct is Wrangellia, which presently is distributed as isolated bodies from the state of Oregon, through British Columbia to southern Alaska, a latitudinal spread of nearly 24 degrees. Paleomagnetic data however, indicate that the original latitudinal spread was likely less than 4 degrees. The Santa Lucia-Orocopia composite terrane is currently

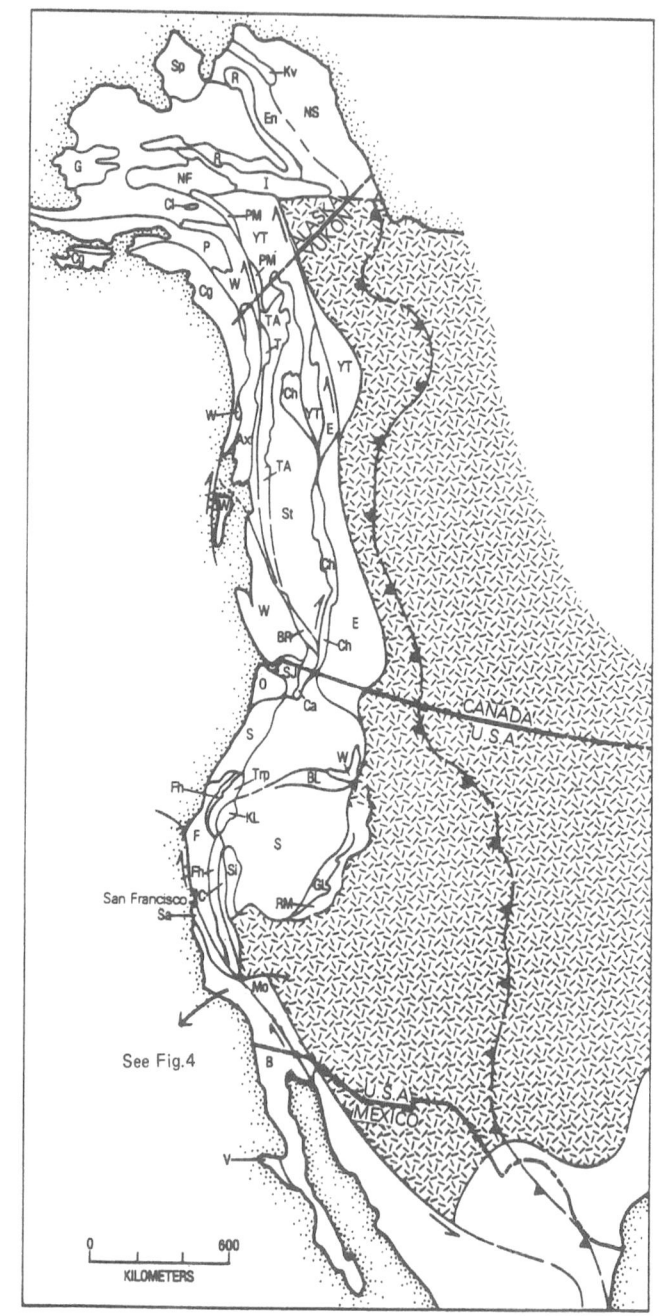

Fig. 1.   Generalized distribution of selected terranes throughout the Cordillera of western North America. Dashed pattern is cratonic basement of authochthonous North America; barbed line represents the eastern limit of Cordilleran Mesozoic-Cenozoic deformation, and arrows show sense of major strike-slip movements. Terranes are described below, from CONEY *et al.* (1980).

Alaska

Sp. Seward Peninsula—structurally complex assemblage of Precambrian metamorphic and sedimentary rocks, and Palaeozoic carbonate rocks.

Ns. North Slope—Precambrian, Palaeozoic, and Mesozoic clastic and carbonate sequence—part of North America, but may have moved from original position.

Kv. Kagvik—Thin sequence of radiolarian chert, argillite, shale, and minor volcanics, Mississippian to Triassic in age.

En. Endicott—metamorphosed Lower to Upper Palaeozoic clastic and carbonate rocks intruded by Palaeozoic granitic rocks.

R. Ruby—composite terrane comprising at least three separate units, including Precambrian metamorphic rocks, mid to Upper Palaeozoic volcanic and sedimentary rocks, and thick piles of Lower Mesozoic basalt and chert.

I. Innoko—structurally deformed sequence of Upper Palaeozoic to early Mesozoic chert, argillite, graywacke, and basic to intermediate volcanics.

NF. Nixon Fork—Precambrian metamorphic rocks overlain by Palaeozoic and Mesozoic carbonate, clastic, and cherty rocks.

G. Goodness (composite)—includes three terranes: (1) a complex assemblage of deformed Upper Palaeozoic volcanics, chert, and graywacke with blocks of older limestone, (2) Precambrian gneisses and schist: and (3) Mesozoic arc-derived volcanic flows, tuff, and graywacke, with interbedded chert.

Cl. Chulitna (composite)—includes three terranes: (1) Devonian ophiolite overlain by Palaeozoic chert, volcanic conglomerate, limestone, and flysch, and Mesozoic limestone, redbeds, flysch, and chert: (2) Mesozoic chert, argillite, crystal tuff, and conglomeratic sandstone: (3) Upper Palaeozoic tuff and chert, volcanic graywacke, with blocks of Lower Palaeozoic limestone.

PM. Pingston & McKinley (composite)—includes three terranes: (1) Upper Palaeozoic phyllite and Triassic thin-bedded limestone and sooty black shale; (2) Upper Palaeozoic chert, Triassic pillow basalt, and Upper Mesozoic flysch and conglomerate; (3) Lower Palaeozoic limestone: tuff and flysch of unknown ages.

YT. Yukon-Tanana (composite)—includes regionally metamorphosed schist and gneiss of Precambrian(?) age. Devonian limestone,

Upper Palaeozoic silicic metavolcanic rocks, Permian ophiolite, and foliated granitic rocks of unknown age.

W. Wrangellia—Upper Palaeozoic arc complex composed of flows, breccias, and volcaniclastic rocks overlain by limestone, clastics, and chert, and Mesozoic pillowed and subaerial basalt flows succeeded by limestone, cherty limestone, and clastic rocks.

P. Peninsular—rare Palaeozoic limestone, Triassic basalt, argillite, and limestone. Lower Jurassic volcanic and volcaniclastic rocks, younger clastics.

Cg. Chugach (composite)—includes (1) deformed Upper Mesozoic flysch and melange units, and (2) deformed Lower Cenozoic flysch and volcanic rocks.

Ax. Alexander—complex terrane of Precambrian (?) and Palaeozoic volcanic rocks, clastics, and limestone, and Mesozoic volcanics, limestone, and clastic rocks.

T. Taku—structurally complex assemblage of Upper Palaeozoic volcaniclastics, limestone, flysch (?) and Lower Mesozoic basalt, limestone, and flysch.

TA. Tracy Arm—structurally complex assemblage of marble, pelitic gneisses, and schist of unknown ages.

Canada

Ch. Cache Creek terrane—Mississippian to Middle (Upper?) Triassic, highly disrupted radiolarian chert, argillite, basalt, alpine-type ultramafics, large shallow-water carbonates, and local blueschist metamorphism.

BR. Bridge River terrane—Middle Triassic to Lower Middle Jurassic, highly disrupted radiolarian chert, argillite, basalt, alpine-type ultramafics and minor carbonate.

St. Stikine terrane—Mississippian and Permian volcaniclastics, basic to acidic volcanics and carbonates, locally deformed and intruded in middle to late Triassic time, overlain by Upper Triassic to Middle Jurassic volcanogenic strata.

E. Eastern assemblage (composite)—includes possible late Precambrian-early Palaeozoic metamorphic terranes, of possible continental affinity, together with Mississippian to Triassic basalt, ultramafics and chert and volcaniclastics and carbonates, overlain unconformably by Middle Triassic to Lower Jurassic volcanogenic strata.

Fig. 1   (continued).

Washington and Oregon

SJ. San Juan (composite)—includes highly deformed Mesozoic chert, argillite, graywacke, and volcanic rocks, partly in melanges, with blocks of lower Palaeozoic plutonic rocks. Palaeozoic chert, carbonates, and volcanic rocks. Permian limestone blocks contain Tethyan fusulinids.

Ca. Northern Cascades (composite)—includes crystalline and pelitic genesses, and thrust sheets composed of (1) Upper Palaeozoic andesitic volcanics and associated sedimentary rocks: (2) greenschist and blueschist; and (3) Jurassic ophiolite.

O. Olympic—Lower Cenozoic volcanic rocks and associated deep and shallow water sedimentary rocks. Basement unknown, but presumed to be oceanic.

S. Lower Cenozoic volcanic and sedimentary rocks lying west of the Cascade Range. Palaeomagnetic data imply post-Eocene clockwise rotation of 70 .

BL. Blue Mountains (composite)—includes melange with blocks of Palaeozoic ophiolite, limestone, and chert, and Mesozoic chert and sandstone, structurally overlain by Triassic and Jurassic volcanic sandstone, conglomerate, and argillite.

California

Fh. Foothills—Upper Jurassic andesitic volcanic and volcaniclastic rocks associated with phyllite, slate, and graywacke, and Upper Jurassic ophiolite.

Trp. Triassic and Palaeozoic of Klamath Mountains (composite)—includes a structurally complex assemblage of Lower Mesozoic ophiolite, chert, basalt, Jurassic andesitic rocks, and associated sedimentary rocks.

KL. Eastern Klamath Mountains—Middle to Upper Palaeozoic clastic, volcanic, and carbonate rocks, overlain by Triassic and Jurassic volcanics and minor limestone.

Si. Northern Sierra—Lower Palaeozoic clastic sedimentary rocks. Upper Palaeozoic and Lower Mesozoic volcanic and associated sedimentary rocks.

C. Calaveras (composite)—including a western belt of melange with ophiolite and Mesozoic chert, and an eastern belt of quartzose clastic rocks, argillite, and minor Permian limestone.

F. Franciscan (composite)—includes Upper Mesozoic Great Valley sequence with ophiolite at base, and structurally underlying disrupted and partially metamorphosed rocks of the Franciscan Complex.

Sa. Salinia—includes metamorphosed pelitic rocks, marble, and graywacke of unknown ages, intruded by Cretaceous granite plutons.

Mo. Mojave (composite)—juxtaposed and disrupted Palaeozoic sedimentary sequences. Lower Mesozoic sedimentary and volcanic rocks intruded by Mesozoic plutons.

Mexico

B. Baja—includes scattered localities of Upper Palaeozoic limestone and Lower Mesozoic clastic rocks, overlain by a thick pile of Upper Mesozoic volcanic and volcaniclastic rocks, capped by latest Cretaceous quartzofeldspathic sandstone.

V. Vizcaino (composite)—includes Triassic basalt, chert, and limestone, Upper Jurassic arc-derived volcanic and volcaniclastic rocks, Upper Jurassic and Cretaceous clastic rocks, ophiolite, and structurally underlying Upper Mesozoic blue schist and distupted rocks similar to the Franciscan Complex.

Nevada

S. Sonomia (composite)—includes Upper Palaeozoic volcanics in the south, and Lower Mesozoic volcanics in the north. Si and Kl. terranes originally included in Sonomia.

Gl., Golconda—structurally deformed assemblage of chert, argillite, minor limestone, and volcanics of Mississippian to Permian age.

RM. Roberts Mountains—structurally complex assemblage of chert, argillite, sandstone, basalt, and minor limestone of Cambrian to latest Devonian or early Mississippian ages.

being disrupted into northwest-oriented slivers owing to Neogene displacement along the San Andreas fault system.

3.1 Terrane boundaries

By definition, all terranes must be separated from adjoining terranes by major faults, fault zones, or complex sutures. In practice one may be confronted with two nearby areas with different stratal units yet the boundary conditions are not known. In such cases the most conservative tack is to assume a fault until other linking relations can be demonstrated.

Confusion may arise in discriminating between fault bounded terranes and successive tectonic elements within a particular terrane. For example, in our definition, the development of a volcanic arc assemblage on top of an earlier-formed sedimentary or igneous package does not constitute formation of a new terrane. This definition is crucial to terrane analysis and failure to discriminate between terranes and contrasting lithogenetic elements within terranes is a potential source of confusion.

3.2 Amalgamation to form composite terranes

Two similar but unrelated events in the history of terranes should be distinguished. An early event referred to as *amalgamation*, is the joining together of separate terranes to form a composite terrane prior to the addition of the amalgamated terrane to a cratonal margin. The later event, termed *accretion*, is the collision and welding of a terrane (either composite or individual) to the craton. These events may be widely separated in time, or closely follow one other. Timing of amalgamation and accretion can be established by three main criteria (Fig. 2): 1) overlap assemblages that depositionally overlie two distinctive, juxtaposed stratigraphic sequences (e.g. the Bowser Basin deposits of British Columbia that link the Stikine and Cache Creek terranes and the Gravina-Nutzotin belt of southeastern Alaska that links Wrangellia to the Alexander composite terrane); 2) sudden appearance of detritus in one terrane derived from its dissimilar neighbor (e.g. in California granite boulders in paralic facies of the Stanley Mountain terrane indicate an Upper Cretaceous suturing to the Salinia terrane); 3) welding together of unlike sequences by granitic intrusions (e.g. 60 m.y. old plutons stitch all the terranes between the Denali and Border Ranges faults of southern Alaska).

By means of a flow chart we can schematically depict successive amalgamation stages of a variety of terranes. Figure 3 demonstrates the suturing sequence for some of the terranes in Alaska and Fig. 4 depicts suturing events in Southern California. In both instances episodes of accretion and amalgamation are inferred from the geologic relations of overlap sequences, plutonic intrusions or debris from a distinctive provenance.

4. Accretion of Terranes

The major tectonic event experienced by most terranes in the Cordillera is their accretion to North America. In many cases, this collisional event has produced intense folding, thrust faulting, penetrative deformation, and recrystallization under green-

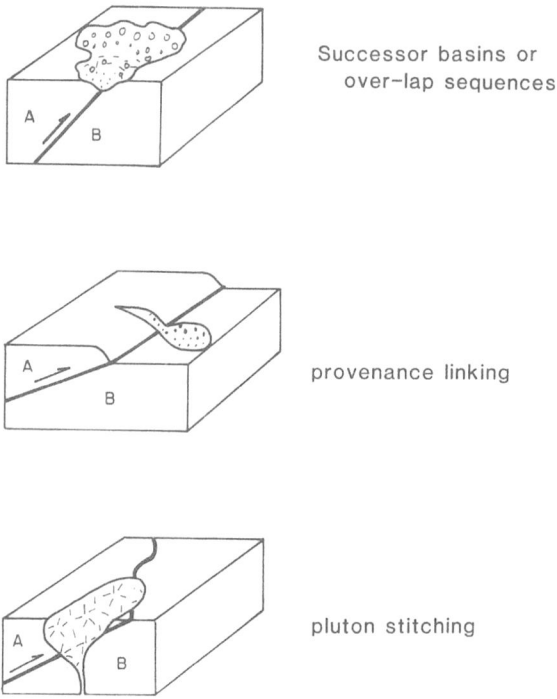

Successor basins or
over-lap sequences

provenance linking

pluton stitching

Fig. 2.   Three geologic relations that help establish the timing of terrane amalgamation and accretion.

schist, blueschist, amphibolite facies conditions. These resulting tectonite fabrics in some instances are regional in scale while in other cases only thin or narrow zones are highly tectonized. Igneous activity directly related to accretion seems to be rare but low temperature alteration is widespread, and local instances of high temperature alteration, including anatexis, have been documented (HUDSON *et al.*, 1979).

Structural styles in accreted terranes vary widely; isoclinal folding is common, but vergence of folds is not consistent even within a single terrane. Thrust faults are ubiquitous, although most fault surfaces have themselves been folded. Identification of the initial accretionary structures that have not been altered by later movements is extremely rare.

The mechanisms of terrane accretion and amalgamation remain obscure; nonetheless, these phenomena are not to be confused with the concept of subduction accretion which relate to off-scraping of soft unconsolidated pelagic or trench deposits during subduction. Many terranes of the Cordillera are composed of sedimentary units which indicate past involvement in subduction accretion, yet their present tectonic position reflects emplacement mainly as coherent, strongly lithified masses, above either the continental margin or on earlier accreted terranes. This implies that the terranes are obducted flakes that are mere remnants of once much larger plates that have now mostly been subducted.

We suspect the major differences between the accretionary wedge-model and the

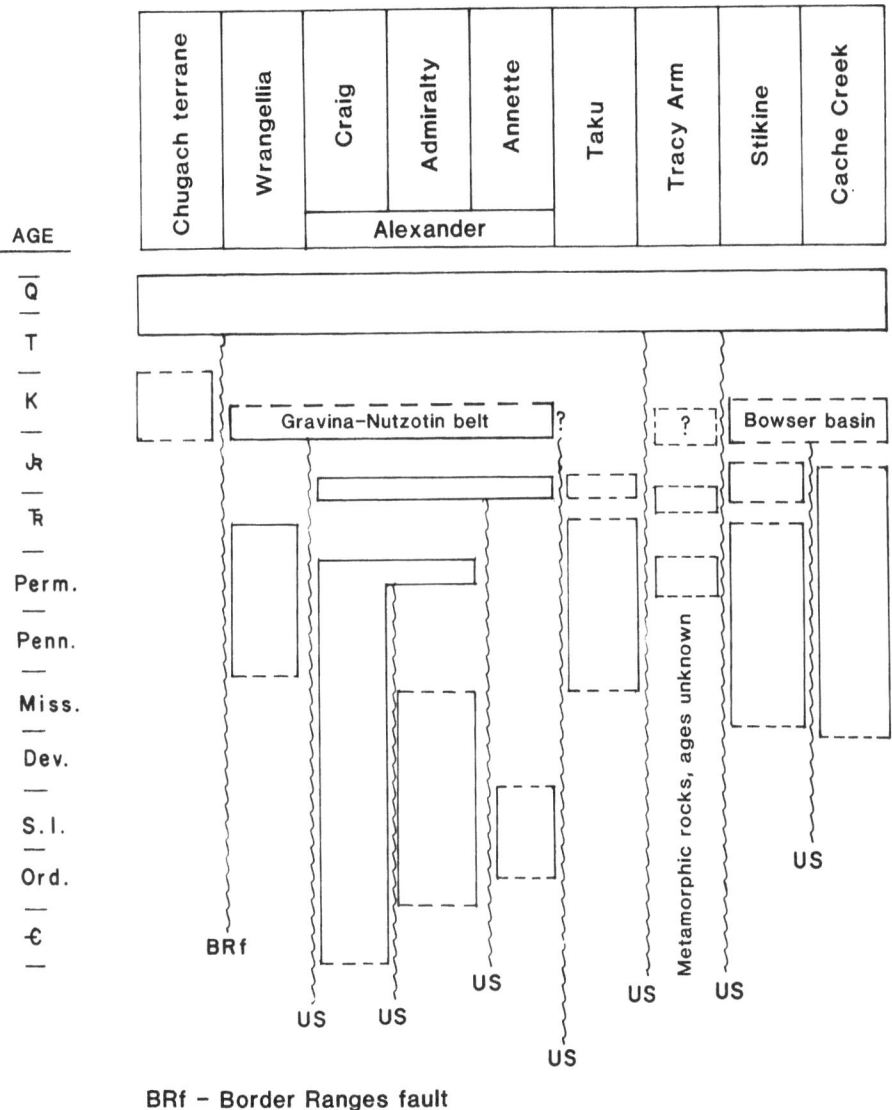

BRf − Border Ranges fault
US − Unnamed suture

Fig. 3. Generalized cross sections of presently contiguous terranes in southern Alaska showing inferred ages of amalgamation and accretion.

accretionary terrane setting are due to differences in crustal thickness of material being subducted. Sediment subduction involves oceanic crust with a thin cover of pelagic deposits being subducted without the formation of a growing accretionary wedge; the formation of the latter may be dependent on the presence of thick clastic trench deposits above a zone of over-pressured pore water (SCHOLL *et al.*, 1980). Thick crustal sections, such as seamounts, oceanic plateaus and ridges, and continental fragments,

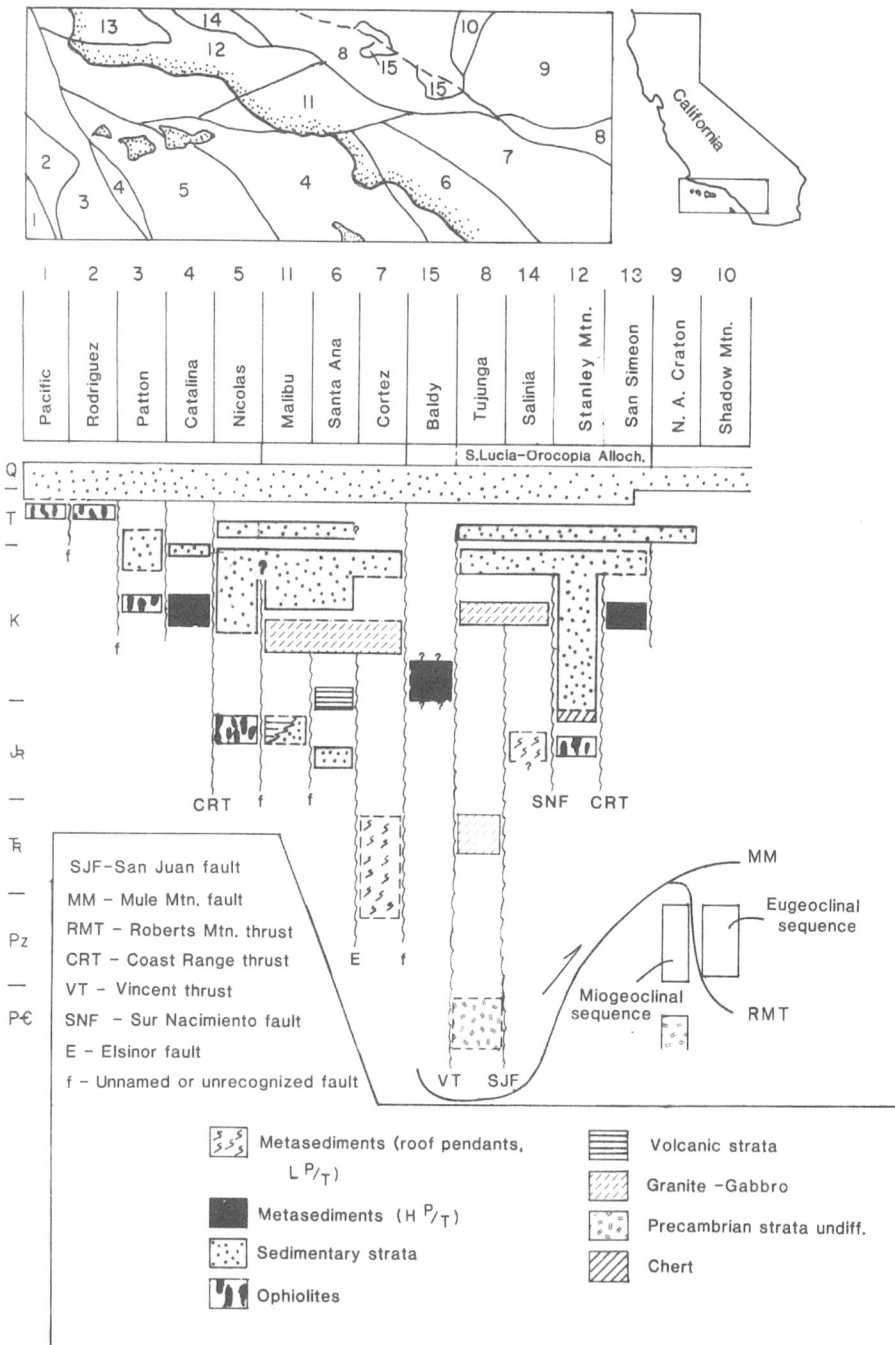

Fig. 4. Generalized cross sections of presently contiguous terranes in southern California showing inferred ages of amalgamation and accretion.

appear to be more difficult to subduct, and are instead, accreted as intact blocks (BEN-
AVRAHAM *et al.*, 1981). Complete kinematic and dynamic explanations of terrane
accretion processes, however, remain to be elucidated.

4.1 Post accretionary dispersion

The main period of accretionary activity ended by early Tertiary time in the
Cordillera. These accretionary episodes have been followed by a long and still poorly
known history of complex strike slip faulting, folding, and thrust faulting resulting in
the breakup of some terranes. This crustal rearranging still is locally active and
represents continuing interactions between oceanic plates of the Pacific and the
continental plate of North America. The results of this neotectonic activity is to
disperse the terranes and to further complicate the structures formed during accretion.
Large scale right slip faults, such as the San Andreas, Fairweather, Denali, Fraser
River, and Tintina, all have minimum displacements of a few hundred kilometers, and
some may have much more. The cumulative relative movement on all of these, plus
innumerable subsidiary faults, must amount to several thousand kilometers. Such a
kinematic history agrees well with the plate tectonic motions between the Pacific and
North American plates. Some of these structures originated during the late Cretaceous
to early Cenozoic Laramide Orogeny, which profoundly affected much of western
North America, including the Precambrian cratonal basement. The Laramide orogeny
has been interpreted as a final "tightening" of a poorly consolidated crust composed of
earlier accreted allochthonous terranes.

4.2 Measuring terrane displacements

The fact that fault-bounded terranes in the Cordillera have different stratigraphic
sequences implies that they have moved relative to each other and to the craton. The
amount of movement may not be large, but must be sufficient to juxtapose dissimilar
rocks and to completely disrupt original facies trends.

Much effort is now being expended in order to establish the kinematic history of
the various terranes. The principal methods of determining relative movements are: (1)
offset of linear geologic elements, e.g. shorelines, dike swarms, fold axes etc.; (2)
matching offset but similar stratigraphic sequences or distinctive rock types; (3)
matching offset biogeographical provinces; (4) matching climatically controlled
lithologic features (e.g., red beds, sabhkas, etc.) with their world-wide regional extent;
(5) determination of paleolatitude through paleomagnetic investigations.

The method (1) permits precise analyses of slip but offsets greater than 500 km are
rarely determined whereas method (5) gives quantitative measurements for displacements that have a very large latitudinal component. Offset in a longitudinal sense is not
possible with paleomagnetic data unless detailed polar-wandering paths are available.
The other methods are mainly qualitative for large displacements, although direct
matching of offset stratigraphic sequences can be quite precise for displacement ranges
of a few tens to a few hundreds of kilometers. Paleomagnetic measurements are now
available from several terranes of the Western Cordillera and these data clearly
substantiate minimum large scale northward displacements in some instances on the
order of thousands of kilometers. Coupled with these right-slip translations are

clockwise rotations commonly of 60 or more.

Thorough terrane analysis involves all of the above parameters to assess relative differential movements of terranes. If two or more methods of estimating original paleolatitude are in essential agreement, much more confidence can be placed on palinspastic and paleogeographic reconstructions. A good example of concordance of geophysical and geologic data is afforded by Wrangellia, where paleomagnetic measurements from Triassic basaltic rocks show equatorial paleolatitudes (HILLHOUSE, 1977) and Triassic carbonates overlying the basalt are Sabhka deposits characteristic of shallow tropical supratidal conditions (ARMSTRONG and MACKEVETT, 1977). These rocks are now at latitudes as high as 62 N, and are completely out of place with respect to cratonal Triassic sequences.

5. Paleogeographic Reconstructions

The aims of terrane analysis are to (1) identify, characterize, and portray on terrane maps all major allochthonous terranes; (2) to relate their faunal and floral characteristics through time to major paleobiogeographic provinces; (3) to establish paleolatitudes through time; (4) and finally, in the case of the Cordillera and Pacific region, to attempt paleogeographic reconstructions of the paleo-Pacific region, to attempt paleogeographic reconstructions of the paleo-Pacific Ocean (Panthalassa) and surrounding cratonal regions. Deep-sea drilling has established that no part of the present Pacific Ocean floor is older than Middle Jurassic—thus, pre-Jurassic history of the Pacific basin can only be gleaned from scraps and fragments plastered around the Pacific Margin in the form of accreted, allochthonous terranes.

REFERENCES

References cited and recommended reading germane to terrane analysis of western Cordillera of North America.

ALBERS, J. P., A Lithologic-tectonic framework for the metallogenic provinces of California, *Econ. Geol.*, **76**, 765–790, 1981.

ALVAREZ, Walter, D. V. KENT, I. PREMOLI-SILVA, R. A. SCHWEICKERT, and R. A. LARSON, Franciscan complex limestone deposited at 17° south paleolatitude, *Geol. Soc. Am. Bull.*, **91**, (pt. 1, no. 8), 476–484, 1980.

ARMSTRONG, A. K. and MACKEVETT, E. M., Jr., The Triassic Chitistone Limestone, Wrangell Mountains, Alaska, stressing detailed descriptions of Sabkha facies and other rocks in the Chitistone and their relations to Kennecott-type copper deposits, U. S. Geol. Survey Open File Report, 77–217, 72p., 1977.

BAILEY, E. H., M. C. BLAKE, and D. L. JONES, Onland Mesozoic oceanic crust in California Coast Ranges, *U.S. Geol. Survey Prof. Paper*, 700-C, C70–C81, 1970.

BECK, M. E., JR., Discordant paleomagnetic pole positions as evidence of regional shear in the western Cordillera of North America, Am. *J. Sci.*, **276**, 694–712, 1976.

BECK, M. E., JR., The paleomagnetic record of Plate-margin tectonic processes along the western edge of North America, *J. Geophys. Res.*, in press, 1982.

BEN-AVRAHAM, Z., A. NUR, D. JONES, and A. COX, Continental accretion and orogeny: From ocean plateaus to allochthonous terranes, *Science*, **213**, 47–54, 1981.

BERG, H. C., D. L. JONES, and P. J. CONEY, Map showing pre-Cenozoic tectonostratigraphic terranes of southeastern Alaska and adjacent areas, *U.S. Geol. Survey Open-file Report*, 78–1085, Scale 1:1,000,000, 1978.

BERG, H. C., D. L. JONES, and D. H. RICHTER, Gravina-Nutzotin belt-Tectonic significance of an upper Mesozoic sedimentary and volcanic sequence in southern and southeastern Alaska, *U.S. Geol. Survey Prof. Paper*, 800-D, D1–D24, 1972.

BLAKE, M. C. and D. L. JONES, Origin of Franciscan melanges in northern California, in *Modern and Ancient Geosynclinal Sedimentation: Soc. Econ. Paleontologists and Mineralogists Spec. Pub.*, **19**, edited by R. H. DOTT, Jr. and R. H. SHAVER, 345–357, 1974.

BLAKE, M. C. Jr. and D. L. JONES, The Franciscan assemblage and related rocks in northern California: a reinterpretation, in *The Geotectonic Development of California*, edited by W. G. Ernst, pp. 307–328, 1981.

BLAKE, M. C. Jr., D. L. JONES, and C. H. LANDIS, Active continental margins: Contrasts between California and New Zealand, in *The Geology of Continental Margins*, edited by C. A. Burk and C. L. Drake, pp. 853–872, Springer-Verlag, 1974.

CHAMPION, D. E., D. G. HOWELL, and C. S. GROMMÉ, Paleomagnetic and geologic data indicating 2,500 km of northward displacement for the Salinian, Sur Obispo, Tujunga and Baldy (?) terranes, California, *J. Geophys. Res.*, in press, 1982.

CHURKIN, Michael Jr., Claire CARTER, and J. H. TREXLER, Jr., Collision-deformed Paleozoic continental margin of Alaska-Foundation for microplate accretion, *Geol. Soc. Am.*, pt. 1, **91**, 648–654, 1980.

CHURKIN, Michael JR. and G. D. EBERLEIN, Ancient borderland terranes of the North American Cordillera: Correlation and microplate tectonics, *Geol. Soc. Am. Bull.*, **88**, 769–786, 1977.

CHURKIN, Michael Jr., W. J. NOKLEBERG, and Carl HUIE, Collision-deformed Paleozoic continental margin, western Brooks Range, Alaska, *Geology*, **7**, 379–383, 1979.

CONEY, P. J., D. L. JONES, and J. W. H. MONGER, Cordilleran suspect terranes, *Nature*, **288**, 329–333, 1980.

DAVIS, G. A., J. W. H. MONGER, and B. C. BURCHFIEL, Mesozoic construction of the Cordilleran "collage," central British Columbia to Central California, in *Mesozoic Paleogeography of the Western United States*, edited by D. G. Howell and K. A. McDougall, Pacific Coast Paleogeography Symposium 2, Pacific Coast Section, Soc. Econ. Paleont. and Mineral., pp. 1–32, Los Angeles, California 1978.

HAMILTON, Warren, Mesozoic tectonics of western United States, in *Mesozoic Paleogeography of the Western United States*, edited by D. G. Howell and K. A. McDougall, Pacific Coast Paleogeography Symposium 2, Pacific Coast Section, Soc. Econ. Paleont. and Mineral., pp. 33–70, Los Angeles, California, 1978.

HILLHOUSE, John, Paleomagnetism of the Triassic Nikolai greenstone, south-central Alaska, *Can. J. Earth Sci.*, **14**, 2578–2592, 1977.

HILLHOUSE, J. W. and C. S. GROMMÉ, Paleomagnetism of the Triassic Hound Island Volcanics, Alexander Terrane, southeastern Alaska, *J. Geophys. Res.*, **85**, No. B5, 2594–2602, 1980.

HUDSON, T., George PLAFKER, and Z. E. PETERMAN, Paleogene anatexis along Gulf of Alaska Margin, *Geology*, **7**, 12, 573–577, 1979.

IRVING, E., J. W. H. MONGER, and R. W. YOLE, New paleomagnetic evidence for displaced terranes in British Columbia, in *The Continental Crust and Its Mineral Deposits*, edited by D. W. Strongway, *Geol. Assoc. Canada Spec. Paper*, **20**, 441–456, 1980.

IRVING, E. and R. W. YOLE, Paleomagnetism and the kinematic history of mafic and ultramafic rocks in fold mountain belts, *Ottawa, Earth Physics Branch Pubs.*, **42**, 87–95, 1972.

IRWIN, W. P., Terranes of the western Paleozoic and Triassic belt in the southern Klamath Mountains, California, *U.S. Geol. Survey Prof. Paper*, 800-C, C103–C111, 1972.

IRWIN, W. P., Ophiolitic terranes of California, Oregon, and Nevada, in *North America ophiolites*, edited by R. G. Coleman and W. P. Irwin, *Oregon Dept. Geol. and Min. Ind. Bull.*, **95**, 75–92, 1978.

IRWIN, W. P. and D. L. JONES, Distribution, age, and tectonic significance of ophiolites of western North America, *IGCP Ophiolite Project*, edited by N. Bogdonov, Wiley, London, in press, 1982.

JONES, D. L., M. C. BLAKE, Jr., E. H. BAILEY, and R. J. MCLAUGHLIN, Distribution and character of upper Mesozoic subduction complexes along the west coast of North America, *Tectonophysics*, **47**, 207–222, 1978.

JONES, D. L., M. C. BLAKE, Jr., Claude RANGIN, The four Jurassic belts of northern California and their significance to the geology of the southern California borderland, *Am. Assoc. Petrol. Geologists, Pacific Sect., Misc. Pub.*, **24**, 343–362, 1976.

JONES, D. L. and N. J. SILBERLING, Mesozoic stratigraphy-the key to tectonic analysis of southern and central Alaska: Frontiers in AAAS, Pac, Div. Sym., in *Frontiers of Western Exploration*, edited by P. Rodola, (Also available as *U.S. Geol. Survey Open-file Report*, 79–1200), 1979.

JONES, D. L., N. J. SILBERLING, H. C. BERG, and George PLAFKER, Tectonostratigraphic terrane map of Alaska, *U.S. Geol. Survey Open-file Report*, 81–792, Scale 1:2,500,000, 1981.

JONES, D. L., N. J. SILBERLING, Bela CSEJTEY, Jr., W. H. NELSON, and C. D. BLOME, Age and structural significance of ophiolite and adjoining rocks in the Upper Chulitna district, south-central Alaska, *U.S. Geol. Suvey Prof. Paper*, 1121–A, 21, 1980.

JONES, D. L., N. J. SILBERLING, Wyatt GILBERT, and Peter CONEY, Character, distribution, and tectonic significance of accretionary terranes in the central Alaska Range, *J. Geophy. Res*, in press, 1982.

JONES, D. L., N. J. SILBERLING, and J. HILLHOUSE, Wrangellia—A displaced terrane in north-western North America, *Can. J. Earth Sci.* **14**, 2565–2577, 1977.

JONES, D. L., N. J. SILBERLING, and J. W. HILLHOUSE, Microplate tectonics of Alaska—Significance for the Mesozoic history of the Pacific Coast of North America, in *Mesozoic Paleogeography of the Western United States*, edited by D. G. Howell and K. A. McDougall, Pacific Coast Paleogeography Symposium 2, Pacific Coast Section, Soc. Econ. Paleont. and Mineral., pp. 71–74, Los Angeles, California, 1978.

MONGER, J. W. H., Correlation of Eugeosynclinal Tectono-stratigraphic belts in the North American Cordillera, *Geo. Sci. Can.*, **2**, 4–10, 1975a.

MONGER, J. W. H., Upper Paleozoic rocks of the Atlin terrane, northwestern British Columbia and south-central Yukon, *Geol. Survey Can.*, paper 74–47, p. 63, 21 figs, 1975b.

MONGER, J. W. H., Upper Paleozoic rocks of the western Canadian Cordillera and their bearing on Cordilleran evolution, *Can. J. Earth Sci.*, **14**, 1832–1859, 1977.

MONGER, J. W. H. and E. IRVING, Northward displacement of north-central British Columbia, *Nature*, **285**, 289–294, 1980.

MONGER, J. W. H. and R. A. PRICE, Geodynamic evolution of the Canadian Cordillera—progress and problems, *Can. J. Earth Sci.*, **16**, 770–791, 1979.

MONGER, J. W. H., R. A. PRICE, and D. J. TEMPELMAN-KLUIT, Tectonic accretion and the origin of the two major metamorphic and plutonic welts in the Canadian Cordillera, *Geology*, in press, 1982.

MONGER, J. W. H. and C. A. ROSS, Distribution of fusulinaceans in the western Canadian Cordillera, *Can. J. Earth Sci.*, **8**, 259–278, 1971.

PACKER, D. R. and D. B. STONE, An Alaska Jurassic paleomagnetic pole and the Alaskan Orocline, *Nature Phys. Sci.*, **237**, 25–26, 1972.

PACKER, D. R. and D. B. STONE, Paleomagnetism of Jurassic rocks from southern Alaska and the tectonic implications, *Can. J. Earth Sci.*, **11**, 976–997, 1974.

PLAFKER, George, D. L. JONES, and E. A. PESSAGNO, Jr., A Cretaceous accretionary flysch and melange terrane along the Gulf of Alaska Margin, *U.S. Geol. Circ.*, 751-B, B41–B43, 1977.

RICHTER, D. H. and D. L. JONES, Structure and stratigraphy of eastern Alaska Range, Alaska, *Arctic Geology, AAPG, Mem.*, **19**, 408–420, 1973.

SCHOLL, D. W., T. VALLIER, R. E. VON HUENE, and D. G. HOWELL, Concepts about tectonic processes that affect large sediment bodies at under-thrust ocean margins, *Geology*, **8**, 564–568, 1980.

STONE, D. B., Tectonic history of the Continental Margin of southwestern Alaska: Late Triassic to earliest Tertiary, in *The Relationship of Plate Tectonics to Alaskan Geology and Resources*, edited by H. Sisson, pp. K1–K29, Alaska Geol. Soc., 1979.

STONE, D. B. and D. R. PACKER, Tectonic implications of Alaska Peninsula paleomagnetic data, *Tectonophysics*, **37**, 183–201, 1977.

STONE, D. B. and D. R. PACKER, Paleomagnetic data from the Alaska Peninsula, *Geol. Soc. Am. Bull.*, **90**, 545–560, 1979.

TEMPELMAN-KLUIT, D. J., The Yukon crystalline terrane: enigma in the Canadian Cordillera, *Geol. Soc. Am. Bull.*, **87**, 1343–1357, 1976.

TEMPELMAN-KLUIT, D. J., Transported cataclasite, ophiolite, and granodiorite in Yukon: evidence of arc-continent collision, *Geol. Survey Can. Paper*, 79–14, 27p., 1979a.

TEMPELMAN-KLUIT, D. J., Five occurrences of transported synorogenic clastic rocks in Yukon Territory,

Geol. Surv. Can. Paper, 79-1A, 1–12, 1979b.

TIPPER, H. W., Offset of an Upper Pliensbachian geographic zonation in the North American Cordillera by Transcurrent faulting, *Can. J. Earth Sci*, in press, 1982.

VAN DER VOO, R., M. JONES, C. S. GROMME, G. D. EBERLEIN, and M. CHURKIN, Jr., Paleozoic paleomagnetism and the northward drift of the Alexander Terrane, southeastern Alaska, *J. Geophys. Res.*, **85**, No. B10, 5281–5296, 1980.

VEDDER, J. G., D. G. HOWELL, and Hugh MCLEAN, Stratigraphy, sedimentation, and tectonic accretion of exotic terranes, southern Coast Ranges, California, *Am. Assoc. Petrol. Geol. Memoir Ser. Hedb*, in press, 1982.

WHETTEN, J. T., D. L. JONES, D. S. COWAN, and R. E. ZARTMAN, Ages of Mesozoic terranes in the San Juan Islands, Washington, in *Mesozoic Paleogeography of the Western United States*, edited by D. G. Howell and K. A. McDougall, Pacific Coast Paleogeography Symposium 2, Pacific Section, Soc. Econ. Paleont. and Mineral., pp. 117–132, Los Angeles, California, 1978.

YANCEY, T. E., Permian marine biotic provinces in North America, *J. Paleon*, **40**, 758–766, 1975.

YOLE, R. W. and E. IRVING, Displacement of Vancouver Island: paleomagnetic evidence from the Karmutsen Formation, *Can. J. Earth Sci.*, **17**, 1210–1228, 1980.

Accretion Tectonics in the Circum-Pacific Regions, edited by M. Hashimoto and S. Uyeda, 37–42.

Tectonostratigraphic Terranes of Alaska and Northeastern USSR
—A Record of Collision and Accretion

Michael CHURKIN, Jr.

ARCO Alaska, Inc., P.O. Box 360, Anchorage, Alaska 99510, U.S.A.

Extended Abstract

The accretionary structure of the northern Pacific consists of collision deformed continental plates, large accreted terranes, and packets of appressed microterranes along suture zones (Fig. 1). The mosaic of allochthonous terranes forming Alaska and northeastern USSR and much of the rest of the Cordillera suggests that: (1) various

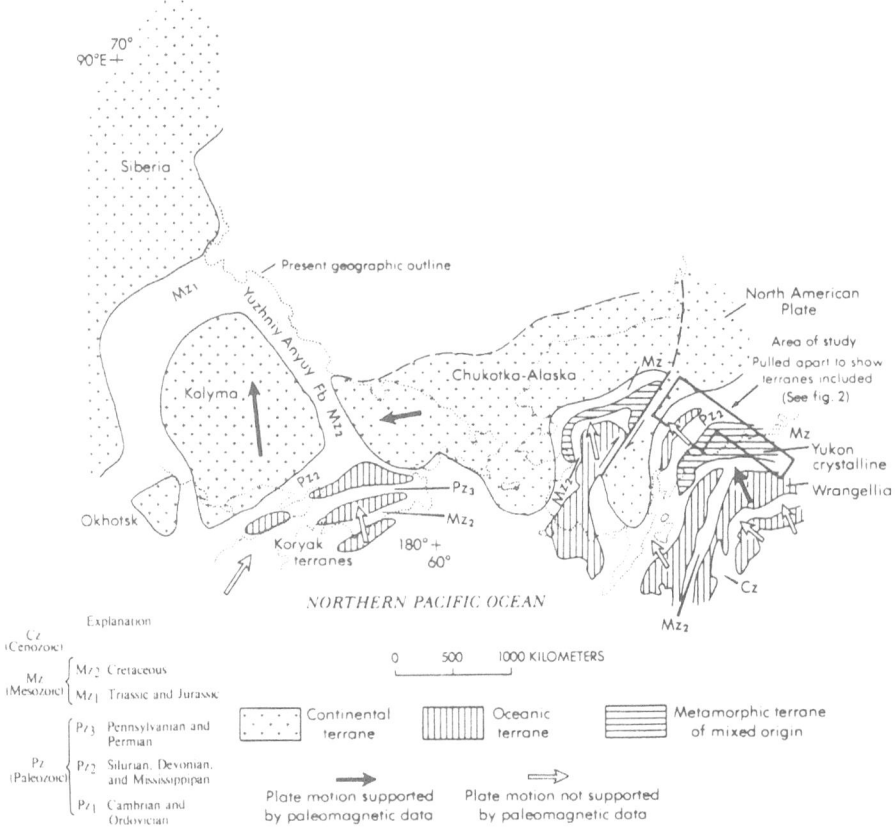

Fig. 1. Pull-apart map of northern Pacific showing major plates and accreted terranes. Individual terranes have been pulled apart by minor reversal of plate movement (From CHURKIN *et al.*, 1981).

types of oceanic features resistant to subduction collided and accreted themselves to Alaska-Chukotka, a peninsular extension of North America; (2) accretion is an old process extending back in time at least into the middle Paleozoic; (3) early Paleozoic continental margins were the oldest docking facilities for migrating plates, and accreted terranes are successively younger toward the modern Pacific; (4) prior to the Cretaceous there was no separate Arctic Ocean basin, and Alaska-Chukotka was separated from the Siberian platform by a wide part of the proto-Pacific plate that extended into the Arctic and Bering Sea; (5) by a process of circumpolar drift and microplate accretion, fragments of the proto-Pacific including fragments less resistant to subduction were cut off and captured in the Arctic; and (6) major strike-slip faults (Tintina and Denali) developed after collision along sutures where the last stages of convergent drift have been oblique to the terrane margins. Where terranes met head on, their leading edges are along multiple thrusts forming packets of microterranes.

Good examples of structures produced by collision and accretion are fault-bounded microterranes and thrust faults in east-central Alaska and the western Brooks Range. In east-central Alaska 17 terranes occur ranging in origin from continental and continental margin to volcanic seamount, arc and ocean floor (Fig. 2). The Yukon crystalline terrane, the largest terrane, is a composite of four terranes juxtaposed across the Tintina fault with the Tatonduk terrane, a northwestern extension of the North American plate in Alaska.

Inboard of the Yukon crystalline terrane are packets of appressed microterranes separated from the Tatonduk terrane and other terranes belonging to North America by splays of the Tintina fault system (Fig. 3). These microterranes define several suture zones separating the mosaic of accretionary terranes of southern Alaska from North America. The most obviously allochthonous terranes within these suture zones are the Woodchopper Canyon terrane, an Early Devonian basaltic seamount, and the White Mountains terrane, an Ordovician volcanic arc terrane capped by Silurian and Devonian carbonate bank deposits. The nearest counterpart to these terranes is the Alexander terrane in southeastern Alaska.

The Tintina fault of Cenozoic age, like other major strike-slip faults of Alaska, follows old suture zones separating terranes. Strike-slip faulting developed after collision in places where the last stages of convergent drift have been oblique to the terrane margins. Where terranes met head on, their leading edges lie along a multiple set of faults that form packets of microterranes.

In the western Brooks Range the allochthonous Kagvik terrane of upper Paleozoic and Triassic argillaceous and cherty rocks lies structurally below thrust-faulted sheets of coeval carbonate rocks, including the Lisburne Group, and structurally above quartz mica schist and greenstone of the southern Brooks Range (Fig. 4). Sedimentary features of the shelly fossil-rich carbonate strata of the Lisburne in the upper thrust sheets indicate deposition in a shallow-water shelf environment. The underlying shale and chert of the Kagvik is an attenuated section about 500 m thick repeated by imbricate thrusts. Ubiquitous pelagic fossils and volcanic material locally forming andesitic tuff and flows point to an oceanic environment of deposition. The presence of some limestone turbidites interbedded with the ocean-floor sediments of the Kagvik suggests that a carbonate shelf was nearby and that a south-facing

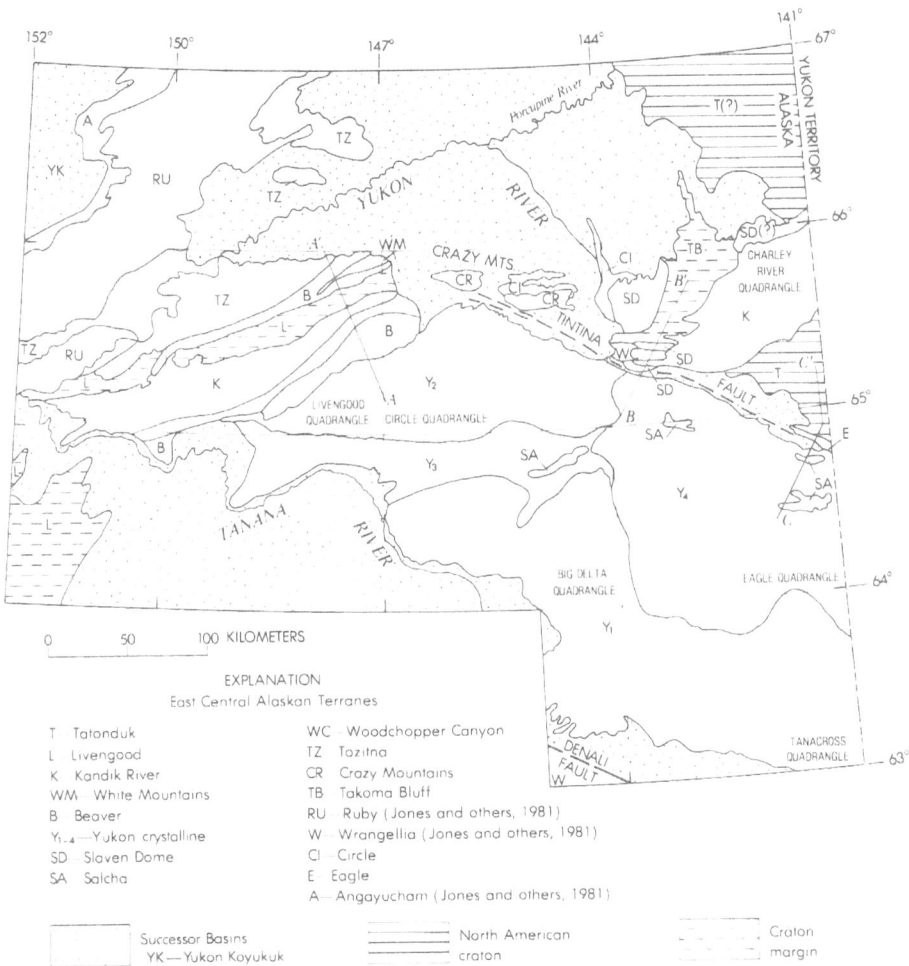

Fig. 2.   Terranes of east-central Alaska and locations of sections shown in Figure 3. (From CHURKIN *et al.*, 1982).

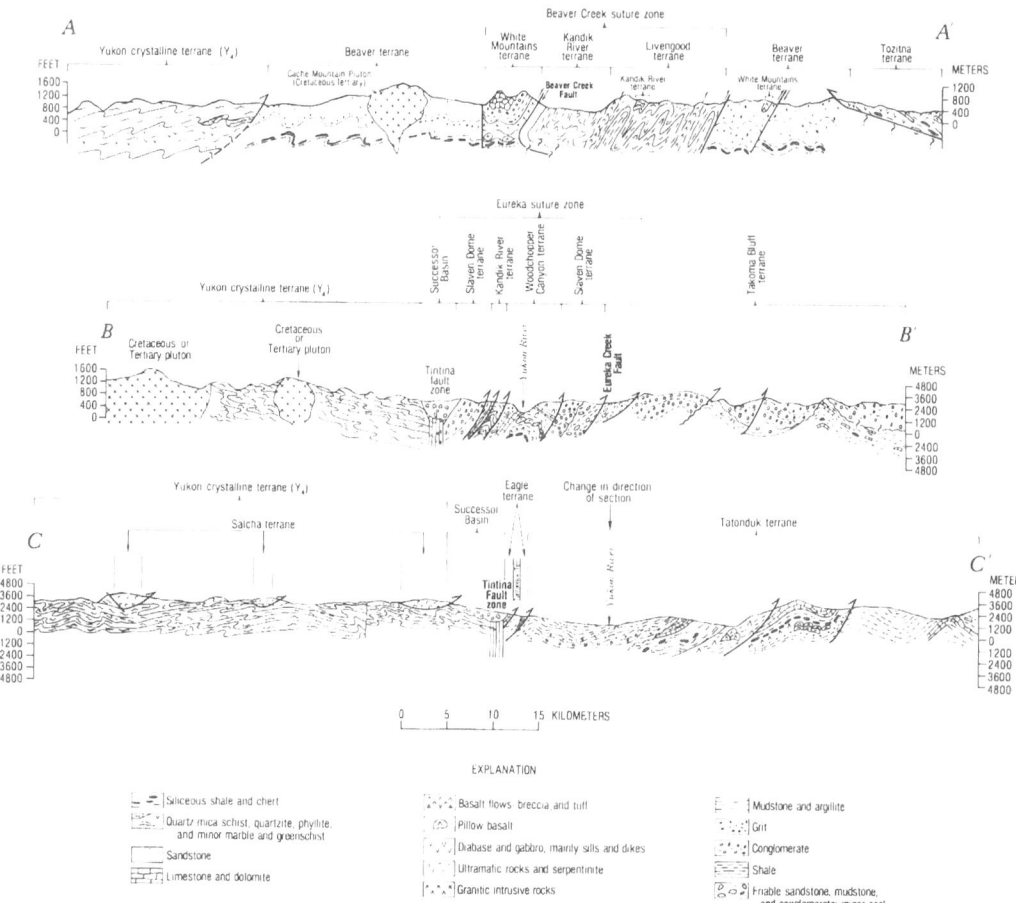

Fig. 3.   Cross sections showing accretionary structures in suture zones (From Churkin *et al.*, 1982).

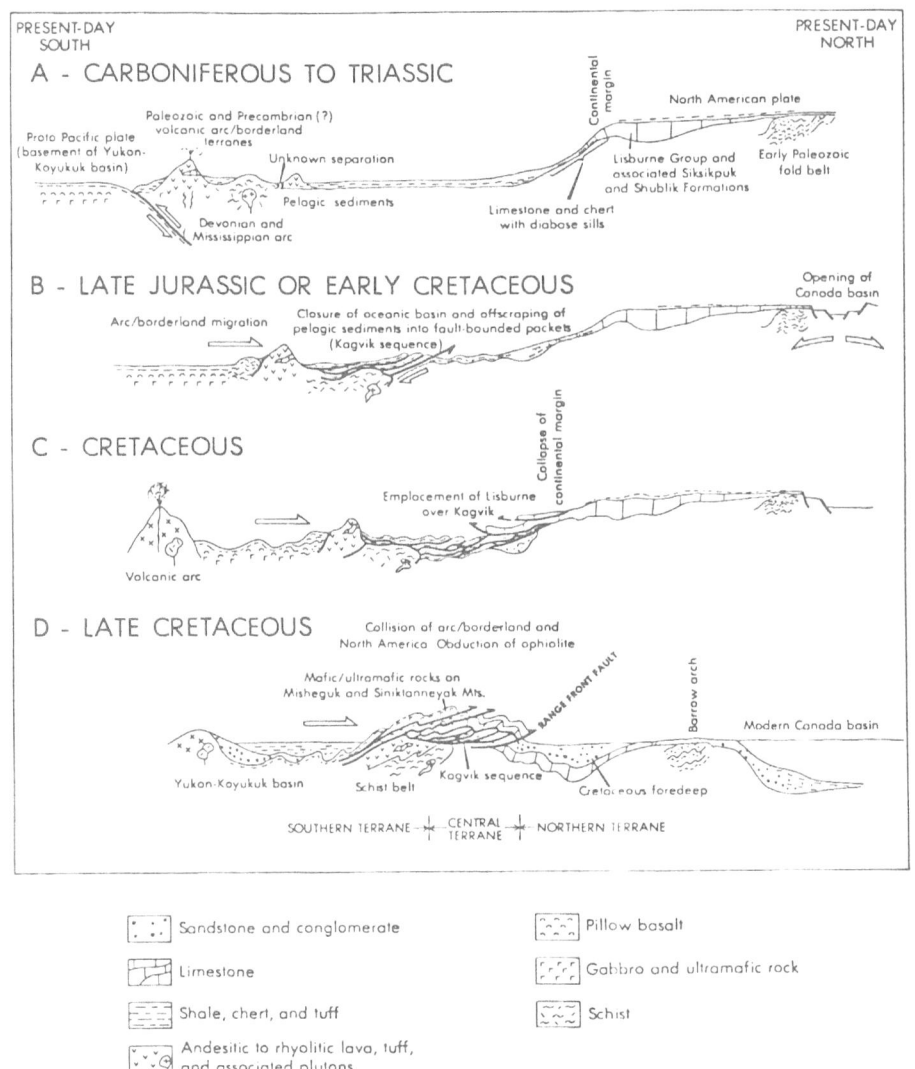

Fig. 4. Tectonic history of northwestern Brooks Range, Alaska, showing collision and accretion (From CHURKIN *et al.*, 1979).

continental margin existed during late Paleozoic and early Mesozoic time near the present-day southern Brooks Range. Collapse of this longstanding continental margin appears to be related to collision and accretion of oceanic and microcontinental blocks now represented by metamorphic rocks along the south flank of the Brooks Range. The structurally higher thrust slices of mafic and ultramafic rocks emplaced on top of the collapsed continental margin represent ophiolitic basement of the Yukon-Koyukuk basin.

REFERENCES

1. Churkin, M., Jr. and G. D. Eberlein, Ancient borderland terranes of the North American Cordillera: Correlation and microplate tectonics, *Geol. Soc. Am. Bull.*, **88**, 769–786, 1977.
2. Churkin, M., Jr., W. J. Nokleberg, and Carl Huie, Collision-deformed Paleozoic continental margin, western Brooks Range, Alaska, *Geology*, **7**, 379–383, 1979.
3. Churkin, M., Jr. and J. H. Trexler, Jr., Circum-Arctic plate accretion—Isolating part of a Pacific plate to form the nucleus of the Arctic basin, *Earth Planet. Sci. Lett.*, **48**, 356–362, 1980.
4. Churkin, M., Jr., C. Carter, and J. H. Trexler, Jr., Collision-deformed Paleozoic continental margin of Alaska—Foundation for microplate accretion, *Geol. Soc. Am. Bull.*, **91**, 648–654, 1980a.
5. Van der Voo, R. M. and others, Paleozoic paleomagnetism and northward drift of the Alexander terrane, southeastern Alaska, *J. Geophys. Res.*, **85**, 5281–5296, 1980.
6. Churkin, M., Jr., H. L. Foster, R. M. Chapman, and F. R. Weber, Terranes and suture zones in east central Alaska, *J. Geophys. Res.*, **87**, 3718–3730, 1982.

Accretion Tectonics in the Circum-Pacific Regions, edited by M. Hashimoto and S. Uyeda, 43–57.

Accretionary Terranes and Tectonic Evolution of Northeast Siberia

Kazuya FUJITA and James T. NEWBERRY

*Department of Geology, Michigan State University,
East Lansing, Michigan 48824, U.S.A.*

The northeastern USSR is composed of allocthonous terranes of both continental and oceanic origin which were accreted onto the Siberian platform in post-Triassic time. Lithologic similarities between the Cherskiy and Novosibirsk terranes and the Siberian platform indicate that the Late Paleozoic to mid-Mesozoic plate margin existed near the Alazeya uplift in the center of the previously postulated Kolyma plate. The Omolon and Prikolymsk terranes were accreted to the eastern margin of the Siberian plate in mid-Jurassic time terminating subduction in the Alazeya region. The enlarged Siberian plate collided with the Arctic Alaska plate in Hauterivian time with the intervening oceanic crust being preserved as the South Anyui terrane. The Koryak highlands are composed of terranes of oceanic origin. Oceanic crust of Late Paleozoic, Middle Mesozoic, and Late Cretaceous age are found, the first of which was trapped behind a post-Jurassic island arc. The ophiolites were obducted onto the adjacent continent and the island arc in four episodes: Late Jurassic, Hauterivian, Early Senonian, and Eocene. The last is associated with the collision of the Shirshov Ridge and the jump in the locus of subduction from the Koryak region to the Aleutians. The presence of Paleozoic oceanic crust in both Koryak and Chulitna, Alaska, suggest a large oceanic plate of Paleozoic age subducted in this area in the Mesozoic.

1. Introduction

Plate tectonic paleoreconstructions of the continents for the Paleozoic and the Mesozoic have ignored the northeastern Soviet Union. Many workers have treated this area as an appendage of Siberia (e.g., SMITH *et al.*, 1981) or of North America (CHURKIN, 1972). More recently, the existance of ultramafic rocks and ophiolite complexes in the South Anyui fold belt of southwestern Chukotka have led to the postulation of a Phanerozoic Kolyma plate encompassing the region between the Verkhoyansk Mountains and the South Anyui fold belt (ZIEGLER *et al.*, 1977; FUJITA, 1978; CHURKIN and TREXLER, 1980).

SESLAVINSKIY (1979) and ANDREWS-SPEED (1981), however, noted that although evidence for a suture in the South Anyui fold belt was excellent, the diagnostic features of a subduction zone were absent in the Cherskiy fold belt or the adjacent Inyali-Deba synclinorium (Fig. 1). In contrast, oceanic and convergent margin type facies were discovered in the central part of the "Kolyma plate" and oceanic-type lithologies were encountered eastward of the Cherskiy Mountains (FUJITA and NEWBERRY, 1982).

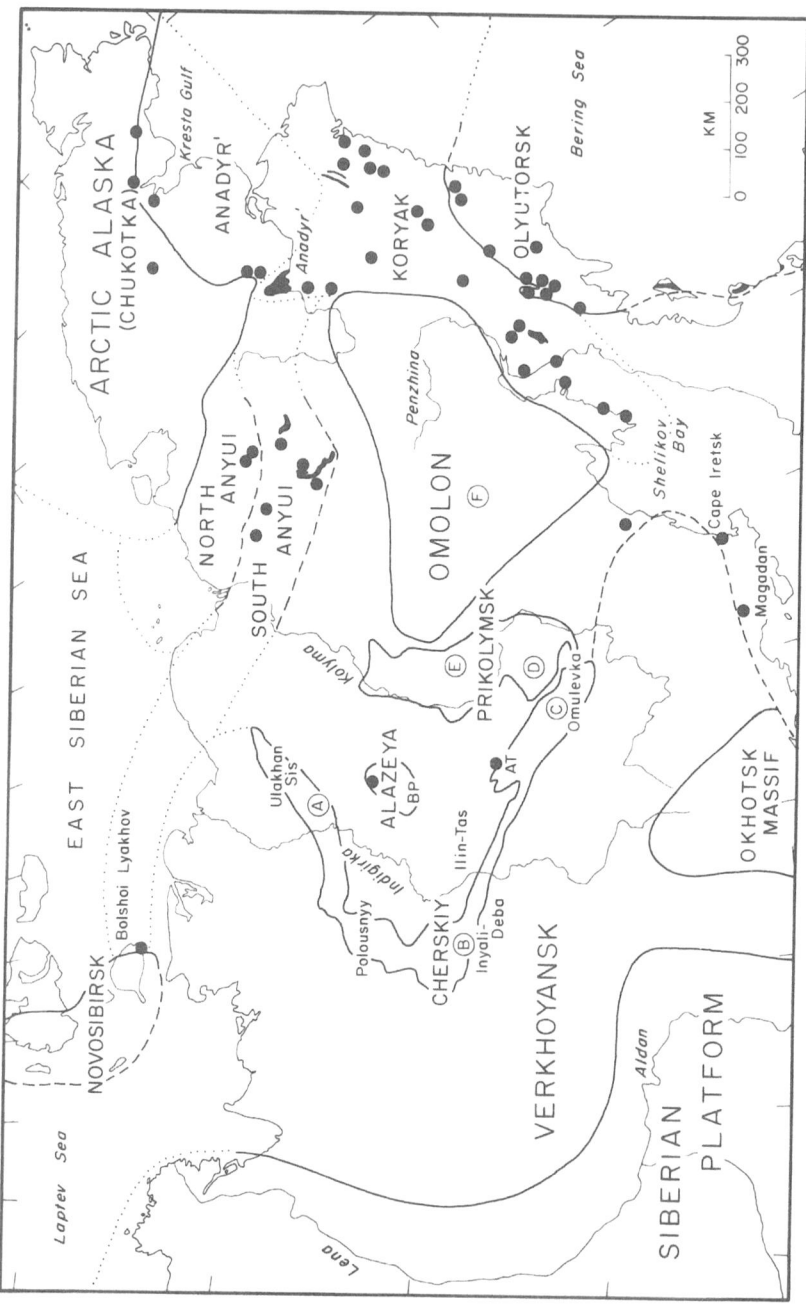

Fig. 1. Index and preliminary tectonostratigraphic terrane map of northeast Siberia. Dashed lines indicate uncertain terrane boundaries and dotted lines indicate speculative boundaries. Dots and solid regions show locations of ophiolite and melange complexes. Circled letters denote locations of stratigraphic columns shown in Fig. 2. AT is the Arga Tas range and BP is the Badyarikha projection.

In this paper, we summarize the geology of the northeastern USSR and construct a preliminary map of tectonostratigraphic terranes for the region. The terranes are primarily based on stratigraphic data using the criteria of JONES and SILBERLING (1979) and CONEY et al. (1980). Zones of ophiolites are used to identify fragments of oceanic crust and, in conjunction with stratigraphic and tectonic data, to determine plate boundaries and postulate a tectonic evolution for the area.

Data have been obtained primarily from Soviet literature in translation (see FUJITA and DRETZKA, 1978) and from paleogeographic and lithologic maps (e.g., VINOGRADOV, 1967, 1968a, b, 1969). Due to the paucity of data in many sections of the northeastern USSR, the resolution of terranes and plates is much poorer than in similar studies on the western United States (e.g., CONEY et al., 1980; JONES et al., 1977). It is probable that many terranes identified here are actually composite terranes.

2. Terranes of the Northeastern USSR

On the basis of available data, we have divided northeast Siberia east of the Lena River into the terranes shown in Fig. 1. The Siberian platform consists of Precambrian basement overlain by a thin veneer of Paleozoic and Mesozoic sediments. Immediately to the east lies the Verkhoyansk complex, a thick pile of deformed Permian to Jurassic clastic sediments. There is no evidence for oceanic crust under the complex and it appears to represent a foreland basin lying on an extension of the Siberian platform (FUJITA and NEWBERRY, 1982). The southern portions of the terrane are covered by the Cretaceous extrusives of the Okhotsk-Chukotsk volcanic belt and much of the terrane is intruded by granitic intrusives of Early Cretaceous age (NENASHEV, 1979). The eastern boundary of the Verkhoyansk terrane with the adjacent Cherskiy terrane lies along the Darpir fault system. Verkhoyansk complex sediments are also believed to extend north of the Ulakhan Sis Range and partially into the Laptev Sea.

The Cherskiy terrane consists primarily of Paleozoic carbonates which thicken to the west (BOGDANOV, 1960) and become associated with igneous formations to the east (FUJITA and NEWBERRY, 1982). The similarity of the stratigraphic columns A-C in Fig. 2 suggest that the Ulakhan Sis Range, the Polousnyy uplift, the Cherskiy Mountains, and the Omulevka uplift all form part of the same terrane. Scraps of obducted oceanic crust are preserved in the Arga Tas Mountains (SAVOSINA et al., 1976). These consist of two thrust sheets, consisting of spilite, basalt, and gabbro, separated by a fault zone containing serpentinized ultramafics. A second zone of terrigenous volcanic rocks developed on continental crust is found 30 km to the southeast (MERZLYAKOV, 1971), thus suggesting a progression towards the north from continent to ocean. These interpretations, however, have been questioned by GULYAYEV (1980).

The Alazeya terrane consists of the Alazeya uplift and the Badyarikha projection. The Alazeya uplift has been known to be the site of intermediate island arc type volcanism throughout the Late Paleozoic and Early Mesozoic (NEKRASOV, 1961; GULYAYEV, 1975; RUSAKOV et al., 1977). The base of the complex, however, has been the source of great controversy. The oldest rocks consist of siliceous clayey rocks, tuffaceous mudstone, spilite, basalt, quartz keratophyres, and andesite porphyry (RUSAKOV et al., 1975). There are also lenses of marmorized limestones and minor

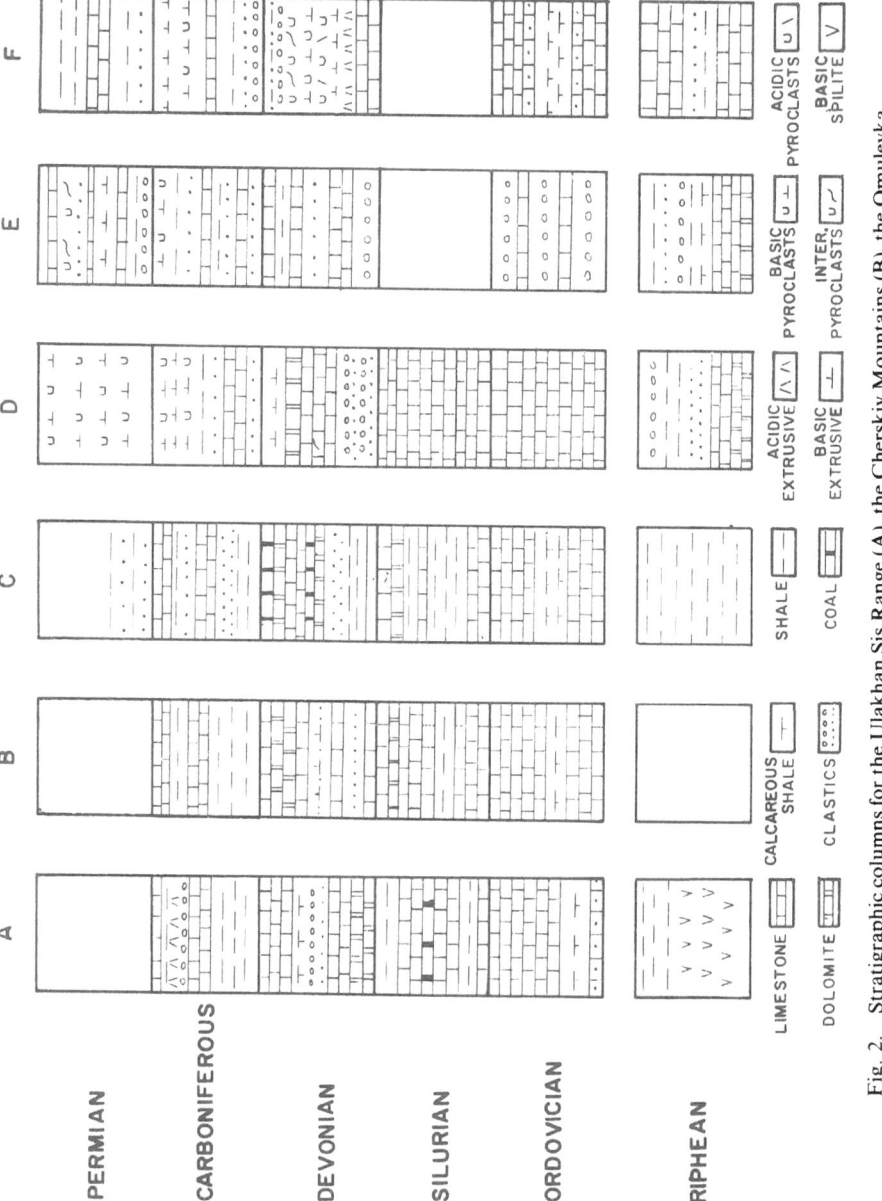

Fig. 2. Stratigraphic columns for the Ulakhan Sis Range (A), the Omulevka Uplift (C), southern Prikolymsk (D), central Prikolymsk (E), and the Omolon massif (F). Data taken from Vinogradov et al. (1968b, 1969).

ultramafic bodies, primarily serpentinites, in fault contact with the host rocks (SHILO *et al.*, 1973b). The above can be interpreted as portions of an ophiolite complex overlain by an island arc. GULYAYEV (1975) questions this interpretation and suggests that the ultramafics are "volcanic rocks of an ancient chert-volcanic sequence" while the spilites are spherical partings in fault zones. He also identifies the keratophyres as trachydacite and rhyolite and claims the jaspers are tuffs. Either interpretation, however, suggests the presence of a convergent plate margin in the immediate vicinity. The GULYAYEV (1975) interpretation places the Alazeya rocks on a continental margin while the SHILO *et al.* (1973b) one indicates a forearc environment. The Badyarikha projection is similar to the lower parts of the Alazeya complex (SHILO and TIL'MAN, 1981) and has been included in the Alazeya terrane.

There is some geophysical evidence to support the more oceanic interpretation. Gravimetric and magnetic data have been collected and interpreted by SHILO *et al.* (1978) and suggest a thin (5 km) layer of sediments overlying a basaltic layer near the exposed part of the Alazeya terrane while the Prikolymsk block has sediments and granitic crust to a depth of 16 km. Elsewhere in central Kolyma, a 9 km pile of sediment overlying a basaltic layer fit the data. Due to the overlying Cenozoic sedimentary cover, the relationship of the Alazeya terrane to adjacent Paleozoic terranes is uncertain.

FUJITA and NEWBERRY (1982) identified a separate Prikolymsk terrane on the basis of a higher amount of clastics in the Devonian section. Columns D and E of Fig. 2 suggest that this difference exists throughout the geologic column and is accentuated in the Late Paleozoic by the presence of pyroclastics. In addition, if the Alazeya uplift and the Arga Tas range represent regions of subduction of oceanic crust through the Paleozoic and into the Mesozoic, geometric considerations preclude the Prikolymsk block from being in its present relative position prior to Jurassic time. A series of faults separate the Prikolymsk terrane from the Cherskiy terrane.

A major fault zone separates the Prikolymsk terrane from the Omolon terrane. The Omolon terrane is very distinct from the Cherskiy and Prikolymsk terranes both in lithology (column F, Fig. 2) and in having a very distinct polar wandering curve which is different from that of the Siberian platform (FUJITA and NEWBERRY, 1982). The Omolon terrane is shown to include the Avekova uplift of northern Taygonos Peninsula and is bounded on the north by the South Anyui suture. The eastern boundaries of the Omolon terrane are indeterminate since they are buried under the Okhotsk-Chukotsk volcanic belt.

The South Anyui suture consists primarily of oceanic crust overlain by deep-sea sediments (SESLAVINSKIY, 1970, 1979). This terrane may extend westward to an ultramafic outcrop on Bolshoi Lyakhov Island (PUSHCHAROVSKIY, 1976) along a band of magnetic anomalies trending west northwest from the mouth of the Kolyma River (SHILO *et al.*, 1973a; IVANOV and BELYAYEV, 1973). The South Anyui suture closed about Hauterivian time (SESLAVINSKIY, 1970) and, from the continuity of the Okhotsk-Chukotsk volcanic belt (KHRENOV *et al.*, 1976), must have done so by Aptian time. Associated with this suture is the North Anyui fold belt consisting of uplifts with volcanics separated by fault bounded slate filled basins (BELYAYEVA *et al.*, 1973). It is possible that these may represent accreted island arcs.

The Novosibirskiye Ostrova (New Siberian Islands) consist of an Early Paleozoic carbonate platform similar to that in the Cherskiy terrane. The Early Paleozoic of the Cherskiy and Novosibirsk terranes is very similar to that of the margins of the Siberian Platform and thus probably represent parts of the same plate (FUJITA and NEWBERRY, 1982). Thus, the boundary of the Siberian plate probably lies along the eastern side of the Novosibirsk terrane, along the western end of the South Anyui suture, near the Alazeya uplift, and between the Verkhoyansk and Omolon terranes. The Prikolymsk terrane and the Omolon terrane appear to be separate microcontinents which attained their present relative position to Siberia in Jurassic time.

3. Other Plate Margins

In addition to the ophiolite associations of the South Anyui and Alazeya terranes, there exist several groups of ophiolites and melanges along the fringes of the Late Mesozoic continent. A series of linear bands of melange and obducted oceanic crust lie in the Koryak highlands region (encompassing the Anadyr', Koryak, and Olyutorsk terranes of Fig. 1) while isolated small exposures of what may be sections of ophiolites are found under the Okhotsk-Chukotsk volcanic belt near Magadan and along the shores of Kresta Gulf. It is therefore likely that these locations also lie adjacent to paleo-convergent plate boundaries.

In the Magadan region, the largest ultramafic complex is found on Iret'sk Cape along the Chelomdzha-Yamsk deep fault. The fault is a southward dipping thrust separating the miogeosynclinal Verkhoyansk terrane from the intermediate-volcanic island arc facies of the Siglan geosyncline, which can be interpreted as an accreted arc (FUJITA, 1978), to the south (YUDIN and IZMAILOV, 1966). The Iret'sk ultramafics consist primarily of peridotites and pyroxenites which are in tectonic contact with host gabbro-amphibolites and greenschist. These ultramafics are similar in chemical composition to those found in the Koyak highlands (UMITBAYEV, 1977). Along with geophysical evidence (BELYAYEV et al., 1966), it is suggested that the Taygonos Peninsula facies extend into this region. Several other ultramafic bodies are found in the region (UMITBAYEV, 1977).

Although Jurassic ophiolites were found in the western Brooks Range of Alaska (PATTON et al., 1977), oceanic formations were unknown in Chukotka until quite recently. KOSYGIN et al. (1974) and VOYEVODIN et al. (1978) have reported the presence of serpentinite, harzburgite, and pyroxenite in tectonic contact with gabbro, basalt, spilite, and siliceous rocks in several parts of southern Chukotka near Kresta Gulf. Although the ages of these formations are not known with certainty, the overlying siliceous-volcanic deposits are between Late Triassic and Late Jurassic in age and the ultramafics are cut by Cretaceous granodiorites. Thus, an emplacement age at the end of the Jurassic is suggested. Since the Chukotkan and Brooks Range ophiolites appear to have been emplaced at about the same time, FUJITA and NEWBERRY (1982) have suggested that they may have been emplaced by the same tectonic event; perhaps the inception of the rotation of Arctic Alaska.

4. Koryak Highlands

The Koryak highlands represent the most seaward portion of northeastern Siberia (Fig. 1). The Koryak highlands are bordered on the west by the Okhotsk-Chukotsk volcanic belt overlying the Paleozoic deposits of the Omolon and Arctic Alaskan terranes. Many of the tectonic elements of the Koryak region extend southward into Kamchatka, however, we restrict our discussion here to the portion north of 61°N.

The Koryak highlands have long been known to be of accretionary origin. ZIEGLER *et al.* (1977) and CHURKIN and TREXLER (1981) suggested that the inner regions, encompassing the Ust' Belaya and Penzhina ophiolites (Fig. 3) were accreted in the Carboniferous while the outer regions (Ekonay, Maynitsk; Fig. 3) were accreted in Jurassic time. Recent geologic data suggest, however, that the accretion of the Koryak terranes is much more recent, of Cretaceous age. We first briefly summarize the ophiolite and ultramafic complexes.

Ust' Belaya: Ultramafic rocks with a radiometric age data of 380 m.y. are overlain by basalt, spilite, gabbro, red chert, and tuffs. The overlying sediments contain a Middle Devonian fauna. The entire complex is thrust over Valanginian clastic sediments (ALEKSANDROV, 1974).

Vayegi: The basal unit consists of undated (Paleozoic?) greenstone, silicilyte, altered extrusives, and tuffs. These are in tectonic contact with a melange containing breccia with spilite, diabase, siliceous rock fragments, and blocks of Paleozoic (Silurian to Devonian?) limestone. The melange is dated to be of Triassic age and is thrust over Jurassic to Hauterivian graywackes, fine clastics, and volcanics (ZINKEVICH, 1978; IVANOV and IL'CHENKO, 1978). This record suggests a subduction zone in the Triassic which consumed Paleozoic oceanic crust.

Khatyrka: Blocks of pyroxenite and gabbro, overlain by Late Silurian to Early Permian limestones, are found in a melange. Triassic clastics and Jurassic siliceous sediments are also mixed into the melange. The complex has been thrust over Jurassic to Valanginian cherty rocks containing mafic volcanics. Ophiolite fragments are found in the overlying Senonian conglomerate (IVANOV and BARATOV, 1974). These deposits suggest the subduction of Late Paleozoic oceanic crust in a Mesozoic subduction zone with the whole complex tectonically emplaced in pre-Senonian time.

Ekonay: The Ekonay complex consists of two thrust plates overlying a Kimmeridgian to Aptian sequence of clastics, flysch, and tuff. The first sheet consists of a Cretaceous olistostrome with olistoliths of Paleozoic rocks. On top lies a relatively intact section of Paleozoic ocean floor consisting of ultramafics and gabbro overlain by basalt, greenstones, jaspers, and limestones with a Tournasian (Carboniferous) fauna. Further up the section are cherts, basic lavas, and limestones of Late Paleozoic age. The entire thrust sheet complex is overlain by a Campanian to Paleogene volcanic-terrigenous complex (ALEKSANDROV *et al.*, 1975). Thus, it is suggested that Paleozoic oceanic crust was emplaced in mid-Cretaceous time in a subduction complex which was active throughout the Early Cretaceous.

Maynitsk: Two types of basal complex exist in this region. The southern (Chirynay) melange consists of crushed serpentinite, gabbros, ultramafics, limestones, jaspers, clastics, and tuffs with occasional large tabular blocks of ultramafics, gabbro,

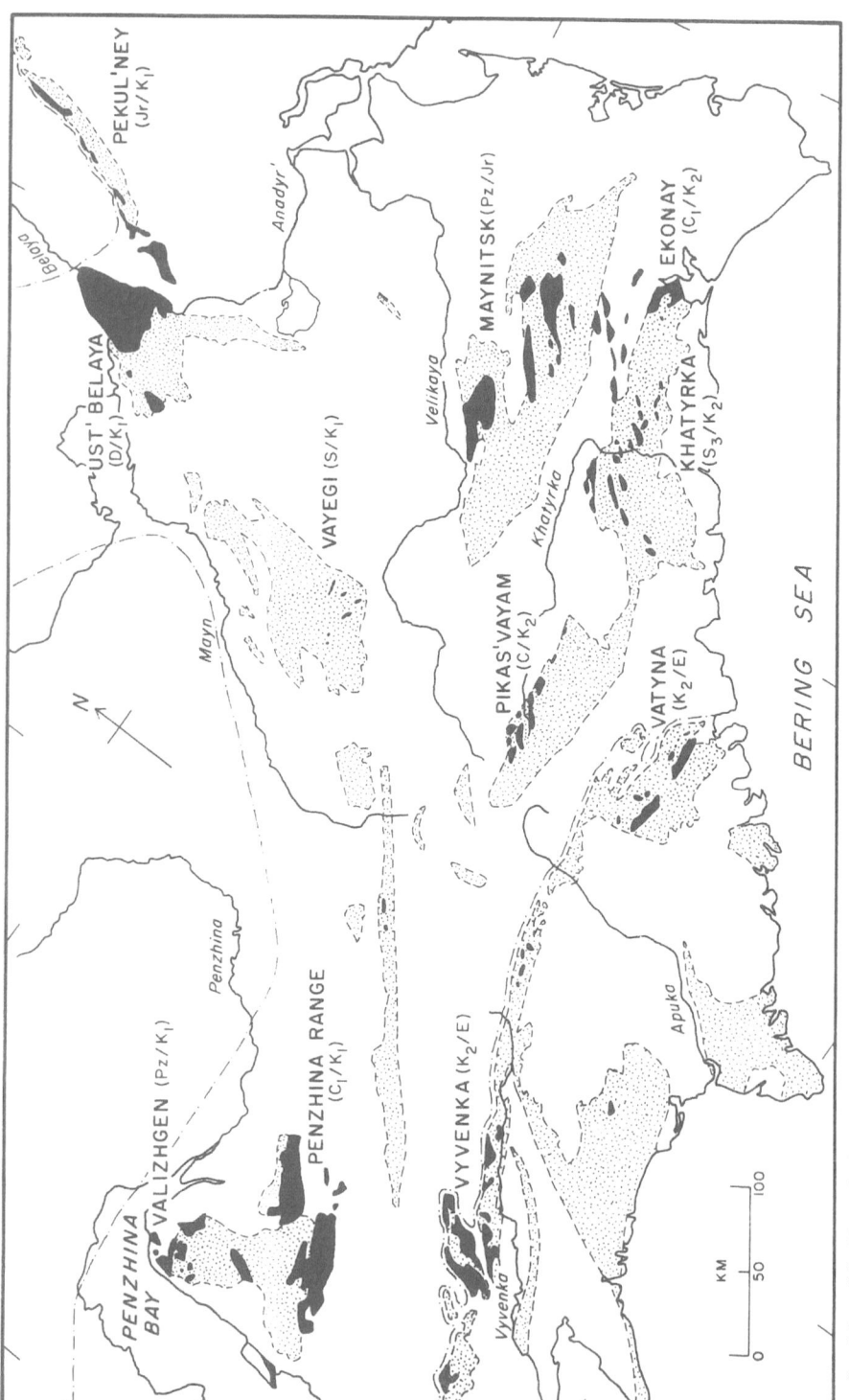

Fig. 3.   Map of the Koryak highlands showing outcrops of ultramafics and melange (solid) and deep marine sediments (dotted). Dash-dot lines show the northern limit of the Olyutorsk terrane and approximate extent of the Omolon and Arctic Alaska terranes. After the name of each ophiolite, the ages of the oceanic crust and emplacement are given in parentheses separated by a slash. Pz is Paleozoic, S is Silurian, D is Devonian, C is Carboniferous, Jr is Jurassic, K is Cretaceous, and E is Eocene. The subscripts 1, 2, and 3 correspond to early, middle, and late, respectively.

and Paleozoic effusives with limestones, and the northern (Yagel'naya) melange of peridotites, gabbros, and tuffs, without any Paleozoic rocks, and with massive blocks of dunite, peridotite, pyroxenite, spilite, and tuff. These are overlain by coarse clastics, with pebbles of the ultramafics and gabbros, and radiolarian chert of Late Jurassic age. Towards the southwest, the cherts are replaced by terrigenous sediments with isolated basalts and tuffs containing Cretaceous faunas (ALEKSANDROV et al., 1975). These formations suggest a Jurassic subduction complex which has been overlain by an island arc. From the differences in the composition of the two melanges, two different ages of oceanic crust may be represented here, a Paleozoic one and a Mesozoic one.

Penzhina Range: Ultramafic bodies and gabbros are found here in tectonic contact with blocks of limestones of Late Carboniferous to Permian age, jaspers, and limestones containing Paleozoic crinoids. The ultramafics consist of serpentinized peridotites and serpentinites. A radiometric age of 320–350 m.y. has been obtained by IVANOV and IL'CHENKO (1978). The ultramafics are overlain by Early Cretaceous coarse clastics and volcanic deposits (PINUS et al., 1973; DOBRETSOV and PONOMAREVA, 1965).

Pekul'ney: Late Jurassic to Valanginian basalt, spilite, jasper, and siltstone overly a melange including blocks of wehrlite, dunite, and lherzolite. The melange includes blocks of Barremian to Hauterivian volcano-clastic rocks as well as some schists. The melange complex is thrust over a Proterozoic to Vendian schist belt (NEKRASOV, 1978). A Late Hauterivian to Early Barremian emplacement of Jurassic oceanic crust is suggested.

Vatyna and Vyvenka: The Vatyna and Vyvenka complexes preserve ultramafics and deep-marine sediments of a much younger age than in any of the previously described complexes. The main thrust sheet consists of a section with pillow lavas at the base with lenses of red siliceous rocks. Higher in the section, a siliceous tuffaceous sequence is encountered containing red and green cherts. Overlying these in thrust contact are ultramafics consisting of dunite, pyroxenite, and gabbros. Radiometric ages of about 30 to 65 m.y. have been obtained. Fragments of the ultramafics are found in Oligocene deposits (MITROFANOV et al., 1980; ALEKSEYEV, 1979; ALEKSANDROV et al., 1980). The emplacement of Late Cretaceous oceanic crust in Paleocene or Eocene time is indicated.

The above data indicate three ages for oceanic crust which was subducted or obducted in the region: Paleozoic (Ust' Belaya, Vayegi, Penzhina Range, Ekonay, Khatyrka, and Chirynay), Early to Middle Mesozoic (Pekul'ney, Yagel'naya, and Kresta Gulf), and Late Cretaceous (Vatyna and Vyvenka). On this basis, we have divided the Koryak highlands into three terranes, Anadyr,' Koryak, and Olyutorsk, in order of decreasing oceanic crustal age. Four distinct emplacement ages are also found: Late Jurassic (Kresta Gulf), Hauterivian (Pekul'ney, Ust' Belaya, and Penzhina Range), Early Senonian (Khatyrka and Ekonay), and Eocene (Vatyna and Vyvenka).

The tectonic evolution of the Koryak highlands may be summarized as follows. Sometime in Triassic time, subduction began under the eastern margin of the Omolon block. This episode is represented by the melange of the Vayegi region. In the mid-Jurassic, an island arc developed at the current site of the Maynitsk and Ekonay zones. The initial subduction at this arc is reflected in the Chirynay ophiolites and the overlying shallow water and tuffaceous formations. Continued subduction under this arc developed the melanges and olistostromes of the Ekonay and Khatyrka zones.

The formation of this island arc trapped the oceanic crust now forming the basement of the Koryak terrane. The back-arc basin was in-filled with Jurassic to Cretaceous volcanic-sedimentary rocks of the so-called Okhotsk complex. Subduction in the Vayegi region ceased sometime in the Jurassic.

In the latest Jurassic or the earliest Cretaceous, Arctic Alaska started to rift off of the Canadian Arctic and the Kresta Gulf and Brooks Range ophiolites were emplaced on its leading margin. In Hauterivian time, Arctic Alaska collided with the Omolon block. With this event, the relative convergence directions changed and the Ust' Belaya and Pekul'ney ophiolites were emplaced. The presence of the younger crust of the Anadyr' terrane along the southern margin of the Arctic Alaska plate suggests that either the crust subducting under Arctic Alaska was part of a different plate than that under Omolon or that an episode of back-arc spreading occurred in the Mesozoic to create this younger section.

Subduction continued through the Jurassic and Early Cretaceous under the Maynitsk-Ekonay arc. The subduction angle was very shallow resulting in a wide arc-trench gap and the formation of the Okhotsk-Chukotsk volcanic belt which was active from Aptian through Maastrichtian time (e.g., KHRENOV and BUKHAROV, 1973; KHRENOV et al., 1976).

In Early Senonian time, a change in relative motions or the collision of an aseismic ridge or seamount chain emplaced the Ekonay ophiolite onto the preexisting arc. Simultaneously, the Khatyrka ophiolite, which is probably an extension of the Ekonay ophiolite, was emplaced. The emplacement age of the Maynitsk ophiolite is not certain, however, it must predate this time. Subduction diminished towards the end of the Cretaceous. In the Early Cenozoic, Shirshov ridge, which may be an oceanic plateau (BEN-AVRAHAM and COOPER, 1981), came into proximity. This plateau was attached to a Late Cretaceous oceanic plate (Kula?) and upon collision with Asia emplaced the Olyutorsk terrane. Due to the bouyancy of the ridge, the site of subduction shifted to the Aleutian trench isolating the Commander and Aleutian basins of the Bering Sea. Much of the basalt in the Vatyna and Vyvenka ophiolites are alkaline and may indicate that much of the Olyutorsk terrane represents a seamount province.

5. Conclusions

We suggest that the whole of northeastern Siberia east of the Alazeya uplift is composed of a mosaic of accreted continental (Omolon, Prikolymsk, and Arctic Alaska) and oceanic (e.g., Koryak, South Anyui) terranes. Figure 4 shows the tectonic evolution of northeastern Siberia as proposed by FUJITA and NEWBERRY (1982) with the addition of the Koryak terranes. The present configuration of the continental fragments and a summary of the convergent margins between them is shown in Fig. 5.

The exact timing and direction of thrusting of the Koryak emplacements is still uncertain. The presence of two distinct episodes of siliceous sedimentation (Paleozoic and Jurassic; IGUMENSHCHEV et al., 1976) suggests a long oceanic history followed by an island arc-back-arc basin stage. It is possible that the Paleozoic oceanic crust of the

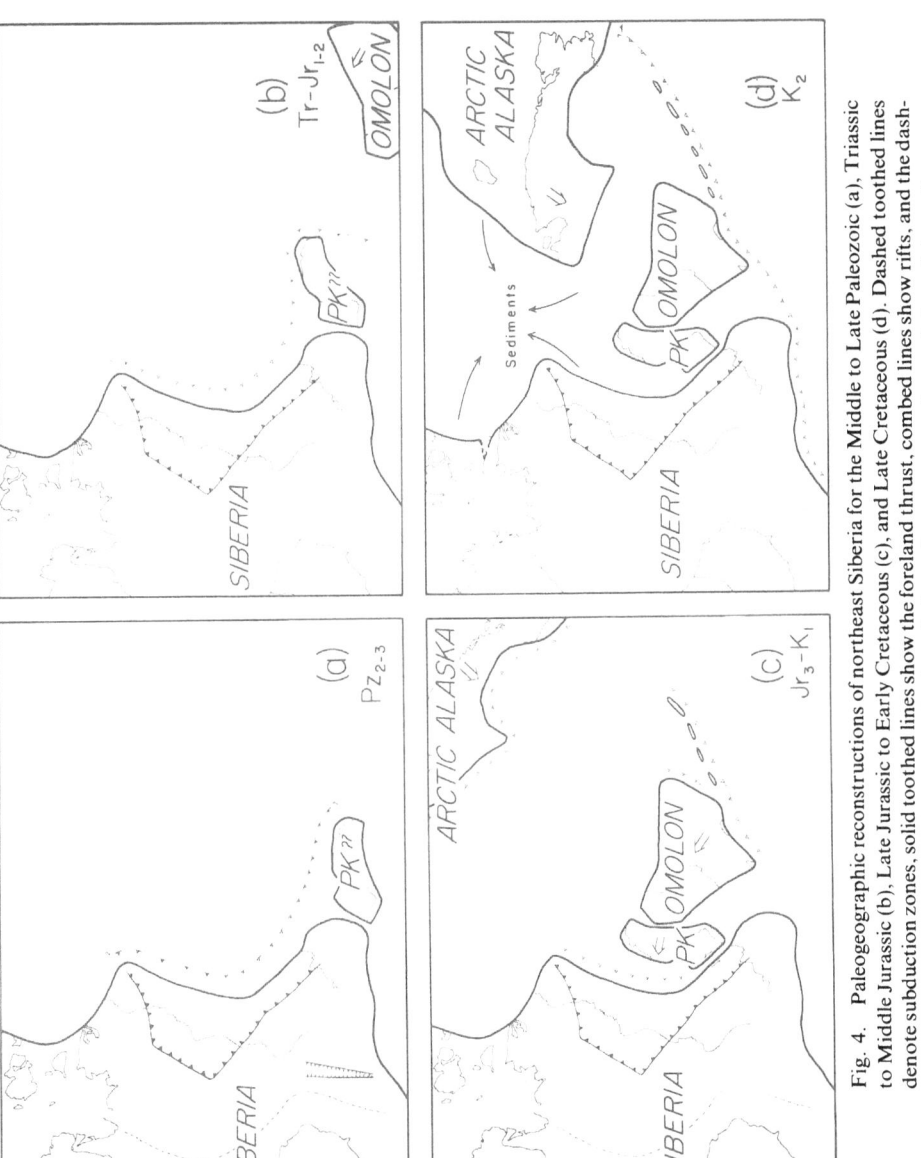

Fig. 4. Paleogeographic reconstructions of northeast Siberia for the Middle to Late Paleozoic (a), Triassic to Middle Jurassic (b), Late Jurassic to Early Cretaceous (c), and Late Cretaceous (d). Dashed toothed lines denote subduction zones, solid toothed lines show the foreland thrust, combed lines show rifts, and the dash-dot line shows the eastern limit of the Siberian platform. Large arrows show relative plate motion.

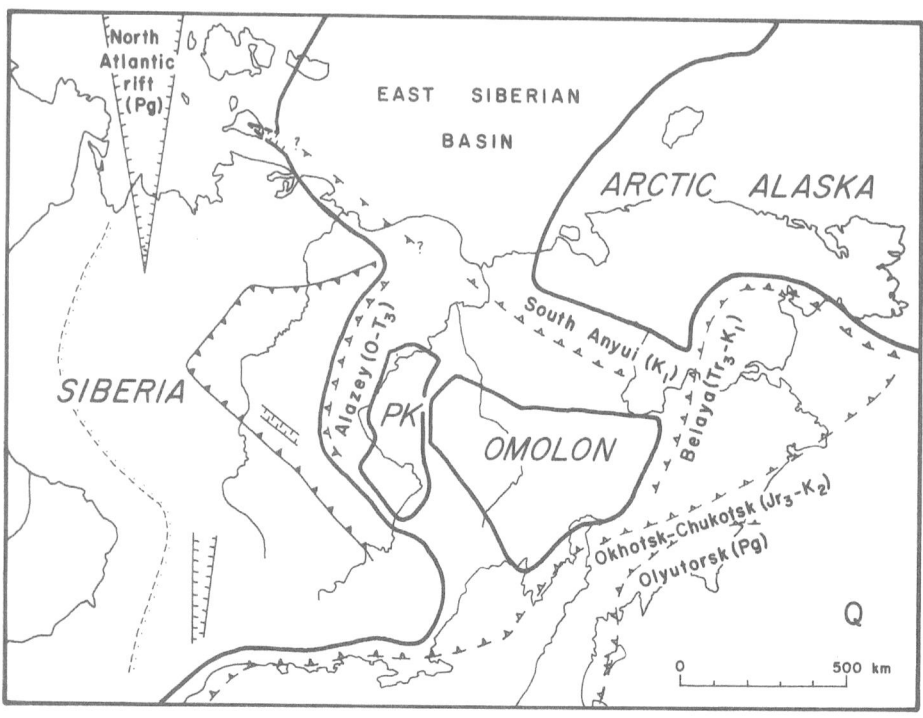

Fig. 5. Current positions of paleoplates in northeastern Siberia. Symbols are as in Fig. 4 with names and times of activity for subduction zones abbreviated as in Fig. 3 with the addition of Pg for Paleogene.

Ekonay zone was first emplaced on the Maynitsk arc and then the whole complex re-thrust as suggested by RUZHENTSEV et al. (1978). However, this seems unlikely due to the intact thrusting of the Paleozoic crust on top of the Cretaceous olistostrome (ALEKSANDROV et al., 1975). Thus, we prefer the continued subduction of Paleozoic crust under the Maynitsk arc until mid-Cretaceous time. It is also possible that some or all of the ophiolites may have been thrust from the north and represent crust of the back-arc basin (RUZHENTSEV et al., 1978).

In addition, we interpret the radiometric ages on the Penzhina Range and Ust' Belaya ophiolites to be the age of the oceanic crust since they correlate extremely well with the faunas in the overlying sediments. The presence of large amounts of limestones over the basaltic section in many of the ophiolites suggests that many of them may have been parts of seamounts or oceanic plateaus.

Finally we note that Paleozoic oceanic crust has also been found in the Chulitna terrane of south-central Alaska (CLARK et al., 1972; JONES et al., 1980) and that the Brooks Range ophiolites may have crustal ages as old as Carboniferous (ROEDER and MULL, 1978). This indicates the possibility of the subduction of Paleozoic oceanic crust in the Koryak-Alaska region as recently as the Late Mesozoic.

We are greatly indebted to helpful discussions with Seiya Uyeda, David Jones, David Howell, Zvi Ben-Avraham, Michael Churkin, and other participants of the Oji seminar. We also thank David LaClair, Arthur Granz, David Scholl, and Alfred M. Ziegler for helpful comments and criticisms. This research was supported by NSF grant EAR 80–25267.

REFERENCES

ALEKSANDROV, A. A., Ophiolites of the Ust' Belaya Mountains, Koryak Highlands, *Acad. Sci. USSR, Doklady, Earth Sci. Sec.*, **219**, 18–20, 1974.

ALEKSANDROV, A. A., N. A. BOGDANOV, S. G. BYALOBZHESKIY, M. S. MARKOV, S. M. TILMAN, V. Y. KHAIN, and A. D. CHEKOV, New data on the tectonics of the Koryak Highlands, *Geotectonics*, **9**, 292–299, 1975.

ALEKSANDROV, A. A., N. A. BOGDANOV, S. A. PALANDZHYAN, and V. D., CHEKHOVICH, Tectonics of the northern part of the Olyutorsk zone of the Koryak Highlands, *Geotectonics*, **14**, 241–248, 1980.

ALEKSEYEV, E. S., Fundamental features of the evolution and structure of the southern part of the Koryak Highlands, *Geotectonics*, **13**, 57–64, 1979.

ANDREWS-SPEED, C. P., The case against a Phanerozoic Kolyma plate in the northeastern USSR, *Geology*, **9**, 174–177, 1981.

BELYAYEV, I. V., O. D. KORSAKOV, B. M. CHIKOV, and A. Y. YUNOV, Tectonic zoning of Shelekhov Gulf and adjacent areas (from geophysical data), *Acad. Sci. USSR, Doklady, Earth Sci. Sect.*, **171**, 109–112, 1966.

BELYAYEVA, D. N., I. V. BELYAYEV, B. M. BRONSHTEYN, Y. P. KIM, and A. P. KUKLIN, Origin of the metallogenic zoning of western Chukotka, *Acad. Sci. USSR, Doklady, Earth Sci. Sect.*, **211**, 83–85, 1973.

BEN-AVRAHAM, Z. and A. K. COOPER, Early evolution of the Bering Sea by collision of oceanic rises and North Pacific subduction zones, *Geol. Soc. Am. Bull., Part I*, **92**, 485–495, 1981.

BOGDANOV, N. A., General features of the Paleozoic trough in the southwestern section of the Kolyma central massif, *Acad. Sci. USSR, Doklady, Earth Sci. Sect.*, **132**, 492–494, 1960.

CHURKIN, M., Jr., Western boundary of the North American continental plate in Asia, *Geol. Soc. Am. Bull.*, **83**, 1027–1036, 1972.

CHURKIN, M., Jr., and J. H. TREXLER, Jr., Circum-Arctic plate accretion—isolating part of a Pacific plate to form the nucleus of the Arctic basin, *Earth Planet. Sci. Lett.*, **48**, 356–362, 1980.

CHURKIN, M., Jr., and J. H. TREXLER, Jr., Continental plates and accreted oceanic terranes in the Arctic, in The Ocean Basins and Margins, v. 5, The Arctic Ocean, edited by A. E. M. Nairn *et al.*, pp. 1–20, Plenum Publishing, New York, 1981.

CLARK, A. L., S. H. B. CLARK, and C. C. HAWLEY, Significance of Upper Paleozoic oceanic crust in the Upper Chulitna district, west-central Alaska Range, *U. S. Geol. Surv. Prof. Pap.*, 800-C, C95–C101, 1972.

CONEY, P. J., D. L. JONES, and J. W. H. MONGER, Cordilleran suspect terranes, *Nature*, **288**, 329–333, 1980.

DOBRETSOV, N. L. and L. G. PONOMAREVA, Lawsonite-glaucophane metaschists of the Penzha Range, northwestern Kamchatka, *Acad. Sci. USSR, Doklady, Earth Sci. Sect.*, **160**, 196–199, 1965.

FUJITA, K., Pre-Cenozoic tectonic evolution of northeastern Siberia, *J. Geol.*, **86**, 159–172, 1978.

FUJITA, K., and E. E. DRETZKA, Bibliography of northeast Siberian and Arctic geology, *Northwestern Univ. Dep. Geol. Sci. Technical Report Nr. 1 (Tectonophysics Ser.)*, pp. 2–58, 1978.

FUJITA, K., and J. T. NEWBERRY, Tectonic evolution of northeastern Siberia and adjacent regions, *Tectonophysics*, **89**, 337–357, 1982.

GULYAYEV, P. V., Contributions to the tectonics of the Alazey uplift, *Geotectonics*, **9**, 350–357, 1975.

GULYAYEV, P. V., Eugeosynclinal formations in the zone between the Momsk and Arga-Tas Ranges (Soviet Northeast), *Geotectonics*, **14**, 383–390, 1980.

IGUMENSHCHEV, S. P., I. M. MIGOVICH, G. P. TEREKHOVA, and O. G. EPSHTEYN, Two epochs of silica accumulation in the eastern part of the Koryak Mountains, *Acad. Sci. USSR, Doklady, Earth Sci. Sect.*, **230**, 56–58, 1976.

Ivanov, O. N., and S. K. Baratov, Serpentinite mélange of the Khatyrka River basin in the Koryak Mountains, *Acad. Sci. USSR, Doklady, Earth Sci. Sect.*, **214**, 56–58, 1974.

Ivanov, O. N., and L. N. Il'chenko, Greenstone-altered metamorphic rocks of the Anadyr'—Koryak fold system, *Acad. Sci. USSR, Doklady, Earth Sci. Sect.*, **238**, 78–80, 1978.

Ivanov, V. V., and I. V. Belyayev, Tectonics and oil and gas potentials of the Kolyma and Primor'ye lowlands and adjacent shelves, *Intern. Geol. Rev.*, **15**, 526–533, 1973.

Jones, D. L., and N. J. Silberling, Mesozoic stratigraphy—the key to tectonic analysis of southern and central Alaska, *U. S. Geol. Surv. Open File Report* 79–1200, 41 p., 1979.

Jones, D. L., N. J. Silberling, and J. Hillhouse, Wrangellia—a displaced terrane in northwestern North America, *Can. J. Earth Sci.*, **14**, 2565–2577, 1977.

Jones, D. L., N. J. Silberling, B. Csejtey, Jr., W. H. Nelson, and C. D. Blome, Age and structural significance of ophiolite and adjoining rocks in the Upper Chulitna district, south-central Alaska, *U. S. Geol. Surv. Prof. Pap.*, 1121-A, A1–A21, 1980.

Khrenov, P. M. and A. A. Bukharov, Marginal volcano-plutonic belts in the North Asian craton, *Intern. Geol. Rev.*, **15**, 688–697, 1973.

Khrenov, P. M., A. A. Bukharov, and Y. A. Nekrasova, Metallogenic features of the volcanic belts of eastern Asia, *Intern. Geol. Rev.*, **18**, 585–596, 1976.

Kosygin, Y. A., V. N. Voyevodin, N. G. Zhitkov, and V. A. Solov'yev, The East Chukotka volcanic zone and the tectonic nature of volcanic belts, *Acad. Sci. USSR, Doklady, Earth Sci. Sect.*, **216**, 79–81, 1974.

Merzlyakov, V. M., Terrigenous volcanic Ordovician section of the Cherskiy Range: *Acad. Sci. USSR, Doklady, Earth Sci. Sect.*, **201**, 123–124, 1971.

Mitrofanov, N. P., A. M. Podol'skiy, N. Y. Kostin, M. A. Talalay, and S. D. Sheludchenko, The Koryak volcanic-plutonic complex, *Intern. Geol. Rev.*, **22**, 1335–1345, 1980.

Nekrasov, G. Y., New data on the structure of the Pekul'ney Range in the left-bank area of the Anadyr' River, *Acad. Sci. USSR, Doklady, Earth Sci. Sect.*, **238**, 87–89, 1978.

Nekrasov, I. Y., Mesozoic volcanism in northeast Yakutia: *Acad. Sci. USSR, Izvestiya, Geol. Ser.*, 10, 64–75, 1961.

Nenashev, N. I., Magmatism and Development of Primary Ore-Magmatism in Eastern Yakutiya, 141 pp. Novosibirsk, Nauka Siberian Section, 1979 (in Russian).

Patton, W. W., Jr., I. L. Tailleur, W. P. Brosgé, and M. A. Lanphere, Preliminary report on the ophiolites of northern and western Alaska, *State of Oregon, Dep. Geol. Mineral Ind. Bull.* **95**, 51–57, 1977.

Pinus, G. V., V. V. Velinskiy, and O. L. Bannikov, On the origin of clastic ultramafic rocks in the north-east of the USSR, *Pacific Geol.*, **6**, 65–72, 1973.

Pushcharovskiy, Y. M., Tectonics of the Arctic Ocean basin, *Geotectonics*, **10**, 85–91, 1976.

Roeder, D. and C. G. Mull, Tectonics of Brooks Range ophiolites, Alaska, *Am. Assoc. Petrol. Geol. Bull.*, **62**, 1696–1702, 1978.

Rusakov, I. M., A. G. Kats, N. S. Bondarenko, G. A. Vasil'yeva, G. P. Koren'kov, and Y. T. Nikolayev, New data on the Paleozoic stratigraphy of the Alazeya plateau in the northeastern USSR, *Acad. Sci. USSR, Doklady, Earth Sci. Sect.*, **223**, 41–43, 1975.

Rusakov, I. M., Z. B. Florova, N. S. Bondarenko, G. A. Vasil'eva, A. G. Kats, G. P. Koren'kov, and Y. T. Nikolaev, Stratigraphy of the Mesozoic deposits of the Alazeya plateau, *Soviet Geol. Geophys.*, **18**, 104–107, 1977.

Ruzhentsev, S. V., S. G. Byalobzheskiy, and S. D. Sokolov, Ophiolite sheets of the Koryak Range, *Acad. Sci. USSR, Doklady, Earth Sci. Sect.*, **239**, 72–74, 1978.

Savosina, A. K., L. M. Natapov, A. I. Sidyachenko, and M. B. Sharkovskily, Spilite-diabase association of the Argatas Range, northeastern USSR, *Acad. Sci. USSR, Doklady, Earth Sci. Sect.*, **230**, 63–65, 1976.

Seslavinskiy, K. B., Structure and development of the South Anyui fault trough, West Chukotka, *Geotectonics*, **4**, 311–317, 1970.

Seslavinskiy, K. B., South Anyui suture (western Chukotka), *Akademii Nauk SSSR, Doklady*, **249**, 1181–1185, 1979 (in Russian).

Shilo, N. A. and S. M. Til'man, The tectonic zones of northeastern USSR and the formation of its continental crust, in *The Ocean Basins and Margins, v. 5, The Arctic Ocean*, edited by A. E. M. Nairn *et al.*, pp. 413–438, New York, Plenum Publishing, 1981.

SHILO, N. A., V. M. MERZLYAKOV, M. I. TEREKHOV, and S. M. TIL'MAN, The Alazeya-Oloy geosynclinal system, a new structure in the Mesozoides of the northeastern USSR, *Acad. Sci. USSR, Doklady, Earth Sci. Sect.*, **210**, 99–101, 1973a.

SHILO, N. A., M. L. GEL'MAN, V. M. MERZLYAKOV, M. I. TEREKHOV, and S. M. TIL'MAN, New zone of glaucophane metamorphism in the Circum-Pacific belt, *Acad. Sci. USSR, Doklady, Earth Sci. Sect.*, **213**, 111–113, 1973b.

SHILO, N. A., Y. Y. VASHCHILOV, T. P. ZIMNIKOVA, and N. M. MIGOVICH, The subsurface structure and nature of the Kolyma-Indigirka interfluve, as revealed by geophysical surveys, *Acad. Sci. USSR, Doklady, Earth Sci. Sect.*, **243**, 52–55, 1978.

SMITH, A. G., A. M. HURLEY, and J. C. BRIDEN, *Phanerozoic Paleocontinental World Maps*, Cambridge Univ. Press, Cambridge, 102 pp., 1981.

UMITBAYEV, R. B., Structural position and some features of ultramafic rocks in the northern Okhotsk region, *Geotectonics*, **11**, 210–214, 1977.

VINOGRADOV, A. P., ed., *Atlas of the Lithological-Paleogeographical Maps of the USSR, v. 4, Paleogene, Neogene and Quaternary*, Ministry of Geology of the USSR, Moskva, 1967.

VINOGRADOV, A. P., ed., *Atlas of the Lithological-Paleogeographical Maps of the USSR, v. 3, Triassic, Jurassic and Cretaceous*, Ministry of Geology of the USSR, Moskva, 1968a.

VINOGRADOV, A. P., ed., *Atlas of the Lithological-Paleogeographical Maps of the USSR, v. 1, Cambrian, Ordovician and Silurian*, Ministry of Geology of the USSR, Moskva, 1968b.

VINOGRADOV, A. P., ed., *Atlas of the Lithological-Paleogeographical Maps of the USSR, v. 2, Devonian, Carboniferous and Permian*, Ministry of Geology of the USSR, Moskva, 1969.

VOYEVODIN, V. N., N. G. ZHITKOV, and V. A. SOLOV'YEV, The eugeosynclinal complex of the Mesozoides of Chukchi Peninsula, *Geotectonics*, **12**, 472–477, 1978.

YUDIN, S. S., and L. I. IZMAILOV, The Chelomdzha-Yamsk deep fault, *Acad. Sci. USSR, Doklady, Earth Sci. Sect.*, **166**, 93–95, 1966.

ZIEGLER, A. M., K. S. HANSEN, M. E. JOHNSON, M. A. KELLY, C. R. SCOTESE, and R. VAN DER VOO, Silurian continental distributions, paleogeography, climatology, and biogeography, *Tectonophysics*, **40**, 13–51, 1977.

ZINKEVICH, V. P., Upper Triassic olistostromes of the Mukarylyan River basin, Koryak Mountains, *Acad. Sci. USSR, Doklady, Earth Sci. Sect.*, **241**, 21–23, 1978.

Accretion Tectonics in the Circum-Pacific Regions, edited by M. Hashimoto and S. Uyeda, 59–68.
Copyright © 1983 by Terra Scientific Publishing Company (TERRAPUB), Tokyo.

Accretional and Collisional Eugeosynclinal Folded Systems of the Northwestern Pacific Rim

B. A. Natal'in* and L. M. Parfenov**

*Institute of Tectonics and Geophysics,
Academy of Science, U.S.S.R., Khabarovsk, U.S.S.R.
**Geological Institute, Siberian Branch,
Academy of Science, U.S.S.R., Yakutsk, U.S.S.R.

Eugeosynclinal folded systems of the northwestern Pacific Rim can be divided into accretional and collisional. Their modern structure and secondary tectonic transformations are different but a pre-deformational history is similar. Both types of the systems include elements similar to those of the modern convergent margins of island arc and andian types. In the accretional systems spatial relations of such elements are unbroken and primary structures of the accretionary wedges and forearc basins are preserved. Collisional systems are narrower and separate large sialic megablocks. Spatial relations of the tectonic elements of the convergent margins in these systems are broken.

1. Introduction

In the northwestern Pacific Rim there are Mesozoic eugeosynclinal systems which occur both parallely to the modern boundaries of the Pacific and within the continent in narrow bands. The parallel systems are accretional and the intra-continental ones are collisional. The modern structure and secondary tectonic transformations of the both types of the systems are different, though the pre-deformational history is similar.

2. Accretional Folded Systems

The Mesozoic Koryak, Shikhote-Alin, Eastern Sakhalin, and Cenozoic Olyutorka-Kamchatka systems belong to the accretional folded systems. The Mesozoic systems have completed their development at the end of Cretaceous and appear to be connected according to the aeromagnetic data (Fig. 1). The Olyutorka-Kamchatka system has completed its development at the end of Miocene and is oceanwards of the Mesozoic systems. Some tectonic elements of this system appear to extend southward to the South-Kuril basin of the Sea of Okhotsk, but its relationship to Japanese structures is unclear.

The accretional systems are up to 500 km wide and are composed primarily of greywackes, cherts, argillites, and basic volcanics. Paleozoic and sometimes Upper Precambrian rocks are present side by side with Mesozoic and Cenozoic ones.

The tectonic elements of these systems are similar to those found in the modern

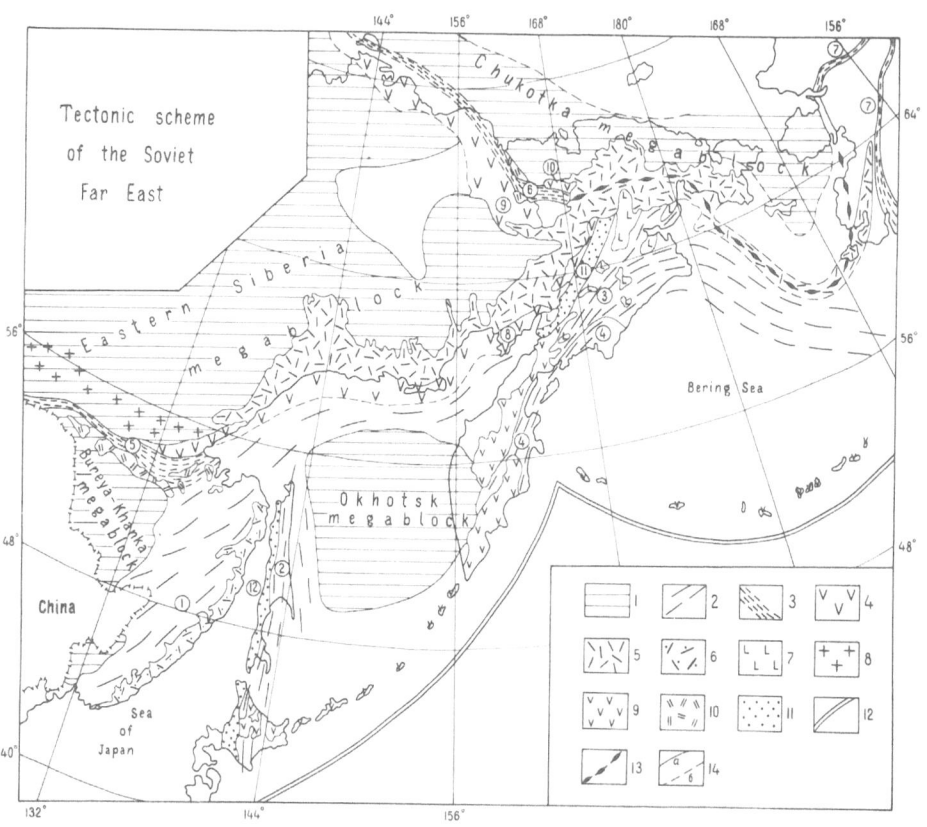

Fig. 1. Tectonic scheme of the Soviet Far East. 1—megablocks with Precambrian basement; 2—
accretional folded systems (1—Sikhote-Alin, 2—Eastern Sakhalin, 3—Koryak, 4—Olyutorka-
Kamchatka); 3—collisional folded systems (5—Mongolia-Okhotsk, 6—South-Anyuy, 7—Early Mesozoic
eugeosynclinal zones of the Yukon-Koyukuk province); 4—Mesozoic volcanic island arcs (8—Uda-
Murgal, 9—Oloy, 10—Nutesyn); 5—Okhotsk-Chukotka volcano-plutonic belt; 6—Eastern Sikhote-Alin
volcano-plutonic belt; 7—Penzhina- Eastern Kamchatka volcanic belt; 8—granodiorite batholiths belt of
the Stanovoy Range; 9—Cenozoic volcanic island arcs; 10—collisional volcanic belts; 11—forearc basins
(11—Penzhina, 12—Eastern Sakhalin); 12—axes of the trenches; 13—hypothetical joining of the South-
Anyuy system with eugeosynclinal zones of Yukon-Koyukuk province of Central Alaska.

convergent plate margins (PARFENOV et al., 1978). The lateral zonation of the tectonic
elements of island arc margin includes: (1) back-arc basin which has an assymmetric
structure characterized by the oceanward increase of thickness and quantity of marine
rocks; (2) volcanic island arc which is usually synchronous to the back-arc basin; (3)
forearc basin which is younger than the volcanic arc; (4) accretionary wedge
incorporating ophiolites which are older than rocks in other zones because the
formation of the convergent margins is more favourable under the subduction of the
older oceanic crust having a large negative buoyancy (USHAKOV and GALUSHKIN,
1979).

The lateral zonation of the tectonic elements of andian margins consists of: (1) marginal volcano-plutonic belt on the continental crust; (2) forearc basin; (3) accretionary wedge. Each accretional system includes both types of convergent plate margins of different age. On the South of the Far East (Fig. 2) Senonian to Paleogene andian type margin includes the Eastern Sikhote-Alin marginal volcano-plutonic belt, the Western Sakhalin forearc basin and the Eastern Sakhalin accretionary wedge. They are superimposed on the Cretaceous island arc margin (PARFENOV *et al.*, 1978, 1981). In the western part of the Sikhote-Alin Range thick (up to 10 km) Hauterivian-Turonian greywackes and shales were formed in the back-arc basin and overlap both conformably and unconformably the older formations. The volcanic island arc is represented by the Apt-Turonian coastal-marine and continental clastic rocks and andesitic volcanics which outcrop in erosional windows of the Eastern Sikhote-Alin volcano-plutonic belt. The volcanic arc rocks are deformed into linear folds while volcanics of the belt dip at low angles. For the margins of the both types the forearc basin existed in the Western Sakhalin. This basin is expressed by up to 6 km Upper Cretaceous marine argillites and greywackes which overlain by Paleogene coal-bearing rocks. Accretionary wedges for the margins of the both types are found in the Eastern Sakhalin and

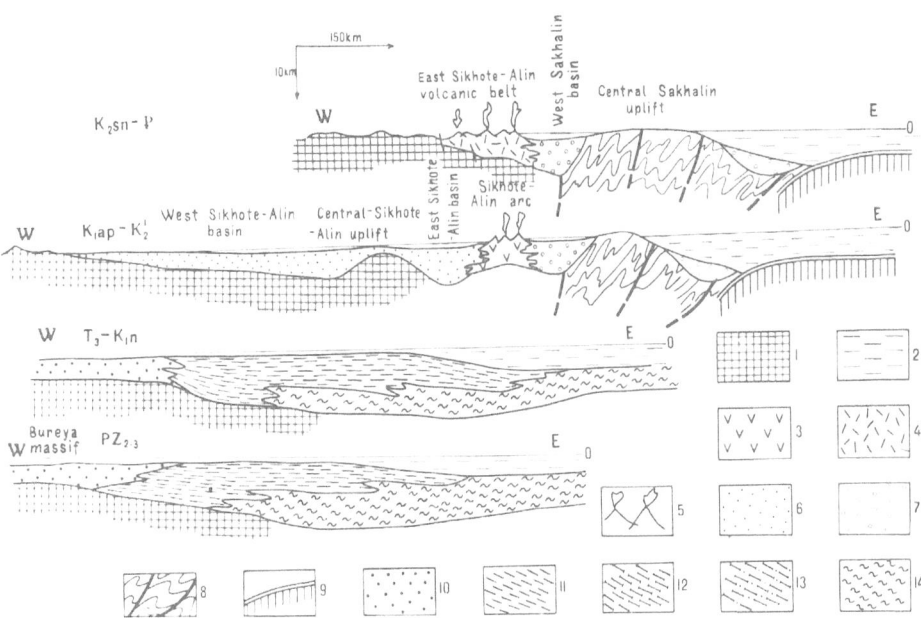

Fig. 2.   Lateral zonation of tectonic elements of the Sikhote-Alin and Eastern Sakhalin folded systems. 1 — continental crust; 2—water; 3–9—tectonic elements of convergent plate margins (3—volcanic island arcs, 4—volcano-plutonic belts on the continental crust, 5—volcanoes, 6—back-arc basins, 7—forearc basins, 8—accretionary wedges, 9—oceanic crust); 10–14—passive continental margins (10—continental and coastal-marine terrigenous rocks of the upper part of the shelf, 11—sandy-argillaceous rocks sometimes with basaltoids of the lower part of the shelf and continental slope, 12—mainly carbonate shelf rocks, 13—terrigenous and carbonate rocks of the upper part of the shelf, 14—volcano-silliceous rocks of the foot of continental slope and oceanic floor).

are represented by Permian to Triassic blueschists and eclogites and Jurassic to Necomian volcano-siliceous and schist-greywacke formations. The Upper Cretaceous volcano-siliceous and flysch rocks with alpine hyperbasites and gabbroids are also found.

In the Koryak folded system three periods of convergent plate margins development are identified—Upper Paleozoic to Necomian, Cretaceous, and Paleogene (Fig. 3). Upper Paleozoic to Neocomian is characterized by the Uda-Murgal volcanic arc. It extends some 2,500 km from the Koryak system to the Uda gulf along the Sea of Okhotsk (Fig. 1). The arc is composed of coastal-marine, rarely continental terrigenous rocks and calc-alcaline volcanics of basic, intermediate and acidic composition (PARFENOV et al., 1978). On the continental side there are found up to 10 km greywacke-shale rocks with basic lava layers and tufogene flysch. They are interpreted as a back-arc basin. Oceanward of the arc there is a forearc basin composed mainly of Carboniferous to Neocomian terrigenous rocks which are exposed along the southeastern edge of the Upper Cretaceous Penzhina basin. The Talovka-Main uplift is located further to the southeast. It is characterized by reduced thickness of the Upper Paleozoic and Masozoic formations. We can regard them as outer non-volcanic arc. On the southeastern slope of the uplift Late Precambrian, Early and Middle Paleozoic, Late Jurassic to Neocomian ophiolites with glaucophane schists, and zones of serpentine melange occur in thrust sheets (ALEXANDROV, 1978; DOBRETSOV, 1974; NEKRASOV, 1976). They belong to the accretionary wedges.

The Cretaceous of the Koryak system was the andian type margin. It includes the Okhotsk-Chukotka volcano-plutonic belt which is continentward and parallel to the Uda-Murgal volcanic arc. The Penzhina forearc basin is oceanward of the volcano-plutonic belt. Within the basin there are up to 6 km of Albian marine aleurite-argillaceous rocks with sandstone and tuff layers (AVDEIKO, 1968; IVANOV and POKHIALAINEN, 1973). In the axial part of the basin Albian rocks are conformable on

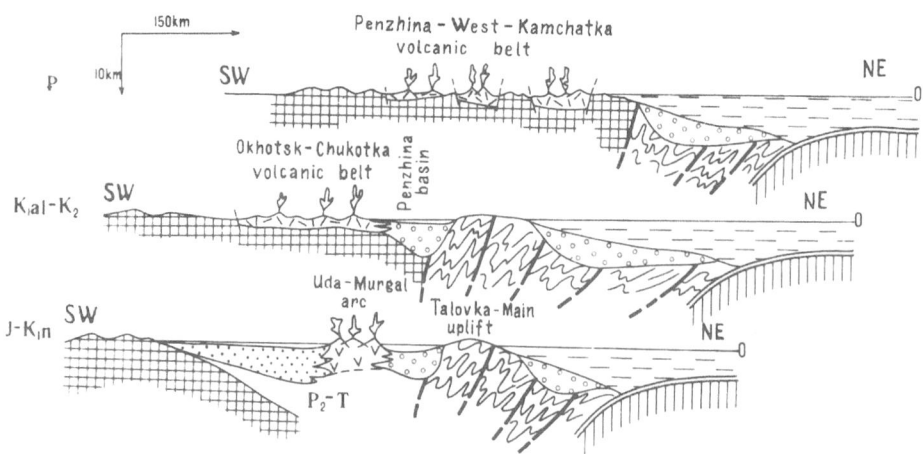

Fig. 3.   Lateral zonation of tectonic elements of the Koryak folded system. Explanatory notes are the same as for Fig. 2.

the Aptian ones however they are unconformable on Valanginian rocks in the Talovka-Main uplift and are thinner. Upper Albian and Upper Cretaceous deposits are unconformable over the aleurites and argillites and are composed of up to 4 km of coastal-marine and continental rocks. The Penzhina basin inherited the forearc basin of the Uda-Murgal arc in a manner similar to that in the Western Sakhalin.

In Cretaceous the Talovka-Main uplift appears also to be a non-volcanic outer arc. Within the uplift Cretaceous thrusts are found (ALEXANDROV, 1978; FILATOVA, 1979). To the southeast of the uplift there are thick (7 km) complexly folded sequences of greywackes, aleurites, argillites with cherty slates, jaspers and basalts in the lower part of the section. They are probably continental slope deposits.

The Paleogene margin includes the Penzhina-Western Kamchatka volcanic belt. The belt is composed of a chain of isolated volcanic fields which extend from the Koryak highland to the northwestern Kamchatka. It can be further geophysically traced along the Western Kamchatka shelf to the Academy of Science Uplift of the Sea of Okhotsk. The belt is composed of calc-alkaline lavas and tuffs of intermediate composition (FILATOVA, 1979) and disordant to the folded structures of underlying formations. To the southeast the volcanics are replaced by thick flyschlike folded sequences of Maastrichtian and Paleogene age which uncomformably overlap older rocks (VOLOBUEVA, and KRASNY, 1979). These deposits become more terrigenous both to the northwest and as they become younger. They are considered to be forearc basin deposits. Multi-folded Paleogene volcano-siliceous and sandstone-shale formations of the Olyutorka region are involved into thrusts with Late Cretaceous ophiolites and correspond to the accretionary wedge.

In the accretional folded systems spatial relations between the lateral zones of the ancient convergent margins are preserved. The volcanics of andian type volcano-plutonic belts are flat-lying and brachyfolded but linear simple folds are typical for the volcanic arcs. The forearc basins are also simple folded and separeted by thrusts from the outer non-volcanic arcs. Primary thrust structure of the accretionary wedges is essentially unchanged. The large nappes have recently been identified in the accret-ionary wedges of the Koryak system (RUZHENTSOV et al., 1977; ALEXANDROV, 1978). They were probably formed by the collision of the accretionary wedge with the microcontinent which was represented by the Early Mesozoic island arc.

3.   Collisional Folded Systems

The Mongolia-Okhotsk and South-Anyuy systems are collisional. Both separate the large sialic megablocks which have been developed independently for the large periods of geological time. These systems are narrow up to tens km and have the same tectonic elements as the accretional systems, but include the axial zones of very complex dislocations. Magmatic belts associated with the development of the Benioff paleozones are superimposed on the bounding megablocks. The accretionary wedges and forearc basins are located in the zones of dislocations and their former structural relations are lost.

The Mongolia-Okhotsk system (Fig. 1) stretches for 2,000 km from the Sea of Okhotsk to Mongolia and separates the Bureya-Khanka and Eastern Siberia megablocks which have the Pre-Riphean basement. The system is divided into some segments. The Galam segment is on the east and 200 km wide. The Tukuringra-Dzhagda segment is opposite to the Bureya-Khanka megablock and its width decreases locally up to 8 km. The Eastern Sabaikal'ie segment which width increases again is westwards of the described region.

In the Tukuringra-Dzhagda segment the Upper Precambrian to Lower Paleozoic greenschists sometimes with glaucophane were formed from basalts, cherts, and greywackes. These schists are associated with altered gabbros with hyperbasites in a zone of 500 km long extending to the Eastern Sabaikal'ie. Cambrian volcano-siliceous rocks are found in the Galam segment. Silurian to Permian formations consist of thick shales, greywackes, and cherts. Within the Silurian to Lower Devonian and Upper Carbon to Lower Permian sections cherty rocks including oceanic basalts are found. From these data we propose a Late Precambrian to Paleozoic oceanic basin located between the Eastern Siberia and Bureya-Khanka megablocks.

The complex dislocations and sometimes high-grade metamorphism of Paleozoic rocks, large thrusts which may overlap some tectonic elements do not allow to distinguish the active or passive character of continental margins. The passive character of the Eastern Siberia margin in the Paleozoic is suggested by: (1) absence of calc-alkaline magmatism; (2) the northward increase of sandstones grain size; (3) the predominance of sandstones over shale cherts and volcanics in the northern part of the Tukuringra-Dzhagda segment; (4) the northward increase of the arkose amount in the Galam segment; (5) the wide distributions of clastics derived from the Stanovoy system.

The main tectonic events happened in the Mesozoic. Upper Triassic and Jurassic mainly sandy-shale formations are preserved as thrust wedges and troughes separated by Paleozoic outcrops. Their original deposition occured in the marine basin connected to the World Ocean (YANSHIN, 1976). Lower and Middle Triassic rocks are absent in the main part of the system and are unconformable on the Paleozoic where they exist (NAGIBINA, 1969).

Mesozoic tectonic movements occured not only within the system but also along the southern margin of the Eastern Siberia megablock—in the Stanovoy folded system and Aldan Shield.

The belt of granitoidal batholiths was formed in the Stanovoy Range. It confined to axial part of the large arched uplift and separeted the geosynclinal marine basin on the south and intermontane basins on the north. This belt extends westward through the Eastern Sabaikal'ie and is everywhere parallel to the Mongolia-Okhotsk system. The belt crosses tectonic elements of the Stanovoy folded system and Kaledonian and Hercinian structures of Sabaikal'ie. The batholiths are poly-phased and include gabbros, diorites, and the later granitoids where epizonal granodiorites are predominant. K-Ar data of 70 to 200 m.y. have been obtained; they are older on the east (GEOLOGY OF THE NORTHEASTERN ASIA, 1973). The batholith belt joins the Uda-Murgal arc on the east and is similar in composition and duration to Mesozoic batholiths of the North America Cordillera. Thus, we consider this belt to mark a cordillerian type margin.

The back arc basin of this margin is represented by Jurassic to Neocomian coal-bearing basins (Chul'man, Tokin, and others) on the southern edge of the Aldan Shield. Precambrian metamorphic rocks are thrusted up to 15 km on the southern part of these basins. Northward within the Aldan Shield there is a belt of small hypabyssal calc-alkaline granitoids, sub-alkaline and alkaline potassic rocks of Mesozoic age (MAKSIMOV, 1975) which is parallel to the belt of the Stanovoy Range.

Forearc basins of this convergent margin are represented by the Uda-Murgal and Tarom basins of the Galam segment composed of simply folded thick (10 km) marine terrigenous Upper Triassic and Jurassic rocks. These rocks are similar to Upper Triassic to Middle Jurassic (4 km) complexly folded rocks of the Lan zone of the Tukuringra-Dzhagda segment. Mesozoic cherts and basic volcanics are found to the south of these basins in the Tugur-Nimelen synclinorium of the Galam segment and in a 350 km long thrust plate in the central part of the Tukuringra-Dzhagda segment. Mesozoic ophiolites are unknown in the Mongolia-Okhotsk system. It may be due to the lack of data or the complete subduction of the heavy Mesozoic oceanic crust.

The closing of Mesozoic marine basins in the Tukuringra-Dzhagda segment occurred in the Late Jurassic when the Bureya-Khanka and Eastern Siberia megablocks had collided. This collision caused reorientation of the previously formed faults and thrusts and origin of new folds, cleavage, transpositional structures and then thrusts displacing the Upper Jurassic and Neocomian molasse. Northeast and northwest strike-slip faults which are widely spread on the Aldan Shield and Stanovoy Range and pass through the Mongolia-Okhotsk system to the northern part of the Bureya-Khanka megablock are formed at the final stage of collision. The sence of motion along these faults is that which would be caused by the clearing of the area in front of the Bureya-Khanka megablock.

The Lower Cretaceous Umlekan-Ogodzha volcano-plutonic belt stretches for 600 km along the southern boundary of the Tukuringra-Dzhagda segment. This belt differs from the andian type belts by another tectonic position and higher alkalinity. We suggest a close relationship of such belts with continental block collision and call them collisional ones.

The South-Anyuy system separates Eastern Siberia and Chukotka megablocks. The basement of the later is composed of Precambrian metamorphic rocks overlapped by Riphean to Early Mesozoic rocks of the Chukotka miogeosynclinal system. Since the Middle Paleozoic the Chukotka megablock has been rigidly bonded to the North American platform by the Brooks-Vrangel folded system. Thus, the South-Anyuy system is the suture between the Early Mesozoic Eurasian and North American plates (PARFENOV, 1975; PARFENOV and NATAL'IN, 1977; NATAL'IN, 1979, 1981; SESLAVINSKY, 1979; FUJITA, 1978).

The Late Jurassic ophiolites composed of dunites, harzburgites, pyroxenites, troctolites, gabbros of eucrit type, volcanics similar to abyssal tholeiites and cherts and Neocomian terrigenous flysch are found in the South-Anyuy system (PINUS and STERLIGOVA, 1973; RADSIVILL and RADSIVILL, 1975; NATAL'IN, 1981). The older rocks of eugeosynclinal type were previously unknown. Recently however Visean fossil-bearing limestones occuring among altered volcanics of basic and intermediate composition were found along the Maliy Anyuy River (SIZYCH et al., 1977). Upper Jurassic conglomerates and breccia contain fragments of basic volcanics, cherts, gabbros and serpentinized hyperbasits (SIZYCH *et al.*, 1977; NATAL'IN, 1981). Poor grading and angular shape of the fragments suggest the proximal origin of this material and indirectly the existence of the older eugeosynclinal complexes.

To the north and south of the South-Anyuy system there are Late Jurassic to Neocomian simply folded volcanic belts which are similar to volcanic island arcs (NATAL'IN, 1981). Northwards from the eastern part of the system the Nutesyn volcanic belt is superimposed on the Chukotka megablock. Southwards from the larger western part of the system is the Oloy volcanic belt. Neocomian flysch of the South-Anyuy system may have been formed in forearc basins and Late Jurassic ophiolites and older formations mark the accretionary wedges of these convergent plate margins.

The structure of the South-Anyuy system is very complex (NATAL'IN, 1981). The early folds and contemporaneous amphibolite facies metamorphism are found in hyperbasits and gabroids. Folds and thrusts of the next generation are observed in the overlying Upper Jurassic and Neocomian rocks. They may have formed over a long period of time in the accretionary wedges of the Nutesyn and Oloy arcs. Folds in Hauterivian greywackes of the Oloy forearc basin are even simpler in morphology than those in underlying Berriasian-Valanginian and Upper Jurassic rocks in spite of the conformity of these stratigraphic units. Tight and isoclinal folds with regionally developed schistosity, transpositional structures sometimes turning into secondary tectonic banding superimpose all the previous folding episodes. All of them were formed in a short time around the Aptian/Neocomian boundary and are connected with collision of the Eastern Siberia and Chukotka megablocks. As in the Mongolia-Okhotsk system all planar structural forms were reorientated and attained subvertical position.

The chain of Post-Neocomian fields of the calc-alkaline andesites and basalts stretches parallel to the South-Anyuy system and was not connected with the Benioff paleozone. It is similar to the Umlekan-Ogodzha belt of the Mongolia-Okhotsk system.

According to aeromagnetic data the South-Anyuy folded system extends west-wards to Bolshoy Lyakhovsky Island. The southeastern end of the system is overlain by the Okhotsk-Chukotka belt but we postulate that system to continue into the Central Alaska (Fig. 1). This suposition is confirmed by discovery of complexly dilocated Jurassic basalts, cherts, schists and greywackes with hyperbasites and banded gabbros which have been found in the Eastern Chukotka (VOEVODIN *et al.*, 1980).

4. Summary

The eugeosynclinal folded systems of the northwestern Pacific Rim can be divided into accretional and collisional. Both types of the systems include the tectonic elements which are similar to those of the modern convergent plate margins of different types. In accretional systems spatial relations of magmatic belts, forearc basins, and accretionary wedges are practically unbroken. The primary structure of the forearc basins and accretionary wedges is usually unchanged. Complex structures, accompanied by nappes appear in the regions of collision of convergent margins with small microcontinents.

Collisional folded systems separate the large sialic megablocks. In comparison with the accretional systems they are narrower. Their primary structures are greatly changed by collision of the sialic megablocks. Spatial relations of the accretionary wedges and forearc basins are broken. All planar structures have subvertical dips. Along the collisional folded systems there are volcanic belts caused by collisional dislocations.

This paper draws the information from the work of many geologists. The authors thank I. P. Voinova, L. I. Popeko, G. S. Gusev and Yu. A. Arkhipov for helpful discussions. We also thank Mrs. I. S. Natal'ina and K. Fujita who help to put this paper into English.

REFERENCES

ALEXANDROV, A. A., Napping and Imbricated-thrust Structures of the Koryak Highland, 122 pp., Nauka, Moscow, (in Russian).

AVDEIKO, G. P., Lower Cretaceous deposits of the North Pacific Rim, 154 pp., Nauka, Moscow, 1968 (in Russian).

DOBRETSOV, N. L., Glaucophane-schist and eclogite-glaucophane compexes of the USSR, 425 pp., Nauka, Novosibirsk, 1974 (in Russian).

FILATOVA, N. I., Cretaceous-Paleogene volcanism of the Verkhoyano-Chukotka and Koryak-Kamchatka transitional zone, *Geotectonics*, **5**, 98–115, 1979 (in Russian).

FUJITA, K., Pre-Cenozoic tectonic evolution of Northeastern Siberia, *J. Geol.*, **86**, 159–172, 1978.

GEOLOGY OF THE NORTHEASTERN ASIA. *Magmatism*, V. 3, 395 pp., Nedra, Leningrad, 1973 (in Russian).

IVANOV, V. V. and V. P. POKHIALAINEN, Cretaceous rocks of the South-Penzhina basin in connection with problem of oil and gas content, in *Problems of Oil and Gas Content on the North-East of the USSR*, pp. 70–107, Magadan, 1973 (in Russian).

MAKSIMOV, E. P., Experience of formational analysis of Mesozoic magmatic formations of the Aldan Shield, *Proc. USSR Acad. Sci., Geol. Ser.*, No. 4, pp. 16–32, 1975 (in Russian).

NAGIBINA, M. S., Stratigraphy and formations of the Mongolia-Okhotsk belt, 399 pp., Nauka, Moscow, 1969 (in Russian).

NATAL'IN, B. A., Eugeosynclinal zones of Chukotka and Alaska, *XIV Pacific Scientific Congress, Comm. B, Sect. B II*, Abstr., pp. 34–35, Moscow, 1979.

NATAL'IN, B. A., Structure and tectonic evolution of the South-Anyuy eugeosynclinal system, Doctor's Thesis, 24 pp., Presidium of the Far East Sci. Cent., Acad. Sci. USSR, Khabarovsk, 1981, (in Russian).

NEKRASOV, G. E., Tectonics and magmatism of the Taigonos Peninsula and Northwestern Kamchatka, 157 pp., Nauka, Moscow, (in Russian).

PARFENOV, L. M., Tectonic scheme of the Soviet Far East and some problems of magmatism, in *Problems of Magmatism and Tectonics of the Far East*, pp. 3–25, Vladivostok, 1975 (in Russian).

PARFENOV, L. M. and B. A. NATAL'IN, Tectonic evolution of the Northeastern Asia in Mesozoic and Cenozoic, *Reports of the USSR Acad. Sci.*, **235**, 1132–1135, 1977 (in Russian).

PARFENOV, L. M., I. P. VOINOVA, B. A. NATAL'IN, and D. F. SEMENOV, Geodynamics of the Northeastern Asia in Mesozoic and Cenozoic time and the nature of volcanic belt, *J. Phys. Earth*, **26**, Suppl., S503–S526, 1978.

PARFENOV, L. M., B. A. NATAL'IN, I. P. VOINOVA, and L. I. POPEKO, Tectonic evolution of the active continental margins of the Northwestern Pacific Rim, *Geotectonics*, **1**, 85–104, 1981 (in Russian).

PINUS, G. V. and V. E. STERLIGOVA, New belt of alpine hyperbasits on the North-West of the USSR and some geological peculiarities of formation of hyperbasits belts, *Geol. Geoph.*, **12**, 109–111, 1973 (in Russian).

Radsivill, A. Ya. and V. Ya. Radsivill, Late Jurassic magmatic formations of the South-Anyuy basin, in *Magmatism of the Northeastern Asia, Part II*, pp. 71–80 Magadan, 1975 (in Russian).

Ruzhentsev, S. V., S. G. Belobzhesky, A. D. Kasimirov, and D. S. Sokolov, Specific features of napping structures development of the Akonay zone in Koryak, *Reports of the USSR Acad. Sci.*, **233**, 1181–1185, 1977 (in Russian).

Seslavinsky, K. B., The South-Anyuy suture (Western Chukotka), *Reports of the USSR Acad. Sci.*, **249**, 1181–1185, 1979 (in Russian).

Sizych, V. I., V. A. Ignat'ev, L. D. Shkol'ny, D. G. Berlimble, V. P. Fomin, R. S. Redyuk, and R. S. Sukhina, New data on stratigraphy and tectonics of the Maliy Anyuy left bank, in *Material on Geology and Minerals of the USSR North-East*, V. 23, Part 1, pp. 29–34 Magadan, 1977 (in Russian).

Ushakov, G. A. and Yu. I. Galushkin, *Lithosphere of the Earth, V. 2, Continental Lithosphere, Physics of the Earth*, V. 4, p. 258, Moscow, 1979 (in Russian).

Voevodin, V. N., N. G. Zhitkov, and B. A. Solov'ev, Eugeosynclinal complex of mesozoid of Chukotka Peninsula, *Geotectonics*, 101–109, 1978 (in Russian).

Volobueva, V. I. and L. L. Krasny, Maastrichtian-Neogene deposits of the Eastern Koryak highland, 131 pp., Nauka, Moscow, 1979 (in Russian).

Yanshin, A. A., Problems of post-geosynclinal tectonic development of the Central Asian folded belt in *Tectonics of Siberia*, pp. 156–158, Nauka, Moscow, 1976 (in Russian).

Accretion Tectonics in the Circum-Pacific Regions, edited by M. Hashimoto and S. Uyeda, 69–88.
Copyright © 1983 by Terra Scientific Publishing Company (TERRAPUB), Tokyo.

Space-Time Distribution of Late Mesozoic to Early Cenozoic Magmatism in East Asia and Its Tectonic Implications

Masaki TAKAHASHI

Department of Earth Sciences, Ibaraki University, Mito 310, Japan

Space-time distribution of magmatism in late Mesozoic to early Cenozoic East Asia is briefly summarized; the type of magmatism is classified into five groups, predominantly calcalkaline (island arc, continental margin, intra-continent, and collision) and alkaline. The evolution of tectonic frameworks in late Mesozoic to early Cenozoic East Asia is discussed on the basis of the distribution of various types of magmatism for five periods: 190–160 Ma, 160–130 Ma, 130–100 Ma, 100–70 Ma, and 70–40 Ma. The tectonic history reconstructed is very complex; various tectonic events such as the collision of continental blocks, subduction of ordinary cold oceanic lithospheres, subduction of young hot oceanic lithospheres, and formation of strike-slip mobile belts are supposed to have occurred.

1. Introduction

The nature and distribution of a group of igneous rocks occurring within a set span of time as well as space over the earth's surface strongly reflect the tectonic setting of that particular province in particular time. Enormous amount of volcanic and intrusive rocks of late Mesozoic to early Cenozoic time are extensively distributed in East Asia (Fig. 1). This paper gives a brief summary of the nature and space-time distribution of these igneous rocks and then tries to reconstruct the tectono-magmatic evolution of East Asia in the respective geologic time.

In this paper, the igneous rock associations are classified into two groups: predominantly calcalkaline and alkaline. The former is further divided into four types: island arc, continental margin, intra-continent and collision types; this classification is mainly based on the particular chemical features (K_2O, $K_2O + Na_2O$ contents, and K_2O/Na_2O ratio) and paleogeographic settings as described in the literature searched. The calc-alkaline island arc type igneous rocks are low in K_2O, $K_2O + Na_2O$ contents and K_2O/Na_2O ratio. Behind the island arcs, the back-arc basins were developed. On the other hand, moderate to high K_2O, $K_2O + Na_2O$, and K_2O/Na_2O are peculiar to the igneous rocks of the continental margin type magmatism, which took place at the active continental margin area and not associated with the back-arc basins. The intra-continent type magmatism includes rocks with high K_2O, $K_2O + Na_2O$ and K_2O/Na_2O; this type of magmatism is developed within continent far from the continent-ocean border. Furthermore, the collision type magmatism is observed along the colliding boundary (suture zone) between the two continental blocks; the rocks of this type usually have moderate to high K_2O, $K_2O + Na_2O$, and K_2O/Na_2O, though

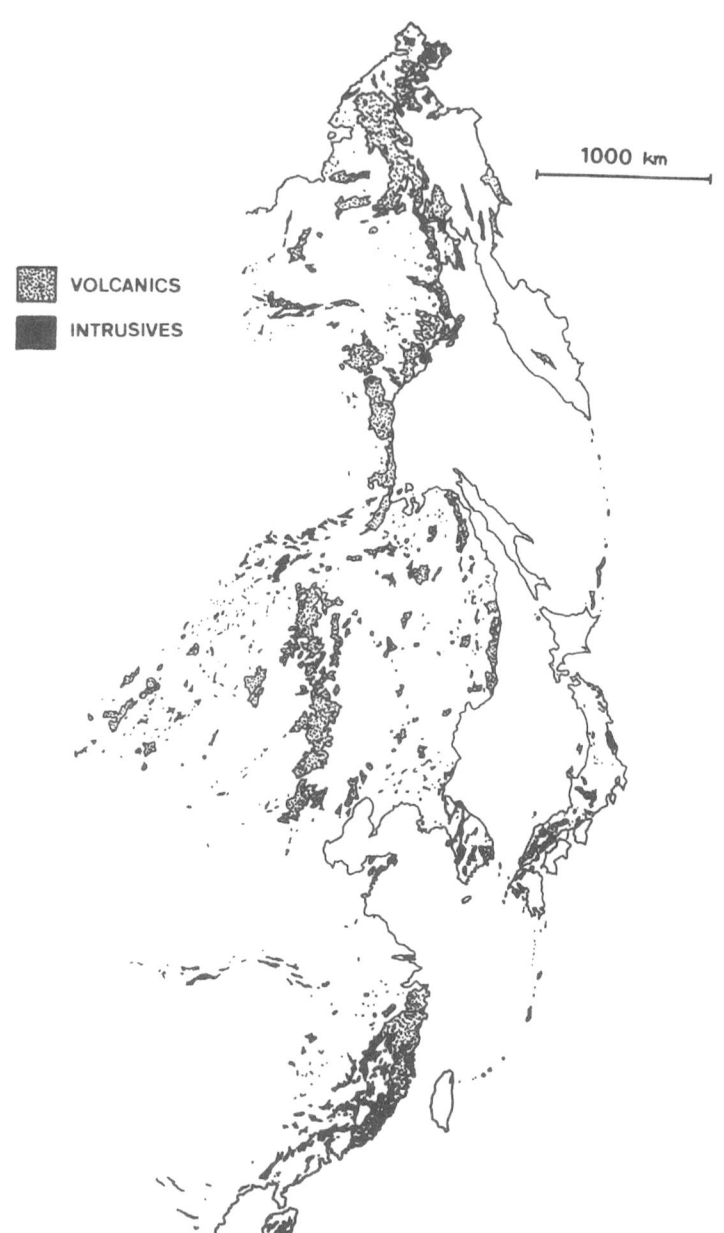

Fig. 1.   Map showing the distribution of Mesozoic to early Cenozoic volcanics and intrusives in East Asia
(ACADEMIA SINICA, 1975).

their chemical features cannot yet be clearly defined. The predominantly alkaline type of magmatism is characterized by the occurrence of abundant alkaline basalt, trachybasalt, trachyandesite, trachyte, alkaline rhyolite, monzonite, syenite and alkaline granite.

2. Space-Time Distribution of Various Types of Magmatic Belt

In the following sections, the areal distribution of previously-defined various types of magmatic belt in late Mesozoic to early Cenozoic East Asia is briefly described for five periods: 190–160 Ma, 160–130 Ma, 130–100 Ma, 100–70 Ma, and 70–40 Ma. The localities referred in the next descriptions are shown in Fig. 2.

2.1   190–160 Ma (Fig. 3A)

Island arc type: this type of magmatic belts are distributed in the Okhotsk (Uda-Murgal arc), north and south areas of the south Anyuy (Nutesyn and Oloy arcs respectively), Kolyma (Polousnensky-Yasachnaya arc) and Alazeya regions (PARFENOV et al., 1978; FUJITA and NEWBERRY, 1981; NATAL'In and PARFENOV, 1981; TIL'MAN and KRASNEY, 1981). The Uda-Murgal island arc type magmatic belt extends for more than 2,500 km. This belt mainly (>80% in volume) consists of mafic to intermediate volcanics characterized by low K_2O, $K_2O + Na_2O$, and K_2O/Na_2O; the K_2O content of volcanics increases inward from the present continental margin (PARFENOV et al., 1978). Continental margin type: the continental margin type magmatic belts are present in the southeast China, Korean peninsula, and southwest Japan (the Hida province). These belts are composed of felsic granitic intrusives, mainly biotite granite; the rocks have moderate to high K_2O, $K_2O + Na_2O$, and K_2O/Na_2O (LEE, 1971; GEOLOGICAL SURVEY of JAPAN, 1977; ACADEMIA SINICA, 1979). Collision type: the magmatic belts of this type extend along the Tsinling and Mongolian-Okhotsk suture zones. In the Tsinling region, felsic biotite granite is predominant (ACADEMIA SINICA, 1973), while moderately felsic to subalkaline granitic intrusives, such as granodiorite, biotite granite and syenite are predominant in the Olekma-Stanovoy region of the Mongolian-Okhotsk suture zone (KHRENOV et al., 1976). They are characterized by high K_2O, $K_2O + Na_2O$, and K_2O/Na_2O. Predominantly alkaline type: this type of magmatic belt is present in the Mongolia region of the Mongolian-Okhotsk suture zone. It comprises trachybasalt trachyandesite, trachyte, alkaline granite, and syenite (KHRENOV et al., 1976).

2.2   160–130 Ma (Fig. 3B)

Island arc type: the magmatism of this type in the Uda-Murgal, Nutesyn and Oloy arcs continued their activity also in this time interval. Continental margin type: the continental margin type magmatic belts occur in the southeast China (Nanling granites etc.) and Korean peninsula (Daebo granites etc.). At present, they mainly consist of felsic biotite granite with minor hornblende biotite granodiorite, which show high K_2O, $K_2O + Na_2O$, and K_2O/Na_2O (LEE, 1971; ACADEMIA SINICA, 1979). Intra-continent type: the magmatic belts of this type are widely distributed in the northeast China; Ta-Khinganling region, Liaodong and Shandong peninsulas. They are mainly

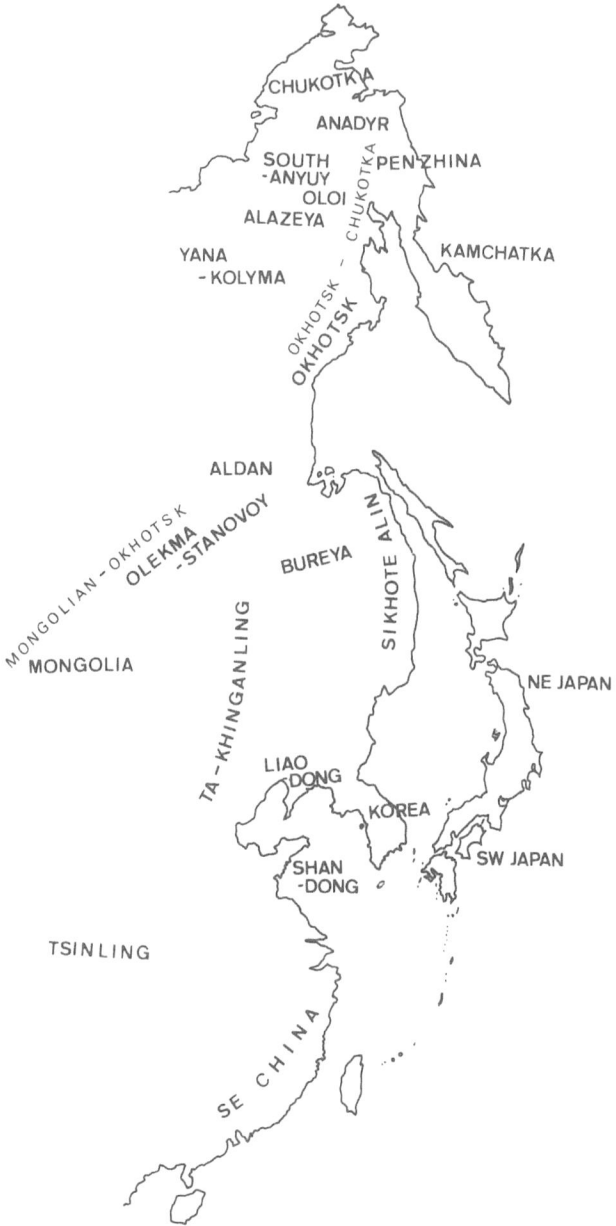

Fig. 2. The name of localities referred in the paper.

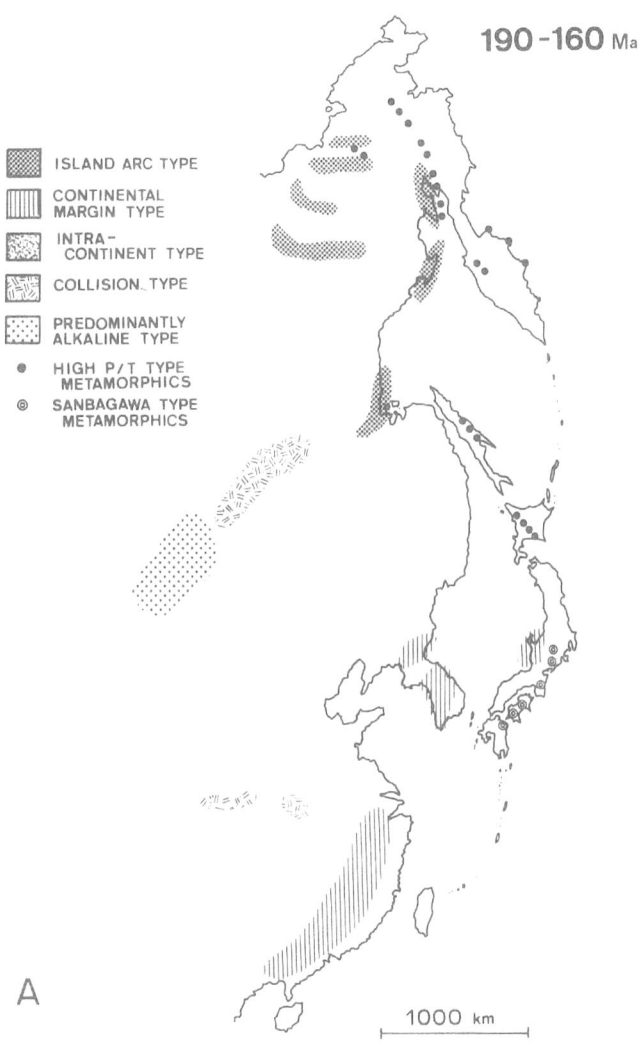

190-160 Ma

ISLAND ARC TYPE

CONTINENTAL
MARGIN TYPE

INTRA-
CONTINENT TYPE

COLLISION TYPE

PREDOMINANTLY
ALKALINE TYPE

● HIGH P/T TYPE
METAMORPHICS

◎ SANBAGAWA TYPE
METAMORPHICS

A

1000 km

Fig. 3. Map showing the space-time distribution of various types of magmatic belt in late Mesozoic to early Cenozoic East Asia. A: 190–160 Ma; B: 160–130 Ma; C: 130–100 Ma; D: 100–70 Ma; E: 70–40 Ma.

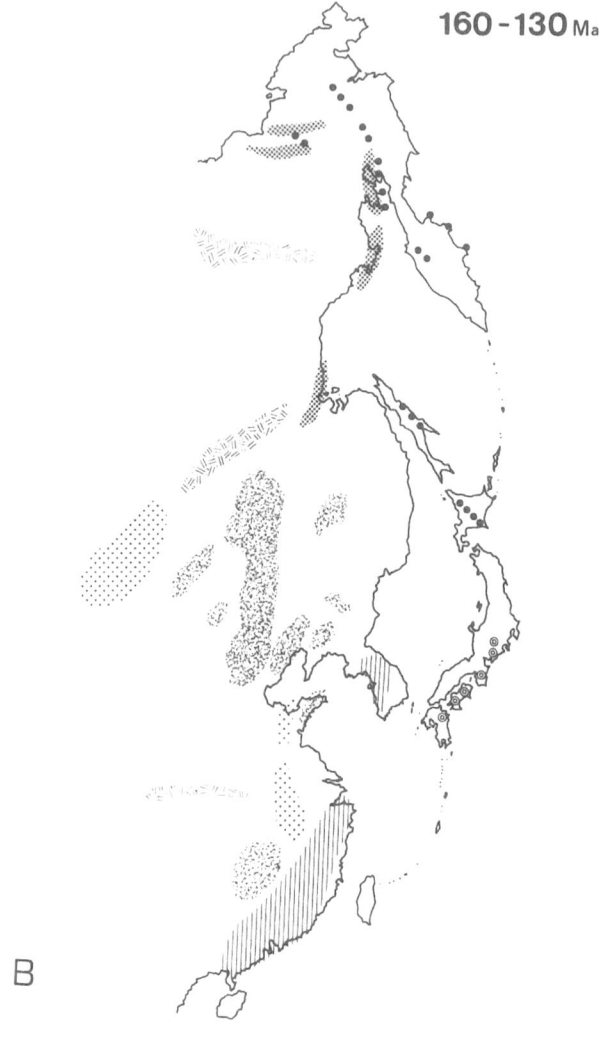

160-130 Ma

B

Fig. 3

composed of voluminous felsic to intermediate welded ash-flow tuffs with subordinate amount of intermediate to mafic lava and pyroclastics (andesite, basalt, trachyandesite, and trachybasalt) and felsic granitic intrusives, which are characterized by higher K_2O, $K_2O + Na_2O$, and K_2O/Na_2O than those of continental margin type (ACADEMIA SINICA, 1977). The similar type of magmatic belt is also exposed behind the continental-margin type belt of southeast China. It consists mainly of biotite granite and two-mica granite (ACADEMIA SINICA, 1973). Collision type: this type of magmatic belts extend along the Tsinling, Mongolian-Okhotsk (Olekma-Stanovoy region) and Yana-Kolyma suture zones. The first one is characterized by felsic biotite granite (ACADEMIA SINICA, 1973), the second by rhyolite, dacite, and granitoids of varied composition (granite, granodiorite, and tonalite) (KHRENOV et al., 1976) and the last, including both volcanics and intrusives, by high Al_2O_3, K_2O/Na_2O, and by the occurrence of cordierite, garnet and alumino-silicate (SHILO and MILOV, 1977); these are the typical S-type igneous rocks of CHAPPELL and WHITE (1974). Predominantly alkaline type: the predominantly alkaline type of magmatic belts are present in the Mongolia region of the Mongolian-Okhotsk suture zone, Shandong peninsula, and middle-lower Yangtze valley in southeast China. The first one consists of trachyandesite, trachybasalt, syenite, and granite (KHRENOV et al., 1976), the second and the last ones are located behind the continental margin type magmatic belt of southeast China and are mainly composed of trachyte, trachyandesite, trachybasalt, shoshonite, and related intermediate to felsic small alkaline intrusives, which show very high K_2O and K_2O/Na_2O (some of them are very low in K_2O/Na_2O) (ACADEMIA SINICA, 1977; WU, 1981; XUE, 1981; YU and BAI, 1981).

2.3   130–100 Ma (Fig. 3C)

Island arc type: the magmatic belts of this type occur in the northern Sikhote Alin (PARFENOV et al., 1978) and northeast Japan (western Hokkaido and Kitakami provinces). The northeast Japan belt consist of andesite, dacite, mafic granodiorite, tonalite, and gabbro; the rocks are classified into the calc-alkaline and alkaline to subalkaline types (KATADA et al., 1974); the former is predominant and characterized by moderate to low K_2O, $K_2O + Na_2O$, and K_2O/Na_2O. The northeast Japan region, to the northeast of the Tanakura Tectonic Line (TTL), is inferred to have been a part of an island arc geographically separated by ocean from the southwest Japan which composed the margin of the Asian continent during this period (OTSUKI and EHIRO, 1978; ISHIHARA, 1978). Continental margin type: this type of magmatic belts are distributed in the Okhotsk-Chukotka, southwest Japan, Korean peninsula, and southeast China. The first one is called the Okhotsk-Chukotka magmatic belt, which extends for more than 3,000 km and is divided into two zones, the inner zone and the outer zone (SHILO and MILOV, 1977; PARFENOV et al., 1978; SHILO et al., 1979). The inner zone, running along the ocean side of the belt, consists of mafic to intermediate volcanics (high-alumina basalt and andesite), and intrusives (quartz diorite, tonalite, plagiogranite, and gabbro; they are low in K_2O, $K_2O + Na_2O$, and K_2O/Na_2O. The outer zone, on the continental side, is composed of intermediate to felsic volcanics (andesite, dacite, and rhyolite) and intrusives (granodiorite and granite); the dominant felsic volcanics are welded ash-flow tuffs and the proportion of felsic volcanics to intermediate ones is variable in the regions of the outer zone; the felsic volcanics are about 90% of the total volume in the Central Chukotka region, about 10% in the Anadyr region, about 45% in the Penzhina region and about 30% in the Okhotsk region (SHILO et al., 1979); the magmatism in the outer zone is characterized by moderately high to high K_2O, $K_2O + Na_2O$, and K_2O/Na_2O. In the Okhotsk-Chukotka magmatic belt, the K_2O content increases from the ocean side into continent. The magmatic belts of this type in the southwest Japan, Korean peninsula (Bulgugsa granites) and southeast China consist of felsic to intermediate volcanics (rhyolite, dacite, and andesite) and intrusives (granite and granodiorite) (ACADEMIA SINICA, 1973; GEOLOGICAL SURVEY OF JAPAN, 1977; MURAKAMI, 1977; ISHIHARA,

Fig. 3

1979; MURAKAMI and IMAOKA, 1980; JIN *et al.*, 1981; LEE, 1981). The felsic volcanics occupy nearly 90% of the total volume and predominantly consist of welded ash-flow tuffs, which show high K_2O, $K_2O + Na_2O$, and K_2O/Na_2O. The systematic increase in K_2O content from the oceanic coast into continent, that is observed in the Okhotsk-Chukotka belt, is not manifest in these magmatic belts. The center of igneous activity in these belts are situated at the ocean-side of those of the previous period. Collision type: the collision type magmatic belts are present in the south Anyuy, Yana-Kolyma, Mongolian-Okhotsk, and Tsinling suture zones. The first one consists of andesite and basalt with high alkalinity (NATAL'IN and PARFENOV, 1981). The belt in the Yana-Kolyma region is composed of biotite hornblende granodiorite to biotite granite with high K_2O, $K_2O + Na_2O$, and K_2O/Na_2O (SHILO and MILOV, 1977). That in the Olekma-Stanovoy region of the Mongolian-Okhotsk suture zone consists of moderately felsic to felsic granitoids (granodiorite and granite), andesite, dacite, and rhyolite (KHRENOV et al., 1976). The magmatic belt in the northern Bureya region of the Mongolian-Okhotsk suture zone is called the Umlekan-Ogodzha volcano-plutonic belt which streches about 600 km (NATAL'IN and PARFENOV, 1981). In the Tsinling suture zone, the magmatism is represented mainly by biotite granite (ACADEMIA SINICA, 1973). Intra-continent type: the magmatism of this type in the Ta-Khinganling region, Liaodong and Shandong peninsulas continued their activity during this period. A newly formed magmatic belt of this type extends in the eastern Bureya, consisting of andesite, dacite, rhyolite (dominantly welded ash-flow tuffs), diorite, granodiorite, and granite with high K_2O, $K_2O + Na_2O$, and K_2O/Na_2O (PARFENOV et al., 1978); they are accompanied by minor amount of trachybasalt, trachyandesite, alkaline rhyolite, and alkaline granite. Predominantly alkaline type: this type of magmatic belts are present in the Aldan region, Mongolia region of the Mongolian-Okhotsk suture zone, western Shandong peninsula and middle-lower Yangtze valley. The first one consists of leucite nepheline syenite, monzonite and alkaline granite (KHRENOV et al., 1976). The second one is composed of trachybasalt and alkaline minor intrusives (KHRENOV et al., 1976) and the last two are mainly comprised of trachyte, trachyandesite, trachybasalt, shoshonite, and related intermediate intrusives with very high K_2O, $K_2O + Na_2O$, and K_2O/Na_2O (some of them show very low K_2O/Na_2O) (ACADEMIA SINICA, 1977; WU, 1981; XUE, 1981; YU and BAI, 1981).

2.4   100–70 Ma (Fig. 3D)

Island arc type: the island arc type magmatic belts extinguished in this period. Continental margin type: This type of magmatic belt extends for more than 8,000 km along the continental margin of the East Asia. In the Okhotsk-Chukotka region, the magmatic belt which was active from 130 to 100 Ma continued its activity throughout this period; the magmatic center of this belt migrated from the oceanic coast into continent during this and previous periods (from 130 to 70 Ma). The Sikhote Alin magmatic belt is divided into two pararell subbelts, the eastern and western belts (PARFENOV et al., 1978). The eastern belt is about 1,500 km long and subdivided into two latitudinal sections, northern and southern subzones. The northern subzone consists of andesite, dacite and granodiorite with moderately high to high K_2O, $K_2O + Na_2O$, and K_2O/Na_2O (PARFENOV et al., 1978); the felsic volcanics amount about 30% of the total in this subzone (BELLY et al., 1974). The southern subzone is composed of rhyolite, dacite, and andesite (mainly welded ash-flow tuffs) and granite with moderately high to high K_2O, $K_2O + Na_2O$, and K_2O/Na_2O (PARFENOV et al., 1978); the felsic volcanics comprise about 80% in this subzone (BELYY et al., 1974). The western subbelt of the Sikhote Alin magmatic belt consists of subalkaline basalt, andesite, dacite, potassic rhyolite, trachyandesite, alkaline rhyolite, monzodiorite, and alkaline granite, characterized by high K_2O, $K_2O + Na_2O$, and K_2O/Na_2O (PARFENOV et al., 1978). The K_2O content decreases and magmatic center migrated towards the oceanic coast in this belt. In the Japanese islands, Korean peninsula and southeast China, the continental margin type magmatic belts continued their activity

Fig. 3

in this period. They are composed of rhyolite and dacite (mainly welded ash-flow tuffs) with subordinate amounts of andesite, granite, and granodiorite, which show high K_2O, $K_2O + Na_2O$, and K_2O/Na_2O. Collision type: the magmatic belts of this type are present in the Stanovoy, northern Bureya, and Tsinling regions. Those in the Stanovoy and northern Bureya consist of felsic to intermediate subalkaline minor intrusives (granite porphyry, granodiorite porphyry, diorite porphyry, diorite, syenite porphyry, and monzonite) with high alkalinity (KHRENOV *et al.*, 1976). That in the Tsinling region is mainly composed of biotite granite (ACADEMIA SINICA, 1973). Intra-continent type: the intra-continent type magmatic belt is only distributed near the Liaodong peninsula, which is comprised of minor felsic intrusives (biotite granite etc.)

(ACADEMIA SINICA, 1973). Predominantly alkaline type: this type of magmatic belts are developed in the Shandong peninsula and middle-lower Yangtze valley, which are present behind the continental margin type magmatic belt. They are composed of alkaline basalt, trachyandesite, and andesite with associated intermediate minor intrusives, which are characterized by high alkalinity (ACADEMIA SINICA, 1977; WU, 1981; XUE, 1981; YU and BAI, 1981).

2.5   70–40 Ma (Fig. 3E)

In this period, the island arc, collision and intra-continent type calc-alkaline magmatic belts are completely lacking. Continental margin type: this type of magmatic

Fig. 3

belts are restricted to the Penzhina-western Kamuchatka, eastern Sikhote Alin, southeast Korean peninsula (LEE, 1980) and southwest Japan. The Penzhina-western Kamchatka belt consists of andesite, dacite, and rhyolite (NATAL'IN and PARFENOV, 1981). In the eastern Sikhote Alin, southeast Korean peninsula and southwest Japan regions, the magmatic activity of this type continued also in this period. The rocks consist of rhyolite-dacite (mainly welded ash-flow tuffs), andesite, granite, and granodiorite with high K_2O, $K_2O + Na_2O$, and K_2O/Na_2O (PARFENOV et al., 1978; ISHIHARA, 1979; MURAKAMI, 1977; MURAKAMI and IMAOKA, 1980). Predominantly alkaline type: the predominantly alkaline type magmatic belts are distributed in the Okhotsk-Chukotka, northeast and southeast China regions. That in the Okhotsk-Chukotka region is mainly composed of alkaline basalt lava plateau formed by the fissure eruption, accompanied with subordinate amounts of trachyte, dacite and rhyolite with high alkalinity (PARFENOV et al., 1978). The belts in the northeast and southeast China regions consist of high alkaline tholeite to alkaline basalt, which erupted through fissures to form lava plateau (ZHENG, 1981; CONG et al., 1979).

3. Tectonic Evolution

The tectonic evolution of East Asia in late Mesozoic to early Cenozoic time is discussed on the view-point of the nature and space-time distribution of magmatism in the following sections. The development of tectono-magmatic frameworks in the area under discussion is shown in a series of Figs. 4A to E for five periods: 190–160 Ma, 160–130 Ma, 130–100 Ma, 100–70 Ma, and 70–40 Ma.

3.1   190–160 Ma (Fig. 4A)

The oceanic lithospheres were present between the Kolyma and Chukotka-Alaska continental blocks and between the Kolyma and Siberia continental blocks. The oceanic lithosphere between the Kolyma and Chukotka-Alaska continental blocks was subducted beneath the two blocks to form island arc type magmatic belts. The oceanic lithosphere between the Kolyma and Siberia continental blocks also was subducted under the two continental blocks to produce island arc type magmatic belts. The Uda-Murgal island arc type magmatic belt in the Okhotsk region was formed as a result of northward subduction of another oceanic lithosphere. The typical high P/T metamorphic belts in the Anadyr, Penzhina, and south Anyuy regions formed in Mesozoic time may be the relict of the subduction zone related to these island arcs (DOBRETSOV and CHIKOV, 1979). On the other hand, the oceanic lithosphere subducted northward beneath the Sino-Korea-Bureya and Yangtze continental blocks produced the continental margin type magmatic belts. The collision type magmatic belt was active along the Mongolian-Okhotsk suture zone formed by the collision of the Siberia and Sino-Korea-Bureya continental blocks during latest Paleozoic to early Mesozoic (NATAL'IN and PARFENOV, 1981). That in the Tsinling suture zone was produced by the collision of the Sino-Korea-Bureya and Yangtze continental blocks in late Triassic (LUO, 1979; FENG and ZHU, 1980). An strike-slip mobile zone was formed by the oblique subduction of oceanic lithosphere along the southwest Japan, northeast Japan and Sikhote Alin. In the northeast Japan and Sikhote Alin regions, the magmatism was

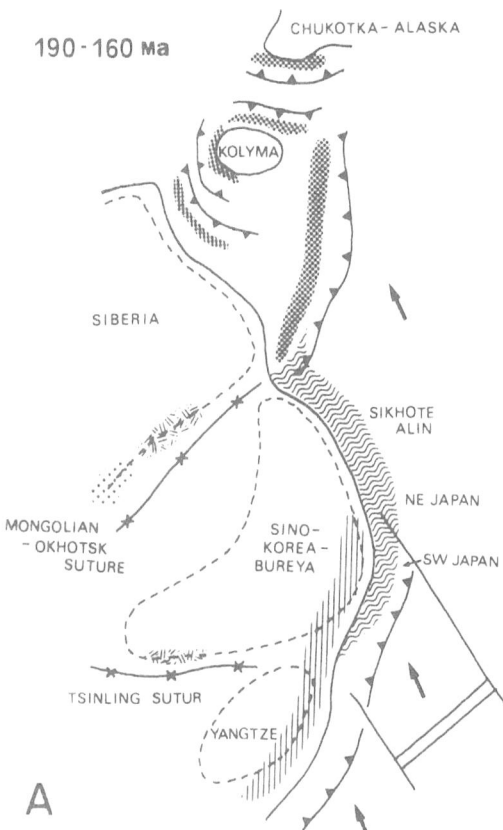

Fig. 4. The evolution of tectono-magmatic frameworks in late Mesozoic to early Cenozoic East Asia. A: 190–160 Ma; B: 160–130 Ma; C: 130–100 Ma; D: 100–70 Ma; E: 70–40 Ma. Waved area: strike-slip mobile belt; line with crosses: suture zone; line with solid triangles: subduction zone; arrow: direction of subduction; double lines: oceanic ridge; others: the same as in Fig. 3.

absent because of the transform nature of the boundary between the oceanic and continental lithospheres. Terrigenous and pelagic sediments, pieces of oceanic lithosphere and micro-continent were accreted to these regions owing to the oblique subduction and strike-slip tectonic movement (TAIRA *et al.*, 1981; SAITO and HASHIMOTO, 1982).

3.2   160–130 Ma (Fig. 4B)
    The oceanic lithosphere continued to be subducted under the Nutesyn, Oloy, and Uda-Murgal island arc type magmatic belts. The oceanic lithosphere also continued to underthrust northward beneath the Sino-Korea-Bureya and Yangtze continental blocks to form the continental margin type magmatic belts. In the Sino-Korea-Bureya continental block, areas of intra-continent type magmatism are widespread (Ta-

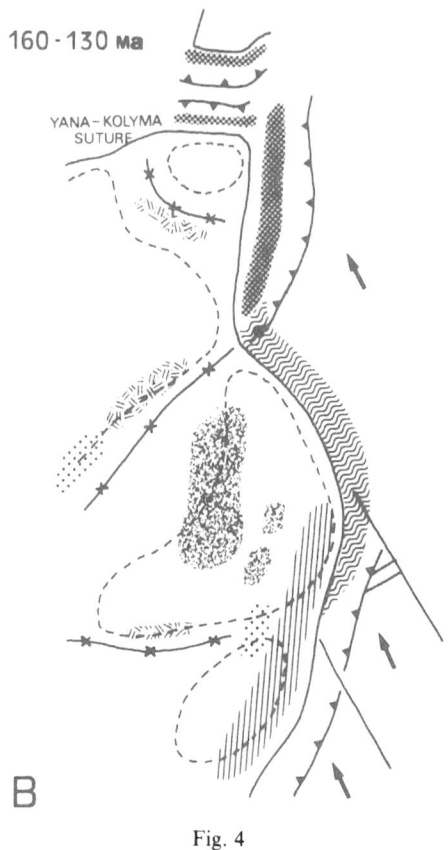

160 - 130 Ma

YANA – KOLYMA
SUTURE

B

Fig. 4

Khinganling region, Liaodong and Shandong peninsulas). This magmatism is very similar to that of western United States during the period 65–20 Ma in its chemical composition, mode of occurrence and space distribution (Fig. 5). The intra-continent magmatism in western U.S., called ignimbrite "flareup" (CONEY, 1978), is considered to be the result of shallow angle subduction of a rather young hot oceanic lithosphere beneath this region. The similar interpretation may apply with the late Jurassic intra-continent type magmatism in East Asia. UYEDA and MIYASHIRO (1974) proposed a similar mechanism, the subduction of oceanic ridge and young hot oceanic lithosphere, as the cause of this magmatism, though they considered that this event occurred in late Cretaceous. The Sanbagawa metamorphic belt, a type II high P meatmorphism (MIYASHIRO, 1973) or high P-intermediate type metamorphism (HASHIMOTO, 1972), represents a relatively high geothermal gradient in comparison to that of the typical high P type metamorphic belts. The latter are characterized by the occurrence of jadeite + quartz, lawsonite, and aragonite. According to BANNO et al., (1981), the maximum pressure and temperature of the Sanbagawa metamorphic belt were about 4 kb and 500°C respectively. This relatively high geothermal gradient may be caused by the subduction of a young hot oceanic lithosphere and an oceanic ridge, not of the ordinary cold oceanic lithosphere. The fact that the extent and age of the Sanbagawa metamorphic belt is nearly the same as the width of the regions and age of the intra-continent type magamtism in the Sino-Korea-Bureya continental block suggests that such a model may be applicable in this case. In this period, the Kolyma continental block collided with the Siberia continental block to produce the Yana-Kolyma collision type magmatic belt (PARFENOV *et al.*, 1978). The northeast Japan and Sikhote Alin regions were still the transform boundary during this period.

3.3 130–100 Ma (Fig. 4C)

The oceanic lithosphere began to be subducted under the northeast Japan and Sikhote Alin regions, where the island arc type magmatic belts were formed. The oceanic lithosphere continued to underthrust beneath the Okhotsk-Chukotka, southwest Japan, Korean peninsula and southeast China during this period. In this period, the Okhotsk-Chukotka region ceased to be an island arc and changed to an active continental margin with continental margin type magmatic belts. The effect of subduction of oceanic ridge and young hot lithosphere in the previous period was still

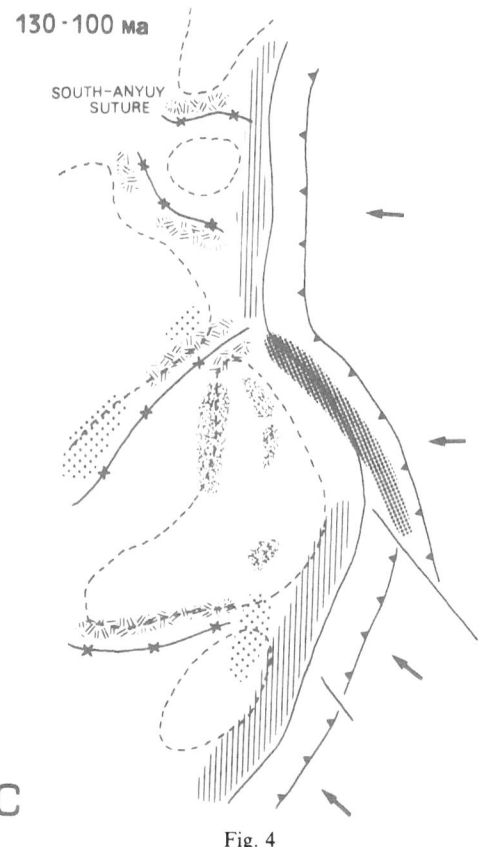

Fig. 4

manifest in northeast China (eastern Bureya, Ta-Khinganling, Liaodong and Shandong peninsulas), where the intra-continent type magmatic belts were active. The Chukotka-Alasks continental block collided with the Siberia-Kolyma block to form the south Anyuy suture zone, where the collision type magmatic belt came to be active (FUJITA, 1981; NATAL'IN and PARFENOV, 1981). A small-scale collision, which probably caused the collision type magmatism in northern Bureya region, occurred in the eastern-most portion of the Mongolian-Okhotsk suture zone (NATAL'IN and PARFENOV, 1981).

3.4   100–70 Ma (Fig. 4D)

The oceanic lithosphere was subducted along the margin of the Eurasian continental block to produce the continental margin type magmatic belt continuously extending for more than 8,000 km from the Chukotka to southeast China. The Okhotsk continental block approached the Eurasian continent from the east in this period.

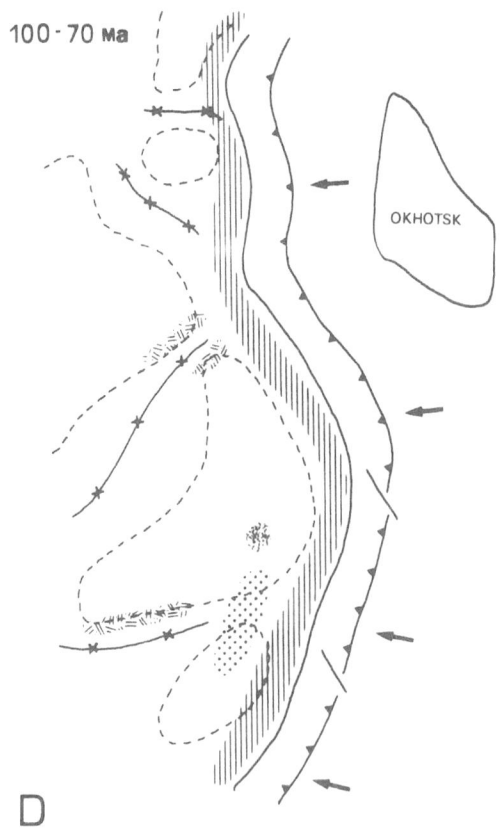

Fig. 4

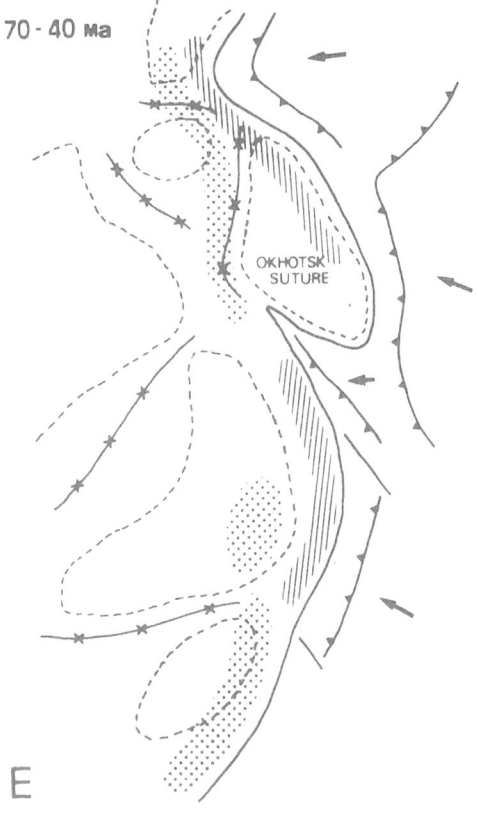

Fig. 4

3.5 70–40 Ma (Fig. 4E)

The Okhotsk continental block collided with the Eurasian continental block to form the Okhotsk suture zone (DICKINSON, 1978; PARFENOV *et al.*, 1978; NATAL'IN and PARFENOV, 1981), where the collision type magmatism (predominantly alkaline type) was active. Then the subduction zone jumped to the eastern side of the Okhotsk block and the continental margin type magmatic belt was formed there (DICKINSON, 1978; NARAL'IN and PARFENOV, 1981). The oceanic lithospheres were still subducted under the Sikohte Alin, Japanese islands, and southeast portion of the Korean peninsula to produce the continental margin type magmatism. On the other hand, the subduction ceased in the southeast China region, where only the predominantly alkaline type magmatism was active.

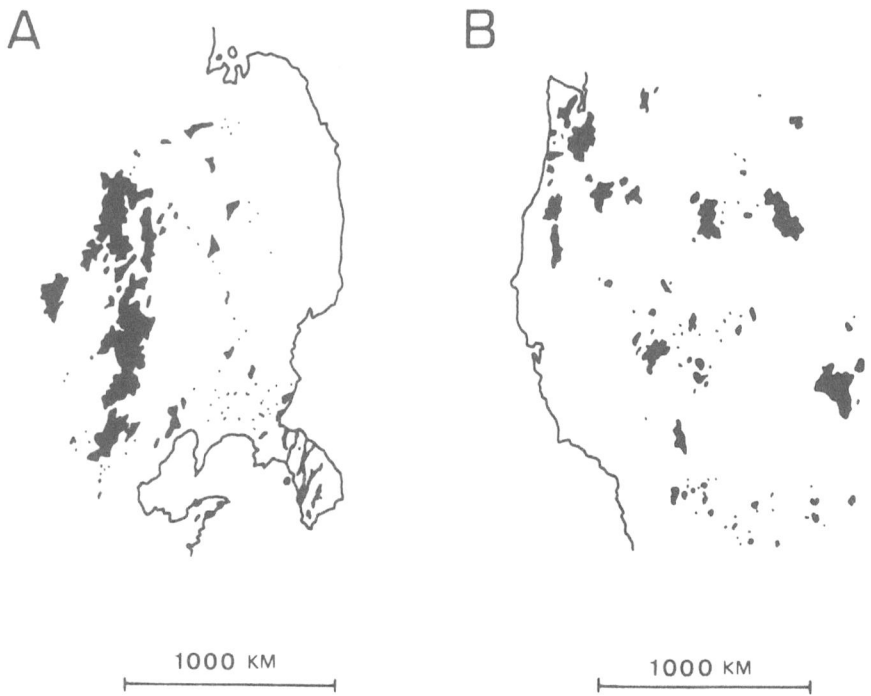

Fig. 5. The distribution of intra-continent type magmatism. A : That in East Asia for the time interval 160 to 130 Ma (ACADEMIA SINICA, 1975); B : That in western U.S. during the period from 65 to 20 Ma (EATON, 1980).

I wish express my thanks to Profs. Hashimoto M. (Ibaraki University) and Aramaki S. (University of Tokyo) for their helpful advices and critical reading of the manuscript.

REFERENCES

ACADEMIA SINICA (Institute of Geology), *Geologic Maps of Peaple's Republic of China*, 1973 (in Chinese).

ACADEMIA SINICA (Institute of Geology), *Geologic Map of Asia*, 1975, (in Chinese).

ACADEMIA SINICA (Minero-Petrographic Laboratory, Institute of Geology), A preliminary study of igneous activities in the region around Bohai Sea, *Scientia Geologica Sinica*, pp. 322–342, 1977 (in Chinese).

ACADEMIA SINICA (Institute of Geochemistry), Geochemistry of granitic rocks in south China, *Science Publication*, 421 pp. 1979 (in Chinese).

BANNO, S., T. HIGASHINO, M. OTSUKI, and C. SAKAI, Time-temperature-pressure trajectory of the Sanbagawa metamorphism, *Abstract of Oji International Seminar on Accretion Tectonics*, p. 14, 1981.

BELYY, V. F., I. N. KOTLYAR, A. P. MILOV, and P. P. PAVLOV, On late Mesozoic felsic volcanism in East Asian system of volcanogenic belts, *Geol. Geophys.*, 3–10, 1974 (in Russian).

CHAPPELL, B. W. and A. J. R. WHITE, Two contrasting granitic types, *Pacific Geol.*, **8**, 173–174, 1974.

CONEY, P. J., Mesozoic-Cenozoic Cordilleran plate tectonics: *Geol. Soc. Am. Mem.*, **152**, 33–50, 1978.

CONG, B., W. ZHANG, and D. YE, The study on the Cenozoic basalts in north China fault block, *Acta Geologica Sinica*, 112–124, 1979 (in Chinese).

DICKINSON, W. R., Plate tectonic evolution of north Pacific Rim, *J. Phys. Earth*, **26**, Suppl., S1–S19, 1978.

DOBRETSOV, N. L. and B. M. CHIKOV, Geology of northeast Asia, in *Chikyu Kagaku*, Vol. 16, edited by A. Miyashiro, pp. 263–299, Iwanami, Tokyo, 1979 (in Japanese).

EATON, G. P., 1980, Geophysical and geological characteristics of the crust of the Basin and Range province, in *Continental Tectonics*, edited by B. C. Burchfiel *et al.*, National academy of sciences, pp. 96–113.

FENG, Y., and ZHU, B., Melanges and tectonic development of the west Tsinling mountains, *Acta Geologica Sinica*, 34–43, 1980 (in Chinese).

FUJITA, K. and J. T. NEWBERRY, Plate margins and accretionary terranes in northeastern Siberia, *Abstract of Oji International Seminar on Accretion Tectonics*, p. 8, 1981.

GEOLOGICAL SURVEY OF JAPAN, *Geology and Mineral Resources of Japan*, edited by K. Tanaka and T. Nozawa, 430 pp., 1977.

HASHIMOTO, M., Metamorphic facies series of the high-pressure intermediate group, *24th I.G.C. Section Rep. 2*, pp. 75–80, 1972.

ISHIHARA, S., Metallogenesis in the Japanese island-arc system: *J. Geol. Soc. Lond.*, 135, 389–406, 1978.

ISHIHARA, S., Granite and rhyolite, in *Chikyu Kagaku*, Vol. 15, edited by K. Kanmera, M. Hashimoto, and T. Matsuda, pp. 105–141 Iwanami, Tokyo, 1980 (in Japanese).

JIN, M. S., S. Y. KIM, and J. S. LEE, Granitic magmatism and associated mineralization in the Gyeongsang Basin, Korea, *Mining Geol. Tokyo*, 31, 245–260, 1981.

KATADA, M., M. YOSHII, S. ISHIHARA, Y. SUZUKI, C. ONO, T. SOYA, and H. KANAYA, Cretaceous granitic rocks in the Kitakami mountains—petrography and zonal arrangement, *Rep. Geol. Surv. Japan*, 251, 139 pp., 1974 (in Japanese).

KHRENOV, P. M., A. A. BUKHAROV, and Y. A. NEKRAZOVA, Metallogenic features of the volcanic belts of Eastern Asia, *Intern. Geol. Rev.*, 585–596, 1976.

LEE, D. S., Study on the igneous activity in the middle Ogcheon geosynclinal zone, Korea, *J. Geol. Soc. Korea*, 7, 153–216, 1971.

LEE, Y. J., Granitic rocks from the southern Gyeongsang Basin, southeastern Korea. Part 1. General geology and K-Ar ages of granitic rocks, *J. Japan. Assoc. Min. Pet. Econ. Geol.*, 76, 331–342, 1980 (in Japanese).

LEE, M. S., Geology and metallic mineralization associated with Mesozoic granitic magmatism in south Korea, *Mining Geol. Tokyo*, 31, 235–244, 1981.

LUO, Z., On the occurrence of Yangtze old plate and its influence on the evolution of lithosphere in the southern part of China, *Scientia Geologica Sinica*, 127–138, 1979 (in Chinese).

MIYASHIRO, A., *Metamorphism and Metamorphic Belts*, 492 pp., George Allen & Unwin Ltd. London, 1973.

MURAKAMI, N., Compositional variations of some constituent minerals of the late Mesozoic to early Tertiary granitic rocks of southwest Japan, *Geol. Soc. Malaysia, Bull.* 9, 75–89, 1977.

MURAKAMI, N. and T. IMAOKA, Chemistry of late Mesozoic to Paleogene volcanic rocks in the inner side of southwest Japan with special reference to west Chugoku, *J. Japan. Assoc. Min. Pet. Econ. Geol.*, Special Issue No. 2, 263–278, 1980 (in Japanese).

NATAL'IN, B. A. and L. M. PARFENOV, Accretion and collision eugeosyncline folded systems in the north-west of the Pacific framing, *Abstract of Oji International Seminor on Accretion Tectonics*, 1981.

OTSUKI, K. and M. EHIRO, Major strike-slip faults and their bearing on spreading in the Japan Sea, *J. Phys. Earth*, 26, Suppl., S537–S555, 1978.

PARFENOV, L. M., I. P. VOINOVA, B. A. NATAL'IN, and D. F. SEMENOV, Geodynamics of the north-eastern Asia in Mesozoic and Cenozoic time and the nature of volcanic belts, *J. Phys. Earth*, 26, Suppl., S503–S525, 1978.

SAITO, Y., and M. HASHIMOTO, South Kitakami region: An allochthonous terrane in Japan, *J. Geophys. Res.*, 87, 3691–3696, 1982.

SHILO, N. A., and A. P. MILOV, Late Mesozoic granitic magmatism in geological structures of the U.S.S.R. North-East, *Geol. Soc. Malaysia, Bull.* 9, 117–122, 1977.

SHILO, N. A., N. V. BABKIN, V. F. BELYY, and A. A. SIDOROV, East Asia system of marginal volcanogenic belts (features of structure, magmatism, and metallogenesis), *Intern. Geol. Rev.*, 774–780, 1979.

TAIRA, A., Y. SAITO, and M. HASHIMOTO, The role of oblique subduction and strike-slip tectonics in the evolution of Japan, *Kagaku Tokyo*, 51, 508–515, 1981 (in Japanese).

TIL'MAN, S. M. and L. L. KRASNY, Relict of Mesozoic island arcs in the northeastern Asia, *Abstracts of IAVCEI symposium—Arc Volcanism*, pp. 379–380, 1981.

UYEDA, S. and A. MIYASHIRO, Plate tectonics and the Japanese Islands, *Geol. Soc. Am. Bull.*, **85**, 1159–1170, 1974.

WU, L., Characteristics and genesis of the Mesozoic volcanic rocks in the eastern part of China, *Abstracts of IAVCEI symposium—Arc Volcanism*, pp. 417–418, 1981.

XUE, Y., Mesozoic volcanic rocks in southeast China, *Abstracts of IAVCEI symposium—Arc Volcanism*, pp. 436–437, 1981.

YU, X. and Z. BAI, Latitic series in Lujiang-Zongyang region, *Geochimica*, pp. 57–65, 1981 (in Chinese).

ZHENG, X., On the Cenozoic basalt in east China, *Abstracts of IAVCEI symposium—Arc Volcanism*, pp. 442–443, 1981.

Hokkaido, North Japan

Accretion Tectonics in the Circum-Pacific Regions, edited by M. Hashimoto and S. Uyeda, 91–105.
Copyright © 1983 by Terra Scientific Publishing Company (TERRAPUB), Tokyo.

Collision Orogenesis and Sedimentation in Hokkaido, Japan

Hakuyu OKADA

Institute of Geosciences, Shizuoka University, Shizuoka 422, Japan

The classic notion of a 'Hidaka Orogeny' in Hokkaido is now explained in terms of collision orogenesis, probably multi-phase collision: an Oligocene to middle Miocene Phase and a latest Miocene Phase. Before the first phase of the collision, the Yezo Arc-Trench System and related sediments, mainly Cretaceous, were developed in the west, while oceanic and marginal sediments were developed around the Okhotsk Paleoland and proto-Kuril Arc. The collision process obducted some of the oceanic 'ophiolite sequence,' which was developed between the Yezo Trench and the Okhotsk Paleoland. After the first phase of the collision enormous amounts of deep-water gravity-flow sediments were deposited in front of the collision suture. The second phase of the collision brought about coarse clastic dispersal in shallow-water basins in front of the younger collision suture.

1. Introduction

Studies of the geology of Hokkaido, stimulated by MIYASHIRO's (1961) concept of paired metamorphic belts, have been developed into plate tectonic interpretations of the geologic evolution of Hokkaido, having led us to conclude that the mid-Tertiary tectonic event called the 'Hidaka Orogeny' implies collision between the Eurasian Plate and the Okhotsk Plate (HORIKOSHI, 1972; OKADA, 1978, 1979, 1980, 1982; KIMURA, 1981b).

Since I reviewed the geology of Hokkaido in terms of plate tectonics (OKADA, 1979, 1982), new information has been added to our knowledge. Incorporating these data I would like to summarize the relationship between the collision tectonics and sedimentation in Hokkaido. In this context, OKADA (1982) should be referred to, in which the tectonic evolution of Hokkaido has been outlined. Therefore, many references cited in OKADA (1982) have been omitted in this paper.

2. Geologic Framework of Hokkaido

The geologic framework of Hokkaido is subdivisible into six major tectonic belts (OKADA, 1982; Fig. 1). The geotectonic setting of these belts is briefly summarized below. The newly revised stratigraphic summary is also shown in Fig. 2.

2.1 Oshima Belt (O-Belt in Fig. 1)

The Oshima Peninsula area in this belt is composed of Carboniferous (?) to Jurassic 'eugeosynclinal' sediments as the basement rocks, which are covered by

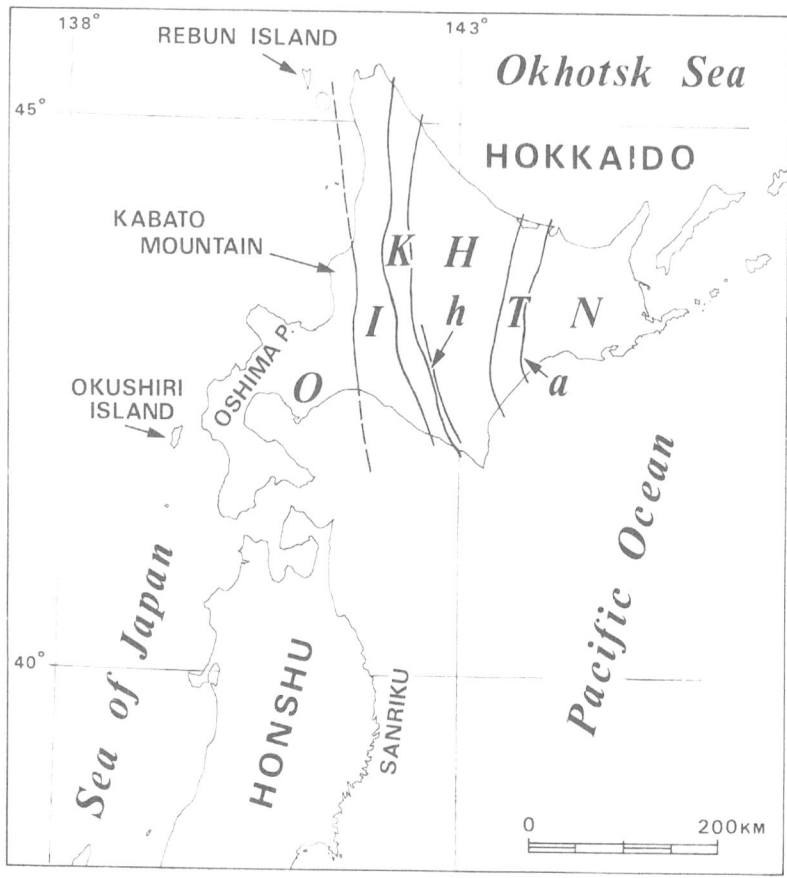

Fig. 1.   Tectonic divisions of Hokkaido (adapted from OKADA, 1982). O: Oshima Belt; I: Ishikari Belt; K: Kamuikotan Belt; H: Hidaka Belt (h: Hidaka Western Margin Tectonic Subbelt); T: Tokoro Belt; N: Nemuro Belt; a: Abashiri Tectonic Line (KIMURA, 1981a).

Neogene to Quaternary sediments mainly composed of thick pyroclastics. These basement rocks and pyroclastic covers are both regarded as the northern extension of the northern Honshu geologic unit. Particularly, the Neogene to Quaternary pyroclastic sequence corresponds to the 'Green Tuff' unit on the Japan Sea side of Honshu.

   Along the eastern margin of this belt, early Cretaceous strata are well exposed, namely the Barremian Rebun Group in Rebun Island (NAGAO et al., 1963) and the Aptian* or earlier Kumaneshiri Group in the Kabato Mountainland (Nakaseko, oral communication). They are correlated with the Harachiyama Formation of the Rikuchu Group exposed along the Sanriku coast of northeastern Honshu (Fig. 1) not only in age and lithology but also in the distribution of magnetic anomalies (OGAWA

---

*In OKADA (1982) the age of the Kumaneshiri Group was erroneously cited as the Valanginian.

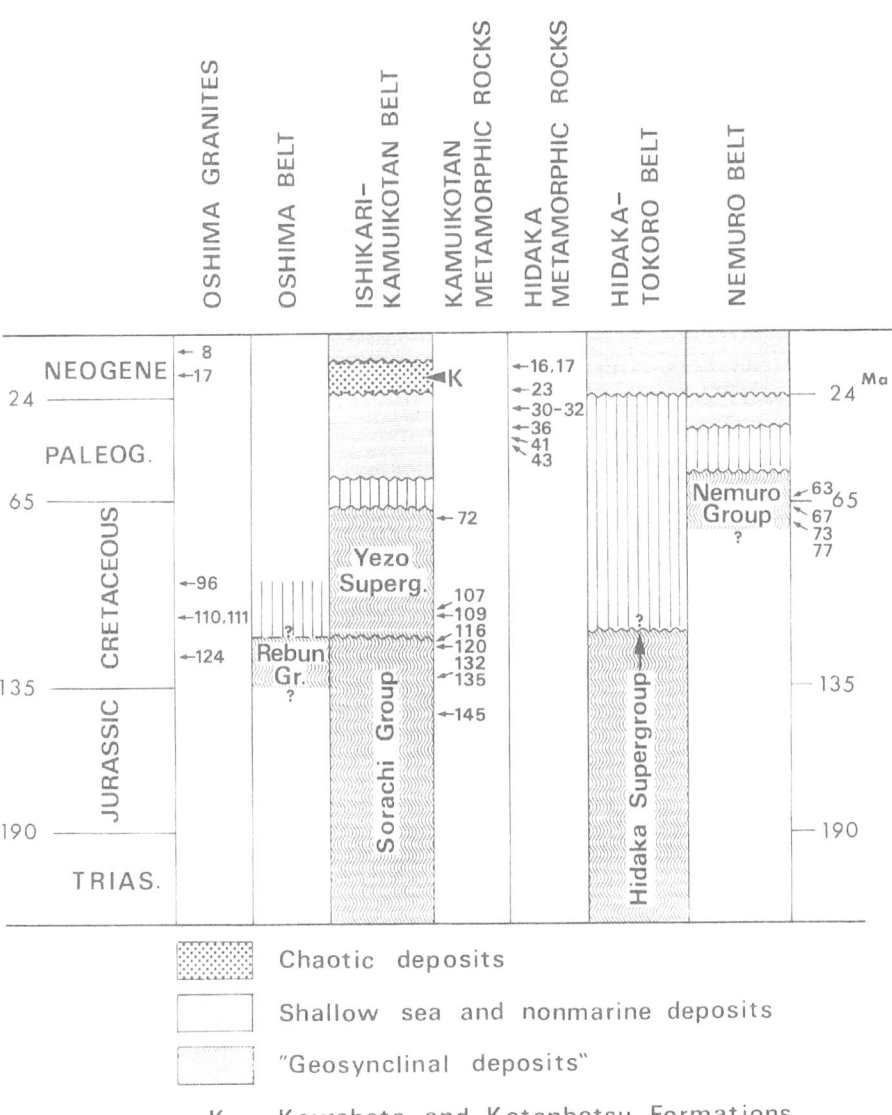

Fig. 2.   Schematic general stratigraphy in Hokkaido. Figures for Oshima granites and Kamuikotan and Hidaka metamorphic rocks indicate the radiometric ages (compiled from KAWANO and UYEDA, 1967a, b; UYEDA and AOKI, 1968; HAMAMOTO *et al.*, 1971; NOZAWA, 1975; SHIBATA and YAMADA, 1978; IMAIZUMI and UYEDA, 1981; SHIBATA and ISHIHARA, 1981); wave line shows unconformity and vertical ruling indicates nondeposition.

and SUYAMA, 1976; SEGAWA and FURUTA, 1978).

Granitic rocks in this belt are divided into two groups according to K-Ar ages: early Cretaceous and Miocene (Fig. 2). It is noteworthy that the early Cretaceous granities correspond well in age to the Kamuikotan metamorphic rocks. Another interesting fact is that granodiorites of 96 Ma are exposed in Okushiri Island together with Cretaceous rhyolitic to andesitic welded tuff (SHIBATA and YAMADA, 1978).

2.2 Ishikari Belt (I-Belt in Fig. 1)

This belt is made up of a thick sedimentary sequence mainly composed of late Mesozoic and Tertiary strata. The late Mesozoic deposits are divided into upper and lower sequences. The former is underlain with unconformity by the lower sequence of the Mesozoic: Valanginian to Jurassic Sorachi Group in the east and pre-Albian Kumaneshiri Group in the west, as revealed by the deep drilling carried out by the Japan Petroleum Exploration Company (ISHIWADA, 1973, p. 159).

The Sorachi Group is mainly composed of an ophiolitic rock association such as chert, micritic limestone, basic pyroclastics, basaltic pillow lava, and diabase, whereas the Kumaneshiri Group is composed of siliceous slate, terrigenous turbidites, andesitic and basaltic pyroclastic rocks, and basaltic pillow lavas. The stratigraphy of the Kumaneshiri Group, however, has not been fully understood.

The upper sequence is composed of terrigenous clastic sediments of the Yezo Supergroup,* ranging in age from Aptian to Maastrichtian. It shows shallow marine coarse-grained facies in the western part and deeper fine-grained facies in the eastern part (MATSUMOTO, 1954; OKADA, 1965, 1974; Matsumoto and Okada, 1971) (Fig. 3). Furthermore, the basal sequence of the Yezo Supergroup is not coeval from place to place, Albian occurring in the west and Aptian in the east (MATSUMOTO, 1978).

The Paleogene, overlying the Cretaceous with unconformity, consists of coal-bearing sediments in the lower and shallow marine sediments in the upper. The Paleogene strata are intensely folded concordantly with the Cretaceous, accompanied by overthrust sheets. This fold structure shows westward vergence (SHIMOKAWARA, 1963). The Paleogene is covered unconformably by very thick Neogene and Quaternary sequences (nearly 10,000 m in maximum thickness; MINATO et al., 1965, p. 271) (Fig. 4), which are composed of alternating marine and brackish water deposits with some pyroclastic rocks and coal seams. The early middle Miocene sequence is, in particular, characterized by marine gravity-flow deposits up to several thousand meters thick, which is represented by the Kawabata Formation in the southern part and the Kotanbetsu and Masuporo Formations in the northern part of this belt. These deposits were once regarded as the post-orogenic 'molasse' related to the 'Hidaka Orogeny' (NAGAO, 1938; MINATO et al., 1965, p. 271; SAKAMOTO, 1977, p. 254). However, the sediments characterized by chaotic deposits and turbidites (OKADA, 1978, 1980), are very different in lithofacies from the classic 'molasse' facies (e.g. VAN HOUTEN, 1973; MIALL, 1978).

---

*I would like to propose here the Yezo Supergroup for the whole sequence comprising the Lower Yezo Group, Middle Yezo Group, Upper Yezo Group, and Hakobuchi Group.

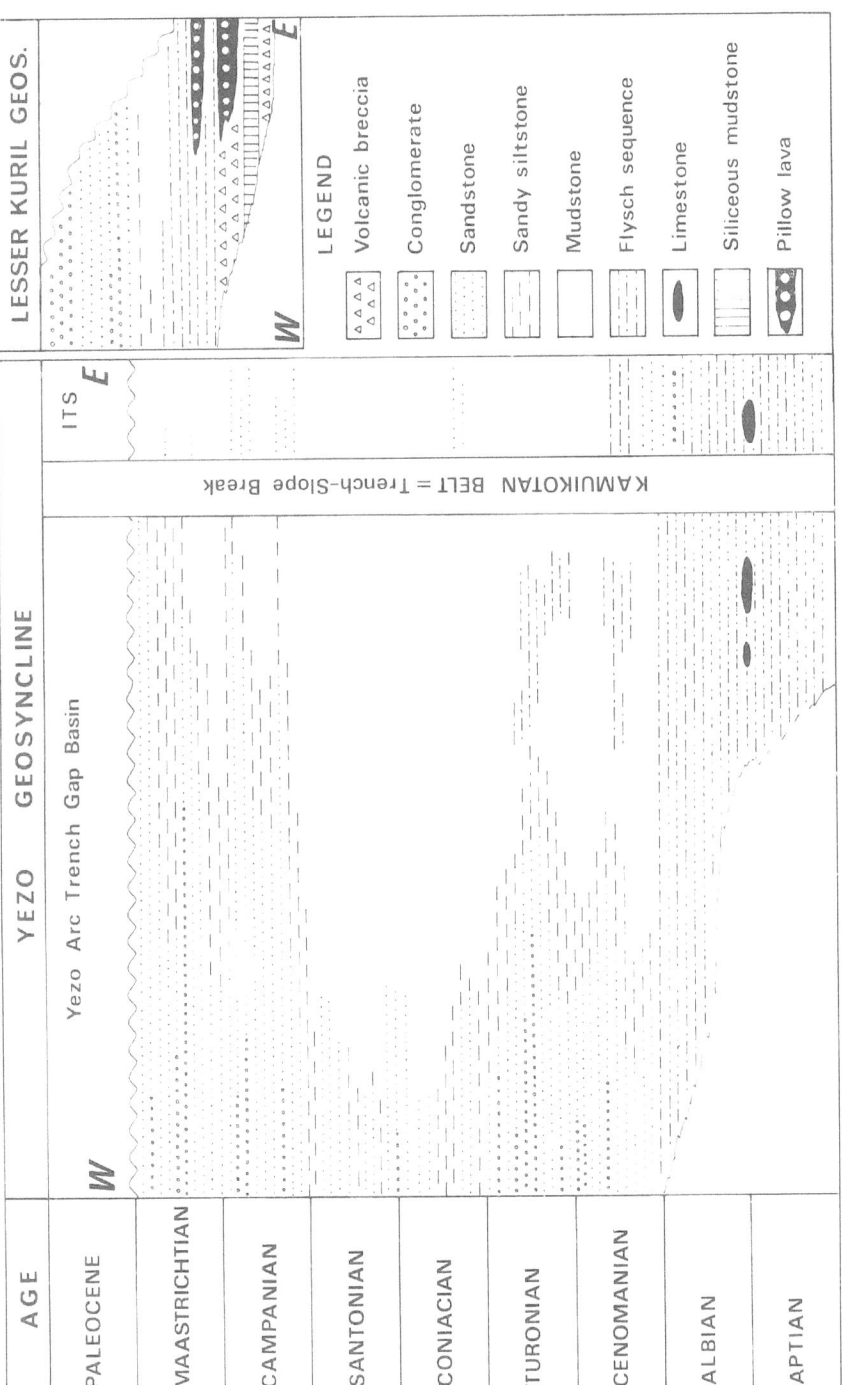

Fig. 3. General lithology of Cretaceous clastic facies in the 'Yezo Geosyncline' and late Cretaceous to Paleocene sequence in the 'Lesser Kuril Geosyncline' (adapted from MATSUMOTO, 1954; MATSUMOTO and OKADA, 1971; OBATA *et al.*, 1973; OKADA, 1974); ITS: trench inner slope basins. Figures not in scale; for the variation in thickness as well as in lithology see Fig. 33 in MATSUMOTO (1978).

2.3   Kamuikotan Belt (K-Belt in Fig. 1)

The Kamuikotan Belt is characterized chiefly by high-P/T-type Kamuikotan metamorphic rocks composed mainly of glaucophane schists, ultramafic and mafic rocks and partly by the Sorachi Group and the Yezo Supergroup. It is interesting to note that the Yezo Supergroup shows, as a whole, much more muddy facies in this belt (Fig. 3).

In the Horokanai area of the Kamuikotan Belt ISHIZUKA (1980) described an ophiolite sequence comprising radiolarian chert (30 to 50 m thick), basaltic pillow lavas (250 to 350 m thick), a complex of basaltic tuff-hyaloclastite-massive lavas (about 700 m thick), banded amphibolites (about 600 m thick), massive amphibolites (about 700 m thick), orthopyroxenite (about 50 m thick), dunite (about 500 m thick), and harzburgite (about 2,000 m thick).

The age of metamorphism of some rocks in this belt has been dated as 107 to 145 Ma (K-Ar ages on muscovites), although there is one value of 70 Ma (IMAIZUMI and UEDA, 1981). The metamorphism may have been completed before the deposition of the Lower Yezo Group (Aptian), as reviewed by OKADA (1982).

2.4   Hidaka Belt (H-Belt in Fig. 1)

The Hidaka Belt consists of the narrow Hidaka Western Margin Tectonic Subbelt (h-Subbelt in Fig. 1) and the main Hidaka Belt. The former is characterized by similar rock association to that of the Sorachi Group, namely basaltic lavas, diabase, basic pyroclastics, siliceous shale, chert and limestone. This narrow belt may be assigned to the collision suture.

The main Hidaka Belt comprises the Hidaka metamorphic rocks and the Hidaka Supergroup. The former, exposed in the axial part, is composed of gneisses, migmatites, green schists, hornfelses, basic plutonic rocks, ultramafic rocks, granitic rocks and others. The K-Ar ages of migmatites and granitic rocks range from 43 to 16 Ma (KAWANO and Ueda, 1967a, b; NOZAWA, 1975; SHIBATA and ISHIHARA, 1981). The age of metamorphism of the Hidaka metamorphic rocks, however, has not yet been dated.

The Hidaka Supergroup is composed of slate, chert, limestone, basic tuff, and flysch-type sediments of great thickness. Limestone and chert, some probably as displaced blocks, in the Hidaka Supergroup yield various kinds of fossils of varying ages such as late Permian fusulines (HASHIMOTO et al., 1975), late Triassic conodonts (IGO et al., 1974, 1980) and bryozoans (SAKAGAMI and SAKAI, 1979), Jurassic calcareous algae and corals (YABE and SUGIYAMA, 1941), and others. As some of these fossil-bearing rocks could be dislocated blocks, it is highly probable that the actual age of the Hidaka Supergroup is mainly Jurassic and perhaps ranges up to early Cretaceous as well as down to late Triassic. In this respect the Hidaka Supergroup is correlated with the Sorachi Group.

A detailed description of the Hidaka Supergroup is given by KIMINAMI and KONTANI (this volume).

2.5   Tokoro Belt (T-Belt in Fig. 1)

This belt is represented by the 'Hidaka Supergroup' composed of three lithostratigraphic units: Yubetsu Group showing flysch facies; Nikoro Group comprising basaltic rocks, basic volcaniclastics, shale, limestone and chert; and the Saroma Group composed of a flysch sequence (KONTANI and SAKAI, 1978). Their stratigraphic relation to each other and their age have not been settled yet, although TERAOKA *et al.* (1962) discovered late Jurassic (Oxfordian to Kimmeridgian) molluscan shells of *Buchia* sp. from the Saroma Group. Other fossils of Triassic and Jurassic ages are also known from limestones and cherts in this belt (IGO *et al.*, 1980).

KIMURA (1981a, b) underlined the importance of a major tectonic line between the Tokoro Belt and the Nemuro Belt following FUJII and SOGABE (1978) and called it the Abashiri Tectonic Line (*a* in Fig. 1). According to his detailed tectonic analysis, this line is the suture caused by the collision of the Kuril Arc with the central Hokkaido massif in the latest Miocene (KIMURA, 1981b).

Furthermore, thick Neogene strata composed of shallow marine to locally non-marine sediments, with a precursor of dacitic to andesitic lavas and welded tuff in the early Miocene sequence, are developed in this belt (Fig. 4). In particular, conglomerates of considerable thickness characterize the latest Miocene sequence, probably intimately related with the collision mentioned above (MIYASAKA and KIKUCHI, 1978; KIMURA, 1981b).

2.6   Nemuro Belt (N-Belt in Fig. 1)

The Nemuro Belt is composed mainly of the Campanian to Paleocene Nemuro Group, coal-bearing Oligocene strata, and Neogene marine beds with andesitic pyroclastic rocks, being covered by Quaternary volcanics.

The Nemuro Group, more than 3,000 m thick, is characterized by alkaline pillow lavas in the lower sequence (YAGI, 1969), flysch facies in the middle, and shallow marine, coarse-grained sediment facies in the upper sequence (OKADA, 1974; KIMINAMI, 1975a, b, c, 1976; KIMINAMI *et al.*, 1978; MATSUMOTO, 1978). The pillowed alkaline basalt is dated as 65–89 Ma (UEDA and AOKI, 1968; HAMAMOTO *et al.*, 1971). The Nemuro Group is also exposed in the outer arc of the Kuril Islands (SASA, 1934; GNIBIDENKO, 1971). Thus, the basin of the Nemuro Group was called the Lesser Kuril Geosyncline (originally Small Kuril Geosyncline) (OKADA, 1974).

The tectonic stress field in this belt was analysed in detail by MABUTI (1962) and KIMURA (1981a, b) among others. They showed that in the outer Kuril Arc area in the Nemuro Belt a nearly E-W compressional stress field prevailed since the Miocene and attained a maximum in latest Miocene to Pliocene times, giving rise to intense thrusting and lateral-slip faulting.

Another important fact is that remarkable positive Bouguer anomalies as high as over +200 milligals characterize the Nemuro area (TSUBOI, 1954; TOMODA, 1973). This value is the highest ever recorded in the Japanese Islands.

3.   Collision Tectonics and Sedimentation

3.1   Sedimentation before the collision Phase I

3.1.1   Yezo Arc-Trench System

In early Cretaceous time, the Oshima Belt extending to northern Honshu constituted a magmatic arc along the eastern margin of the proto-Siberian continent (Eurasian Plate). This magmatism is represented by the 124 to 96 Ma plutonic activity in Oshima Peninsula (Fig. 2) as well as by chiefly andesitic volcanism in the Rebun Group and the Kumaneshiri Group. In addition to them, rhyolitic to andesitic welded tuff in Okushiri Island also characterizes the related magmatism. Particularly it shows that the Okushiri area was a land at that time.

This arc magmatism should have been caused by the westward subduction along the eastern margin of the Kamuikotan Belt or the Hidaka Western Margin Tectonic Subbelt. This tectonic framework is called the Yezo Arc-Trench System according to the modern plate tectonic analogy.

In the arc-trench gap basin in this tectonic setting, the Yezo Supergroup was deposited, onlapping against the basement composed of the Kumaneshiri Group in the west and the Sorachi Group in the east. As already mentioned, the Yezo Supergroup indicates the marginal facies in the western part of the basin while the deeper marine facies is in the eastern part. Particularly the Yezo Supergroup begins with the Lower Yezo Group which is characterized by turbidite facies and is developed near the tectonic high corresponding to the trench-slope break in the modern analogy. Especially the basal sequence of the Lower Yezo Group, the Tomitoi Sandstone Member, contains sand particles of green schists probably derived from the Kamuikotan metamorphic rocks (FUJII, 1958). This means that the Kamuikotan rocks already constituted the tectonic high at time of sedimentation of the Lower Yezo Group.

The Lower Yezo Group thins out westwards (ISHIWADA, 1973), and instead, the Middle Yezo Group directly rests on the basement. The Mikasa Formation of the Middle Yezo Group characterizes the marginal facies in the western part of the basin. That is to say, (a) shallow-marine conglomerates and other coarse-grained sediments are predominant; (b) red-beds and coal seams on a small scale are intercalated; (c) molluscan fossils of shallow-sea and brackish-water elements such as *Pterotrigonia*, *Glycymeris*, *Aphrodina*, *Ostrea* and others are common (OKADA, 1965).

In sharp contrast, the Middle Yezo Group represented by the Shimanoshita Shale Member overlying the Tomitoi Sandstone Member is characterized by a shaly flysch sequence, which contains blocks of 'Orbitolina' limestones yielding rudistid bivalves as well. These limestone blocks were displaced from originally shallow sea areas into much deeper environments.

The Upper Yezo Group of the Yezo Supergroup is generally muddy in lithology and still shows the tendency to become more sandy in the west and much more muddy, with some intercalation by turbidites, in the east. The Hakobuchi Group is predominantly characterized by coarse-grained, andesitic sandstones of neritic to littoral environments associated with acidic to intermediate tuff layers. This sequence is also thinner and coarser in the west and thicker and finer in the east.

The Cretaceous clastic sequence in the arc-trench gap basin shows major paleocurrent patterns of eastward lateral flows and northward or southward axial flows, as summarized by TANAKA and SUMI (1981). This arc-trench gap basin seems to have been filled up gradually and eventually got shallow enough to be exposed above sea-level towards the end of the Cretaceous Period.

3.1.2 Kamuikotan Tectonic High and sedimentation

As mentioned in the preceding section, the Kamuikotan metamorphic rocks might have been formed and exposed partly as tectonic highs before the deposition of the Tomitoi Sandstone Member of the Lower Yezo Group. Therefore, this uplift should have controlled the sedimentation, just as the present-day Oyashio uplift has supplied gravels and has also affected sedimentation on the inner side of the uplift (VON HUENE et al., 1980).

On the western flank of the Kamuikotan uplift, well-defined, large-scale submarine channel structures are recognized near Saku, northwest Hokkaido (see MATSUMOTO and OKADA, 1973, Fig. 6). The structures are characterized by chaotic channel-fill deposits. This feature suggests the tectonically active nature of the uplift. Local unconformities below the Middle Yezo Group pointed out by INOMA (1969) also suggest unstable tectonic setting of this uplift.

East of the Kamuikotan Tectonic High, Aptian to Maastrichtian terrigenous deposits were developed mainly as turbidites on the trench inner slope, as is well exemplified in the Hidaka and Urakawa sections (OBATA et al., 1973; KANIE, 1966) (Fig. 3). As TANAKA and SUMI (1981) showd, general flow patterns in these deposits are represented by eastward lateral flows and northward or southward axial flows. In these deposits displaced blocks of shallow-water limestones are recognized as well (OBATA et al., 1973). This implies that the Kamuikotan Tectonic High was tectonically unstable during the sedimentation of the trench inner slope deposits.

3.1.3 Sorachi Ocean

In the Jurassic to early Cretaceous time the ocean floor outside the trench was the place of deposition of the Sorachi Group, which is lacking in terrigenous material. According to ISHIZUKA (1980), the Sorachi Group exposed in the Kamuikotan Belt shows the ophiolitic succession of harzburgite, dunite, orthopyroxenite, massive amphibolite, banded amphibolite, hyaloclastite-basic tuff, pillowed basalt and radiolarian chert with limestone in ascending order.

Recently IGO et al. (1980) discovered various fossils of latest Permian to Jurassic ages from limestone and chert. However, ISHIZUKA et al. (1982) detected Valanginian radiolarians in greenish gray shale enclosing a Triassic limestone block in the Hidaka Western Margin Tectonic Subbelt. Therefore, chert and limestone containing Permian and some Triassic fossils could be dislocated blocks and the Sorachi oceanic sediments may range only from the Jurassic to early Cretaceous.

3.1.4 Okhotsk Paleoland and its margin

To the east of the Sorachi Ocean should have been the Okhotsk landmass. The reason is threefold: (a) since at least Jurassic times clastic sediments of the Jurassic to early Cretaceous Hidaka Supergroup and the late Cretaceous to Paleocene Nemuro Group are characterized by terrigenous and volcaniclastic rocks. Composition of these clastics is rather similar between the Hidaka Supergroup and the Nemuro Group, but is very different from that of the Yezo sediments (MATSUMOTO and OKADA, 1971; OKADA, 1974; KONTANI and KIMINAMI, 1980); (b) paleocurrent systems show a striking contrast between the Yezo basin and the Hidaka and Nemuro basins (KIMINAMI, 1976; KONTANI and KIMINAMI, 1980; TANAKA and SUMI, 1981). In the Yezo basin, eastward flows and northward or southward axial flows are prevalent, while westward flows occur in the Hidaka basin and southward flows in the Nemuro basin ; (c) the modern Okhotsk Sea is underlain by continental crust (GNIBIDENKO, 1971 ; TUYEZOV, 1971 ; BURK and GNIBIDENKO, 1977), as was also discussed in OKADA (1982). From these lines of evidence, the Hidaka Supergroup and the Nemuro Group belong most likely to the Okhotsk geologic province.

3.1.5   Sedimentation during the Paleogene

On the former forearc areas, an Eocene to Oligocene sequence, more than 3,000 m thick in the Ishikari Belt and an Oligocene series more than 2,000 m thick in the Nemuro Belt were deposited, both characterized by paralic facies, in which important Japanese coal-fields occur. During the Paleogene no significant volcanism is recorded. This means that arc magmatism stopped during the Paleogene, if not throughout the period. Therefore, subduction must have slowed considerably or stopped, as in the case of the Japan Trench area off northeast Honshu (VON HUENE et al., 1982).

3.2   Sedimentation during and after the collision Phases I and II

3.2.1   Collision phase

After the sedimentation of Paleogene strata in the Ishikari Belt, they were folded along with the underlying Cretaceous, showing westward vergence. This phase was caused by southwesterly movement of the Okhotsk Plate colliding with the Eurasian Plate fringed by the Cretaceous Yezo Arc-Trench System [collision Phase I] (OKADA, 1979, 1980, 1982; KIMURA, 1981b). This Phase I collision may have taken place in Oligocene to middle Miocene times.

Based on the stress-field analysis, KIMURA (1981b) discriminated a later phase of collision [Phase II], which was completed by the collision of the Outer Kuril Arc with the central Hokkaido massif in the latest Miocene.

3.2.2   Sedimentation related with the Phase I collision

The 'Hidaka Orogeny' caused by the collisions mentioned above had remarkable influence upon sedimentation. The most notable fact is that very deep basins (more than 1,000 m) was developed in front of the Hidaka collision suture, which might have been strike-slip basins (READING, 1980) related to a convergent stress field. These basins are filled by the Kotanbetsu and Masuporo Formations in the northern part of the Ishikari Belt and the Kawabata Formation in the southern part (Fig. 4).

Sedimentation in these Hidaka frontal depressions has been intensively studied by OKADA (1978, 1980) with reference to the Kotanbetsu Formation. The Kotanbetsu basin was filled by gravity-flow deposits, which were supplied from the tectonic fronts to the east upheaved by the collision. The gravity-flow sedimentary facies is divisible into three subfacies reflecting the proximality : chaotic conglomerate facies nearest the source, then graded-bed facies, and ripple-bed facies furthest from the source (OKADA, 1978, 1980).

The chaotic conglomerate facies, though subdivisible into more sedimentary types, is regarded as an olistostrome. The graded-bed facies and ripple-bed facies are interpreted as turbidity-current deposits (turbidites) and bottom-current deposits (tractionites or contourites), respectively.

Furthermore, TAKAHASHI (1974) showed by petrographical analysis that the middle Miocene Kotanbetsu and Masuporo conglomerates were derived from the

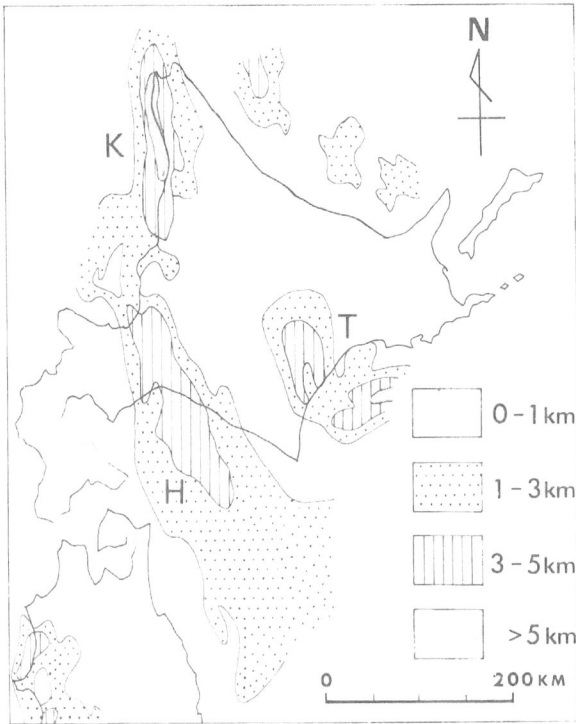

Fig. 4.   Isopach map of Neogene basins on and around Hokkaido (adapted from Ishiwada and Ogawa, 1976). K : Tenpoku (or Kotanbetsu) Basin ; H : Hidaka and Ishikari Basins ; T : Tokachi Basin.

uplift in the northern Hidaka Belt. Minato et al. (1965, pp. 271–276) also stressed that conglomerates in the Neogene formations in central Hokkaido are intimately related with the Hidaka upheaval. These uplifts must have been caused by the Phase I collision.

3.2.3   Sedimentation related with the Phase II collision

It is interesting to point out that in Miocene times a marked subsidence belt was developed in the southern Tokoro Belt, where a fairly thick (3,400 m +) Miocene to Pliocene sequence overlies the basement comprising the Hidaka Supergroup (Fig. 4). This Neogene sequence is represented by thick sandy to muddy facies (shallow marine and bay sediments) in the lower part (middle Miocene), and conglomeratic facies (shallow marine) in the upper part (late Miocene). Miocene to Pliocene sediments (1,800 to 3,000 m thick) with similar lithologies are also developed on the west side of the southern Hidaka Range (Shizunai area). According to Miyasaka and Kikuchi (1978), the upper Miocene conglomerates on both sides of the Hidaka Range are composed of clasts of granite, migmatites, gneissose and schistose gabbroic rocks, hornfelsed sediments, sandstone, and slate. They related these coarse clastics to the very rapid uplift of the southern Hidaka metamorphic belt. This uplift may have been caused by the Phase II collision. Further sedimentological studies of these sediments are necessary for better understanding of the relationship between the Phase II collision and the sedimentation.

4. Concluding Remarks

It may be concluded that the geology of Hokkaido is best explained in terms of multi-phase collisions, as demonstrated by the tectonic sutures between the Kamuikotan and Hidaka Belts (OKADA, 1979, 1980, 1982) and along the Abashiri Tectonic Line between the Tokoro and Nemuro Belts (KIMURA, 1981b). That is to say, the former represents the earlier phase of collision (Oligocene to middle Miocene) and the latter the later one (latest Miocene).

In this paper sedimentation features related to the earlier phase of collision are emphasized, but more data are necessary for a better understanding of the sedimentation related to the later collision phase.

It seems that the notion of the westward subduction tectonics of the Yezo Arc-Trench System has been generally accepted, but a serious and urgent problem to be solved is when and where the eastward subduction took place before the collision.

I would like to thank Professor Emeritus Tatsuro Matsumoto of Kyushu University and Dr. J. H. McD. Whitaker of University of Leicester for their reviews and comments on the manuscript. Thanks are due to Drs. K. Kiminami, G. Kimura and the participants of the Oji Seminar for constructive discussions. I also thank Drs. M. Hashimoto, Y, Saito, and S. Uyeda for inviting me to attend the Oji Seminar on Accretion Tectonics and to present this paper for this volume. This study was partly supported by a Grant-in-Aid for Scientific Researches from the Ministry of Education, Science and Culture (Monbusho), Japan. Many facilities for all phases of my field work for this study were provided by the Japan Petroleum Exploration Company, Limited (Tokyo and Sapporo Offices).

REFERENCES

BURK, C. A. and H. S. GNIBIDENKO, The structure and age of acoustic basement in the Okhotsk Sea, in *Island Arcs, Deep Sea Trenches and Back-arc Basins*, edited by M. Talwani and W. C., Pitman, III, pp. 451–461, American Geophysical Union, 1977.

FUJII, K., Petrology of the Cretaceous sandstones of Hokkaido, Japan, *Memoirs of Faculty of Science, Kyushu University*, Ser. D, **6**, 129–152, 1958.

FUJII, K. and M. SOGABE, Tectonic movement occurred in Hokkaido during the late Miocene and Pliocene time, *Geol. Surv. Japan Bull.*, **29**, 631–644, 1978 (in Japanese with English abstract).

GNIBIDENKO, H. S., Geology and deep structure of Sakhalin, Kuril Islands, and Kamchatka, in *Island Arc and Marginal Sea*, edited by S. Asano and G. B. Udintsev, pp. 5–16, Tokai University Press, (in Japanese with English abstract).

HAMAMOTO, R., M. YAMAGUCHI, and T. YANAGI, Rb-Sr ages of the alkaline rocks from Nemuro, Hokkaido, *Sci. Rep. Dept. Geol.*, *Kyushu Univ.*, **10**, 247–251, 1971 (in Japanese with English abstract).

HASHIMOTO, W., T. KOIKE, and T. HASEGAWA, First confirmation of the Permian System in the central part of Hokkaido, *Proc. Japan Acad.*, **51**, 34–37, 1975.

HORIKOSHI, E., Orogenic belts and plates in Japanese Islands. *Kagaku (Science)*, **42**, 665–673, 1972 (in Japanese).

IGO, Hy., Hh. IGO, S. ADACHI, and Y. SATO, On the geologic age of the Hidaka and Sorachi Groups, in *Studies of Geosynclines and Tectonic Division of the Northern Part of the Japanese Islands*, pp. 69–75, 1980 (in Japanese).

IGO, Hy., T. KOIKE, Hh. IGO, and T. KINOSHITA, On the occurrence of Triassic conodonts from the Sorachi Group in the Hidaka Mountains, Hokkaido, *J. Geol. Soc. Japan*, **80**, 135–136, 1974 (in Japanese).

IMAIZUMI, M. and Y. UEDA, On the K-Ar ages of the rocks of two kinds existed in the Kamuikotan metamorphic rocks located in the Horokanai district, Hokkaido, *J. Jpn. Assoc. Mineral. Petrol. Econ. Geol.*, **76**, 88–92, 1981 (in Japanese with English abstract).

INOMA, A., The intra-Yezo disturbance and sedimentation of the Middle Yezo Group, *J. Jpn. Assoc. Petrol. Technol.*, **34**, 155–161, 235–239, 196 (in Japanese with English abstract).

ISHIWADA, Y., ed., Petroleum industry and technology in Japan, in *Petroleum Technologists Association of Japan*, 430 pp., 1973 (in Japanese).

ISHIWADA, Y. and K. OGAWA, Petroleum geology of offshore areas around the Japanese Islands, *Un, ESCAP, CCIP, Techn. Bull.*, **10**, 23–34, 1976.

ISHIZUKA, H., Geology of the Horokanai ophiolite in the Kamuikotan Tectonic Belt, Hokkaido, Japan, *J. Geol. Soc. Japan*, **86**, 119–134, 1980 (in Japanese with English abstract).

ISHIZUKA, H., M. OKAMURA, and Y. SAITO, Radiolaria-bearing shale from the Sorachi Group at the Pippu area in the Hidaka Western Marginal Tectonic Zone, Hokkaido, *J. Geol. Soc. Japan*, in press, 1982.

KANIE, Y., The Cretaceous deposits in the Urakawa district, Hokkaido, *J. Geol. Soc. Japan*, **72**, 315–328, 1966 (in Japanese with English abstract).

KAWANO, Y. and Y. UEDA, K-A dating on the igneous rocks in Japan (IV)—Granitic rocks, summary, *J. Jpn. Assoc. Mineral. Petrol. Econ. Geol.*, **57**, 177–187, 1967a.

KAWANO, Y. and Y. UEDA, Periods of the igneous activities of the granitic rocks in Japan by K-A dating method, *Tectonophysics*, **4**, 523–530, 1967b.

KIMINAMI, K., Sedimentology of the Nemuro Group (Part I), *J. Geol. Soc. Japan*, **81**, 215–232, 1975a.

KIMINAMI, K., Sedimentology of the Nemuro Group (Part II)—Radiographic investigation of flysch-type sandstones in the Akkeshi Formation of the Nemuro Group, *J. Geol. Soc. Japan*, **81**, 697–708, 1975b (in Japanese with English abstract).

KIMINAMI, K., Sedimentology of the Nemuro Group (Part III)—Sedimentation of the lower Akkeshi Member, *J. Geol. Soc. Japan*, **81**, 755–768, 1975c (in Japanese with English abstract).

KIMINAMI, K., Sedimentology of the Nemuro Group (Part IV)—On the change of the source areas in the Akkeshi Formation, *J. Geol. Soc. Japan*, **82**, 773–782, 1976 (in Japanese with English abstract).

KIMINAMI, K., and Y. KONTANI, Pre-Cretaceous paleocurrents of the northeastern Hidaka Belt, Hokkaido, Japan, *J. Fac. Sci., Hokkaido Univ.*, Ser. IV, **19**, 179–188, 1979.

KIMURA, G., Abashiri Tectonic Line, with special reference to the tectonic significance of the southwestern margin of the Kurile Arc, *J. Fac. Sci., Hokkaido Univ.*, Ser. IV, **21**, 95–111, 1981a.

KIMURA, G., Tectonic evolution and stress field in the southwestern margin of the Kurile Arc, *J. Geol. Soc. Japan*, **87**, 757–768, 1981b (in Japanese with English abstract).

KONTANI, Y. and K. KIMINAMI, Petrological study of the sandstones in the pre-Cretaceous Yubetsu Group, northeastern Hidaka Belt, Hokkaido, Japan, *J. Assoc. Geol. Collab. Japan*, **34**, 307–319, 1980.

KONTANI, Y. and A. SAKAI, The stratigraphic and tectonic problems of the Hidaka Supergroup, *Assoc. Geol. Collab. Japan, Monograph*, **21**, 9–26, 1978 (in Japanese with English abstract).

MABUTI, S., A study on sedimentation and tectogenic history of the Paleogene System of the Kushiro coal field, in *Association of Promotion of Mining in Hokkaido*, Sapporo, 42 pp., 1962 (in Japanese with English abstract).

MATSUMOTO, T., ed., *The Cretaceous System in the Japanese Islands*, Japan Society of Promotion of Science, Tokyo, 324 pp., 1954.

MATSUMOTO, T., Japan and adjoining areas, in *The Phanerozoic Geology of the World. II. The Mesozoic, A*, edited by M. Moullade, and A. E. M. Nairn, pp. 79–144, Elsevier, Amsterdam, 1978.

MATSUMOTO, and H. OKADA, Clastic sediments of the Cretaceous Yezo Geosyncline, *Mem. Geol. Soc. Japan*, **6**, 61–74, 1971.

MATSUMOTO, T., and H. OKADA, Saku Formation of the Yezo Geosyncline, *Sci. Rep. Dept. Geol. Kyushu Univ.*, **11**, 275–309, 1973 (in Japanese with English abstract).

MIALL, A. D., Tectonic setting syndepositional deformation of molasse and other nonmarine-parallic sedimentary basins, *Can. J. Earth Sci.*, **15**, 1613–1632, 1978.

MINATO, M., M. GORAI, and M. HUNAHASHI, eds., *The Geologic Development of the Japanese Islands*, Tsukiji-Shokan, Tokyo, 442 pp., 1965.

MIYASAKA, S. and K. KIKUCHI, The Neogene Tertiary upheaval of the Hidaka Metamorphic Belt, *Assoc. Geol. Collab. Japan, Monograph*, **21**, 139–153, 1978 (in Japanese with English abstract).

MIYASHIRO, A., Evolution of metamorphic belts, *J. Petrol.*, **2**, 277–311, 1961.

NAGAO, T., Tertiary orogeny in Hokkaido, *J. Fac. Sci., Hokkaido Imperial Univ., Ser. IV*, **4**, 23–30, 1938.

NAGAO, S., C. AKIBA, and T. OMORI, *"Rebunto." Explanatory text of the geological map of Japan. Scale 1:50,000*, Geol. Surv. Hokkaido, Sapporo, 43 pp., 1963 (in Japanese with English abstract).

NOZAWA, T., ed., Radiometric age map of Japan—Granitic rocks, in *1:2,000,000 Map Series, No. 16–1*, Geological Survey of Japan, 1975.

OBATA, I., T. MAEHARA, and H. TSUDA, Cretaceous stratigraphy of the Hidaka area, Hokkaido, *Mem. Natl. Sci. Museum, Tokyo*, **6**, 131–145, 1973 (in Japanese with English abstract).

OGAWA, K. and J. SUYAMA, Distribution of aeromagnetic anomalies, Hokkaido, Japan, and its geologic implication, in *Volcanoes and Tectonosphere*, edited by K. Aoki and S. Iizuka, Tokai Univ. Press, Tokyo, pp. 207–215, 1976.

OKADA, H., Sedimentology of the Cretaceous Mikasa Formation, *Mem. Fac. Sci., Kyushu Univ., Ser. D*, **16**, 81–111, 1965.

OKADA, H., Migration of ancient arc-trench systems, in *Modern and Ancient Geosynclinal Sedimentation*, edited by R. H. Dott, Jr. and R. H. Shaver, *Soc. Econ. Paleontol. Mineral., Spec. Pub.* **19**, 311–320, 1974.

OKADA, H., Sedimentary patterns in apparent back-arc basins: a case study of the Neogene sequence in northwestern Hokkaido, Japan, *J. Phys. Earth*, **26**, Suppl., 477–490, 1978.

OKADA, H., The geology of Hokkaido and its plate tectonics, *Earth Monthly*, **1**, 869–877, 1979 (in Japanese).

OKADA, H., Sedimentary environments on and around island arcs: an example of the Japan Trench area, *Precambrian Res.*, **12**, 115–139, 1980.

OKADA, H., Geologic evolution of Hokkaido, Japan: an example of collision orogenesis, *Proc. Geol. Assoc.*, **93**, 201–212, 1982.

READING, H. G., Characteristics and recognition of strike-slip fault systems, *Spec. Pub. Intern. Assoc. Sedimentol.*, **4**, 7–26, 1980.

SAKAGAMI, S. and A. SAKAI, Triassic bryozoans from the Hidaka Group in Hokkaido, *Transact. Proc. Palaeontol. Soc. Japan. New Ser.*, **114**, 77–86, 1979.

SAKAMOTO, T., Neogene System, in *Geology and Mineral Resources of Japan*, 3rd ed., edited by K. Tanaka and T. Nozawa, pp. 233–259, Geological Survey of Japan, 1977.

SASA, Y., A preliminary note on the geology of the Island of Shikotan, southern Tisima (Southern Kuril Islands), *5th Pac. Sci. Congr. Proc.*, **2**, 2479–2482, 1934.

SEGAWA, J. and T. FURUTA, Geophysical study of the mafic belts along the margins of the Japanese Islands, *Tectonophysics*, **44**, 1–26, 1978.

SHIBATA, K. and S. ISHIHARA, K-Ar ages of granitic rocks in the Hidaka Belt, Hokkaido, in *Abstr. 88th Ann. Meet. Geol. Soc. Japan*, 342, 1981 (in Japanese).

SHIBATA, and N. YAMADA, K-Ar age of a granodiorite from Okushiri Island, Hokkaido, *Bull. Geol. Surv. Japan*, **29**, 611–613, 1978 (in Japanese).

SHIMOKAWARA, T., Geology and structural development of the Yubari coal field, Hokkaido, Japan, *Stud. Coal Geol.*, **5**, 1–244, 1963 (in Japanese).

TAKAHASHI, K., Composition of Tertiary conglomerates in northern Hokkaido, *Rep. Geol. Surv. Hokkaido*, **46**, 17–43, 1974 (in Japanese with English abstract).

TANAKA, K. and Y. SUMI, Cretaceous paleocurrents in the central zone of Hokkaido, Japan, *Bull. Geol. Surv. Japan*, **32**, 65–127, 1981 (in Japanese with English abstract).

TERAOKA, Y., K. KURODA, and K. HIRAYAMA, On the unknown Mesozoic formations, south of the Lake Saroma, Hokkaido, Japan, *J. Geol. Soc. Japan*, **68**, 416, 1962 (in Japanese).

TOMODA, Y., *Map of Free-air and Bouguer Gravity-Anomalies in and around Japan*, University of Tokyo Press, Tokyo, 1973.

TSUBOI, C., Gravity survey along the lines of precise levels throughout Japan by means of a Worden gravimeter. Part 4, Map of Bouguer anomaly distribution in Japan based on approximately 4,500 measurements, *Earthq. Res. Inst. Bull., Spec. Vol.*, **4**, 125–127, 1954.

TUYEZOV, J. K., Crustal structure of the Okhotsk and Japanese areas from regional seismic prospecting data, in *Island Arc and Marginal Sea*, edited by S. Asano and G. B. Udintsev, Tokai University Press, Tokyo, 121–136, 1971 (in Japanese).

UEDA, Y., and K. AOKI, K-Ar dating on the alkaline rocks from Nemuro, Hokkaido, *J. Jpn. Assoc. Mineral. Petrol. Econ. Geol.*, **59**, 230–235 1968 (in Japanese with English abstract).

VAN HOUTEN, F. B., Meaning of molasse, *Geol. Soc. Am. Bull.*, **84**, 1973–1976, 1973.

VON HUENE, R., M. LANGSETH, N. NASU, and H. OKADA, Summary, Japan Trench transect, *Initial Rep. Deep Sea Drilling Project*, **56/57**, 473–488, 1980.

VON HUENE, R., M. LANGSETH, N. NASU and H. OKADA, A summary of Cenozoic tectonic history along the IPOD Japan Trench transect, *Geol. Soc. Am. Bull.*, **93**, 829–846, 1982.

YABE, H. and T. SUGIYAMA, Discovery of *Circoporella semiclathrata* HAYASAKA from Hokkaido, *J. Geol. Soc. Japan*, **48**, 38–42, 1941 (in Japanese).

YAGI, K., Petrology of the alkalic dolerites of the Nemuro Peninsula, Japan, *Geol. Soc. Am. Mem.*, **115**, 103–147, 1969.

Note added in proof

After the manuscript went to the press, two papers on the age of the Kumaneshiri Group (OKADA *et al.*, 1982) and the Sorachi Group (OKADA *et al.*, 1982) have been published. The Kumaneshiri Group is correlated to the Aptian by radiolarian fauna. The top of the Sorachi Group exposed in the type area is assigned to the *Sethocapsa trachyostraca* Zone and the lower part of the *Eucyrtis tenuis* Zone according to radiolarian assemblages, which represent the Valanginian.

REFERENCES

OKADA, H., K. ANDO, and K. NAKASEKO, Discovery of Aptian radiolarian fauna from the Kumaneshiri Group, Hokkaido, *News Osaka Micropaleont. Spec.*, **5**, 359–360, 1982 (in Japanese with English abstract).

OKADA, H., A. HATAKENAKA, and K. NAKASEKO, Age of the Sorachi Group in its type area in Hokkaido, *News Osaka Micropaleont. Spec.*, **5**, 353–357, 1982 (in Japanese with English abstract).

Accretion Tectonics in the Circum-Pacific Regions, edited by M. Hashimoto and S. Uyeda, 107–122.
Copyright © 1983 by Terra Scientific Publishing Company (TERRAPUB), Tokyo.

Mesozoic Arc-Trench Systems in Hokkaido, Japan

Kazuo Kiminami* and Yoshihiro Kontani**

*Department of Geology and Mineralogical Sciences, Faculty of Science,
Yamaguchi University, Yoshida, Yamaguchi 753, Japan
**Department of Geology and Mineralogy, Faculty of Science,
Hokkaido University, Sapporo 060, Japan

The Mesozoic stratigraphic-tectonic belts in Hokkaido would have been developed from three arc-trench systems. Tectonic settings of the Oshima-Rebun and Yezo-Sorachi Belts and Hidaka-Tokoro Belt require for westward and eastward subducting plates on both sides of the Paleo-Sorachi-Hidaka ocean respectively. The Nemuro Belt represents a part of the Paleo-Kuril arc-trench system.

1. Introduction

The geologic development of Hokkaido has been explained in terms of geosynclinal concept "the Hidaka Orogeny" during the 1950's and 1960's (Minato et al., 1965, etc.). Since the concept of plate tectonics has been advanced, some authors have tried to apply it to explain the geology of Hokkaido and its neighborhood (Horikoshi, 1972; Dickinson, 1978; Parfenov et al., 1978; Okada, 1979, etc.). These aims, however, have not been achieved successfully because of the lack of enough knowledge on the geology of Hokkaido. In particular, the tectonic setting of the Hidaka Supergroup of Hokkaido has not been fully understood. This paper intends to summarize the characteristics of the Mesozoic stratigraphic-tectonic belts in Hokkaido and to reconstruct Mesozoic arc-trench systems.

2. Geology

The Pre-Cenozoic geologic framework of Hokkaido can be divided into four stratigraphic-tectonic major belts (Fig. 1). They are, from west to east, (i) Oshima-Rebun Belt, (ii) Sorachi-Yezo Belt, (iii) Hidaka-Tokoro Belt, and (iv) Nemuro Belt.

2.1 Oshima-Rebun Belt

The Pre-Cenozoic constituents of this Belt are chiefly late Paleozoic to Triassic sedimentary strata, late Jurassic to early Cretaceous volcanics and early Cretaceous granites. The late Paleozoic to Triassic sedimentary strata are distributed in southwest Hokkaido, and represents the northern extension of the northeast Honshu Geosynclinal Belt. The early Cretaceous granites are distributed in southwest Hokkaido sporadically intruding into late Paleozoic to Triassic strata.

The granites have been dated as 110 and 120 m.y. in the Oshima Peninsula

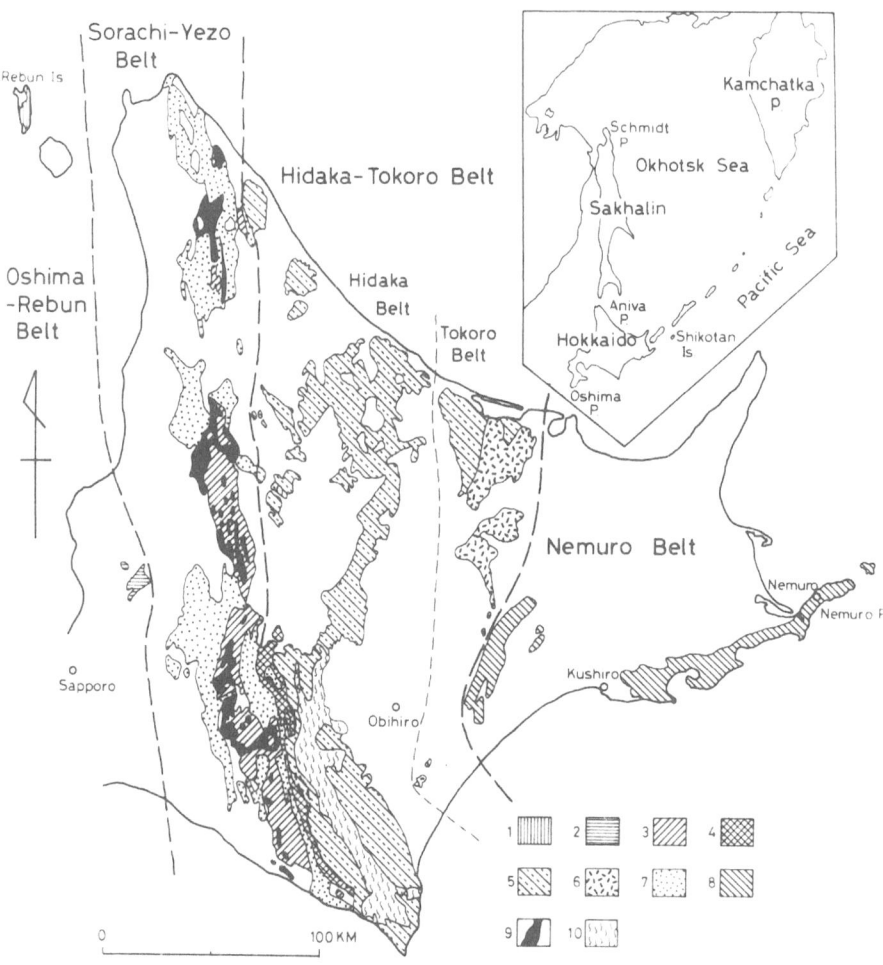

Fig. 1. Tectonic division and the distribution of Mesozoic and the Hidaka metamorphic rocks in Hokkaido. 1: Rebun Group, 2: Kumaneshiri Group, 3: Sorachi Group and Kamuikotan metamorphic rocks, 4–6: Hidaka Supergroup (4: Greenstone in the Hidaka Western Greenstone Belt, 5: "Greenstone-chert-turbidite facies" and "turbidite facies" of the Hidaka Supergroup, 6: Nikoro Group), 7: Yezo Group, 8: Nemuro Group, 9: Serpentinite, 10: Hidaka metamorphic rocks (Tertiary).

(KAWANO and UEDA, 1967), and 96 m.y. in Okushiri Island (SHIBATA and YAMADA, 1978). The late Jurassic to early Cretaceous volcanics are contained in the Kumaneshiri and Rebun Groups, and their northern extension is found in Moneron Island situated 45 km west of southern Sakhalin. The Rebun Group distributed in Rebun Island is mainly composed of andesitic tuff breccia, agglomerate, tuff and clastic sediment. The base of the Group is not exposed, and the Group's thickness is estimated to exceed 2,500 m. On megafossil evidences the Rebun Group is considered to be early Cretaceous in age (NAGAO et al., 1963). The Kumaneshiri Group resembles the Rebun

Group in lithology. The upper part of this Group contains inoceramus fragments, and is assigned to the Jurassic to Cretaceous (HASHIMOTO *et al.*, 1960). The log of a 4,215 m deep hole drilled on Moneron Island shows the existence of the volcanic formation of late Jurassic to early Cretaceous age (PISUKNOV and KHVEDCHUK, 1976); between depths of 1,481 m and 4,215 m occur mafic and intermediate lava and lava breccia with partings of volcanic sedimentary rocks. The uppermost part of the volcanic formation has radiometric ages of 98, 103, and 118 m.y., and the extrusive rocks found at depth of 3,680 m have a radiometric age of 141 m.y.. The Moneron-Rebun-Kumaneshiri zone is clearly expressed in the gravity field as a band of intense high.

2.2   Sorachi-Yezo Belt

This Belt is composed of serpentinite mélange with high-P/T-type Kamuikotan metamorphic rocks and ophiolite (ASAHINA and KOMATSU, 1979; ISHIZUKA, 1980) (Kamuikotan Metamorphic Belt), the Sorachi Group and the Yezo Group.

The Sorachi Group is composed of greenstone (basaltic massive lava, pillow lava, and hyaloclastite), chert, limestone, and a small amount of flysch-type sandstone. The Group is considered to represent the upper part of the ophiolite sequence. Limestones yield the Torinosu fauna that appears to be of late Jurassic to early Cretaceous age (TAKAHASHI and SUZUKI, 1978, etc.). The age given by the radiolarian assemblages in chert indicates the same as the limestone age (NAKASEKO *et al.*, 1979; WATANABE *et al.*, 1981).

Radiometric ages of the Kamuikotan metamorphic rocks are 109 and 120 m.y. (tectonic blocks) (BIKERMAN *et al.*, 1971), and 116, 107 and 72 m.y. (schists), and 145, 135 and 132 m.y. (tectonic blocks) (IMAIZUMI and UEDA, 1981). The age of 72 m.y. may be due to rejuvenation. The Kamuikotan Metamorphic Belt extends north to the Susunai Range in Sakhalin. The Susunai metamorphic rocks have been dated as 210–140 m.y. (DOBRETSOV and KURODA, 1970).

The Aptian to Maastrichtian Yezo Group overlies the Sorachi Group partly with unconformity and partly with conformity (KANIE *et al.*, 1981). This Group is mainly composed of thick turbidite sequence, and is divided into lower, middle, and upper Groups (MATSUMOTO, 1954). The clastics of the Group were mainly supplied from the west (MATSUMOTO and OKADA, 1971; KIMINAMI *et al.*, 1978; TANAKA and SUMI, 1981, etc.). The relationship between the lower and middle Yezo Groups is a conspicuous unconformity near the Kamuikotan Metamorphic Belt (Fig. 2). This unconformity is ascribed to the diastrophism in the Kamuikotan Metamorphic Belt occurred at the end of the Albian (INOMA, 1969; KIMINAMI *et al.*, 1978).

2.3   Hidaka-Tokoro Belt

The Pre-Cenozoic constituent of this Belt is the Hidaka Supergroup. The Hidaka Supergroup can be subdivided into four parts according to the differences of lithofacies and structure: (1) greenstones in the Hidaka Western Greenstone Belt, (2) greenstone-chert-turbidite facies in the Hidaka Belt, (3) turbidite facies in the Hidaka-Tokoro Belt, and (4) pillow lava-hyaloclastite-chert facies in the Tokoro Belt (Nikoro Group).

The Hidaka Western Greenstone Belt consists of basaltic massive lava, pillow lava and hyaloclastite, and a small amount of limestone and chert. This Belt is narrow and is

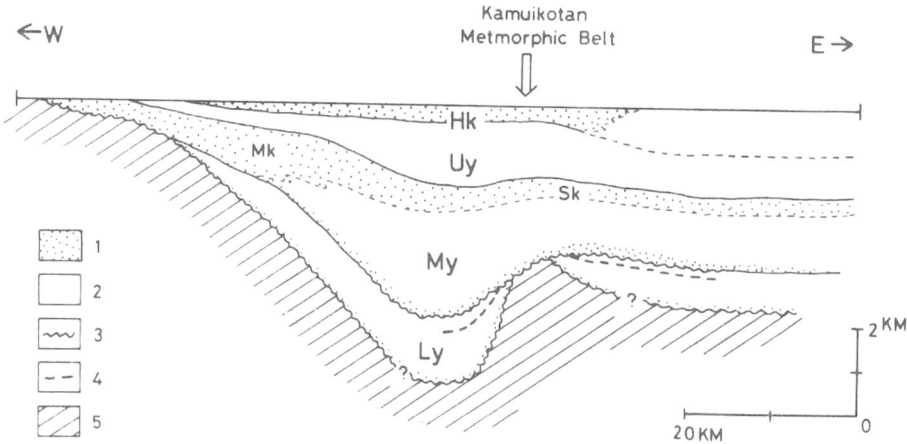

Fig. 2. Schematic cross section in the depositional stage of the Yezo Group. Ly: Lower Yezo Group, My: Middle Yezo Group, Mk: Mikasa Formation, Sk: Saku Formation, Uy: Upper Yezo Group, Hk: Hakobuchi Group. 1: Coarse clastic sediments, 2: Fine clastic sediments, 3: Unconformity, 4: Limestone, 5: Basement of the Yezo Group.

in fault contact with the Yezo Group on the west side and with "greenstone-chert-turbidite facies" of the Hidaka Supergroup on the east. Limestones in this Belt yield late Triassic condonts (IGO *et al.*, 1974; SAKAGAMI and SAKAI, 1979) and Triassic Bryozoan fossils (SAKAGAMI and SAKAI, 1979). The petrochemical study of greenstones in the Belt indicates that the original rocks of these greenstones show transitional affinity between tholeiite and alkaline basalt (ISHIZUKA, 1981).

"The greenstone-chert-turbidite facies" of the Hidaka Supergroup is mostly distributed in the western part of the Hidaka Belt. This facies is represented by the Idonnappu Formation distributed west of the Hidaka Metamorphic Belt.* The Idonnappu Formation consists of greenstone (basaltic pillow lavas and hyaloclastite), chert, micritic limestone, slate and flysch-type alternation. The middle to late Triassic Bryozoans and the late Triassic conodonts were obtained from chert and limestone of the Formation and its equivalents (IGO *et al.*, 1974; ISHIZAKI, 1979; IGO *et al.*, 1980). Figure 3 shows the schematic cross section of the Idonnappu Formation along the River Chiroro where the Formation is exposed typically. This cross section indicates that the Formation is cut by many east dipping reverse faults. Each slab separated by reverse faults has an average thickness of several hundred meters. The structure of most slabs is trending N-S facing to the east. The lithological assemblages of the slabs are chert, limestone, slate and flysch-type sandstone in the western part, chert, chert bearing slate and flysch-type sandstone in the middle part, and hyaloclastite with pillow lava, chert, chert bearing slate and flysch-type sandstone in the eastern part in ascending order. Stratigraphic relationships of these slabs are not yet known. The

---

*The radiometric ages of granites and migmatites from the Hidaka Metamorphic Belt show 36 to 16 m.y. (Oligocene to Miocene) (KAWANO and UEDA, 1967; NOZAWA, 1975). The geologic setting of the Belt will be discussed by KOMATSU *et al.* (1982).

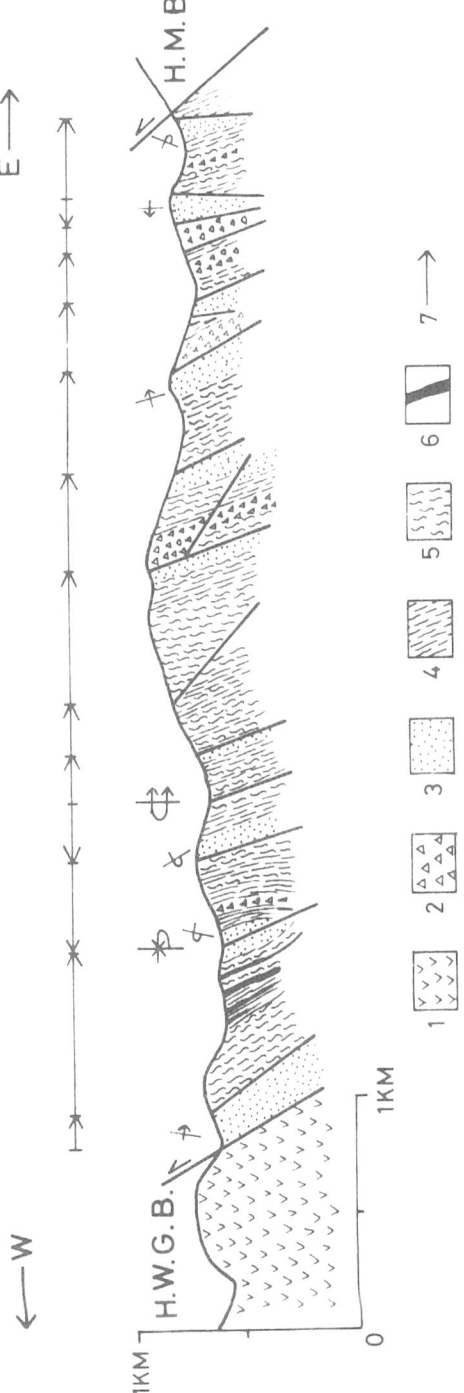

Fig. 3.   Structure of the Idonnappu Formation along the River Chiroro. H.W.G.B.: Hidaka Western Greenstone Belt, H.M.B.: Hidaka Metamorphic Belt. 1: Basaltic massive and pillow lavas, 2: Hyaloclastite, 3: Alternation of sandstone and shale, 4: Slate, 5: Chert, 6: Limestone, 7: Upward direction in each slab.

petrochemical data suggest that the greenstones belonging to the Formation were derived from tholeiitic magma (ISHIZUKA, 1981). The flysch-type sandstones of the Formation are mainly lithic wacke, and their framework grains are composed of quartz, feldspar (plagioclase and K-feldspar), lithic fragment and a small amount of heavy mineral. The lithic fragments are composed of basic to acid volcanic rocks, acid plutonic rock, and small amounts of semi-schist and clastic sediment. The provenance of the clastics must be a paleo-land with active volcanism.

"The turbidite facies" of the Hidaka Supergroup is mostly distributed in the eastern part of the Hidaka-Tokoro Belt. This facies is represented by the Nakanogawa Group exposed on the eastern flank of the Hidaka Metamorphic Belt, and by the Yubetsu and Saroma Groups in the northeastern part of the Hidaka-Tokoro Belt. The Nakanogawa Group consists mainly of turbidite, massive sandstone and slate intercalated with conglomerate and tuff. The Group is estimated to be 8,000 m thick (KONTANI, 1978). IGO et al. (1980) reported the occurrence of the Triassic conodonts from limestones in the lower part of this facies (Kamui Group) of the southwestern flank of the Hidaka Metamorphic Belt. The Yubetsu and Saroma Groups are regarded as of the same geologic age and resemble the Nakanogawa Group in lithology. The middle part of the Yubetsu Group yields middle Cretaceous radiolarias. The total thickness of the Yubetsu Group is estimated to be over 10,000 m, and the structure is trending N-S facing to the east and forms a homocline. The base of this Group is not known. The Saroma Group conformably overlies the Nikoro Group and is about 3,000 m in thickness. Figure 4 shows the compositions of the sandstones of the Nakanogawa, Yubetsu and Saroma Groups based on the studies of 150 samples. Most sandstones of the three Groups belong to lithic wacke and are rich in feldspar, lithic

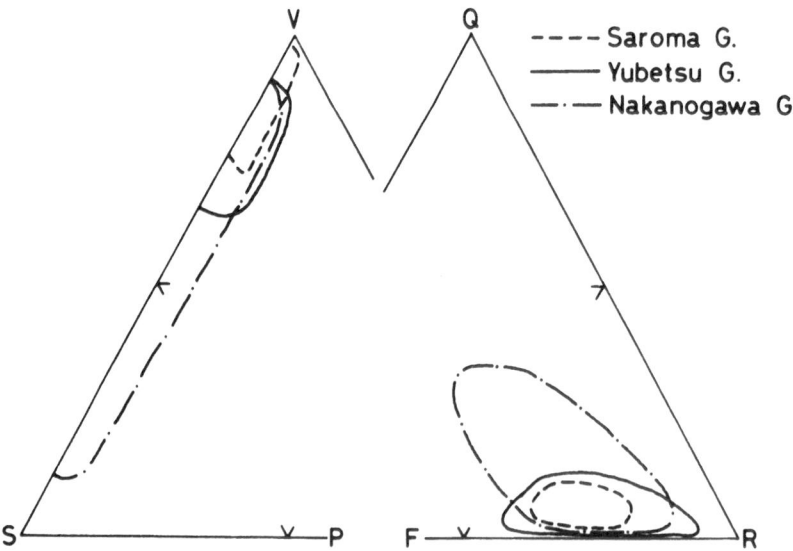

Fig. 4. Ternary diagram showing the composition of sandstones from the Nakanogawa, Yubetsu and Saroma Groups. Q: Quartz, F: Feldspar, R: Rock fragment, V: Volcanic rock fragment, S: Sedimentary rock fragment, P: Plutonic rock fragment.

fragment and pyroxene and/or hornblende. The lithic fragments are mainly composed of volcanic rock with acid plutonic rock, semi-schist, chert, and clastic sediment. The volcanic fragments consist chiefly of basic to acid volcanic rocks with welded tuff and scoria. Provenance of the material of these Groups must be a paleo-land with active volcanoes of basic to acid rocks. Figure 5 shows the paleocurrent directions of the Yubetsu and Saroma Groups examined by a three dimentional analysis of internal sedimentary structures utilizing the soft X-ray technique and the measurements of sole markings. From the data obtained in this study, the prevailing current direction in the

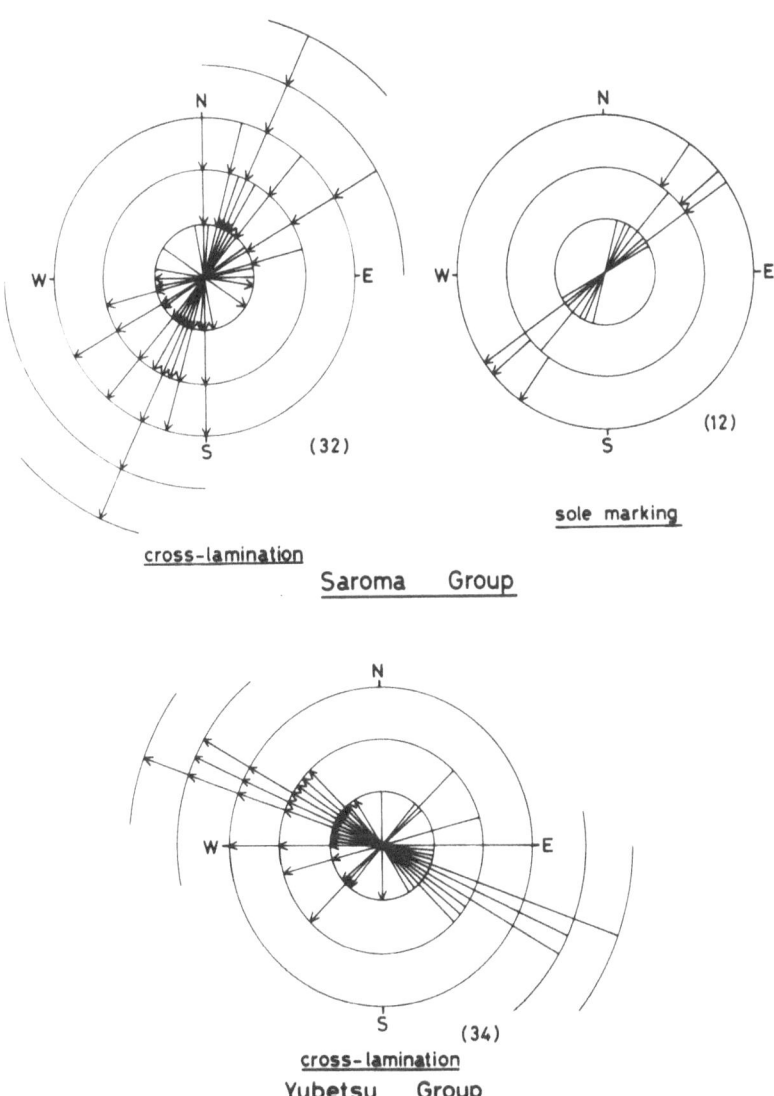

Fig. 5. Paleocurrent directions of the Yubetsu and Saroma Groups.

turbidites seems to be a lateral current from east to west with minor occurrence of axial currents along N-S direction. Therefore, the provenance called Paleo-Okhotsk Land (HASHIMOTO, 1958; KIMINAMI et al., 1978; KIMINAMI and KONTANI, 1979; OKADA, 1979; KONTANI and KIMINAMI, 1980) was presumably located to the east. "The turbidite facies" must have been deposited in the western margin of the Paleo-Okhotsk Land during the late Triassic to middle Cretaceous.

"The pillow lava-hyaloclastite-chert facies" in the Tokoro Belt is peculiar and is called the Nikoro Group. The base of the Nikoro Group is not exposed, and the lowermost part observed in the field is composed of massive sandstone of volcanic origin. The clastic rocks are believed to have accumulated in relatively shallow waters. A thick sequence consisting of pillow lavas and hyaloclastite with red chert, limestone and shale conformably overlies the massive sandstone. The sequence is estimated to be 3,000 m thick (TERAOKA et al., 1962). Fossils in the limestones of the Group resemble those from the upper Jurassic Torinosu Limestone (YAMADA et al., 1963). The age given by radiolarias in the cherts indicates the late Jurassic to early Cretaceous. The Group has the form of an open synclinal fold trending N ∼ NE.

2.4 Nemuro Belt

The Pre-Cenozoic constituent of this Belt is the late Cretaceous to Paleocene Nemuro Group distributed in the southern part. The Group is unconformably overlain by coal-bearing Oligocene sediments, and its lower limit is not known. The total thickness exceeds 3,000 m. The Group distributed from Kushiro to Nemuro extends eastward from Suisho Island through Shibotsu Island to Shikotan Island. This zone extends farther to the east, to the Vityaz submarine rise over several hundreds kilometers long, and represents non-volcanic outer arc of the Kuril arc-trench system. The Nemuro Group in this zone consists of turbidite, shale, conglomerate, acidic tuff, tuff breccia, pillow lava, and dolerite sheet (KIMINAMI, 1978; KOCHERGIN and KRESNYY, 1977; TROFIMUK, 1976; FROLOV et al., 1980). The general strike and dip of the beds are E-W and 10°–20° S respectively, and in many areas the rocks have a homocline structure. The sedimentary facies, paleocurrents and many submarine slumpings indicate that the sedimentary basin of the Group in this zone extended in an E-W direction. In the eastern part, many pillow lavas and dolerite sheets which are alkalic in nature are contained in the lower and middle parts of the Group. These pillow lavas are found in turbidite or in shallow marine sandstone and shale. The data of paleocurrents and submarine slumpings indicate that the main paleo-landmass existed to the north of the basin, and subordinately to the south. The northern landmass seems to be a part of the Paleo-Okhotsk Land (KIMINAMI et al., 1978). From the petrological study of the sandstones, the northern paleo-landmass is inferred to have been an active basic to intermediate volcanic chain in the early and middle stages of the Group (KIMINAMI, 1979). The southern paleo-landmass supposed temporarily supplied the sediments derived from basic to intermediate volcanic rocks. The volcanic activities are supposed to have been comparatively violent in the eastern part of the "heimat" and basin. The deposition and igneous activity must have occurred in the southern margin of the Paleo-Okhotsk Land. Schematic sedimentary history of the Nemuro Group is shown in Fig. 6.

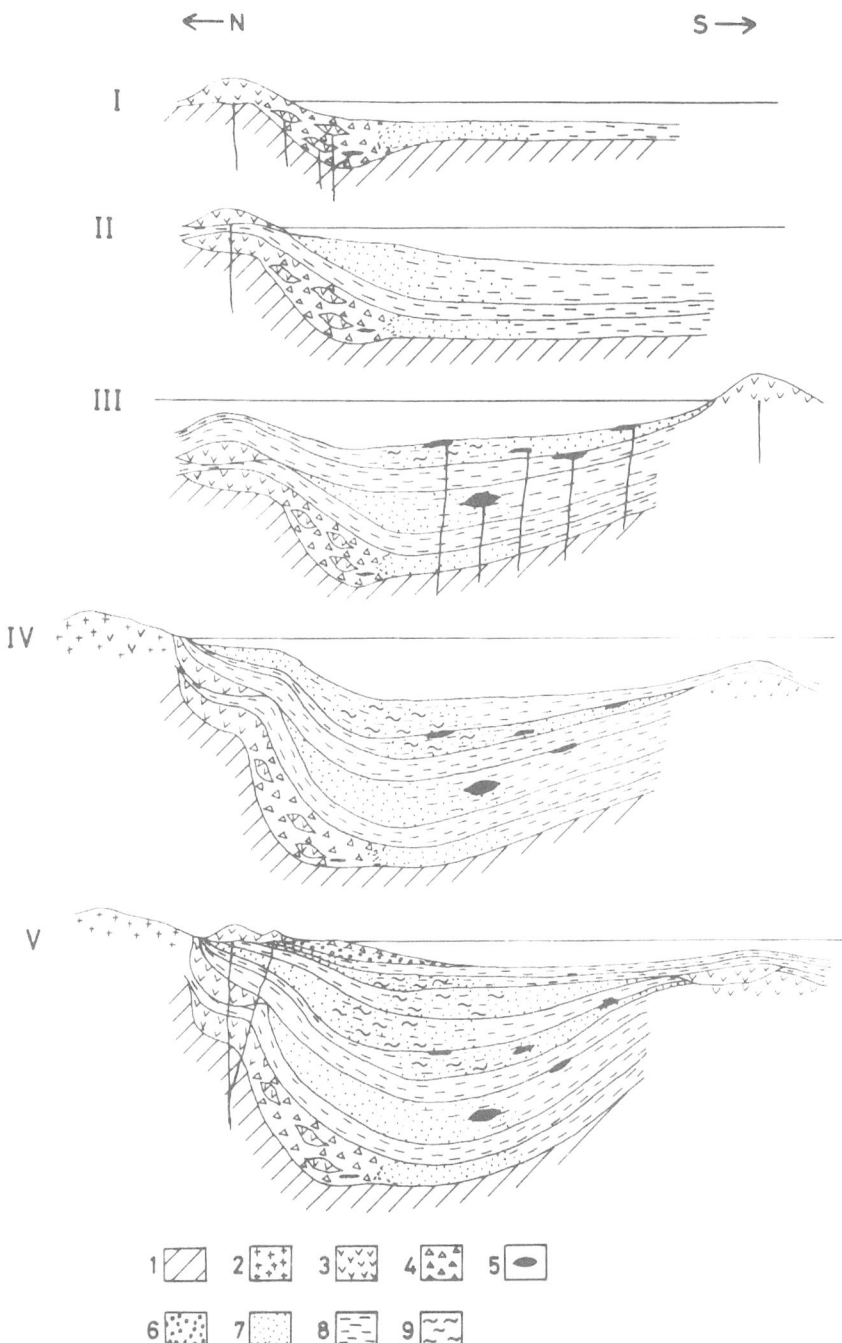

Fig. 6. Schematic cross sections showing the depositional stages of the Nemuro Group. I: Campanian, II · III · IV: Maastrichtian, V: Danian. 1: Basement of the Nemuro Group, 2: Granitic rock, 3: Volcanic rock, 4: Tuff breccia, 5: Dolerite, 6: Conglomerate, 7: Massive sandstone and proximal turbidite, 8: Distal turbidite and shale, 9: Slumping.

3. Triassic to Cretaceous Arc-Trench Systems

Summarizing the above description, the tectonic setting of each Belt is discussed in terms of arc-trench system below.

3.1 Jurassic to early Cretaceous arc-trench system in the Oshima-Rebun and Sorachi-Yezo Belts

The Oshima-Rebun Belt consists of late Jurassic to early Cretaceous thick andesitic volcanic sequence and early Cretaceous granites. Although, the lower limit of the volcanic sequence is not known, these volcanics and plutonics are considered to represent the products of active island arc (OKADA, 1979). The Sorachi-Yezo Belt including the Kamuikotan Metamorphic Belt is composed of Jurassic to early Cretaceous sediments, ophiolite and high-P/T-type metamorphic rocks. The sediments must have been deposited in ocean or trench environments. Isotopic ages indicate the metamorphic rocks of the Kamuikotan Belt and Susunai Range to be early Jurassic to early Cretaceous. These metamorphic rocks were presumably formed in connection with subduction in the Paleo-Kamuikotan trench. The tectonic settings of the Oshima-Rebun and Sorachi-Yezo Belts require for a westward subducting plate at least throughout the early Jurassic to early Cretaceous time. The episodes of subduction and volcanism in this arc-trench system ended by obduction of oceanic plate in middle Cretaceous time which caused the diastrophisms in the Kamuikotan Metamorphic Belt shown by the unconformities in the Yezo Group. Triassic history of this arc-trench system is not known.

3.2 Triassic to Cretaceous arc-trench system in the Hidaka-Tokoro Belt

The Hidaka Western Greenstone Belt may represent a part of the basaltic layer of oceanic crust, although the tectonic setting is not clear and the greenstones in this Belt are not typical abyssal tholeiites.

Most sediments and greenstones of "the greenstone-chert-turbidite facies" in the Hidaka Supergroup are oceanic in nature. The composition of the framework grains of the turbidite in this facies indicates that the provenance was a magmatic arc. These turbidites must have been deposited in a trench. "The greenstone-chert-turbidite facies" is considered to be deposited on ocean floor or in the trench environments. The structural features indicate the possibility that the facies represents the accretionary prism ascribed to the east dipping subduction. Although, the geologic age of the facies has not yet been confirmed by fossil evidence except for conodonts, it may range up to the early or middle Cretaceous.

The stratigraphic relationship between "the greenstone-chert-turbidite facies" and "the turbidite facies" of the Hidaka-Tokoro Belt is still not fully clear. The geologic ages of "the turbidite facies" confirmed by fossil evidences are late Triassic to middle Cretaceous. "The greenstone-chert-turbidite facies" and "the turbidite facies" are regarded as roughly of the same geologic age. The composition of the framework grains of the turbidite indicates that the provenance must be an active magmatic arc which is inferred to have existed in the east from the paleocurrent analysis. "The turbidite facies" must have been deposited in a fore-arc basin. This sedimentary basin has migrated to the east as it goes to the later stage of the deposition of the facies. The migration presumably resulted from the growth of outer high (trench-slope break). These data strongly suggest that there was another convergent plate margin dipping to the east during the late Triassic to middle Cretaceous.

The tectonic setting of the Nikkoro Group exposed in the eastern margin of the Hidaka-Tokoro Belt is not still clear. General features of the Group appear to be those of ophiolite. However, the Group includes too much clastic sediment to be regarded as ophiolite and the thick pillow lava-hyaloclastite-chert sequence of the Group is underlain by sandstone with conglomerate. The Group may represent a primitive magmatic arc along the western margin of the Paleo-Okhotsk Land. Main magmatic arc of this arc-trench system seems to have existed farther east.

3.3 Cretaceous arc-trench system in the Nemuro Belt

The sediments of the late Cretaceous to Paleocene Nemuro Group distributed from Kushiro to Shikotan Island are not oceanic in nature. The clastic sediments are derived from basic to intermediate volcanic rocks in the lower and middle parts, and from intermediate volcanic rocks, granitic rocks, and older sedimentary rocks in the upper part. On the basis of the compositional variation of the sandstones, paleocurrent analysis, features of the volcanism in the basin, and reconstruction of the paleogeography, it is assumed that the sediments of the Nemuro Group exposed along the Pacific coast were deposited in a fore-arc basin or intra-arc basin along the southern margin of the Paleo-Okhotsk Land, although OKADA (1974) pointed out that the Group was deposited in a trench. The volcanic activity in the source of supply was comparatively violent in the east part and became gradually felsic as it goes to the later stage of the deposition of the Nemuro Group. This Group must have been emplaced in connection with the Paleo-Kuril arc-trench system.

4. Tectonic Development

As already stated, the two arc-trench systems trending N-S in Hokkaido are inferred to have been present during the major part of the Triassic to Cretaceous. One is represented by the Oshima-Rebun and Sorachi-Yezo Belts, and the other by the Hidaka-Tokoro Belt. These systems possibly extended northward to Sakhalin. The northern extension of the Kamuikotan Metamorphic Belt and Sorachi Group is traceable from the Susunai Range through East Sakhalin Mountains to Nabil Range. The eastern part of the Hidaka-Tokoro Belt must extend from the Tonino-Aniva Range through East Sakhalin block zone (PUSHCHAROVSKIY, 1965) to Schmidt Peninsula, although oceanic sediments and flysch range in age to late Cretaceous in these area (ROZHDESTVENSKIY, 1973; MELANKHOLINA, 1976; RAZNITSIN, 1975; ROZHDESTVENSKIY and RECHKIN, 1975). These two arc-trench systems must have been separated with considerable distance at first, and "the Paleo-Sorachi-Hidaka ocean" have existed between the two. The westward movement of the Paleo-Okhotsk Land which presumably constituted a part of the American plate possibly gave rise to the eastward subduction. The movement may have been related to the opening of the North Atlantic Ocean or the Arctic Ocean. In particular, the opening of the Amerasian

Basin in the Arctic Ocean, whose initiation is inferred to be in Triassic time (JACOBS *et al.*, 1974) or Jurassic time (HERRON *et al.*, 1974) may have a close relation with the movement. The westward subduction in the Kamuikotan trench requires another force, possibly spreading of the Paleo-Sorachi-Hidaka ocean. In the early Cretaceous, the Paleo-Okhotsk Land approached closely to the Sorachi-Yezo arc-trench system. The subduction in the Sorachi-Yezo arc-trench system turned into obduction at that time. On the other hand, the subduction in the Hidaka-Tokoro arc-trench system continued on into the middle Cretaceous. In the middle to late Cretaceous, the Paleo-Okhotsk Land collided with the Sorachi-Yezo arc system. Figures 7 and 8 illustrate these tectonic developments.

The Nemuro Group distributed from Kushiro to Shikotan Island must have been formed in connection with another arc-trench system, that is the Paleo-Kuril arc-trench system. In the outer and inner Kuril arcs, the presence of metamorphic basement is indicated by xenoliths of greenschist, gneiss and granite found in the Cretaceous and Tertiary effusives (BELYAYEVSKI and RODNIKOV, 1972; RESANOV, 1978). This basement is considered to be a southern part of the Paleo-Okhotsk Land. The history of the Paleo-Kuril trench is not fully understood in Hokkaido. Its evidences, however, are found in the Kamchatka-Koryak region. The Ganal and

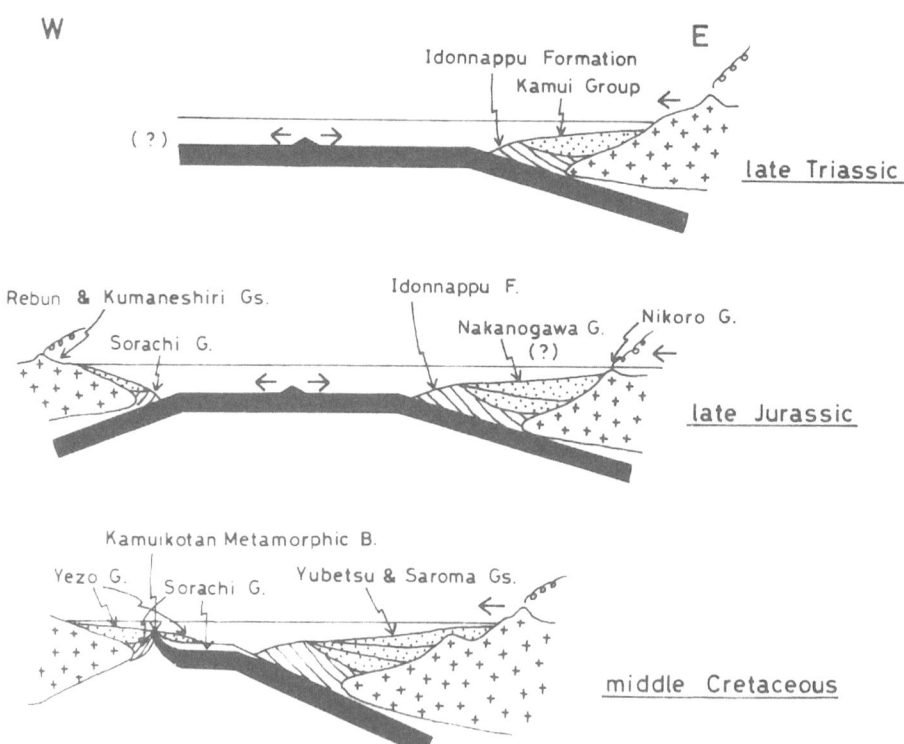

Fig. 7.   The major stages in the Mesozoic geologic development of Hokkaido.

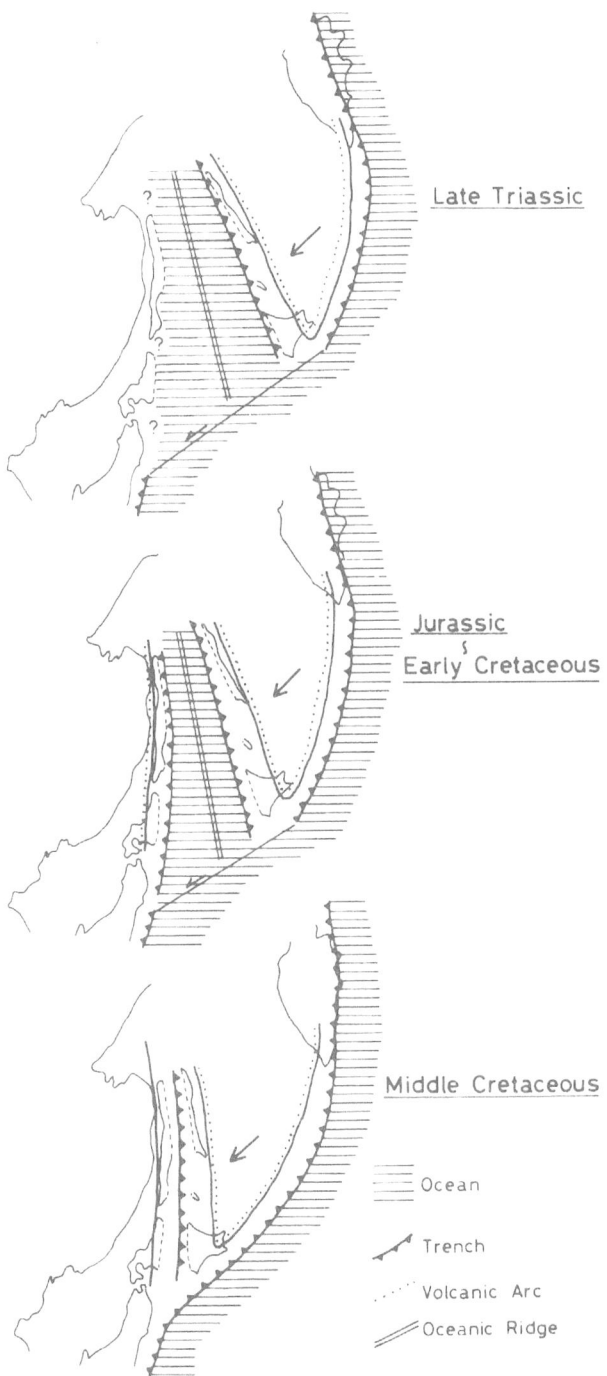

Fig. 8.   Schematic plate reconstruction in the Mesozoic.

Karaginsk-Kronotsk Belts which are composed of high-P/T-type metamorphic rocks and oceanic sediments (RIVOSH, 1964; DOBRETSOV and KURODA, 1970; MISHKIN, 1975; DOBRETSOV, 1975; MARKOV et al., 1973) must represent the Jurassic and late Cretaceous-Paleogene subduction zones. The southwestern extension of these Belts is not be found in Hokkaido. They may be present in the continental slope off the coast of Hokkaido or have been lost by tectonic erosion in the Kuril-trench. The eastern extension of the Nemuro Group is found in the East Kamchatka Range (Valaginsk series, ROTMAN et al., 1973; Drozdov suite, SHAPIRO and SELIVERSTOV, 1975) and Koryak Range (flysch of Ukelayat trough, YERMAKOV and SUPRUNENKO, 1976). The history of the Paleo-Kuril arc-trench system is schematically pictured in Fig. 8.

Our sincere thanks are due to Prof. M. Matsui and Prof. Emeritus S. Hashimoto, Hokkaido University, for their continual encouragement and guidance. Thanks are also due to Prof. S. Igi (Yamaguchi University), Prof. H. Okada (Shizuoka University) and Prof. M. Komatsu (Niigata University) for critical reading of the manuscript and to members of Hokkaido Geotectonics Research Group for their valuable discussions and encouragement. We are indebted to Dr. A. Yao (Osaka City University) and Dr. F. Kumon (Kyoto University) for paleontological work on radioralia.

REFERENCES

ASAHINA, T. and M. KOMATSU, The Horokanai ophiolitic complex in the Kamuikotan tectonic belt, Hokkaido, Japan, J. Geol. Soc. Japan, 85, 317–330, 1979.

BELYAYEVSKI, N. A. and A. G. RODNIKOV, Crustal structure of the island arcs and Far Eastern Seas, Int. Geol. Rev., 14, 171–184, 1972.

BIKERMAN, M., M. MINATO, and M. HUNAHASHI, K-Ar age of the garnet amphibolite of the Mitsuishi district, Hidaka province, Hokkaido, Japan, Earth Sci., 25, 27–30, 1971.

DICKINSON, W. R., Plate tectonic evolution of North Pacific rim, J. Phys. Earth, 26, 1–19, 1978.

DOBRETSOV, N. L., Metamorphic belts of the northwestern Circum-Pacific Region, Geol. Soc. Am. Special Paper, 151, 133–144, 1975.

DOBRETSOV, N. L. and I. KURODA, Geologic laws characterizing glaucophane metamorphism in northwestern part of the folded frame of Pacific Ocean, Int. Geol. Rev., 12, 1389–1407, 1970.

FROLOV, V. T., BURIKOVA, I. A. and GUSHCHIN, A. V., Zone of high magmatic permeability of the southern part of the Lesser Kuril Ridge, Int. Geol. Rev., 22, 1303–1308, 1980.

HASHIMOTO, W., Consideration on the geological history of the Yezo-Saghalien geosynclinal area, in Jubilee Publication in the commemoration of Professor H. Hujimoto, Sixtieth Birthday, pp. 101–112, 1958(in Japanese with English abstract).

HASHIMOTO, W., H. IGO, K. ASAKURA, T. TATENO, and K. NAGASE, On the Kumaneshiri Group, Kabato Mountains, Hokkaido, J. Geol. Soc. Japan, 66, 361, 1960(in Japanese).

HERRON, E. M., J. F. DEWEY, and W. C. PITMAN, Plate tectonics model for the evolution of the Arctic, Geology, 3, 377–380, 1974.

HORIKOSHI, E., Orogenic belts and plates in Japanese Islands, KAGAKU, 42, 665–673, 1972(in Japanese).

IGO, Hy., T. KOIKE, Hh. IGO and T. KINOSHITA, On the occurrence of Triassic conodonts from the Sorachi Group in the Hidaka Mountains, Hokkaido, J. Geol. Soc. Japan, 80, 135–136, 1974 (in Japanese).

IGO, Hy., Hh. IGO, S. ADACHI, and Y. SATO, On the geologic age of the Hidaka and Sorachi Groups, in Studies of geosynclines and tectonic division of the northern part of the Japanese Islands, pp. 69–75, 1980 (in Japanese).

IMAIZUMI, M. and Y. UEDA, On the K-Ar ages of the rocks of two kinds existed in the Kamuikotan metamorphic rocks located in the Horokanai district, Hokkaido, J. Japan. Assoc. Mineral. Petrol. Econ. Geol., 76, 88–92, 1981.

INOMA, A., The Intra-Yezo disturbance and sedimentation of the middle Yezo Group, J. Japan. Assoc.

Petrol. Technol., **34**, 11–25, 1969(in Japanese with English abstract).

ISHIZAKI, S., Discovery of the Triassic Bryozoan fossils from the Esashi Mountain, Hokkaido and its significance, *Earth Sci.*, **33**, 355–359, 1979(in Japanese).

ISHIZUKA, H., Geology of the Horokanai ophiolite in the Kamuikotan tectonic belt, Hokkaido, Japan, *J. Geol. Soc. Japan*, **86**, 119–134, 1980(in Japanese with English abstract).

ISHIZUKA, H., Greenstones from the Idonnappu Formation along the River Oku-Niikappu in the axial zone of Hokkaido, Japan, *Mem. Fac. Sci., Kochi, Univ., Ser. E. Geol.*, **2**, 1–22, 1981.

JACOBS, J. A., R. D. RUSSEL, and J. T. WILSON, *Physics and Geology*, 622 pp., MacGraw-Hill, New York, 1974.

KANIE, Y., Y. TAKETANI, A. SAKAI, and Y. MIYATA, Lower Cretaceous deposits beneath the Yezo Group in the Urakawa area, Hokkaido, *J. Geol. Soc. Japan*, **87**, 527–533, 1981(in Japanese with English abstract).

KAWANO, Y. and Y. UEDA, K-Ar dating on the igneous rocks in Japan (VI)—Granitic rocks, summary, *J. Jpn. Assoc. Mineral. Petrol. Econ. Geol.*, **57**, 177–187, 1967.

KIMINAMI, K., Stratigraphic re-examination of the Nemuro Group, *Earth Sci.*, **32**, 120–132, 1978(in Japanese with English abstract).

KIMINAMI, K., Sedimentary petrography of the Nemuro Group, *Earth Sci.*, **33**, 152–162, 1979(in Japanese with English abstract).

KIMINAMI, K. and Y. KONTANI, Pre-Cretaceous paleocurrents of the northeastern Hidaka Belt, Hokkaido, Japan, *J. Fac. Sci., Hokkaido Univ., Ser. IV*, **19**, 179–188, 1979.

KIMINAMI, K., K TAKAHASHI, and K. MANIWA, The Cretaceous system in Hokkaido: Yezo and Nemuro Groups, *Monogr. Geol. Collab.*, **21**, 111–126, 1978 (in Japanese with English abstract).

KOCHERGIN, Ye. V. and M. L. KRASNYY, Geologic nature of offshore magnetic anomalies near Shikotan Island, *Doklady Akademii Nauk SSSR*, **232**, 181–183, 1977.

KOMATSU, M., S. MIYASHITA, J. MAEDA, Y. OSANAI, and T. TOYOSHIMA, Disclosing of a deepest section of continental type crust up-thrust as the final event of collision of arcs in Hokkaido, noth Japan, this volume, pp. 149–165, 1982.

KONTANI, Y., Geological study of the Hidaka Supergroup distributed on the east side of the Hidaka metamorphic belt (Part I), *J. Geol. Soc. Japan*, **84**, 1–14, 1978 (in Japanese with English abstract).

KONTANI, Y. and K. KIMINAMI, Petrological study of the sandstones in the Pre-Cretaceous Yubetsu Group, northeastern Hidaka Belt, Hokkaido, Japan, *Earth Sci.*, **34**, 307–319, 1980.

MARKOV, M. S., G. Ye. NEKRASOV, and M. Yu. KHOTIN, Basement of the Cretaceous geosyncline on the Kamchatka Capa Peninsula (Eastern Kamchatka), *Geotectonics*, **7**, 246–249, 1973.

MATSUMOTO, T., ed., *The Cretaceous System in the Japanese Islands*, pp. 1–324, Japan Soc. Prom. Sci. Res., Tokyo, 1954.

MATSUMOTO, T. and H. OKADA, Clastic sediments of the Cretaceous Yezo Geosyncline, *Mem. Geol. Soc. Japan*, **6**, 61–74, 1971.

MELANKHOLINA, Ye. N., Formation complexes in structures of Sakhalin and Hokkaido, *Geotectonics*, **10**, 186–193, 1976.

MINATO, M., M. GORAI, and M. HUNAHASHI, eds., *The Tectonic Development of Japanese Islands*, 422 pp., Tsukiji, Tokyo. 1965.

MISHKIN, M. A., Metamorphic associations in transition zone between the Asiatic continent and the Pacific Ocean, *Int. Geol. Rev.*, **17**, 1402–1414, 1975.

NAGAO, S., C. AKIBA, and T. OMORI, *Explanatory Text of the Geological Map (1:50,000) "Rebunto,"* 43 pp. Geol. Surv. Hokkaido, 1963(in Japanese with English abstract).

NAKASEKO, K., A. NISHIMURA, and K. KANNO, Study of radiolarian fossils from the Shimanto Belt with special reference of the Cretaceous radiolaria, *Newslett. Osaka Micropaleont., Spec. Issue*, **2**, 1–49, 1979 (in Japanese).

NOZAWA, T., ed., *Radiometric Age of Japan—Granitic Rocks, 1:2,000,000 Map Series, No. 16–1*, Geol. Surv. Japan, 1975.

OKADA, H., Migration of ancient arc-trench systems, in *Modern and Ancient Geosynclinal Sedimentation*, pp. 311–320, edited by R. H. Dott Jr. and R. H. Shaver, Soc. Econ. Paleont. Mineral., Spec. Publ. No. 19, 1974.

OKADA, H., The geology of Hokkaido and its plate tectonics, *Earth Monthly*, **1**, 869–877, 1979(in Japanese).

Parfenov, L. M., I. P. Voinova, B. A. Matal'in, and D. F. Semenov, Geodynamics of the north-eastern Asia in Mesozoic and Cenozoic time and the nature of volcanics belts, *J. Phys. Earth*, **26**, 503–525, 1978.

Piskunov, B. N. and I. I. Khvedchuk, New data on composition and age of rocks of Moneron Island in the northern part of the Sea of Japan, *Doklady Akademii Nauk SSSR*, **226**, 647–650, 1976.

Pushcharovskiy, Yu. M., The tectonics of Sakhalin, *Int. Geol. Rev.*, **7**, 2184–2196, 1965.

Raznitsin, Yu. N., Comparative tectonics of ultrabasic belts in the Schmidt Peninsula (Sakhalin), Papua (New Guinea) and Sabah (Kalimantan), *Geotectonics*, **9**, 108–115, 1975.

Resanov, I. A., Origin of the deep-water basin of the Sea of Okhotsk and Japan, *Int. Geol. Rev.*, **20**, 1072–1080, 1978.

Rivosh, L. A., Some geophysical data bearing on deep structure of the Central Kamchatka Trough, *Int. Geol. Rev.*, **6**, 2009–2014, 1964.

Rotman, V. K., B. A. Markovskiy, and M. I. Khotina, The Kamchatka ultrabasic volcanic province, *Int. Geol. Rev.*, **15**, 1015–1024, 1973.

Rozhdestvenskiy, V. S., Strike-slip faults in the Vostochnyy range of Schmidt Peninsula on Sakhalin, *Int. Geol. Rev.*, **15**, 1391–1393, 1973.

Rozhdestvenskiy, V. S. and A. N. Rechkin, Serpentinite mélange and some aspects of tectonic evolution of Sakhalin Island, *Doklady Akademii Nauk SSSR*, **221**, 1156–1159, 1975.

Sakagami, S. and A. Sakai, Triassic Bryozoans from the Hidaka Group in Hokkaido, *Trans. Proc. Paleont. Soc. Japan*, **114**, 77–86, 1979.

Shapiro, M. N. and V. A. Seliverstov, Morphology and age of folded structures in eastern Kamchatka at the latitude of the Kronotskiy Peninsula, *Geotectonics*, **9**, 204–244, 1975.

Shibata, K. and N. Yamada, K-Ar age of a granodiorite from Okushiri Island, Hokkaido, *Bull. Geol. Surv. Japan*, **29**, 611–613, 1978(in Japanese).

Takahashi, K. and M. Suzuki, *Explanatory Text of the Geological Map of Japan (1:50,000) "Iwachishi,"* 46 pp., Geol. Surv. Hokkaido, 1978(in Japanese with English abstract).

Tanaka, K. and Y. Sumi, Cretaceous paleocurrents in the central zone of Hokkaido, Japan, *Bull. Geol. Suv. Japan* **32**, 65–127, 1981(in Japanese with English abstract).

Teraoka, T., K. Kuroda, and K. Hirayama, On the unknown Mesozoic formations, south of the Lake Saroma, Hokkaido, Japan, *J. Geol. Soc. Japan*, **68**, 416, 1962(in Japanese).

Trofimuk, A. A., The sedimentary sequence in the Kuril Island, *Int. Geol. Rev.*, **18**, 1149–1154, 1976.

Watanabe, T., S. Sano, H. Yano, and S. Koitabashi, Radioralian age and stratigraphy of weakly metamorphosed rocks of the southern part of the Kamuikotan Metamorphic Belt, in *Abstract of '81 Annual Meetings of "Sanko-Gakkai,"* 118, 1981(in Japanese).

Yamada, K., Y. Teraoka, and M. Ishida, *Explanatory Text of the Geological Map of Japan (1:50,000) "Ikutawara,"* 35 pp., Geol. Surv. Japan, 1963(in Japanese with English abstract).

Yermakov, B. V. and O. I. Suprunenko, Structure and conditions of development of Upper Cretaceous and Miocene flysch of Koryak-Kamchatka, *Int. Geol. Rev.*, **18**, 1139–1148, 1976.

Accretion Tectonics in the Circum-Pacific Regions, edited by M. Hashimoto and S. Uyeda, 123–134.
Copyright © 1983 by Terra Scientific Publishing Company (TERRAPUB), Tokyo.

Collision Tectonics in Hokkaido and Sakhalin

G. Kimura,*† S. Miyashita,* and S. Miyasaka**

*Department of Geology and Mineralogy, Faculty of Science,
Hokkaido University, Sapporo 060, Japan
**Meiji Consultant Co., Sapporo Branch, Sapporo 060, Japan

Hokkaido and Sakhalin in the northwestern Pacific border are situated along the
boundary between the Eurasian and North American (or Okhotsk) Plates. Hokkaido
also is located at the junction of the Tohoku and Kurile Arcs. In Latest Cretaceous or
Paleogene, subduction around Hokkaido and Sakhalin ceased and collision tectonics
started. The structural configuration of Hokkaido and Sakhalin is complicated, and it
contains "two" collision systems: one is the oblique collision of the Eurasian and
North American Plates, and the other is the collision of the Kurile Arc and the Tohoku
Arc. The southern part of central Hokkaido was formed by superposition of these two
collisions. The former collision is related to the opening of the Atlantic Ocean. The
latter one is caused by the westward shift of the frontal Kurile Arc due to the oblique
subduction of the Pacific Plate along the Kurile-Kamchatka Trench.

1. Introduction

Generally, Hokkaido is divided into three geologic provinces; southwestern,
central, and eastern Hokkaido. Southwestern Hokkaido is a northern extension of the
Tohoku Arc. Central Hokkaido, which continues to Sakhalin has a very complicated
geologic association; this geologic province was a subduction and accretion zone
during Mesozoic period. Eastern Hokkaido belongs to the Kurile Arc. The respective
boundaries among these provinces are the Ishikari Low Land and the Abashiri
Tectonic Line (Fig. 1).

Several plate tectonic models of Hokkaido have been proposed until now
(Horikoshi, 1972; Den and Hotta, 1973; Miyashiro, 1977; Dickinson, 1978;
Okada, 1979, 1980, etc.). According to them, Hokkaido was formed by collision of two
different continents, though the vergence of subduction before the collision was not the
same. Somebody insist the eastward subduction (Horikoshi, 1972; Den and Hotta,
1973; Miyashiro, 1977) and the other the westward subduction (Dickinson, 1978;
Okada, 1979, 1980). Recently, Kiminami and Kontani (1982) suggested that both the
eastward and westward subduction systems existed during Mesozoic time.

Moreover, neotectonic models of Hokkaido and its environs have been proposed
(Kaizuka, 1972, 1975, 1980; Fujii and Sogabe, 1978; Kimura, 1981a, b). All the

---

†Present address: Department of Earth Sciences, Faculty of Education, Kagawa University,
Takamatsu 760, Japan.

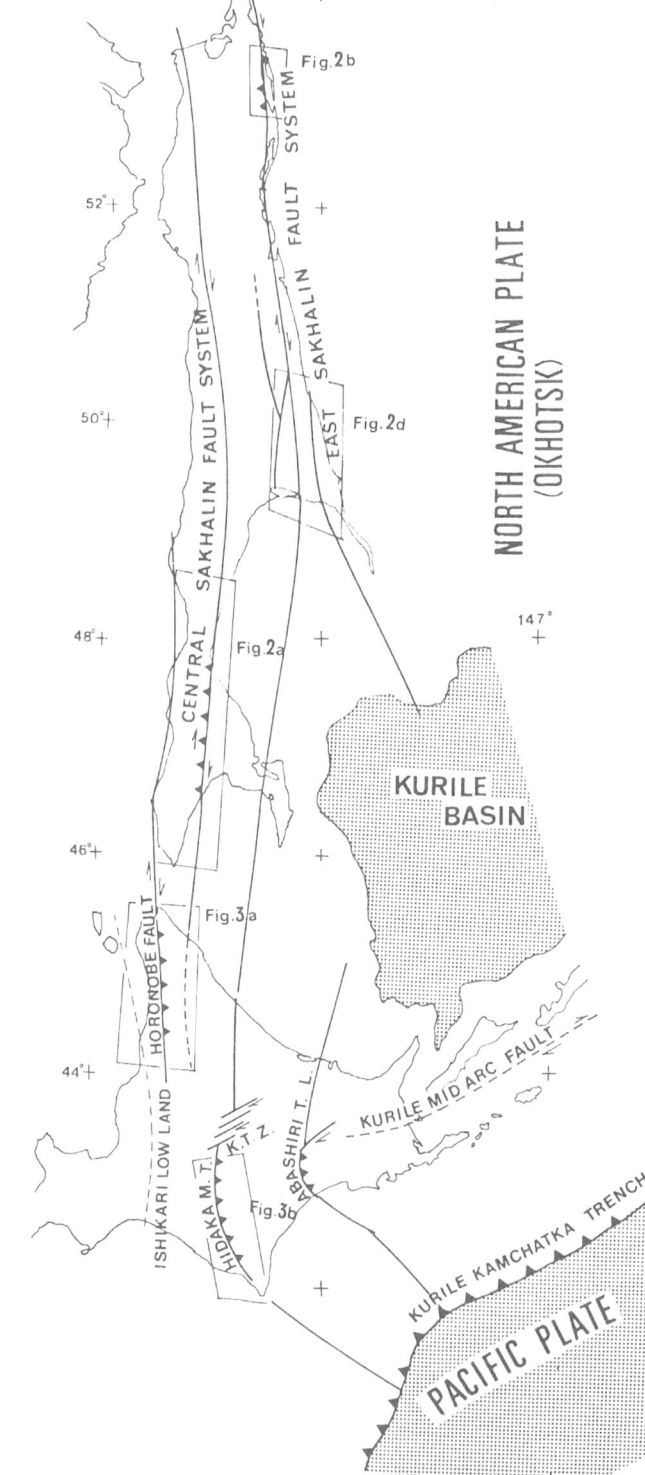

Fig. 1.   Fault systems in Hokkaido and Sakhalin. K. T. Z. : Kamishiyubetsu Tectonic Zone, Hidaka M. T. :
Hidaka Main Thrust, Abashiri T. L.: Abashiri Tectonic Line.

models show that the Kurile Arc collided with the Tohoku Arc in Hokkaido. KIMURA (1981a, b) showed that the collision began in Latest Miocene. The Mesozoic subduction systems probably ceased from developing during Late Cretaceous or Paleogene time (DICKINSON, 1978; OKADA, 1979, 1980; KIMINAMI and KONTANI, this volume) when two different continents collided with each other. In this paper, we will analyze the structure formed since Paleogene and discuss its tectonic evolution.

2. Central Hokkaido and Sakhalin as a Right-Lateral Strike-Slip Mobile Zone

Some N-S trending fault systems can be found in Hokkaido and Sakhalin. These faults were active from Miocene to Pliocene, and some of them are still active at present.

2.1 Sakhalin

We can see two deep seated N-S trending fault systems in Sakhalin; the Central Sakhalin Fault System in the central part, and the Eastern Sakhalin Fault System along the eastern coast of the island (Fig. 1).

The Central Sakhalin Fault System (ZANYUKOV, 1971), which is called the Tym'Poronay Fault in southern Sakhalin, is the boundary between the uplifting western block composed mainly of Cretaceous to Tertiary strata and subsiding eastern low lands. This fault system extends about 600 km. Many investigations have been made on this fault system such as TOKUDA (1926), MELINIKOV (1970), ZANYUKOV (1971), ROZHDESTVENSKIY (1975; 1976), MELANKHOLINA (1976), MELANKHOLINA and MOLCHANOVA (1977), and KHARAKHINOV et al. (1979). This fault was recognized earlier as a west dipping reverse fault, but TOKUDA (1926), and ROZHDESTVENSKIY (1975, 1976) suggested it to be a right-handed strike-slip one.

ROZHDESTVENSKIY (1975, 1976) discussed: Tertiary and Cretaceous strata in the western block of the fault are folded en echelon (Fig. 2a), and the axial traces of fold trend in NW-SE direction. Moreover, a number of volcanic dikes intruding in NE direction during Middle Miocene are found along the fault zone, and in it recent mudvolcanoes also are distributed similarly in NE direction. The NNW trending segment of the fault is a reverse fault while the NNE trending segment dips vertical. These facts indicate that the Central Sakhalin Fault is a right-handed strike-slip one.

MELINIKOV (1970), ZANYUKOV (1971), and MELANKHOLINA and MOLCHANOVA (1977) suggested that the uplift movement of the western block was initiated during Late Paleogene or Early Miocene and the most intensive movement occurred from Late Miocene to Pliocene.

Many N-S trending faults are developed along the eastern coast of Sakhalin. They are of the Eastern Sakhalin Fault System (Fig. 1), which comprises the Tuin, Kheyton, and Longri Faults in the Schmidt Peninsula (Fig. 2c), the East Ekhaba and Pil'Tun Faults in the northern coast of Sakhalin (Fig. 2b), and the Central, Coastal, and Limnask Faults in the Eastern Sakhalin Mountains (Fig. 2d), as described by ROZHDESTVENSKIY (1972, 1975) and ROZHDESTVENSKIY and RECHKIN (1975) in detail. The Longri, East Ekhaba, Pil'Tun, and Coastal Faults are connected to each other (ROZHDESTVENSKIY, 1975).

There are three N-S trending Faults: the Tuin, Kheyton, and Longri Faults

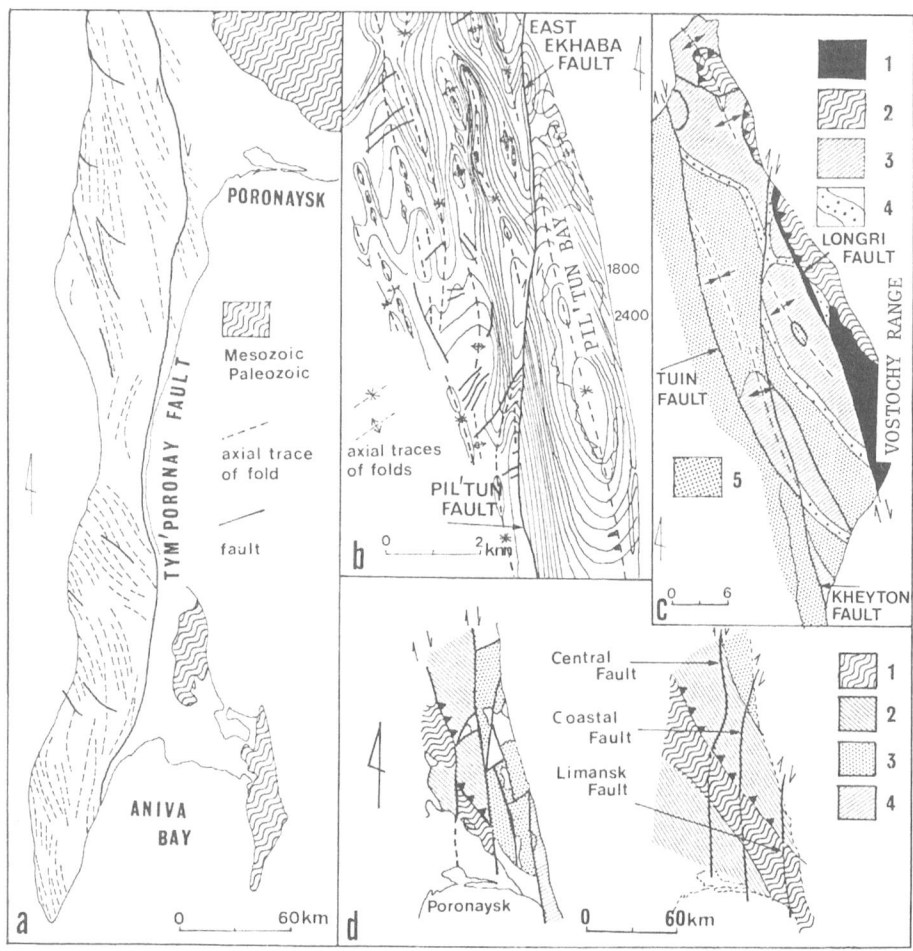

Fig. 2. Structural map of Sakhalin. a: Structural map of the southern Sakhalin (modified after MELINIKOV, 1970), b: Strike line map near the Pil'Tun Bay in the northeast coast of Sakhalin (ROZHDESTVENSKIY, 1975), c: Geologic outline of the Schmidt Peninsula in the northern Sakhalin (ROZHDESTVENSKIY, 1972), 1: serpenitinized peridotite, 2: tectonic breccia and Jurassic to Lower Cretaceous sediments, 3: Upper Cretaceous sediments, 4: Neogene basal sediments, 5: Neogene sediments, d: Geologic outline of the East Sakhalin Mountains (left) and reconstructed structure (right), 1: Paleozoic strata, 2: Mesozoic strata, 3: Mesozoic silicious sediments and volcanics, 4: Upper Cretaceous sediments (modified after ROZHDESTVENSKIY, 1975).

situated on the western side of the Vostochy Range which is composed mainly of serpentinized peridotite, gabbro and Upper Jurassic to Lower Cretaceous greenstones. These rocks thrust up onto the Upper Cretaceous sediments of the western area, but they make contact with the Upper Cretaceous at the vertical Longri Fault in the southern area (Fig. 2b). Moreover, numerous gently inclined striations are observed on the fault plane. ROZHDESTVENSKIY (1972) showed that a displacement along the fault is about 7 km, which he inferred from offset of the Neogene strata. The Kheyton and Tuin Faults cut the Miocene and Pliocene strata. The Upper Cretaceous and Miocene sediments are fairly folded, and the axes of folds are traced from northwest to southeast and are oblique to the Tuin and Kheyton Faults. These folds were formed by the drag associated with the strike-slip displacement of about 20 km in total along the faults. ROZHDESTVENSKIY (1972) suggested that the most intense movement occurred from Late Miocene to Pliocene time.

The Longri Fault in the Schmidt Peninsula extends to the East Ekhaba and Pil'Tun Faults near the Pil'Tun Bay in the northeast coast of Sakhalin. The East Ekhaba and Pil'Tun Faults are respectively the northern and southern segments of the same fault. The East Ekhaba Fault dips eastward whereas the Pil'Tun Fault westward. Neogene strata around the faults are folded en echelon (Fig. 2c). The arrangement pattern of folds, which is similar to that in the Schmidt Peninsula, indicates that the faults are of the right-handed strike-slip type. Moreover, the development of NW trending reverse faults and thrusts and NE trending normal faults near these major faults is consistent with the movement associated with the N-S trending strike-slip faulting.

The East Sakhalin Mountains (Fig. 2d) consist mainly of Mesozoic greenstones, chert, and terrigenous rocks which compose a melange (ROZHDESTVENSKIY and RECHKIN, 1975), and these are displaced by N-S trending faults. ROZHDESTVENSKIY (1975) and ROZHDESTVENSKIY and RECHKIN (1975) reconstructed the tectonic evolution of the mountains as follows. The East Sakhalin Mountains were thrust under western Sakhalin in Late Cretaceous time, and in Latest Cretaceous to Paleogene the underthrusting ceased. Then, the uplift of the mountains occurred and the thrust sheets were strongly deformed and cut by N-S trending right-handed strike-slip faults (Fig. 2d). The Coastal Fault in the East Sakhalin Mountains extends to the East Ekhaba and Pil'Tun Faults in the northern Sakhalin as mentioned above.

2.2 Central Hokkaido

Central Hokkaido is subdivided into four N-S trending geologic belts: the Ishikari-Teshio Belt, the Kamuikotan Belt, the Hidaka Belt, and the Tokoro Belt from west to east (Fig. 3c). Tectonic development in this province since Paleogene can be reconstructed by means of structural analysis of the Ishikari-Teshio Belt and the Hidaka Belt.

In the Ishikari-Teshio Belt, thick Cretaceous, Paleogene, and Neogene sedimentary rocks are accumulated. These strata are folded but the grade of folding in the northern part of the belt is different from that in the southern part. Most of the folds in the northern part are gentle to open, whereas those in the southern part are strongly deformed, tight or isoclinal. Most of the folds and thrusts in this belt show a westward vergence in contrast to those of Sakhalin. The Central Sakhalin Fault extends to the boundary between the Ishikari-Teshio Belt and the Kamuikotan Belt in Hokkaido. The Horonobe Fault in northern Hokkaido is parallel to this boundary and extends to the western coast of Sakhalin (Figs. 1, 3b). The structural arrangement in the northern part of the Ishikari-Teshio Belt is as follows (Fig. 3b). The axial traces of folds, secondary thrusts, and reverse faults trend in NW direction and are arranged en

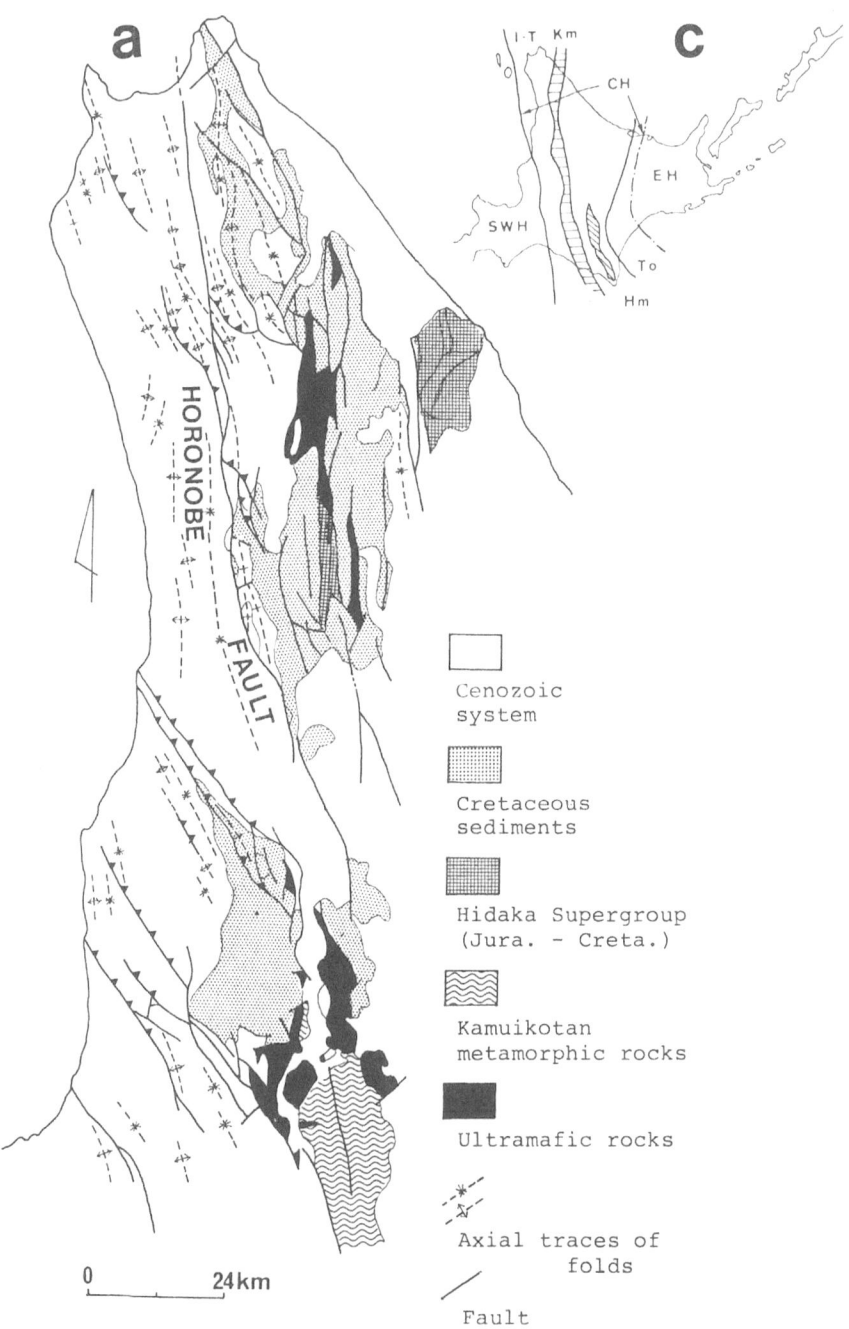

Fig. 3

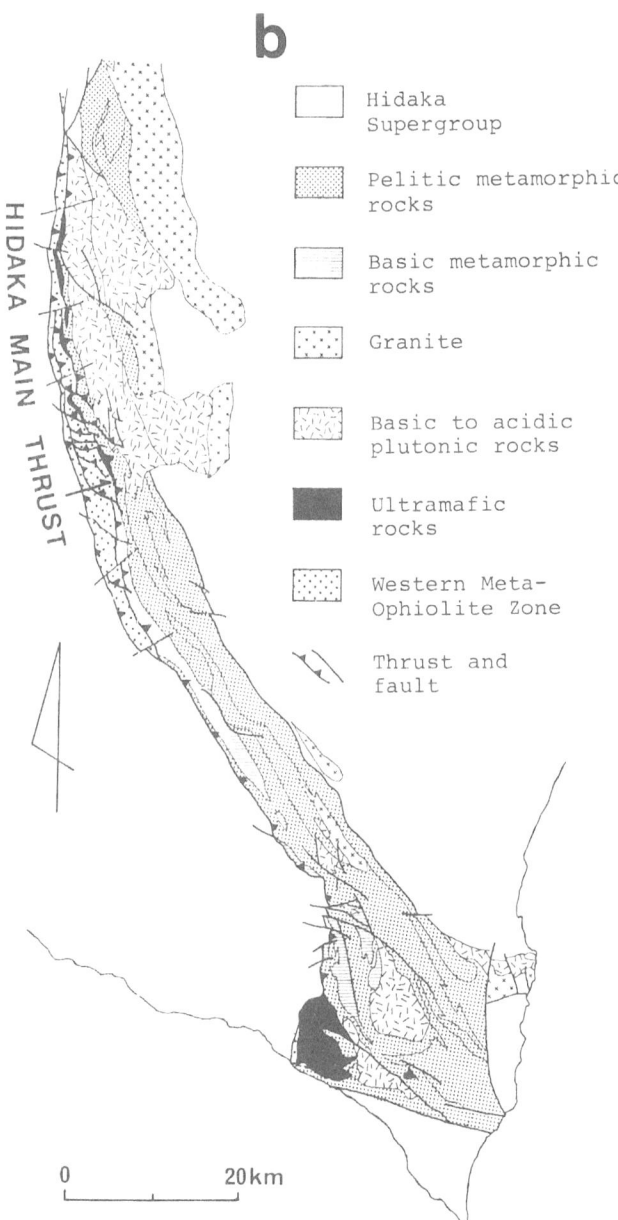

Fig. 3. a: Structural map of the northern Ishikari-Teshio Belt, b: Geologic outline and structure of the Hidaka Metamorphic Belt, c: Geologic division of Hokkaido, SWH: southwestern Hokkaido, CH: central Hokkaido, EH: eastern Hokkaido, I-T: Ishikari-Teshio Belt, Km: Kamuikotan Belt, Hm: Hidaka Metamorphic Belt, Hi: Hidaka Belt, To: Tokoro Belt.

echelon; this fact indicates that the displacement along the Horonobe Fault has a right-handed strike-slip component. The vergence of the folds and thrusts in Sakhalin is different from that in Hokkaido, though the right-handed strike-slip movement picture along them is very similar to each other. These structure were formed in Miocene to Pliocene.

The Hidaka Metamorphic Belt is situated in the southern part of the Hidaka Belt (Figs. 1, 3c). This metamorphic belt is composed of two different geologic units: the Western Zone composed of overturned meta-ophiolite (Miyashita, 1982a, b) and the Main Zone of continental or island arc crust consisting of granulite, gneiss, migmatite, and basic to acidic pultonic rocks (KOMATSU *et al.*, 1982a, b). Both zones are bounded by a thick mylonite zone of the Hidaka Main Thrust. The metamorphism and magmatism in the Main Zone took place from about 40 to 17 Ma: Late Paleogene to Middle Miocene (HASHIMOTO, 1976; SHIBATA and ISHIHARA, 1981). The structure of these zones was analysed by KIZAKI (1972), KIZAKI and HAYASHI (1979), and MIYASHITA (1982a, b). Most of the migmatite domes and granitic rocks in the Main Zone intruded from north to south. The Western Zone is deformed by many NW trending thrusts oblique to the Hidaka Main Thrust (Fig. 3b). The thrusting and metamorphism occurred simultaneously (Miyashita, 1982b). These facts indicate that the Hidaka Main Thrust was formed as a right-handed strike-slip fault during Late Paleogene to Middle Miocene. In the Middle Miocene, when the strike-slip movement and metamorphism occurred, the area of the Hidaka Metamorphic Belt was situated under sea level, and terrigeneous sediments were deposited thereover (MIYASAKA and KIKUCHI, 1978). In Latest Miocene, the Hidaka Metamorphic Belt started to uplift and the Hidaka Mountains appeared (MIYASAKA and KIKUCHI, 1978). This mountain building movement was caused by overthrusting of the Main Zone onto the Western Zone.

3.   Mid-Arc Fault System Formed in Latest Miocene

The southern part of the Ishikari-Teshio Belt and the Hidaka Metamorphic Belt are located in the present frontal arc region close to the southwestern margin of the Kurile Arc. Eastern Hokkaido is bounded on Central Hokkaido by the Abashiri Tectonic Line (Fig. 1). This line is divided into two segments, northern and southern (KIMURA, 1981a, b), the latter being near the volcanic front. The southern segment is a westward convex thrust which was formed in Latest Miocene, and it branches in a NE direction near the volcanic front (Fig. 1). The NE trending branched fault is a right-handed reverse one (KIMURA, 1981a). On the other hand, at the northern end of the Hidaka Metamorphic Belt, there is a wide fault zone of NE trend (Fig. 1), the Kamishiyubetsu Tectonic Zone (KIMURA *et al.*, 1982). This tectonic zone also is located near the boundary between the volcanic and frontal Kurile Arcs. The tectonic zone has been regarded as a normal dip-slip fault zone (HASEGAWA *et al.*, 1961), but recently KIMURA (1981b) and KIMURA *et al.* (1982) have revealed that this is a right-handed strike-slip fault zone which was formed in Latest Miocene. The fault of the Kamish-iyubetsu Tectonic Zone displaces right-laterally the granitic plutons of the Hidaka Belt. KIMURA (1981b) proposed that the NE trending branch of the Abashiri Tectonic

Line and the Kamishiyubetsu Tectonic Zone correspond to the Mid Arc Fault system of KAIZUKA (1972).

4. Collision Tectonics since Paleogene

From the above-mentioned characteristics of the structure, we can reconstruct a tectonic framework of the boundary between the Eurasian and North American (or Okhotsk) Plates, together with the junction of the Kurile and Tohoku Arcs. The vergence of folds, thrusts and reverse faults in Sakhalin is mostly different from that in Hokkaido; however, the right-handed movement picture along the major fault system is very similar. This fact shows that the Sakhalin and the central Hokkaido provinces are right-handed strike-slip mobile zone which was developed since Paleogene or Early Miocene time. The rapid development of Miocene basins in Hokkaido such as the Kotambetsu and Kawabata basins (OKADA, 1980), may have been caused by this right-handed movement. The most active phase of this movement occurred from Middle Miocene to Pliocene. This movement picture of Sakhalin and Hokkaido indicates that the oblique collision tectonics occurred along the boundary between the Eurasian and North American Plates since Late Paleogene or Early Miocene.

Why did the oblique collision occur along the boundary? We can answer this question by observing the position of the relative rotation pole of the Eurasian and North American Plates deduced by PITMAN and TALWANI (1972). They estimated the relative rotation pole to be located at 65°N 133°E from 38 Ma to 9 Ma and at 68°N 137°E since 9 Ma, on the basis of the analysis of fracture zones and magnetic liniation pattern in north Atlantic. The boundary between the Eurasian and North American Plates in northwest Pacific is traced from the Verkhoyansk or Cherskiy Mountains in Siberia to Sakhalin and central Hokkaido (CHURKIN, 1972; CHAPMAN and SOLOMON, 1976), where collision tectonics have been taken place. According to the Tertiary relative rotation pole revealed by PITMAN and TALWANI (1972), the slip vectors of two plates near the plate boundary of Sakhalin and Hokkaido must be oblique to the suture. The right-handed oblique collision movement along the suture deduced from the field evidence, may be ascribed to the above mentioned rotation of the two plates.

Moreover, along the collision suture zone, metamorphism and magmatism took place, for example, those in the Hidaka Belt (MAEDA et al., 1982), Miocene pultonic intrusions in the East Sakhalin Mountains and in the Schmidt Peninsula, and Miocene volcanism along the Central Sakhalin Fault System. Until now, it has been thought that the Tertiary magmatism and metamorphism along this suture zone occurred in the N-S trending island arc which is related to either the westward subduction (OKADA, 1979, 1980) or the eastward subduction (HORIKOSHI, 1972, MIYASHIRO, 1977); however, it is very difficult to find their respective subduction zones in Tertiary. The alternative possibility is that the magmatism and metamorphism along this zone may have been developed by another tectonic framework like as collision. The tectonic significance of the magmatism and metamorphism must be left for future investigation.

There is still another collision tectonic system which was developed since Latest Miocene in Hokkaido, i.e. the collision of the frontal Kurile Arc and Tohoku Arc. The volcanic Kurile Arc is arranged en echelon (TOKUDA, 1926), and this pattern was

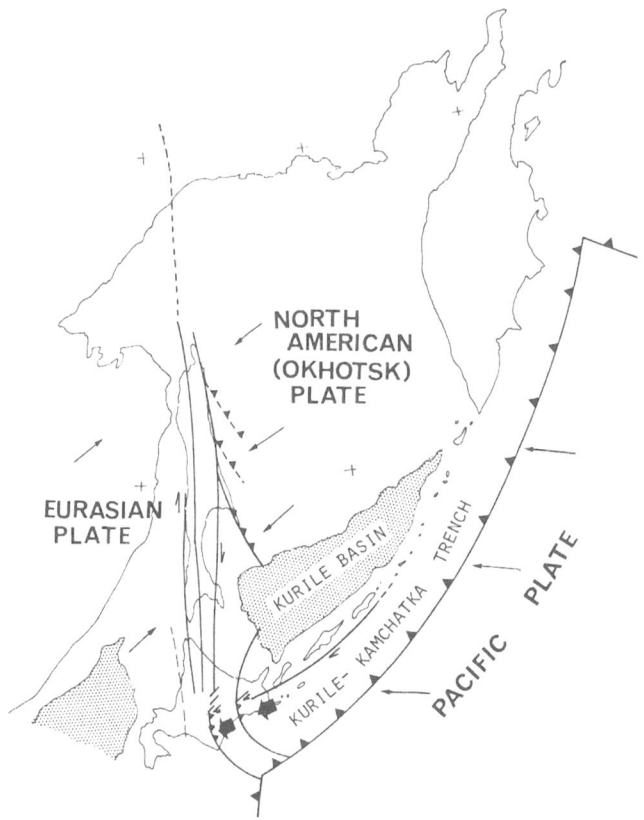

Fig. 4. Tectonic framework near Hokkaido and Sakhalin. The oblique collision of the Eurasian Plate and the North American (Okhotsk) Plate took place along the central Hokkaido and Sakhalin since Paleogene or Early Miocene. At the south western end of the Kurile Arc, the frontal Kurile Arc also collided with the Tohoku Arc after Latest Miocene, caused by the oblique subduction of the Pacific Plate along the Kurile-Kamchatka Trench.

developed by a right-handed strike-slip along the mid-arc fault. This mid-arc fault is recognized as the Kamishiyubetsu Tectonic Zone and the branching fault of the Abashiri Tectonic Line. KIMURA (1981a, b) and KIMURA et al. (1982) pointed out that the right-handed strike-slip movement along the mid-arc fault system commenced in Latest Miocene due to the change in stress field of the volcanic arc. Formation of the echelon ridges of the volcanic arc and the right-handed strike-slip along the mid-arc fault system indicate a westward shift of the frontal Kurile Arc (KAIZUKA, 1980; KIMURA, 1981a, b; KIMURA et al., 1982).

What tectonic event occurred at the southwestern margin of the Kurile Arc? The Hidaka Metamorphic Belt, which crops out only in the southern part of central Hokkaido, consists of a westward vergent pile of two different crusts (KOMATSU et al., 1982a, b). The Hidaka Mountains were created by overthrusting of the eastern crust upon the western crust in Latest Miocene, as discussed in a previous section. The strata in the southern part of the Ishikari-Teshio Belt have been intensely deformed due to a compressional force from the east since Latest Miocene. These intensely deformed structures are limited only around the southwestern margin of the Kurile Arc. Taking into consideration the movement picture along the mid-arc fault system, these phenomena indicate that the frontal Kurile Arc has shifted westward and collided with the Tohoku Arc since Latest Miocene.

The N-S trending right-handed simple shear along the Hidaka Main Thrust occurred in the metamorphic phase. It indicates that the southern part of central Hokkaido was formed by superposition of the two collisions: the collision of the Eurasian and North American Plates, and that of the Kurile and Tohoku Arcs (Fig. 4). The westward shift of the frontal Kurile Arc is ascribed to oblique subduction of the Pacific plate along the Kurile-Kamchatka Trench.

We wish to express our sincere thanks to Dr. M. Komatsu, Dr. T. Watanabe, Dr. K. Kiminami, Dr. Y. Kontani, Dr. K. Niida, and Dr. J. Maeda for their kind advices. We express our deep appreciation to Prof. Y. Katsui for critical reading of the manuscript.

REFERENCES

CHURKIN, M., Western boundary of the North America continental plate in the Asia, *Geol. Soc. Am. Bull.*, **83**, 1027–1036, 1972.

CHAPMAN, M. E., and S. C. SOLOMON, North American-Eurasian plate boundary in Northeast Asia, *J. Geophys. Res.*, **81**, 921–930, 1976.

DEN, N., and H. HOTTA, Seismic refraction and reflection evidence supporting plate tectonics in Hokkaido, *Pap. Meteorol. Geophys.*, **24**, 31–54, 1973.

DICKINSON, W. R., Plate tectonic evolution of North Pacific rim, *J. Phys. Earth*, **21**, Suppl. S1–S9, 1978.

FUJII, K., and SOGABE, M., Tectonic movement occurred in Hokkaido during the Late Miocene and Pliocene time, *Bull. Geol. Surv. Japan.* **29**. 1–14, 1978 (in Japanese with English abstract).

HASEGAWA, K., T. TAKAHASHI, and M. MATSUI, *Explanation Text of the Geological Map "Kamishiyubetsu" (scale 1/50,000)*, Hokkaido Development Agency, 44 pp., 1961.

HASHIMOTO, S., Structural significance of the Western Zone of the Hidaka Metamorphic Belt, Hokkaido, Japan, *Rep. Geol. Mineral., Niigata Univ.*, **4**, 409–414, 1976 (in Japanese with English abstract).

HORIKOSHI, E., Orogenic Belt of Japanese Islands and plate tectonics, *Kagaku*, **43**, 665–673, 1972 (in Japanese).

KAIZUKA, S., Macro-topography of the island arc system and plate tectonics, *Kagaku*, **42**, 573–581, 1972 (in Japanese).

KAIZUKA, S., Tectonic model for the morphology of arc-trench systems, especially for the echelon ridges and mid-arc fault, *Japan J. Geol. Geogr.*, **45**, 9–28, 1975.

KAIZUKA, S., Quaternary tectonic map of the "Outer Izu Bar" and the "Outer Kurile Bar," *Chikyu Monthly*, **2**, 155–156, 1980 (in Japanese).

KHARAKHINOV, V. V., V. E. KONONOV, I. M. AL'PEROVICH, V. M. NIKIFOROV, Yu. G. SLUDNEV, and A. A. TERESHCHENKOV. Deep-seated structure of Sakhalin, *Sovetskaya Geology*, **4**, 50–61, 1979 (in Rusian).

KIMINAMI, K. and Y. KONTANI, Mesozoic arc-trench system in Hokkaido, Japan, this volume, pp. 107–122, 1982.

KIMURA, G., Abashiri Tectonic Line—with special reference to the tectonic significance of the southwestern margin of the Kurile Arc, *J. Fac. Sci., Hokkaido Univ.*, Ser. IV, Geol. Mineral., **20**, 95–111, 1981a.

KIMURA, G., Tectonic evolution and stress field in the southwestern margin of the Kurile Arc, *J. Geol. Soc. Japan*, **87**, 757–768, 1981b (in Japanese with English abstract).

KIMURA, G. S. MIYASAKA, Y. KONTANI, S. MIYASHITA, K. HOYANAGI, and Y. WATANABE, Tectonic significance of the Kamishiyubetsu Tectonic Zone in the uplift of the Hidaka Metamorphic Belt, *Bull. Tectonic Res. Group Japan*, **27**, 167–177, 1982 (in Japanese with English abstract).

KIZAKI, K., Configuration of migmatite dome comparative tectonics of migmatite in the Hidaka Metamorphic Belt, *J. Fac. Sci., Hokkaido Univ.*, Ser. IV. *Geol. Mineral.*, **15**, 157–172, 1972.

KIZAKI, K., and D. HAYASHI, Migmatite tectonics of the Hidaka Metamorphic Belt, Hokkaido, Japan, *Tectonophysics*, **56**, 203–220, 1979.

KOMATSU, M., S. MIYASHITA, J. MAEDA, Y. OSANAI, T. TOYOSHIMA, Y. MOTOSOSHI, and K. ARITA, Petrological constitution of the continental type crust upthrusted in the Hidaka Belt, Hokkaido, *Spec. Pap. Jpn. Assoc. Meneral. Petrol. Econ. Geol.*, **3**, 229–238, 1982a (in Japanese with English abstract).

KOMATU, M., S. MIYASHITA, J. MAEDA, Y. OSANAI, and T. TOYOSHIMA, Disclosing of a deepest section of continental type crust upthrust as the final event of collision of arcs in Hokkaido, north Japan, This volume, pp. 149–165, 1982b.

MAEDA, J., Y. MOTOYOSHI, and T. TAKAHASHI, Magmatism in the Main Zone of the Hidaka Metamorphic Belt, Hokkaido, in *Symposium on the Paired Metamorphism*, Hiroshima, 19–24, 1982.

MELANKHOLINA, Ye. N., Formation complexes in structures of Sakhalin and Hokkaido, *Geotectonics*, **9**, 88–104, 1975 (in Russian).

MELANKHOLIAN, Ye. and T. V. MOLCHANOVA, Tectonic system of the Late Mesozoic Continental Margin, Eastern Asia, *Geotectonics*, **11**, 310–322, 1977 (in Russian).

MELINIKOV, O. A., *Geological development of southern Sakhalin from Paleogene to Neogene*, Nauka, Moscow, 153 pp., 1970 (in Russian).

MIYASHIRO, A., Subduction-zone ophiolites and island-arc ophiolites, in *Energetics of Geological Processes*, edited by S. K. Saxena and Bhattacharji, pp. 188–213, Springer-Verlag, New-York, 1977.

MIYASHITA, S., Geology of the metamorphosed ophiolite, Western Zone of the Hidaka Metamorphic Belt, Hokkaido, *J. Geol. Soc. Japan*, in press, 1982a (in Japanese with English abstract).

MIYASHITA, S., Ophiolite succession and metamorphism of the Western Zone of the Hidaka Metamorphic Belt, Hokkaido: *Symposium on the Paired Metamorphism*, Hiroshima, 25–30, 1982b.

MIYASAKA, S. and K. KIKUCHI, The Neogene Tertiary upheaval of the Hidaka Metamorphic Belt, *Assoc. Geol. Collab. Japan, Monogr.*, **21**, 139–153, 1978.

OKADA, H., Geology of Hokkaido and plate tectonics, *Chikyu Monthly*, **1**, 869–877, 1979 (in Japanese).

OKADA, H., Sedimentary environments on and around island arcs; an example of the Japan trench area, *Precambrian Res.*, **12**, 115–139, 1980.

PITMAN, W. C. and M. TALWANI, Sea-floor spreading in the North Atlantic, *Geol. Soc. Am. Bull.*, **83**, 619–646, 1972.

ROZHDESTVENSKIY, V. S., Strike-slip faults in the Vostochnyy range of Schmidt Peninsula of Sakhalin, *SSSR Geol. Geophys.*, **10**, 131–134, 1972 (in Russian).

ROZHDESTVENSKIY, V. S., Tectonics in northeastern Sakhalin, *Geotectonics*, **9**, 85–97, 1975 (in Russian).

ROZHDESTVENSKIY, V. S., Displacement along the Tym'Poronay fault on Sakhalin island, *Doklady Akademy Nauk, SSSR*, **230**, 678–680, 1976 (in Russian).

ROZHDESTVENSKIY, V. S. and A. N. RECHKIN, Serpentinite melange and some aspects of tectonic evolution of Sakhalin island, *Doklady Akademy Nauk, SSSR*, **221**, 1156–1159, 1975 (in Russian).

SHIBATA, K., and S. ISHIHARA, K-Ar ages of granitic rocks in the Hidaka Belt, Hokkaido, in *86th Annual Meeting of the Geological Society of Japan, Abstract*, 342, pp. 1981 (in Japanese).

TOKUDA, T., On the echelon structure of Japanese archipelagoes, *Japan J. Geol. Geogr.*, **5**, 41–76, 1926.

ZANYUKOV, V. N., The Central Sakhalin Fault and its role in the tectonic evolution of the island, *Doklady Akademy Nauk, SSSR*, **196**, 913–916, 1971 (in Russian).

Accretion Tectonics in the Circum-Pacific Regions, edited by M. Hashimoto and S. Uyeda, 135–148.
Copyright © 1983 by Terra Scientific Publishing Company (TERRAPUB), Tokyo.

From Subduction to Paleosubductions in Northern Japan

Jean-Paul CADET and Jacques CHARVET

*Département des Sciences de la Terre,
Université D'Orléans, 45046 Orleans, France*

In Northern Japan, subduction corresponding to the westward diving of Pacific plate under Japan Arc settled, with its present geometry, in Neogene time; the main feature of its evolution during that period seems to be the westward migration of the trench and volcanic arc front, together with the Japanese margin subsidence.

The paleogeodynamic evolution former stages in this area of Northwest Pacific can be explored in Hokkaido Central Belt, which results from the evolution of a Triasico-Jurassic subduction zone, apparently dipping eastward under a microcontinent (Hidaka volcanic paleoarc and Okhotska paleoland) which has collided since Cretaceous with the Asian margin, giving a remarkable set of nappes with a westward vergence.

This disposition was later separated from the Asian continent by the opening of Japan Sea and sectioned eastward by the initiation of the present Neogene subduction.

Hidaka-Okhotska microcraton can be interpreted either as a fragment of a vast continent scattered through Pacific Ocean ("Pacifica" concept of A. NUR and Z. BEN-AVRAHAM), or, in so much as Hokkaido Central Belt goes on to Sakhaline and perhaps further North, as a plate boundary corresponding to the Mesozoic and Cenozoic trace of Asia-America collision.

Northern Japan is a highly favorable area for the study of the relation between present subduction and paleosubductions for it includes a mountain belt resulting from a subduction-collision process itself cut by a present subduction zone (Fig. 1):

—the present westward dipping subduction corresponds to the underthrusting of the Pacific plate under Japan Arc. Subduction is characterized, beside geophysical evidences linked with high rate convergence (9 cm/year), by trenches (Kuril off Hokkaido then after a bend, Japan off Honshu) bordered by an active volcanic arc, both among the best known and best studied in the world;

—traces of paleosubduction zones can be found in "Hokkaido Central Belt," which has a westward tertiary vergence (thus with an eastward dipping paleosubduction?) cut southward by Japan trench and going northward on to Sakhaline and further.

By using the retro-tectonic method, attempts will be made to make out the different stages that allowed to reach the present setting. The transition between the present subduction setting and the ancient subductions will be specially emphasized.

1. Present Stage to Lower Miocene

Japan convergent margin is better understood since Deep Sea Drilling on IPOD

Legs 56 and 57, prepared by a comprehensive geological and geophysical surveying including multichannel seismic reflection lines that crossed the Japan trench at about 40°N Latitude (JNOC and Ocean Research Institute records, cf. NASU et al., 1980).

Main results can be summarized as follows:

(1) from the multichannel data, the broad deep sea terrace between the shore and the trench seems to be made of reflectors corresponding to Neogene sediments dipping gently seaward. These sediments rest by an angular unconformity on underlying Cretaceous, dipping landward, and are cut by normal probably synsedimentary faults (Fig. 2).

(2) This extentional tectonic is associated to a remarkable subsidence of about 3 km of the deep sea terrace bordering the trench. This subsidence start at the end of Paleogene from an emerged paleoland (OYASHIO landmass, cf. VON HUENE et al., 1978; CHAMLEY and CADET, 1981). This subsidence is very fast (the rate is nearly 50% higher than the normal midoceanic ridge, cf. M. G. LANGSETH et al., 1981) and is mostly due to cooling of the mantle below the Japan fore arc after the end of the arc volcanism known between 22–24 m.y. near the trench slope (cf. infra) to which are added the effects of the mass transport of cold 80 to 100 m.y. old Pacific oceanic lithosphere into the mantle during subduction (LANGSETH et al., 1981).

(3) Drilling in the trench inner wall did not reveal any subduction complexe or major thrust faults and the accretionary wedge, if any, has limited size despite the thousand of kilometers of oceanic crust that have been subducted, most of the sediments that entered the trench having been subducted rather than accreted.

These data confirm that subduction process, here characterized by generalized extension (cf. CADET, 1980), subsidence, and small frictional resistance between the two plates even if it causes important stress, probably cannot build mountain belts and that a collision is necessary to realize this type of setting.

(4) Besides, the discovery on site 439 on the deep sea terrace of a boulder conglomerate, overlying the cretaceous deposits with blocks of rhyolitic to dacitic composition rocks (FUJIOKA, 1980), angular and fresh, suggesting that the source of these rocks was close to site, raises the problem of the meaning of island arc related rocks near the trench slope. The most likely hypothesis is to consider (cf. VON HUENE et al., 1978) that those rocks, dated radiometrically between 22 and 24 m.y. (YANAGISAWA et al., 1980), are remnants of a lower Miocene volcanic arc far seaward of the presently active arc. This hypothesis is strengthened by refraction data (NAGUMO et al., 1980) indicating that another igneous body occurs near by and the probable presence of other plutons suggested by low-amplitude magnetic anomalies (OSHIMA et al., 1975).

As pointed by MOORE and FUJIOKA (1980) the 24–22 m.y. dacite in hole 439 lines up with a belt of granitic and granodioritic plutons that extends northward from Erimo cape across Hokkaido, and are dated from 36–16 m.y. (NOZAWA, 1975). Because of the aligment and of the similar age, we can consider these magmatic rocks all to have all had a common origin and to correspond to a middle Tertiary magmatic arc parallel with the Japan trench, then crossing toward the north present Kuril arc toward Sakhaline (Fig. 3).

The position of the ancient volcanic arc much closer to the trench (90 km) than the

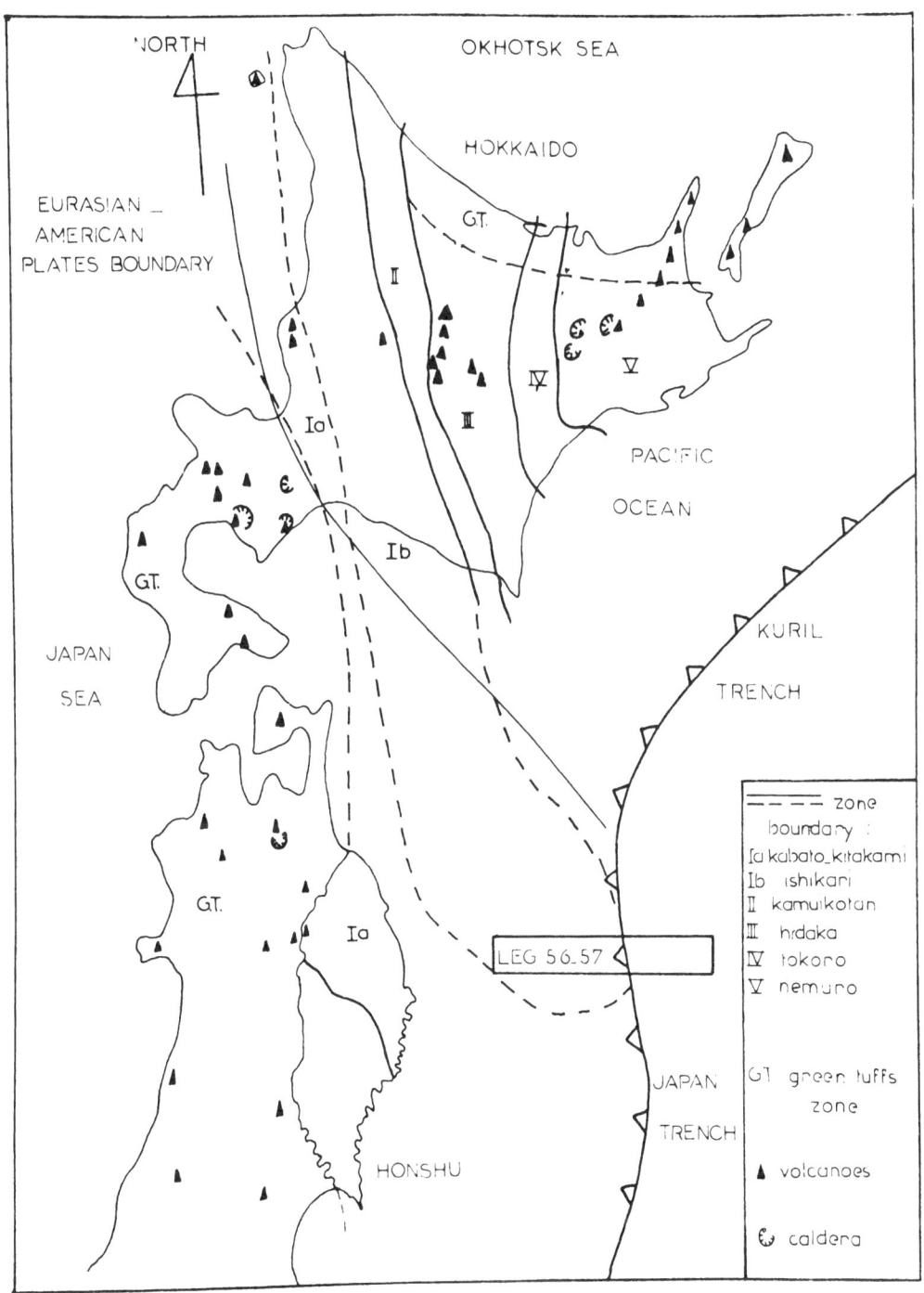

Fig. 1.   General outline of the studied area.

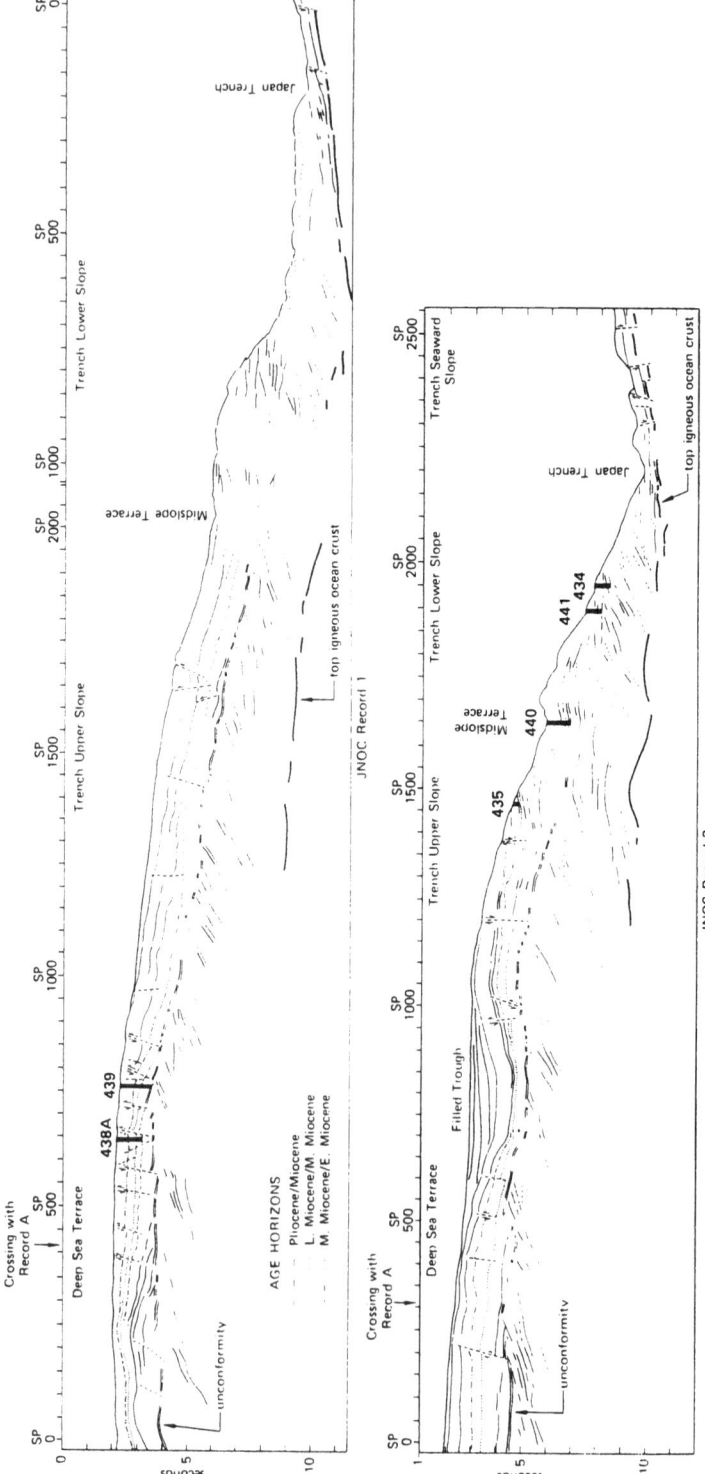

Fig. 2. Japan Deep Sea Terrace and Trench off Kitakami massif. Tracing of selected reflections in JNOC multichannel seismic reflections records and interpretation of faults with location of DSDP-IPOD legs 56 and 57 drill sites (from Nasu et al., 1980).

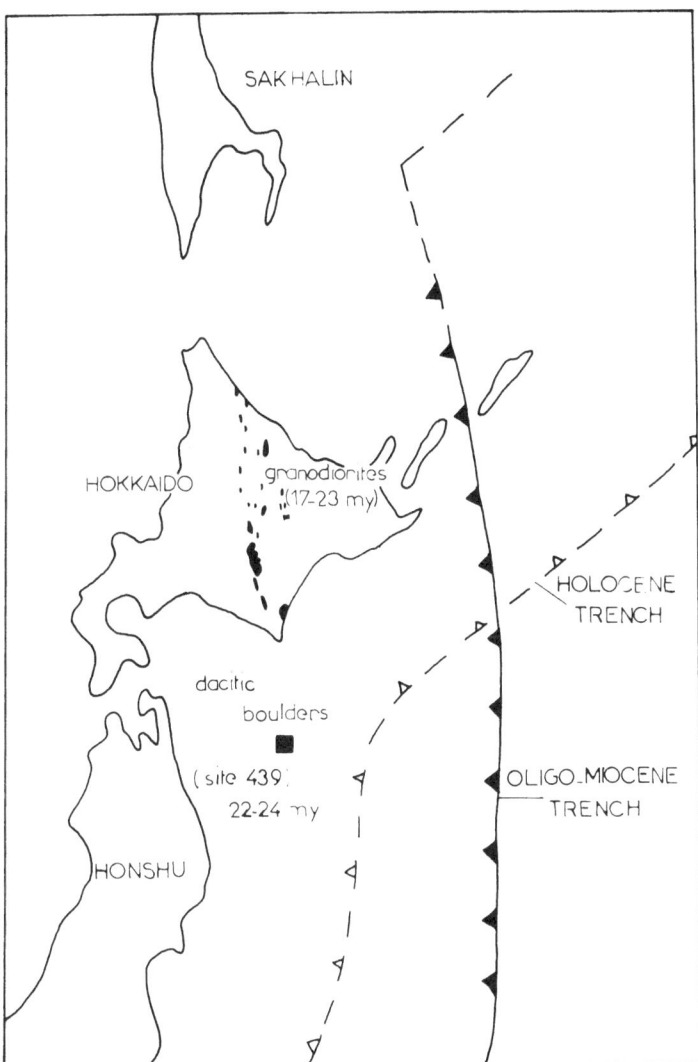

Fig. 3. Position of the Japan and Kuril Trenches infered from the position of the magmatic rocks at the Oligocene-Miocene boundary (modified from MOORE and FUJIOKA, 1980).

present active arc (300 km from the Japan trench) suggests (cf. MOORE and FUJIOKA, 1980; VON HUENE and UYEDA, 1981) that the subduction zone may formely have been farther seaward unless the depth to the top of the subducting oceanic lithosphere was once deeper or fore arc magma are produced by some other process than the melting of subducted materials.

Beside the existence of arc magmatism in the present fore arc area, evidence of massive subsidence in the fore arc and trench slope area during subduction and abrupt truncation of the continental framework on the landward slope of the trench, all suggest that the trench has moved from a further east position since lower Miocene. (VON HUENE and UYEDA, 1981).

This hypothesis seems supported by the simultaneous migration of the Tohoku arc volcanism linked with the present subduction geometry which, after a long quiescence periode, became active in late Oligocene and has seemed to migrate westward since Miocene in the Green Tuffs area in northern Honshu (MASUDA and KITAMURA, 1970; HONZA et al., 1977).

In short, the stage starting at the end of Paleogene corresponds to the initiation of the present subduction, then the migration of the arc-trench system westward and the subsidence of the fore-arc region.

2.   Paleogene and Mesozoic Stages

The former stages of the paleogeodynamic evolution of this area may partly be sketched by the analysis of Hokkaido folded belts (Hokkaido Central Belt).

2.1   Hokkaido Central Belt

This north-south stretching belt forms the backbone of Hokkaido island on over 500 km and goes on to Sakhaline island. It is made up of a pile of tectonic sheets and nappes with tertiary westward vergence. It is classicaly divided into five overthrusting zones from West to East: Ishikari, Kamuikotan, Hidaka, Tokoro, and Nemuro (Fig. 4).

The Ishikari zone is made up of westward thrusting units involving Miocene (HASHIMOTO, 1977). Its essentially detritic stratigraphic series are restricted to upper Cretaceous, Paleogene and Neogene. Upper Cretaceous (Yezo group) with flysch facies supplied from an emerged area situated westward (OKADA, 1965; KIMINAMI et al., 1978), dating from Aptian to Coniacian (TANAYANAGI et al., 1976; OKAMURA, 1977) is overlaid unconformably with Paleogene coal bearing molasses, then Neogene Green Tuffs equally with unconformity (NAGAO, 1938) (Fig. 5).

These sequences, rest, at least in the eastern part of Ishikari zone, upon oceanic series with ophiolitic affinity (Sorachi Group) known by drilling (SEGAWA and FURUTA, 1978).

Ishikari zone corresponds thus to subsiding basin, supplied in detrital material from the west (i.e. from the relative fore-land of Hokkaido central belt: Kitakami massif, mainly southern part) and from the east, the origin being an active orogenic inner zone. Its offshore continuation has been evidenced by oil drilling and cut by leg 57, hole 439 (Neogene hemi-pelagic series resting unconformably on pelagic upper Cretaceous).

The Kamuikotan zone is made of nappe and thrust sheets (NAGAO, 1933) of ophiolitic nature thrust on the whole upon Ishikari zone.

The following terranes are found:

—*Sorachi group* (Triassic, Jurassic, and Neocomian), oceanic series made of weakly metamorphosed tuffs, cherts, basalts, and pillow-lavas;

—*an ophiolitic complex* (ultrabasic bodies, harzburgite, dunite with minor gabbroic rocks, (cf. ASAHINA and M. KOMATSU, 1975; BANNO et al., 1978) sometimes complete (Horokanai massif, KOMATSU et al., 1977) and ending with radiolarian cherts dated from Tithonian to Valanginian;

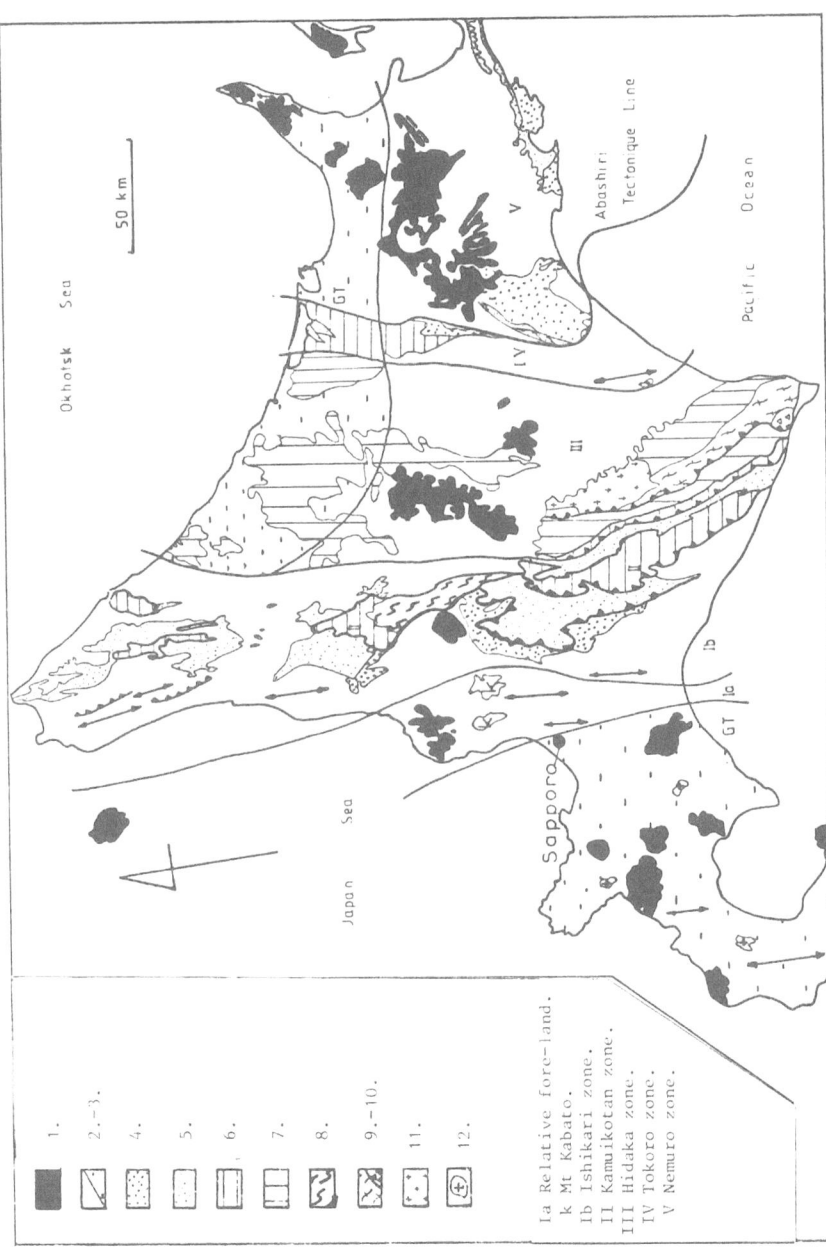

Fig. 4. Simplified structural sketch of Hokkaido island : 1—Quaternary volcanoes, 2— Neogene, 3— Green Tuffs, 4— Paleogene, 5— Cretaceous, 6—Ophiolitic series, 7—Hidaka supergroup, 8—Kamuikotan blue-schists, 9—Migmatites, 10—Lherzolites, 11—Miocene granodiorites, 12—Cretaceous granites. Ia relative fore-land : K—Mt Kabato, Ib—Ishikari zone, II—Kamuikotan zone, III—Hidaka zone, IV— Tokoro zone, V—Nemuro zone.

—*a set of metamorphic rocks* derived from basaltic rocks, sediments and retrograded amphibolites showing several metamorphic types and dated from 145 to 72 m.y. (IMAIZUMI and UEDA, 1981). Among them one is the Kamuikotan high P/T type metamorphism characterized by the occurrence of jadeite + quartz and glaucophane.

These three elements are highly tectonised and separated into incoherents blocks by faults and shear zones so that their relations are difficult to analyse.

These series with oceanic affinities are unconformably overlaid by transgressive middle cretaceous detrital Yezo group.

The eastern part of Kamuikotan zone is involved in the complex front of Hidaka zone (=Hidaka western marginal zone, MIYASHITA and MAEDA, 1978) and is affected by a different (*LP-LT*) metamorphism than Kamuikotan zone stricto sensu.

The Kamuikotan zone thus corresponds to an oceanic series sometimes intercalated with ophiolitic sheets and partly affected by an *HP* polyphased metamorphism.

Hidaka zone overthrust the inner part of Kamuikotan zone (=Hidaka western marginal zone) by a complex front (S. Miyashita) corresponding to fundamental shear split by a right lateral strike-slip fault (Chiroro and Pankenushi valleys).

At the basis of Hidaka series outcrops, besides probably old gneiss, peridotitic massifs (Horoman massif, cf. HASHIMOTO, 1975; HASHIMOTO *et al.*, 1975) with dunite, lherzolite and gabbros (NIIDA, 1974, 1975; S. HASHIMOTO *et al.*, 1979), recalling the Lanzo type peridotites (continental lithosphere upper mantle).

The bulk of Hidaka zone is made up by a several kilometers thick pile of graywackes rich in volcanic debris, sandstones and schists: it is the Nakanogawa group dated from Triasic and Jurassic. These series are cut through by migmatites (ARITA *et al.*, 1978) whose age is controversial, but probably quite recent (post Cretaceous?), which in turn are cut again by 22–24 m.y. granodiorites (YANAGISAWA *et al.*, 1980).

The Hidaka zone thus shows clear continental affinities and the abundance of volcanic material reworked in Nakanogawa group recalls the proximity of an active volcanic arc in Trias and Jurassic time (Cf. GARCIA, 1978).

Tokoro zone. It is situated east of Hidaka zone but their contact is hidden by Neogene and Quaternary volcanism (Daisetsuzan massif).

It includes:

—*a non metamorphic series* with oceanic affinities (radiolarian cherts, pillow lavas (OKADA, 1980), dating from upper Jurassic (Nikoro group, very similar to Sorachi group);

—*a set of detritic series* (Yubetsu and Saroma groups) rich in calc-alkaline volcanic debris, sometimes coarse, (KONTANI and KIMINAMI, 1980) supplied from the east (KIMINAMI, 1979; KONTANI *et al.*, 1980).

The Tokoro zone seems to correspond to an oceanic, ophiolitic type series partly covered (at least as concerns Saroma group) by a detritic series derived from a close eastern paleo-continent. This raises the problem of the relations between this oceanic serie and Kamuikotan zone.

Nemuro zone. It is separated from Tokoro zone by Abashiri tectonic line (KIMURA, 1981). There is again Nikoro group (basic tuffs, cherts and slate) covered by

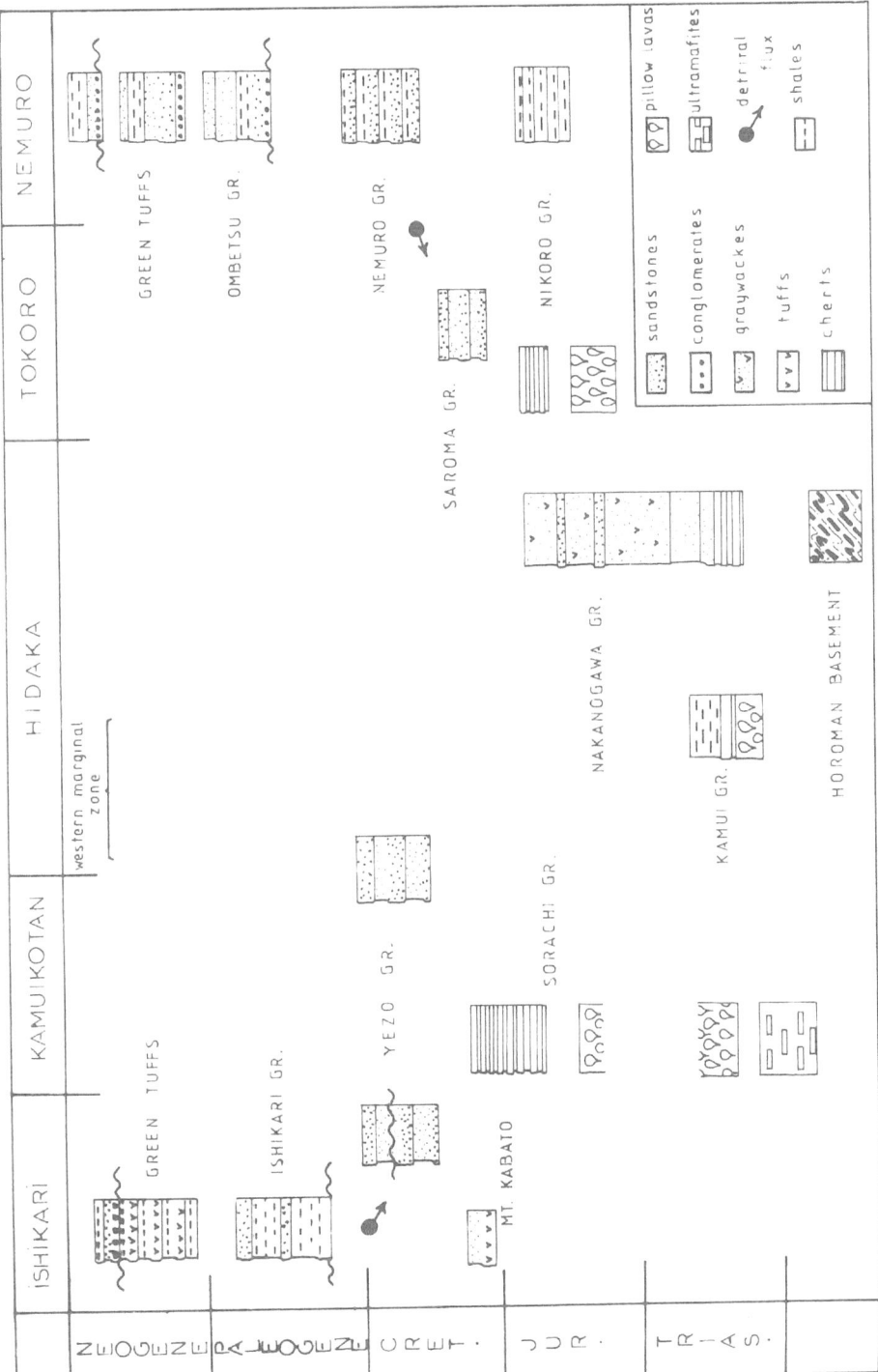

Fig. 5. Simplified stratigraphic columns of Hokkaido Central Belt zones.

Nemuro group, dating upper Cretaceous with flysch facies (KIMINAMI *et al.*, 1978).

These two groups are unconformably covered by molasse type Paleogene and neogene Green Tuffs, the whole being cut through by neogene and quaternary kuril arc volcanism.

The bulk of detritic material supplied since Cretaceous has been issued from the east and their continental nature and granulometry recall an emerged eroding continent (notion of paleo-Okhotsk continent or Okhotska, cf. HASHIMOTO, 1977; KONTANI *et al.*, 1980; etc.)

However, there are major questions left concerning tectonics and paleogeography. About tectonics, the problem of how important are tangential displacements is raised: for example, the existence of a parautochtonous nappe made of Yezo flysch overlaying Ishikari zone Neogene and Paleogene and overlaid by a kamuikotan nappe preceeded by klippes is hypothetically considered by some (cf. NAGAO, 1933). In the same way the early tectonic phases vergence is unknown. For, if tertiary vergences are obviously westward, nothing shows for sure that early vergences have the same orientation though the tertiary shear orientation fits in with an eastward dipping subduction zone. Besides, due to the lack of paleomagnetic data, it is difficult to determine the importance of the longitudinal displacement for which we have clues (Hidaka front for example). The paleogeographic organisation itself is equally questionable: the main point is the meaning of Tokoro zone where the description of a pelagic serie with ophiolitic affinities, if checked, would induce the existence of two oceanic zones (Tokoro and Kamuikotan) on each side of Hidaka continental zone. In short the question would be the original position of Kamuikotan zone in comparison with Hidaka zone, that is to say whether this zone is originally west of Hidaka zone or thrust over during an early phase and later overlaid by Hidaka zone during tertiary movements.

2.2   *Paleogeodynamic evolution*

Starting from field data, the evolution of Hokkaido Central Belt can be summed up as follows (Fig. 6):

—an upper Jurassic—lower Cretaceous stage resulting in ophiolites setting and high-pressure metamorphism. We can consider, as an hypothesis, that the ophiolites are tectonically emplaced portion of oceanic material obducted on the Asian margin while an ocean corresponding to Kamuikotan zone and bordered eastward by Hidaka paleo-volcanic arc and paleo-Okhotska (or Okhotsk paleoland) micro-continent was closing. Kamuikotan high pressure metamorphic rocks partly could been formed by thrusting of oceanic series on continental margin, following the model sketched in Corsica by MATTAUER and PROUST (1976) as relatively young radiometric ages (72–145 m.y. K-Ar age on muscovites, IMAIZUMI and UEDA, 1981) seems to show.

The progressive closing of the oceanic domain during Cretaceous and early Tertiary is witnessed by the deposit of Yezo group inconformable detritic series and the Paleogene molasses of the Ishikari basin. It would end by the collision between the Asian continental margin and the Okhotsk paleoland finally giving the Hokkaido Central Belt mountain building in middle Miocene (Hidaka orogenesis) while the Japan sea was opening and a proto-Kuril trench (cf. OKADA and KIMINAMI, *OJI*

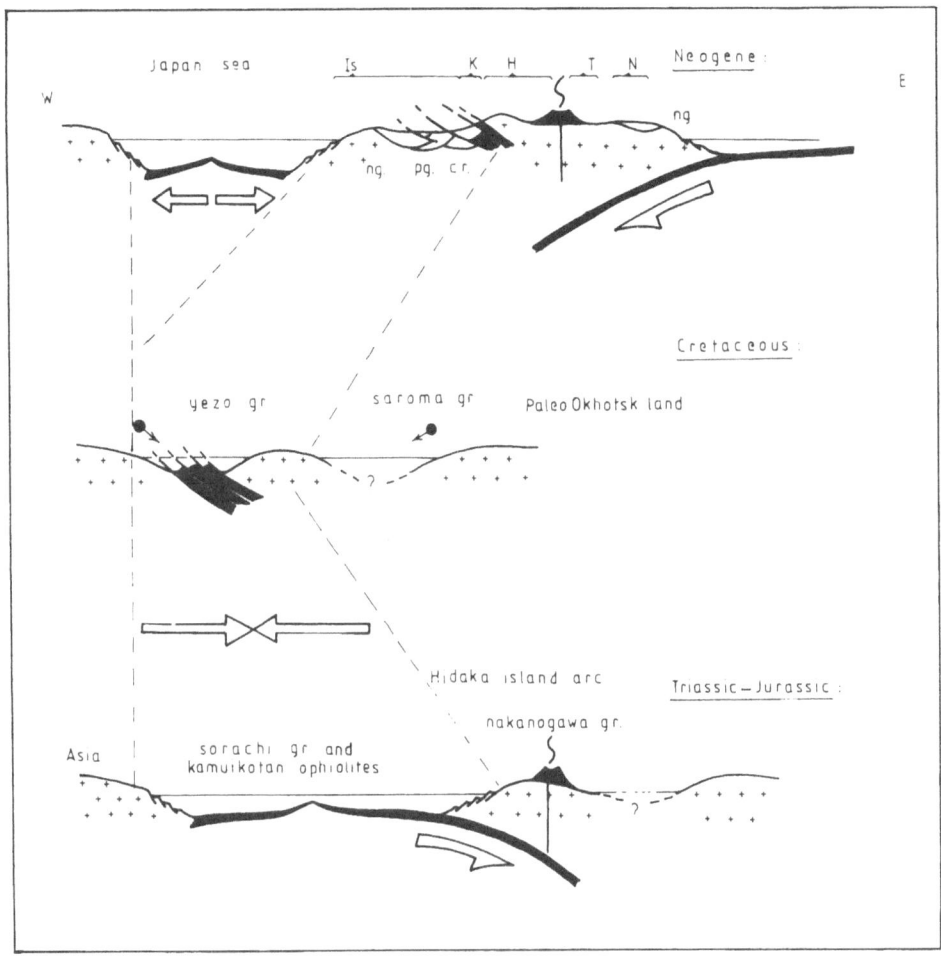

Fig. 6.   Hokkaido Central Belt paleogeodynamic evolution attempt.

Symposium) was forming.

3.   Conclusion

Hokkaido Central Belt results from a subduction-collision process probably more or less oblique between the proto-Siberian margin and a continental bloc. Two main types of alternates hypotheses can be considered as regards the origin of this continental domain:

(1)   either it is a microcontinent drifting with the Pacific plate from a more southern position, this hypothesis fit with NUR and BEN-AVRAHAM (1978) hypothesis on the scattering of a lost Pacifica continent (may be a former oceanic plateau).

Up to 40 m.y. (cf. LANCELOT and LARSON, 1975), subduction then collision between this microcontinent and the Asian margin would have been essentially oblique and corresponding to a left lateral displacement. From Oligocene onward, this movement turn into a convergence movement where from the final collision, still partly oblique, will result together with the initiation of a new subduction zone with a westward vergence (the precise mechanism of this accretion remaining speculative . . .)

The reference to a continental paleodomain to account for the collision process that begot Hokkaido Central Belt is supported by stratigraphic and sedimentologic data but suffers from the lack of paleomagnetic informations or paleontological data that would demonstrate its allochtonous origin comparable to the allochtonous terranes known along many active margin (western north America, etc . . .)

(2) or as much as Hokkaido Central Belt goes on to Sakhaline (HASHIMOTO, 1977) and perhaps further in Koriak peninsula (PARFENOV et al., 1979), we can consider that the suture between the oceanic domain (or domains) of Kamuikotan (and Tokoro?) and paleookhotsk land correspond to a plate boundary and in that case Hokkaido Central Belt is the Mesozoic and Cenozoic trace of the Asia-America collision.

The transition between the two subduction systems, the first having worked during Mesozoic and Cenozoic and a second, actual one, having cut it since Neogene corresponds to a key period marked by several important events besides initiation of a new subduction trench:

—end of collision in Hokkaido Central Belt,

—beginning of subsidence of the fore arc area of the present Japan trench,

—start of active colcanism of the present volcanic arc Green Tuffs episode),

—at least partial opening of Japan sea (late Oligocene, early Miocene, as deduced from heat flow data (cf. LANGSETH et al., 1981) or extrapolated from leg 31 information).

Relations between these phenomena raise the badly elucidated question of the links between back-arc spreading, subduction and arc volcanism: it seems that northern Japan because of the amount of collected geological and geophysical data is a hightly favorable place for the study of this fundamental problem.

The authors express their grateful thanks to Professors T. Bamba and S. Hashimoto and Drs. Y. Fujiwara and S. Miyashita for acquainting them with the problems of Hokkaido Central Belt during field trips.

This work benefited from discussions with Drs. J. Aubouin, S. Uyeda and R. Von Huene and we thank them for the help. We also acknowledge Drs. S. Uyeda and M. Hashimoto for their invitation to participate to OJI Symposium where discussions were very enriching. Thanks are due to Laurent Jolivet and Jacqueline Martinez for technical assistance. This research was partly supported by a CNRS-INAG grant.

REFERENCES

ARITA, K., H. MORI, M. OKAZAKI, K. OGURA, and Y. MOTOYOSHI, The metamorphic rocks and migmatites of the Southern part of the Hidaka metamorphic belt, Assoc. Geol. Collab. Japan, Monogr., 21, 27–41, 1978.

ASAHINA, T. and M. KOMATSU, The Horokanai ophiolitic complex in the Kamuikotan tectonic belt, Hokkaido, Japan, J. Geol. Soc. Japan, 85, 317–330, 1979.

BANNO, S., H. ISHIZUKA, N. GOUCHI, and M. IMAIZUMI, Kamuikotan belt in Hokkaido: the tectonic contact of high-pressure metamorphic belt and low-pressure ophiolite succession, in *Inter. Geody. Conf., Tokyo*, pp. 14–15, 1978.

CADET, J. P., Tectonique et subduction dans l'Ouest Pacifique: la fosse du Japon, *C. R. Soc. Geol. France*, **5**, 178–181, 1980.

CHAMLEY, H. and J. P. CADET, Tectonique, volcanisme, morphologies et climats cenozoiques au large du Japon d'après la sédimentation argileuse marine, *C. R. Acad. Sci. Paris*, **292**, 219–224, 1981.

FUJIOKA, K., Conglomerate of volcanic rocks of the Deep Sea Drilling Project, site 439, in *Initial Reports of the Deep Sea Drilling Project, Vol. 57*, pp. 1075–1082, U.S. Govt. Printing Office, Washington, D. C., 1980.

GARCIA, M. O., Criteria for the identification of ancient volcanic arcs, *Earth Sci. Rev.*, **14**, 147–165, 1978.

HASHIMOTO, S., The basic plutonic rocks of the Hidaka metamorphic belt, Hokkaido, part I, *J. Fac. Sci., Hokkaido Univ., Ser. IV*, **16**, 367–420, 1975.

HASHIMOTO, S., Mesozoic and Cenozoic orogenic belts in northern Japan, in *Colloques Internationaux du C.N.R.S., 268: Ecologie et Géologie de l'Himalaya*, 1977.

HASHIMOTO, S., S. MIYASHITA, and J. I. MAEDA, The basic plutonic rocks of the Hidaka metamorphic belt, Hokkaido, Part II. The Ameyama layered gabbros in the Tokashi Province, *J. Fac. Sci., Hokkaido Univ., Ser. IV*, **19**, 241–255, 1979.

HONZA, E., H. KAGAMI, and N. NASU, Neogene Geological history of the Tohoku Island Arc system, *J. Oceanogr. Soc. Japan*, **33**, 297–310, 1977.

IMAIZUMI, M. and Y. UEDA, On the K-Ar ages of the rocks of two kinds existing in the Kamuikotan metamorphic rocks located in the Horokanai district, *J. Japan Assoc. Mineral. Petrol. Econ. Geol.*, **76**, 88–92, 1981.

KIMINAMI, K., Pre-cretaceous paleocourants of north eastern Hidaka belt, Hokkaido, *J. Fac. Sci. Hokkaido Univ. Ser. IV*, **19**, 179–188, 1979.

KIMINAMI, K., K. TAKAHASHI, and K. MANIWA, The cretaceous system in Hokkaido-Yezo and Nemuro groups, *Assoc. Collab. Geol. Japan, Monogr.*, **21**, 111–126, 1978.

KIMURA, G., Abashiri tectonic line with special reference to the tectonic significance of the southwest margin of the kuril arc, *J. Fac. Sci. Hokkaido Univ.*, **20**, 95–111, *Ser. IV*, p. 317, 1981.

KITAMURA, N., N. E. Japan arc and the Hidaka arc since late cretaceous, pp. 161–168, 1978.

KOMATSU, M., K. TAZAKI, and Y. KURODA, Ophiolite suite rocks in some ophiolite belt in Japan, *Int. Symp. on Geodynamic in South-West Pacific. Ed. Technip. Paris*, pp. 3–14, 1977.

KONTANI, Y. and K. KIMINAMI, Petrological study of the sandstones in the pre-cretaceous yubetsu group, northeastern Hidaka belt, Hokkaido, Japan, *Earth Science; J. Assoc. Geol. Collab. Japan*, **33**, 307–319, 1980.

LANCELOT, Y. and R. LARSON, Sedimentary and tectonic evolution of the northwestern Pacific, in *Initial Reports of the Deep Sea Drilling Project, Vol. 32*, in edited by R. L. Larson, R. Moberly *et al.*, pp. 925–940, U.S. Govt. Printing Office, Washington, D.C. 1975.

LANGSETH, M. G., R. VON HUENE, N. NASU, and H. OKADA, Subsidence of the Japan Trench forearc region of northern Honshu, in *Ocean. ACTA, Proc. 26th Int. Geol. Congr., Geol. Contin. Marg. Symp. Paris*, pp. 173–179, 1981.

MATSUDA, T. and N. KITAMURA, Northeast Japan, Mesozoic-Cenozoic Orogenic belt, *Geol. Soc. London, Special Pub.*, **4**, 543–552, 1974.

MATTAUER, M. and F. PROUST, La Corse alpine: un modèle de génèse du métamorphisme haute pression par subduction de croûte continentale sous du matériel océanique, *C.R. Acad. Sci. Paris*, **282**, 1249–1252, 1976.

MIYASHITA, S. and J. I. MAEDA, The basic plutonic and metamorphic rocks from the northern Hidaka metamorphic belt, Hokkaido, *Assoc. Geol. Collab. Japan, Monogr.*, 43–60, 1978.

MOORE, G. W. and K. FUJIOKA, Argon 40/39 age and origin of Dacite 90 km from the Japan Trench, in *Initial Report of the Deep Sea Drilling Project, Vol. 57, Part II*, pp. 1083–1088, U.S. Govt. *Printing Office*, Washington, D.C., 1980.

NAGAO, T., "Nappes" and "Klippes" in central Hokkaido. *Proceeding of the Imperial Academy of Japan, Vol. 9*, pp. 101–104, 1933.

NAGAO, T., Tertiary orogeny in Hokkaido, *J. Fac. Sci. Hokkaido Imp. Univ. Ser. IV*, **4**, 23–30, 1938.

NAGUMO, S., J. KASAHARA, and S. KORESAWA, OBS airgun seismic refraction survey near sites 441 and 434, 438 and 439 and proposed site J-2B-Legs 56 and 57, Deep Sea Drilling Project, in *Initial Reports of the*

Deep Sea Drilling Project, Vol. 57 Part I, pp. 459–462, U.S. Govt. Printing Office, Washington, D.C., 1980.

Nakano, N., Metamorphism of the greenstones in the Kamuikotan zone and the Hidaka western marginal tectonic zone in the Shizunai-Mitsuishi district, Hokkaido, pp. 211–224, 1981.

Nasu, N., R. von Huene, Y. Ishiwada, M. Langseth, T. Bruns, and E. Honza, Interpretation of multichannel seismic reflection data legs 56 and 57 Japan Trench Transect, in *Intial Report of the Deep Sea Drilling Project, Vol. 56–57, part I,* pp. 489–504, U.S. Govt. Printing Office, Washington D.C., 1980.

Niida, K., Stucture of the Horoman ultramafic massif of the Hidaka metamorphic belt in Hokkaido, Japan, *J. Geol. Soc. Japan,* **1**, 31–44, 1974.

Niida, K., Textures and olivine fabrics of the Horoman ultramafic rocks, Japan, *J. Jpn. Assoc. Mineral Petrol Econ. Geol.,* **70**, 265–285, 1975.

Nozawa, T., Radiometric age map of Japan; granitic rocks, in *Geol. Surv. Japan Map ser. 16–1, 1:2,00 0,000,* 1975.

Nur, A. and Z. Ben K Avraham, Speculations on mountain building and the lost Pacifica continent, *J. Phys. Earth,* **26**, Suppl., 521–537, 1978.

Okada, H., Serpentine sandstone from Hokkaido, *Mem. Fac. Sci. Kyushu Univ. Ser. D, Geologie,* **15**, 23–38, 1964.

Okada, H., Sedimentology of the cretaceous Mikasa formation, *Mem. Fac. Sci. Kyushu Univ. Ser. D, Geology,* **16**, No. 1, 1965.

Okada, H., Sedimentary environments on and around islands arcs: an example of the Japan Trench area, *Precambrian Res.,* **12**, pp. 115–139, 1980.

Okamura, M., Geology and microfossils of the cretaceous strata of the Saku area, Teshio district, Hokkaido, *Mem. Fac. Educ. Kumamoto Univ., Natural Sci.,* **26**, 145–161, 1977.

Oshima, S., T. Kondo, T. Tsukamoto, and K. Onodera, Magnetic anomalies at sea around the northern part of Japan, *Rep. Hydro. Res.,* **10**, Hydrographic office of Japan, Tokyo, 1975.

Parfenov, L. M., I. P. Voinova, B. A. Natalin, and D. F. Semenov, Geodynamics of the Northeastern asia in mesozoic and cenozoic time and the nature of volcanic belts, in *Advances in Earth and Planetary Sciences, Vol. 6, (J. Phys. Earth. Suppl.), Geodynamic of the Western Pacific,* edited by S. Uyeda, R. W. Murphy, and K. Kobayashi, pp. 503–525, Center for Academic Publications Japan/Japan Scientific Sciences Press, Tokyo, 1979.

Segawa, J. and T. Furuta, Geophysical study of the mafic belts along the margins of the Japanese Islands, *Tectonophysic,* **44**, 1–26, 1978.

Takayanagi, Y. and M. Okamura, Mid-cretaceous planktonic microfossils from the obiro area, Rumoi, Hokkaido, in *Paleontol. Soc. Japan, Mid-cretaceous Events-Hokkaido Symposium, 1976, Spec. Papers,* **21**, pp. 31–39, 1977.

von Huene, R. and S. Uyeda, S. A summary of results from the IPOD transects across the Japan, Mariana and middle-America convergent margins, Oceanol ACTA, in *Proc. 26th Int. Geol. Congr., Geol. Contin. Marg. Symp., Paris,* pp. 233–239, 1981.

von Huene, R., M. Langseth, M. Nasu, and M. Okada, Summary Japan Trench transect, in *Initial Reports of the Deep Sea Drilling Project Vol. 56–57, Part I,* pp. 473–488, U.S. Govt. Printing office, Washington, D.C., 1980.

von Huene, R., N. Nasu, M. Arthur, J. P. Cadet, B. Carson, G. W. Moore, E. Honza, K. Fujioka, J. A. Barron, G. Keller, R. Reynolds, B. Shaffer, S. Sato, and G. Bell, Japan Trench transected on Leg 57, *Geotimes,* **23**, 4, 16–21, 1978.

Yanagisawa, M., Y. Takigami, M. Ozima, and I. Kaneoka, 40 Ar/39 Ar age of boulders drilled at site 439, Leg 57, in *Deep Sea Drilling Project, Initial Report of the Deep Sea Drilling Project, Vol. 57, Part II,* pp. 1281–1284, U.S. Govt. Printing Office, Washington, D.C., 1980.

Accretion Tectonics in the Circum-Pacific Regions, edited by M. Hashimoto and S. Uyeda, 149–165.

Disclosing of a Deepest Section of Continental-Type Crust Up-Thrust as the Final Event of Collision of Arcs in Hokkaido, North Japan

Masayuki Komatsu,* Sumio Miyashita,** Jinichiro Maeda,**

Yasuhito Osanai,* and Tsuyoshi Toyoshima*

*Department of Geology and Mineralogy,
Niigata University, Niigata 950–21, Japan
**Department of Geology and Mineralogy,
Hokkaido University, Sapporo 060, Japan

The metamorphic sequence in the main zone of the Hidaka metamorphic belt represents almost the whole section through the continent-type crust which was obducted on the accretion complex along the collision zone of two island arcs. The exposed crust section has about 23 km in thickness, consisting of granulite facies gneisses at the base, amphibolite facies gneisses in the middle to upper and greenschist facies rocks and unmetamorphosed sediments at the top. The crust might have been developed in an island arc environment during the Paleogene to Early Miocene and thrust up from east to west, having been affected by the westward collision of the third island arc, the Kuril arc, in the Late Miocene to Pliocene time.

1. Introduction

The Hidaka metamorphic belt has been considered as the axial zone of sedimentation in the Hidaka Geosyncline and of metamorphism, plutonism and mountain building of the Hidaka Orogenesis of Alpine phase (Hunahashi and Hashimoto, 1951; Hunahashi, 1957; Minato et al., 1965). The previous authors are proper to have stressed that the metamorphic zone has thrust up from east to west at the time of mountain building, which is inferred from the overturned structures and many thrust faults facing to west observed in the region to the west of the Hidaka metamorphic belt. In this region is there the Kamuikotan belt characterized by blueschists and serpentinite, and its formation has been also ascribed to the resultant westward influence of the thrust-up movement of the Hidaka metamorphic zone (e.g. Minato et al., 1965). This interpretation, however, is not acceptable, because the Kamuikotan blueschists and metaophiolites were formed in the Early Cretaceous (Bikerman et al., 1971; Imaizumi and Ueda, 1981), which is the age long before the metamorphism (Paleogene to Early Miocene) and upthrusting (Late Miocene to Pliocene) of the Hidaka belt. In this connection, the concept of paired metamorphic belt by Miyashiro (1961) is not valid ragarding to the 'appearent pair' of the Hidaka and Kamuikotan belts.

On the structure and evolution of the Hidaka metamorphic belt itself, we have the opinion different from those of the previous authors; we interpret that the Hidaka metamorphic belt is composed of two tectonic units, the western metaophiolite and the up-thrust continental type crust block in the eastern Main Zone (KOMATSU et al., 1982a). The latter was formed in an island arc environment during the Paleogene to Early Miocene and thrust up from east to west under the influence of the westward collision of Kuril arc in Late Miocene to Pliocene (KIMURA, 1981; KIMURA et al., 1982).

2.   Tectonic Settings

Three major provinces are recognized in Hokkaido; West Hokkaido, Central Hokkaido, and East Hokkaido (Fig. 1). West Hokkaido is the northern extension of Northeast Honshu (Kitakami belt). Along its eastern margin occurs the andesitic volcanic terrain (Rebun-Kabato range) of Early to Middle Cretaceous in age (MINATO, 1978; OKADA, 1979). It extends southward to the Iwaizumi belt in the outer Kitakami region. East Hokkaido (Nemuro belt) belongs to the Kuril arc and is divided into outer and inner arcs. The outer arc consists of Late Cretaceous to Paleogene turbidites and andesitic volcanic rocks (KIMINAMI, 1975). Central Hokkaido is a composite terrain consisting of the Kamuikotan belt, Hidaka belt, and Tokoro belt from west to east. The Hidaka belt is subdivided into two zones, the western melange zone and the Hidaka sedimentary zone. The Hidaka metamorphic belt appears along the boundary between the western melange zone and the sedimentary zone in the southern half of the Hidaka belt. The structural trends of these belts are traceable in the adjacent submarine region and northward to Sakhalin.

In Central Hokkaido, there are two fossil subduction zones; one is the Kamuikotan belt and the other is the Hidaka western melange zone (KOMATSU et al., 1982b; KIMINAMI and KONTANI, 1982). The former was active in the Late Jurassic to Early Cretaceous, while the latter was in the Late Triassic to Late Jurassic. The Rebun-Kabato range corresponds to the volcanic arc related to the Kamuikotan trench where the subduction was directed westward (OKADA, 1979; ISHIZUKA et al., 1981; KIMINAMI and KONTANI, 1982). This arc-trench system is the northern extension of that in Northeast Honshu.

Another set of arc-trench system with the direction opposite to the former could have occurred in Central Hokkaido, inferred from the facts that the rocks in the Hidaka western melange are older than the Kamuikotan rocks and that the related fore-arc basin (Hidaka sedimentary zone, after KONTANI et al., 1982; KIMINAMI and KONTANI, 1982) and volcanic arc (the Tokoro belt, although it is a part of the arc) are arranged from west to east (Fig. 1).

The arc-trench gap between the supposed southern extension of the Kamuikotan belt (Jurassic to Cretaceous subduction zone) and the volcanic arc (the Iwaizumi belt) remains relatively wide, more than 150 Km in width in the off-Kitakami region, while the gap is seen very narrow in Hokkaido (Fig. 1). In this regard, it is worth of noting that in Central Hokkaido the basement beneath the Cretaceous Yezo Group has been reveald by deep drilling to consist of porphyritic rocks at 3,713 m deep; the porphyritic rock belongs to the volcanic group of the Rebun-Kabato range and the boundary

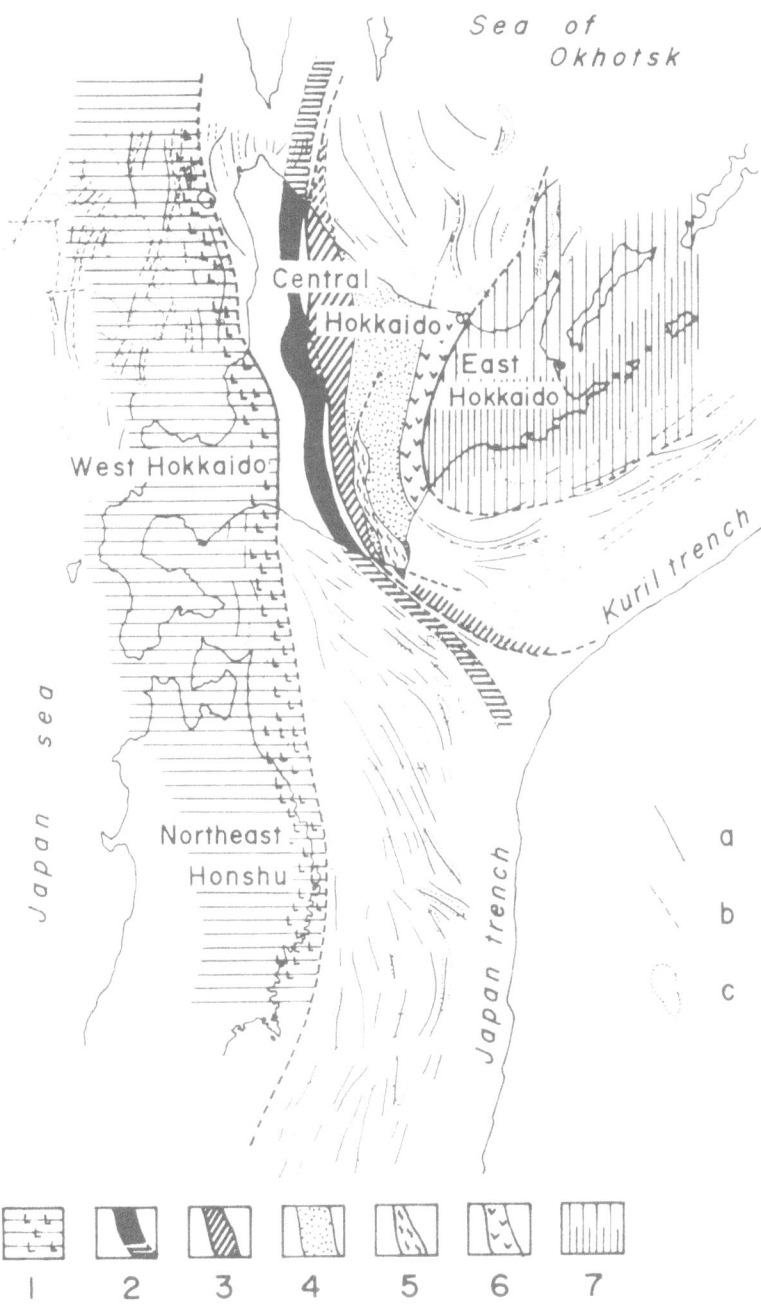

Fig. 1. Tectonic division and structures of Hokkaido and the adjacent areas, 1: Rebun-Kabato range and Iwaizumi belt, 2: Kamuikotan belt and its extension, 3: Hidaka western melange zone, 4: Hidaka sedimentary zone, 5: Hidaka metamorphic belt, 6: Tokoro belt, 7: Kuril arc (outer arc). a: fold axes, b: faults, c: axial part of acoustic basement high. The structures in the adjacent sea area are modified from those given by GEOLOGICAL SURVEY OF JAPAN (1978; 1979).

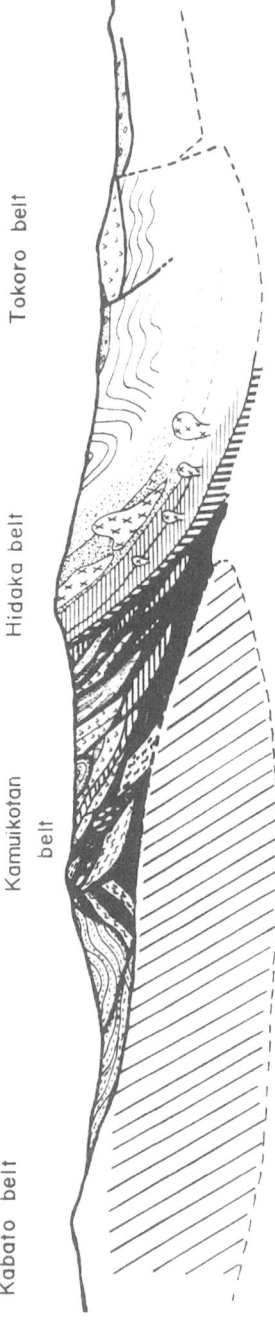

Fig. 2.   Schematic geological profile of Hokkaido (not scaled) showing overthrusting of Central Hokkaido on West Hokkaido.

surface between the basement and the Yezo sediments gently dips to the east (MINATO, 1978; MITANI, 1978). These evidences suggest that the large part of the zone corresponding to the arc-trench gap is concealed under the tectonic belts in Central Hokkaido (Fig. 2); in other words, the Yezo group sediments, the Kamuikotan subduction complex and part of the Hidaka belt are overthrust on the marginal realm of West Hokkaido. This tectonic relation is considered to be related to the convergence of two trench-arc systems and more effectively to the collision of Kuril arc to Central Hokkaido in Late Miocene to Pliocene.

3.   Petrological Composition of the Hidaka Metamorphic Belt

The Hidaka metamorphic belt consists of two zones; the Western Zone and the Main Zone (Fig. 3). The former is composed of metaophiolite, which was faulted into slices and overturned in appearence (MIYASHITA and MAEDA, 1978; MIYASHITA, 1982b). On the other hand the Main Zone is regarded to represent the crustal section of a continent or an island arc type (KOMATSU et al., 1981; OSANAI et al., 1982). The boundary between the Western Zone and the Main Zone is a large thrust fault (Hidaka main thrust) associated with mylonite over the whole extension of the metamorphic belt.

3.1   Ophiolite succession of the western zone
The Western Zone is exposed typically in the northern to middle part of the Hidaka metamorphic belt for 70 km long with a range from 1 to 4 Km in width. This Zone is bounded on the west by thrust faults against weakly metamorphosed sediments (slate, chert, and greenstones in the western melange zone), and on the east against mylonitized high-grade metamorphic rocks of the Main Zone.

The Zone is largely composed of nine lighofacies: pelitic schist, greenschist, blastoporphyritic amphibolite, metagabbro, metamorphosed basic to ultrabasic cumulates, peridotite tectonite, and quartz-biotite amphibolite. The former five lithofacies are generally distributed successively from west to east with N-S trend. Metagabbro frequently appear as lenticular bodies in the blastoporphyritic amphibolite. Peridotite tectonites composed of dunite and harzburgite occur along the eastern margin with tectonic contacts. The arrangement of the lithofacies is disturbed by thrust faults with a NW-SE trend and dipping to north east. Over the thrust faults, metagabbros, metamorphosed cumulates, and tectonites are exposed.

The original rocks of the metamorphic rocks were evaluated on the basis of their field relations and petrographic features (MIYASHITA, 1982b). The geologic sections based on the original rocks are shown along the four main routes (Fig. 4). From these sections, the ophiolite succession of the Western Zone is reconstructed as follows;

The layered basic complex consisting of three units (ultrabasic to basic cumulate, olivine gabbro and pyroxene gabbro) is more than 3 km thick. Above the layered series occurs the composite body consisting of differentiated gabbro and dolerite dike swarm (500 m), followed by dolerite or massive lava (700 m), and hyaloclastite and pillow lavas (?) (600 m) in ascending order. At the upper horizon occurs alternation of greenstone and pelitic to silicious sediments (20 m +); the greenstone seems to be

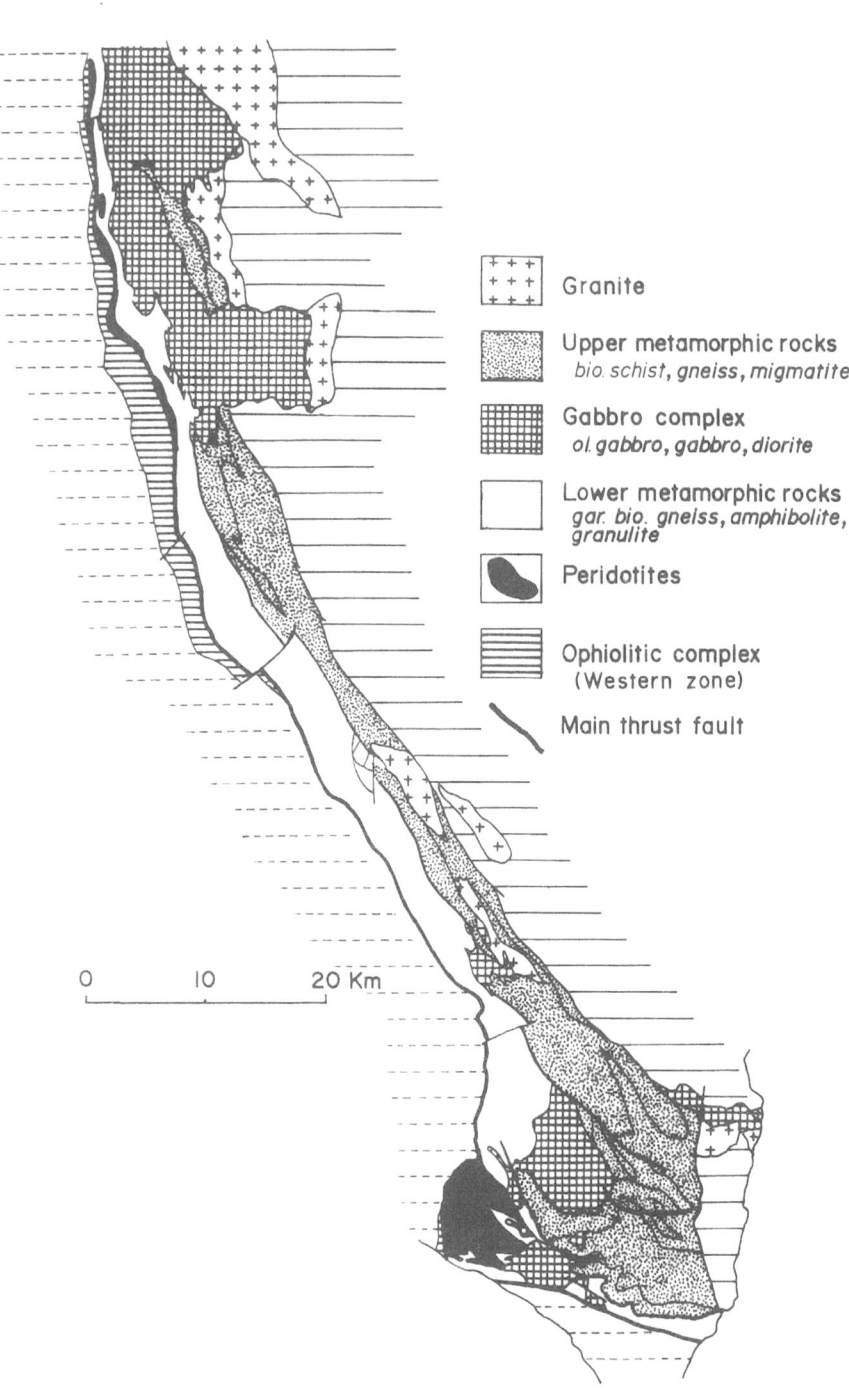

Fig. 3.   Simplified geological map of the Hidaka metamorphic belt.

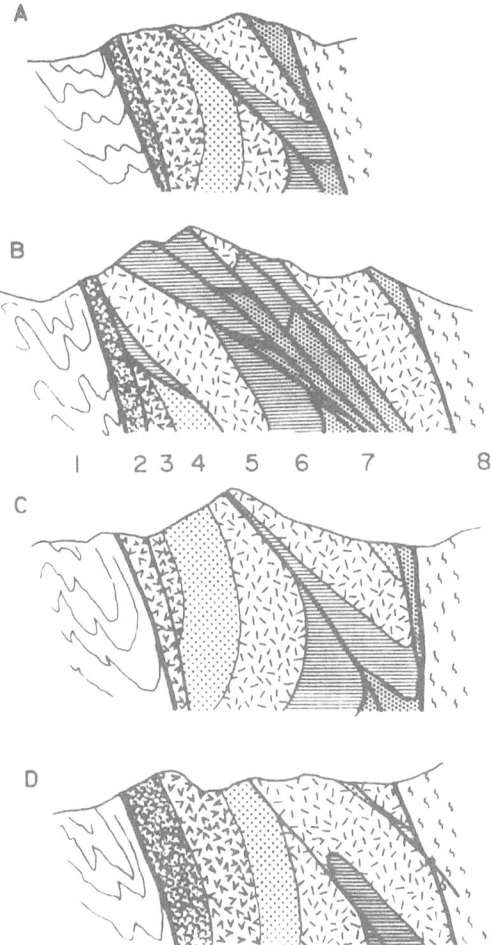

Fig. 4. Schematic cross sections of the Western Zone along the representative routes showing successions of primary ophiolitic rocks. 1: sediments in the western melange zone, 2: basaltic rocks, 3: massive lave(?) and dolerite, 4: dolerite and gabbro, 5: pyroxene gabbro and olivine gabbro, 6: layered basic-ultrabasic cumulates, 7: peridotite tectonite, 8: mylonites in the Main Zone.

effusive in origin. The maximum total thickness of the ophiolite succession excluding peridotite tectonite seems at least 5.3 Km.

Plots of FeO^*/MgO versus TiO_2, FeO^*, and SiO_2 for the rocks of the Western Metaophiolite are shown in Fig. 5. With increasing FeO^*/MgO, TiO_2 and FeO^* increase, and SiO_2 is nearly constant. These compositional relations show a trend due to fractional crystallization and correspond well to those of oceanic ridge tholeiite (MIYASHIRO, 1973).

3.2   Main Zone

The Main Zone is composed of granulite, amphibolite, thin alternation of

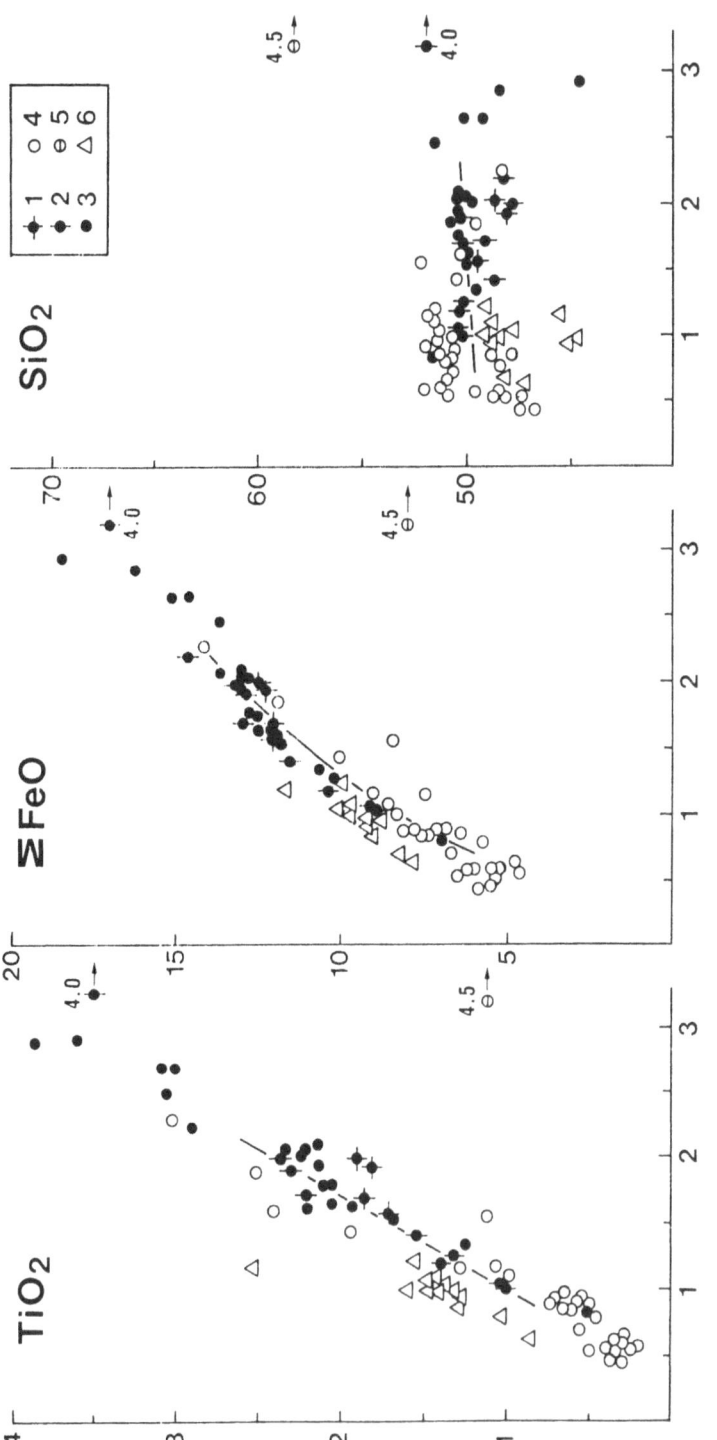

Fig. 5.   Variations of TiO₃, FeO* and SiO₂ contents in relation to the FeO*/MgO ratios of basic rocks in the western metaophiolite. 1: greenschists, 2: epidote amphibole schists, 3: schistose amphibolite, 4: blastoporphyritic amphibolite, 5: quartz amphibolite, 6: meta-dolerite, Solid lines: variations of mid-oceanic ridge tholeiite (MIYASHIRO, 1973).

amphibolite and biotite gneiss, and of biotite hornblende gneiss and biotite gneiss, biotite muscovite gneiss, schist, hornfels, and unmetamorphosed sediments from west to east i.e., from lower to upper horizon. Basic metamorphic rocks are dominated in the lower portion and pelitic, psammitic metamorphic rocks in the upper. The Main Zone, therefore, can be devided into the lower sequence where basic rocks are predominant and the upper sequence without basic rocks (Fig. 6). The spinel and plagioclase lherzolite bodies are associated within the granulite unit.

Various kinds of igneous rocks (HASHIMOTO, 1975; MIYASHITA and MAEDA, 1978; MOTOYOSHI, 1981; MAEDA et al., 1982) plentifully intrude in selective horisons, basic rocks being, in general, predominant in the lower part and acidic ones in the upper. "Migmatite" intrudes into the biotite muscovite gneiss and the schist and hornfels units (KIZAKI and HAYASHI, 1979).

3.2.1 Lower metamorphic sequence

The granulite unit in the basal part of the lower sequence can be divided into three parts; 1. garnet-orthopyroxene migmatitic rock, 2. garnet-orthopyroxene gneiss, and 3. orthopyroxene-plagioclace gneiss and amphibolite from lower to upper (OSANAI et al., 1982). Mylonite which occurs along the main thrust fault was derived from the rocks of the granulite unit.

The amphibolite unit consists essentially of brown hornblende amphibolite with minor garnet amphibolite in the basal part of the unit and some thin intercalations of biotite-gedrite-cordierite gneiss. The biotite-hornblende gneiss unit is largely composed of thin alternation of biotite gneiss and biotite amphibolite.

The regular arrangement of lithofacies from lower to upper as described above is most prominent at the Shizunai river district in the middle part of the metamorphic belt. The arrangement, however, is disturbed in the northern part due to gabbroic intrusions and in the southern part due to thrust faults with right hand lateral slip.

3.2.2 Upper sequence

The metamorphic rocks of the upper sequence are divided into biotite-muscovite gneiss, schist and hornfels, in general, from lower to upper, but the regular arrangement is not clear due to intrusions of migmatite and granite. The upper metamorphic rocks might have been derived from the sedimentary rocks of the lowest formation of the Hidaka Group (Nakanogawa Group).

The migmatites have been classified into biotite-migmatite, cordierite-migmatite and granitic migmatite (KIZAKI, 1964). Although the genesis of the migmatites has not sufficiently studied, we suppose that there are two types of migmatite: so-called injection gneiss and anatectic granite (S-type granite by WHITE and CHAPPEL, 1977) derived from lower crustal materials.

3.2.3 Metamorphic zoning

The Main Zone can be divided into five metamorphic zones according to the changes of mineral assemblage of pelitic rocks. The upper sequence comprises the zones I and II, and the lower sequence, the zones II, III, IV, and V. The critical mineral assemblages of each zone are as follows;

I: Chl + Mus
II: Mus + Bi
III: Bi + Sill + Gar + Kfs

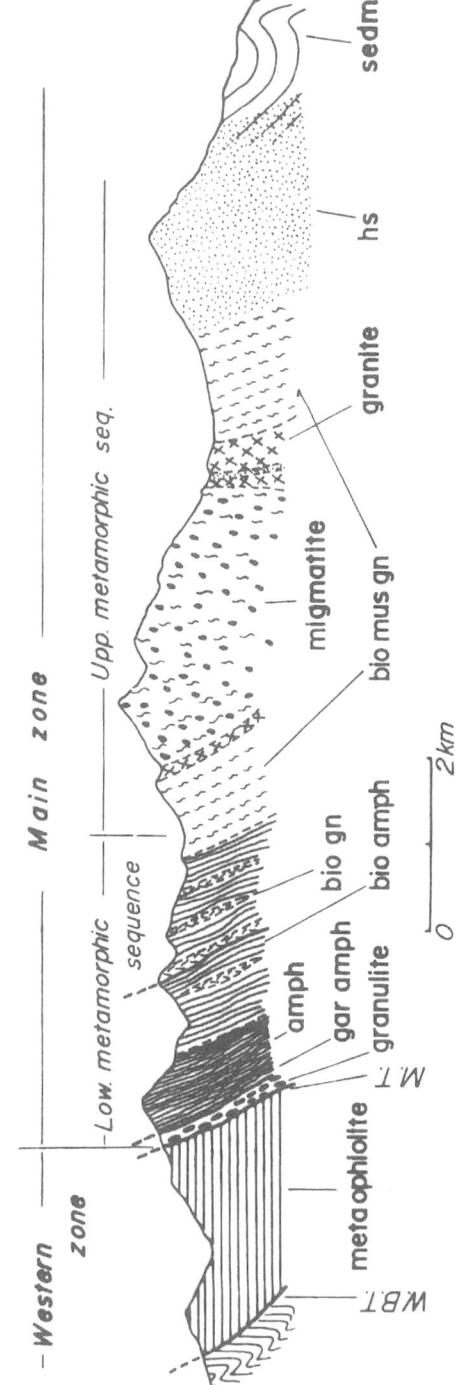

Fig. 6. Cross section of the Hidaka metamorphic belt in the Koibokushu-shibichari-gawa to Satsunai-gawa area. W.B.T.: western boundary thrust fault, M.T.: Hidaka main thrust fault.

$$
\begin{array}{ll}
& \text{Bi + Sill + Cord + Kfs} \\
\text{IV:} & \text{Gar + Sill + Cord + Kfs} \\
& \text{Gar + Bi + Opx + Kfs} \\
& \text{Gar + Cord + Bi + Kfs} \\
\text{V:} & \text{Gar + Sill + Cord + Kfs} \\
& \text{Gar + Opx + Cord + Kfs} \\
& \text{(each includes quartz and plagioclase)}
\end{array}
$$

The mineral assemblages of basic metamorphic rocks change as follows;

$$
\begin{array}{ll}
\text{II, III:} & \text{Hb + Bio + Pl} \\
\text{III:} & \text{Hb + Pl} \\
\text{IV, V:} & \text{Hb + Pl, Hb + 2px + Pl, Opx + Pl}
\end{array}
$$

The change of chemical composition of minerals of pelitic metamorphic rocks from zone II to V is shown in the A'FM diagram (Fig. 7). The metamorphic grade in the Main Zone increases progressively from the greenschist facies of the upper (east) part to the granulite facies of the lower (west).

P-T conditions calculated by means of the garnet-biotite geothermometer (THOMPSON, 1976), garnet-cordierite geothermometer and barometer (HENSEN and GREEN, 1973) and garnet-orthopyroxene geobarometer (WOOD, 1974) for these mineral pairs from the zones III, IV, and V range 5.5 to 6.0 Kb, 620° to 650°C in the zone III, 6.0 to 6.5 Kb, 680° to 750°C in the zone IV, and 6.5 to 7.0 Kb, 750° to 800° in the zone V, respectively. The results are consistent with a change of metamorphic grades inferred from the mineralogical changes.

3.3 Igneous activities

The igneous rocks in the Main Zone are largely composed of olivine gabbro and gabbro (olivine gabbro group), hornblende-pyroxene gabbro and diorite (diorite group) and biotite granite and migmatite (granite group). Besides these, small bodies of alkalic rocks occur (MAEDA et al., 1982). The isotopic ages of the igneous rocks in the Main Zone range from 43 to 17 Ma (KAWANO and UEDA, 1967; SHIBATA and ISHIHARA, 1979, 1981).

The Pankenushi gabbro intrusion (MAEDA, 1981) is typical of the olivine gabbro group. It consists of troctolite, olivine gabbro and ferrogabbro. The estimated liquid line of the pankenushi gabbros plotted in the AFM diagram is nearly parallel to the Skaergaard liquid trend (Fig. 8). The Memurodake plutonic complex and Oshirabetsu mass are representative of the diorite group (MAEDA, 1981; MAEDA et al., 1982). Main constituent minerals in the dioritic complexes are plagioclase, hornblende, clinopyroxene, orthopyroxene, and biotite. A small amount of quartz, alkali feldspar, sphene, ilmenite, and sulphide are also included. The bulk compositions of the diorite group rocks are shown in the AFM diagram of Fig. 8. The plots show the trend of the calc-alkalic rock series, as there is no sign of iron enrichment. SiO_2 contents range from about 50 to 65%. Granitic rocks are also calc-alkalic in nature.

Conclusively, the igneous rocks in the Main Zone is characterized by the association of calc-alkalic and tholeiitic series with a small amount of alkalic series. All these rocks belong to the ilmenite series (ISHIHARA: 1977; MAEDA et al., 1982).

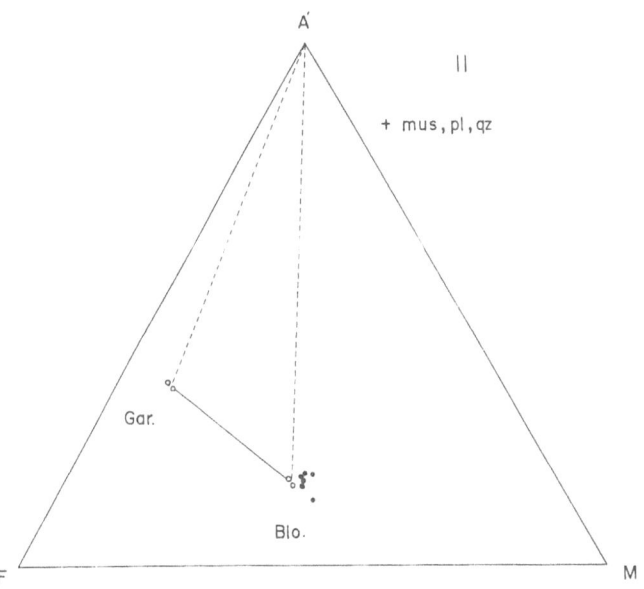

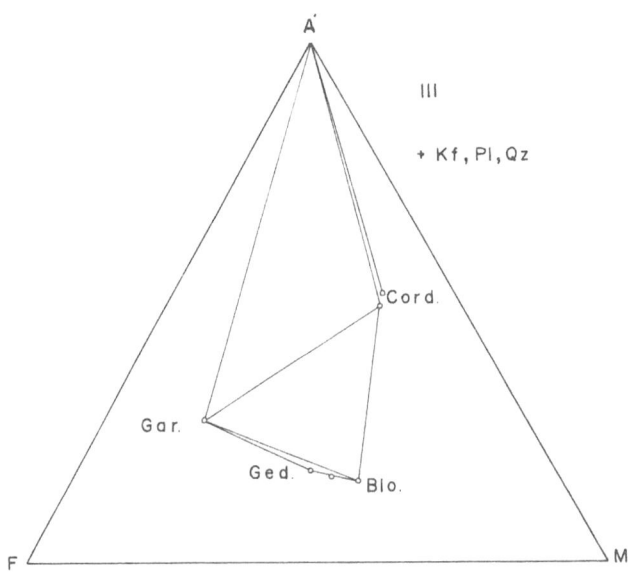

Fig. 7. A'FM diagrams of pelitic metamorphic rocks showing changes of mineral assemblages and mineral

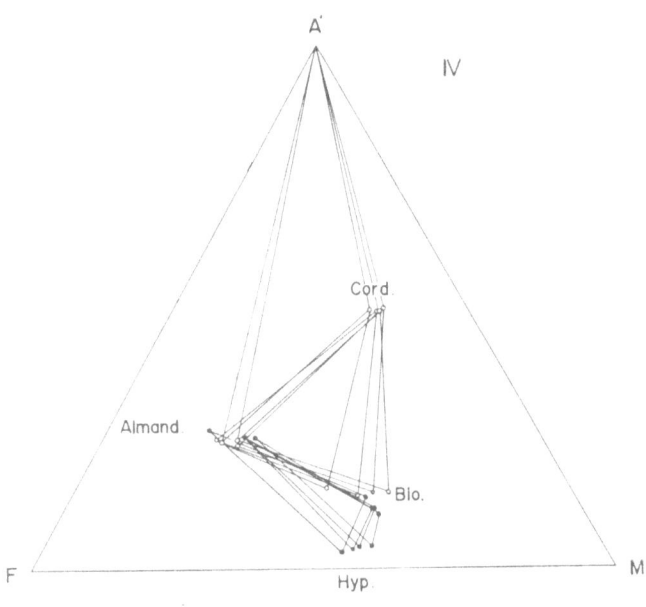

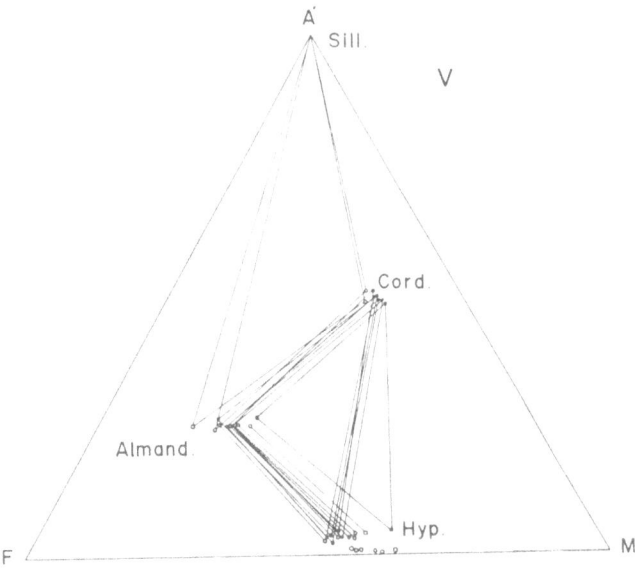

of the Main Zone in the Hidaka metamorphic belt,
compositions from zone II to V.

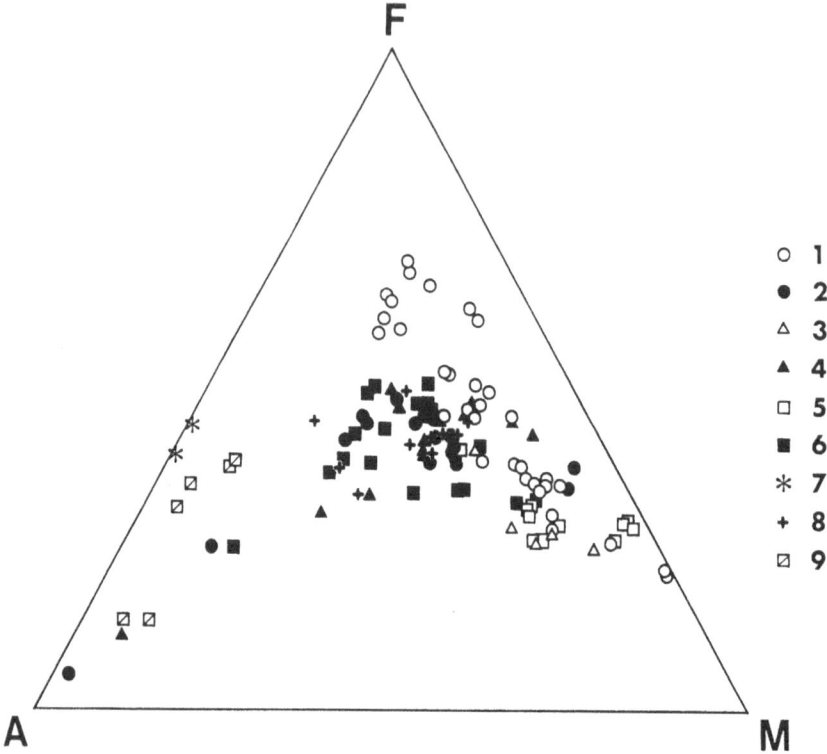

Fig. 8.   AFM diagram of igneous rocks in the Hidaka metamorphic belt. 1: Pankenushi intrusion, 2:
Memurodake complex, 3: olivine gabbro group in the Horoman complex, 4: diorite group in the Horoman
complex, 5: olivine gabbro group in the Oshirabetsu mass, 6: diorite group in the Oshirabetsu mass, 7:
Pipairo quartz monzonite, 8: Meguro dolerite, 9: granitic rocks.

4.   Evolution of the Hidaka metamorphic belt

The rock sequence of the Main Zone represents the profile of the continental or
island arc type crust with small flakes of upper mantle lherzolite (Fig. 9). This crust
block thrusts on the overturned ophiolite (Fig. 6). The thrustup movement of the crust
would have taken place in Middle to Late Miocene time and reached the climax in
Pliocene, being inferred from the amounts of igneous and metamorphic rock pebbles in
conglomerates of these ages distributed around the Hidaka mountains (MIYASAKA and
KIKUCHI, 1978). Being accompanied with the thrust movement, a thick mylonite zone
has developed along the thrust fault.

The radiometric ages of biotite-muscovite gneiss and migmatite indicate that the
metamorphism of the Hidaka belt took place during Paleogene to Early Miocene
(KAWANO and UEDA, 1967; SHIBATA and ISHIHARA, 1979). In the same period, igneous
activity occurred (SHIBATA and ISHIHARA, 1979). It dose not mean that the metamorphism was caused by igneous intrusives, because the metamorphic grade is not
connected with the intrusives but varies in regional scale. The metamorphic grade

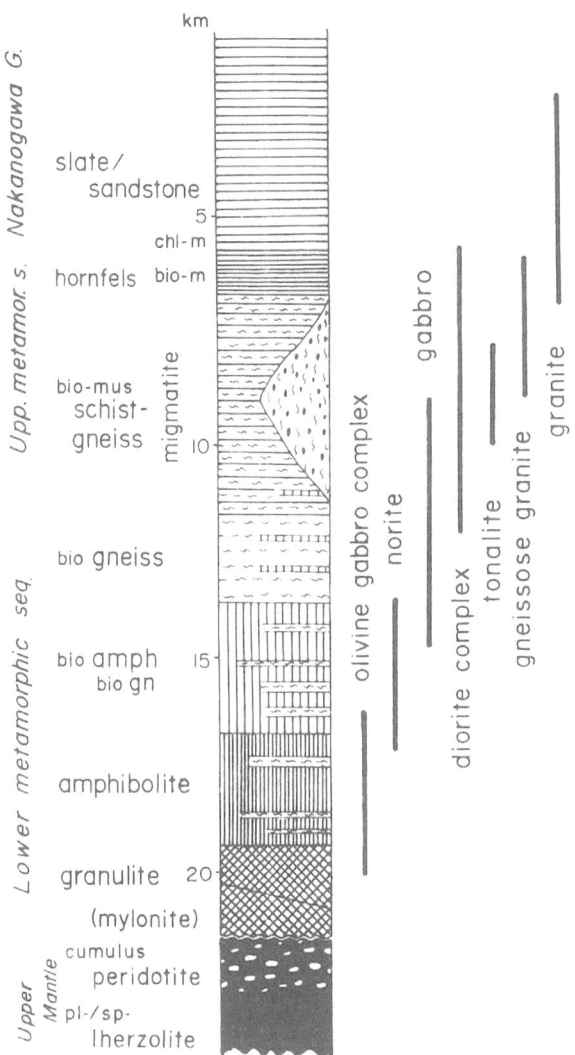

Fig. 9.   Generalized columner section of the island arc-type crust of the Hidaka belt. Vertical bars show the ranges in horizons of igneous rock intrusions.

decreasing from the lower to upper horizons can be ascribed to the elevation of geothermel gradient during the Tertiary period. At the same time it caused the generation of tholeiitic and calc-alkalic magmas in the upper mantle. Judging from the high geothermal gradient and nature of igneous rocks, the tectonic environment where these events occurred is considered to have been an island arc, although its corresponding subduction zone of Tertiary age can not be specified in Hokkaido.

Before the metamorphism and igneous activity took place, the Hidaka belt was the zone of sedimentary rocks which had been presumably deposited in the fore arc basin of the Triassic to Jurassic arc trench system (Kiminami and Kontani, 1982). The lower part of the sedimentary sequence and its basement formations were suffered, as described in the proceeding section, the metamorphism of the Tertiary event and constitute respectively the upper and lower metamorphic sequences of the Hidaka up-thrust crust. The original nature of the basement rocks, however, is not known. Two possibilities we have: (1) it was the crystalline basement of an island arc in continental margin of Triassic to Jurassic, or (2) it was composed of oceanic complex accreted in an early stage of the development of Triassic to Jurassic arc trench system.

We would like to express our appreciation to Drs. K. Arita, Y. Kontani, G. Kimura, K. Niida, T. Watanabe, K. Kiminami, T. Uemura, T. Uda, and M. Toriumi and Messrs. Y. Motoyoshi, and T. Takahashi for discussions.

REFERENCES

Bikerman, M., M. Minato, and M. Hunahashi, K-Ar age of the garnet amphibolite of the Mitsuishi district, Hidaka province, Hokkaido, Japan, *Earth Sci.*, **25**, 27–30, 1971.

Geological Society of Japan, Geological map of the Japan and Kuril trenches and the adjacent areas, *Marine Geology Map Ser.*, **11**, 1978.

Geological Society of Japan, Geological map of the Japan and Okhotsk Seas around Hokkaido, *Marine Geology Map Ser.*, **14**, 1979.

Hashimoto, S., The basic plutonic rocks of the Hidaka metamorphic belt, Hokkaido. Part 1, *J. Fac. Sci. Hokkaido Univ.*, Ser. IV, **16**, 367–420, 1975.

Hensen, B. J. and D. H. Green, Experimental study of the stability of cordierite and garnet in pelitic compositions at high pressures and temperatures, III. Synthesis of experimental data and geological applications, *Contrib. Mineral. Petrol.*, **38**, 151–166, 1973.

Hunahashi, M., Alpine orogenic movement in Hokkaido, Japan, *J. Fac. Sci. Hokkaido Univ.*, Ser. IV, **9**, 415–469, 1957.

Hunahashi, M., and S. Hashimoto, Geology of the Hidaka zone, Hokkaido, *Assoc. Geol. Collab. Japan Monogr.*, **6**, 1–38, 1951.*

Ishihara, S., The magnetite-series and ilmenite-series granitic rocks, *Mining Geol.*, **27**, 293–305, 1977.

Ishizuka, H., T. Hirajima, M. Imaizumi, and S. Banno, Guidebook for field trip to Horakanai area in *Oji International Seminar on Accretion Tectonics, Tomakomai, 1981*, pp. 1–28, 1981.

Imaizumi, M., and Y. Ueda, On the K-Ar ages of the rocks of two kinds existed in the Kamuikotan metamorphic rocks in the Horokanai district, Japan, *J. Jpn. Assoc. Mineral. Petrol. Econ. Geol.*, **76**, 88–92, 1981.*

Kawano, Y., and Y. Ueda, K-Ar dating on the igneous rocks in Japan (VI). Granitic rocks, summary, *J. Jpn. Assoc. Mineral. Petrol. Econ. Geol.*, **57**, 177–187, 1967.

Kiminami, K., Sedimentalogy of the Nemuro Group (part 1), *J. Geol. Soc. Japan*, **81**, 215–232, 1975.

Kiminami, K., and Y. Kontani, Mesozoic arc-trench systems in Hokkaido, Japan, this Volume, pp. 107–122, 1982.

Kimura, G., Tectonic evolution and stress field in the southwest margin of the Kuril arc, *J. Geol. Soc. Japan*, **87**, 757–768, 1981.*

Kimura, G., S. Miyashita, and S. Miyasaka, Collision tectonics in Hokkaido and Sakhalin after Paleogene, this Volume, pp. 123–134, 1982.

Kizaki, K., On migmatites of the Hidaka metamorphic belt, *J. Fac. Sci. Hokkaido Univ.*, Ser. IV, **12**, 111–169, 1964.

Kizaki, K., and D. Hayashi, Migmatite tectonics of the Hidaka metamorphic belt, Hokkaido, Japan, *Tectonophys.*, **6**, 203–220, 1979.

Komatsu, M., S. Miyashita, J. Maeda, Y. Osanai, T. Toyoshima, Y. Motoyoshi, and K. Arita,

Petrological constitution of the continental type crust upthrust in the Hidaka belt, Hokkaido, *J. Jpn. Assoc. Mineral. Petrol. Econ. Geol., Spec. Paper*, No. **3**, 220–230, 1982b.*

KOMATSU, M., G. KIMURA, and K. KIMINAMI, Tectonics of Hokkaido, with special reference to the Hidaka metamorphic belt, in *Geotectonics of the Paired Metamorphic Belts in Japan*, edited by I. Hara, 1982a.

KONTANI, Y., K. KIMINAMI, and A. SAKAI, Hidaka belt as a fore arc basin, in *Geotectonics of the Paired Metamorphic Belts in Japan*, edited by I. Hara, 1982.

MAEDA, J., Petrology of the Pankenushi gabbroic intrusion, Hidaka metamorphic belt, Hokkaido, Dr. thesis, Fac. Sci. Hokkaido Univ., pp. 1–114, 1981.

MAEDA, J., Y. MOTOYOSHI, and T. TAKAHASHI, Magmatism in the Main Zone of the Hidaka metamorphic belt, Hokkaido, in *Geotectonics of the Paired Metamorphic Belts in Japan*, edited by I. Hara, 1982.

MINATO, M., Submarine volcanic rocks of the older age (Pre-Aptian) in Hokkaido, *Assoc. Geol. Collab. Japan Monogr.*, **21**, 193–197, 1978.*

MINATO, M., M. GORAI, and M. HUNAHASHI, eds., *The Geologic Developement of the Japanese Islands*, Tsukiji-shokan, Tokyo, 1965.

MITANI, K., Changing of the Tertiary sedimentary basins in the western flank of the axial belt of Hokkaido, *Assoc. Geol. Collab. Japan Monogr.*, **21**, 127–137, 1978.*

MIYASAKA, S. and K. KIKUCHI, The Neogene Tertiary upheaval of the Hidaka Metamorphic Belt, *Assoc. Geol. Collab. Japan Monogr.*, **12**, 139–153, 1978.*

MIYASHIRO, A., Evolution of the metamorphic belts, *J. Petrol.*, **2**, 277–311, 1961.

MIYASHIRO, A., The Troodos ophiolitic complex was probably formed in an island arc, *Earth Planet. Sci. Lett.*, **19**, 218–224, 1973.

MIYASHITA, S., Ophiolite succession and metamorphism of the Western Zone of the Hidaka metamorphic belt, Hokkaido, in *Geotectonics of the Paired Metamorphic Belts in Japan*, edited by I. Hara, 1982a.

MIYASHITA, S., Geology of metamorphosed ophiolite in the Western Zone of the Hidaka metamorphic belt, Hokkaido, I; Reconstruction of the ophiolite succession, *J. Geol. Soc. Japan*, in press, 1982b.*

MIYASHITA, S., and J. MAEDA, The basic plutonic and metamorphic rocks from the northern Hidaka metamorphic belt, Hokkaido, *Assoc. Geol. Collab. Japan, Monogr.*, **21**, 139–153, 1978.*

MOTOYOSHI, Y., Fe-Ti oxide minerals in the Horoman Plutonic complex of the Hidaka metamorphic belt, *J. Fac. Sci. Hokkaido Univ., Ser. IV.* **20**, 87–94, 1981.

OKADA, H., The Geology of Hokkaido and its plate tectonics, *Earth monthly*, **1**, 869–877, 1979.*

OSANAI, Y., T. TOYOSHIMA, and M. KOMATSU, Constitution of the Hidaka metamorphic belt; its metamorphism and structure, in *Geotectonics of the Paired Metamorphic Belt in Japan*, edited by I. Hara, 1982.

SHIBATA, K., and S. ISHIHARA, Rb-Sr whole rock and K-Ar mineral ages of granitic rocks in Japan, *Geochem. J.*, **13**, 113–119, 1979.

SHIBATA, K., and S. ISHIHARA, K-Ar ages of granitic rocks in the Hidaka belt, Hokkaido, in *86th Ann. Meet. Geol. Soc. Japan, Abstr.*, p. 342, 1981.*

THOMPSON, A. B., Mineral reactions in pelitic rocks. II. Calculation of some P-T-X (Fe-Mg) phase relations, *Am. J. Sci.*, **276**, 425–456, 1976.

WHITE, A. J. R. and B. W. CHAPPEL, Ultrametamorphism and granitoid genesis, *Tectonophys.*, **43**, 7–22, 1977.

WOOD, B. J., The stability of alumina in orthopyroxene coexisting with garnet, *Contrib. Mineral. Petrol.*, **46**, 1–15, 1974.

*In Japanese with Englich abstract.

Southwest Japan

Accretion Tectonics in the Circum-Pacific Regions, edited by M. Hashimoto and S. Uyeda, 169–178.
Copyright © 1983 by Terra Scientific Publishing Company (TERRAPUB), Tokyo.

Hida and Mino: Tectonostratigraphic Terranes in Central Japan

Shinjiro MIZUTANI* and Isamu HATTORI**

*Department of Earth Sciences, Nagoya University, Nagoya 464, Japan
**Geological Laboratory, Fukui University, Fukui 910, Japan

Litho- and biostratigraphic analyses have shown that the sedimentary complex of the Mino area includes disorganized Permian, Triassic, and Jurassic sedimentary and volcanic rocks of an oceanic affinity. The paleomagnetism of these rocks has revealed their long-distance northward drift after the deposition in the equatorial or sub-equatorial regions. On the contrary, the Hida area, underlain by granitic metamorphic complex of Precambrian to Paleozoic age, is characterized in the Mesozoic history by granitic plutonism and by subsequent sedimentation of coarse clastic rocks. This evolution differs profoundly from that of the Mino area, suggesting that the Mino and the Hida areas were two independent terranes during Mesozoic time. Paleomagnetic, paleontologic, and lithologic contrasts provide the most compelling evidence in favour of the hypothesis of allochthonous origin of the Mino terrane. In late Cretaceous time, extensive outpouring of acidic volcanic rocks took place overlapping both terranes, indicating that they were amalgamated in this time.

1. Introduction

Recent biostratigraphic works on the sedimentary complex of the Mino area, central Japan, have disclosed that radiolaria-bearing Jurassic siliceous sediments are dominantly developed in this area (MIZUTANI et al., 1981a), and paleomagnetic investigations (Hattori, 1982) unravelled that the Jurassic siliceous sediments are magnetized in the directions nearly parallel to bedding planes. To the north of the Mino area, thick piles of Jurassic to Cretaceous coarse clastic sedimentary rocks deposited on land and under the shallow sea are distributed unconformably covering the Hida metamorphic complex. This outstanding contrast in lithology between the coeval sedimentary rocks in the Mino and the Hida areas, reinforced by the paleomagnetic data, shows that they were deposited in two quite different geologic environments. This paper is concerned with the tectonic division of central Japan on the basis of new lines of biostratigraphic and paleomagnetic evidence derived from the Mino and the Hida areas, and discusses the Mesozoic history in terms of the accretion tectonics in central Japan.

2. Geology of Central Japan

The Hida metamorphic complex consisting principally of quartzofeldspathic gneiss and crystalline limestone with a subordinate amount of amphibolite is believed to be by far the oldest geologic unit in Japan. A Rb-Sr whole rock age of about 600 Ma has been determined for gray granite intruded into the central part of this complex (Shibata and Nozawa, 1980). Younger bimodal isotopic ages centering on 240 Ma and 180 Ma, which have been claimed in many reports, support the geologic interpretation that the Precambrian Hida gneiss underwent repeated metamorphism.

Recently, Hiroi (1981) has subdivided the Hida metamorphic complex into the Hida gneiss region and the Unazuki zone, chiefly on the basis of his discovery of fossils of late Carboniferous age in bedded limestones intercalated within the Unazuki schist. This zone can apparently be traced along the entire length of the southern boundary of the Hida gneiss region. Further outward to the Unazuki zone occurs the Circum-Hida tectonic zone where glaucophane schists and garnet amphibolites are embedded within serpentinites together with non-metamorphosed Ordovician (Igo et al., 1980) to Permian rocks in a complicated fashion.

The Kuruma Group of early to middle Jurassic age and the Tetori Group of late Jurassic to Cretaceous age, both made up of conglomerate, sandstone and shale deposited subaerially and in shallow sea environments, unconformably overlie the rock formations of the Hida gneiss region, the Unazuki zone, and the Circum-Hida tectonic zone. These Mesozoic sediments commonly carry orthoquartzite-cobbles of Precambrian to early Paleozoic age (Shibata, 1979), but their provenances are utterly unknown so far in any part of the present Japanese Islands. Abundant fossil flora and fauna have been reported from the Tetori Group, and they are collectively called the Tetori flora and fauna.

In the Mino area, a thick sedimentary complex consisting of shale, sandstone, chert, greenstone, and limestone with minor intercalation of conglomerate are extensively distributed, all of them having been believed to be mostly of late Paleozoic age on the basis of fusulinid fossils in limestone. During the past decade, however, it has become accepted from the paleontologic evidence of conodont that many chert members are Triassic in age, and recent radiolarian biostratigraphy has disclosed the ubiquitous occurrence of Jurassic formations in this area (Mizutani et al., 1981a). Most of the paleontologically dated Permian and Triassic rocks and Jurassic siliceous shales are now considered to be exotic blocks included in the middle Jurassic and late Jurassic to early Cretaceous clastic facies. Of much importance in reconstructing the paleogeography relevant to the sedimentary complex of the Mino area is the occurrence of the Jurassic Kamiaso conglomerate (Adachi, 1971) containing clasts of Precambrian gneiss and orthoquartzite (Shibata et al., 1971; Shibata and Adachi, 1974). The source of these clasts has not been identified in the Japanese Islands.

In late Cretaceous time, the Nohi rhyolite unconformably covered all the geologic units in the Hida and the Mino areas (Fig. 1), with which the Mesozoic evolution terminated in central Japan.

3. Mesozoic Terranes in Central Japan

3.1 Stratigraphy

Radiolarian fossils extracted from siliceous shales in the Mino area, are divisible into many assemblages of different ages ranging from the early Jurassic to the late

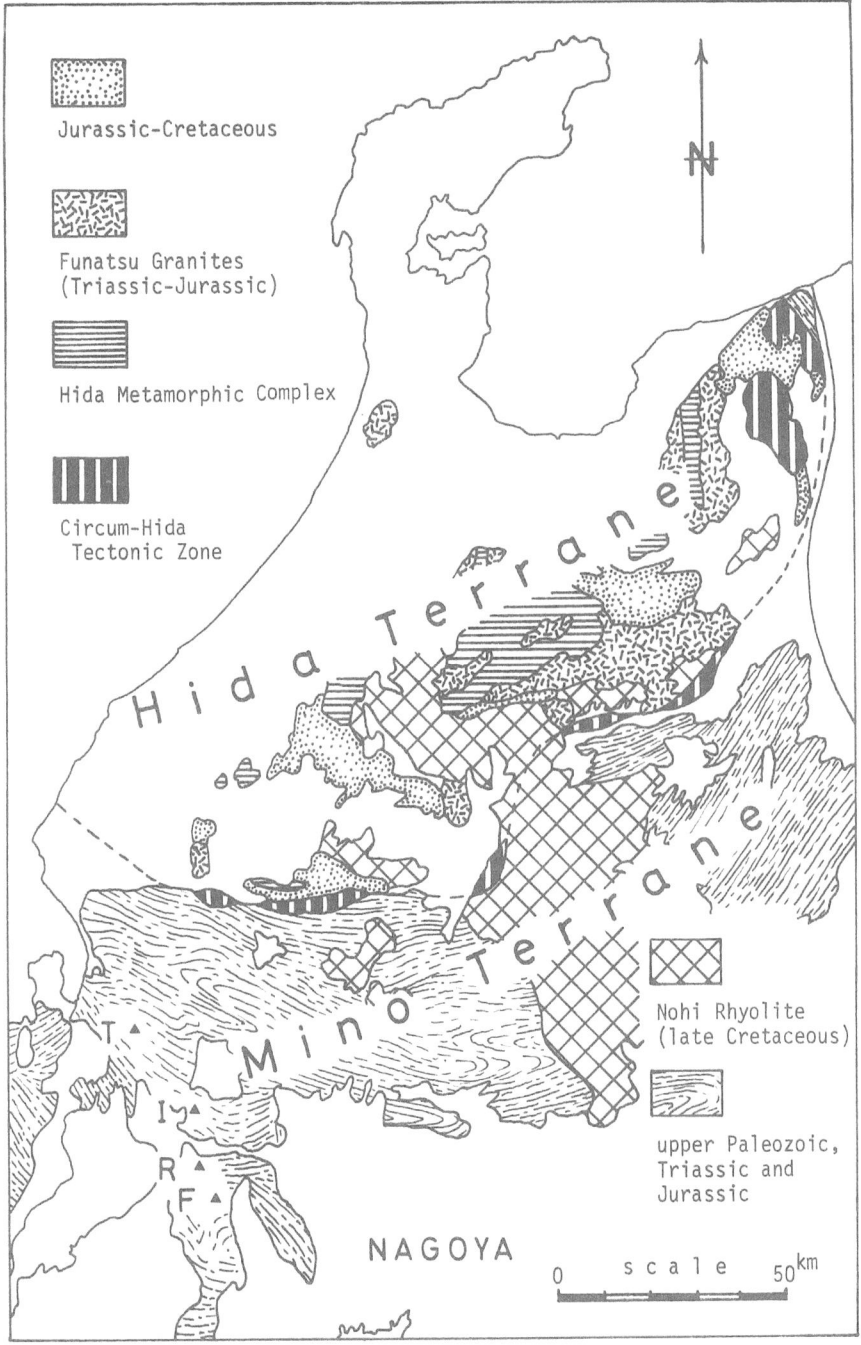

Fig. 1.   Geologic sketch map of central Japan. F(Fujiwara)-R(Ryozen)-I(Ibuki): Permian limestone-greenstone complex. T(Tsuchikura): bedded cupriferous pyrite ore deposits.

Jurassic (Yao et al., 1980; Mizutani et al., 1981a). Each member of the siliceous shales is usually several tens of meters thick and is intercalated sometimes together with underlying Triassic cherts within a coarse clastic formation, taking the form of an allochthonous sheet of olistolith origin. Among the paleontologically dated siliceous shales, the youngest one is Tithonian in age (Mizutani, 1981). Since this is also found to be interbedded apparently in black shale of an olistostromal nature, it is very likely that extensive post-depositional mixing occurred after the Tithonian. This sedimentary complex, generally dipping steeply, is sharply differentiated from the Circum-Hida tectonic zone to the north. This geologic entity, actually homogeneous in its lithologic diversity, is defined here as the Mino Terrane, but its internal structure is very complicated owing to syn- and post-depositional deformation so that the successive process of formation of this terrane is poorly understood.

The lithofacies of the Mesozoic sedimentary complex in the Mino Terrane is in striking contrast to that of the Hida area despite the close proximity of their present locations as illustrated in Fig. 2. As compared with an oceanic affinity of the sediments in the Mino Terrane, the Jurassic to Cretaceous strata in the Hida area, in which pelagic sediments are absent, are characterized by non-marine deposits with a large amount of conglomerate; no olistolith derived from the Permian or Triassic formation is found. Most of these Jurassic to Cretaceous formations are intersected by faults, and their fold structure is much gentler than that of the Mino Terrane. The geologic entity defined by the Jurassic to Cretaceous groups and their metamorphic basement forms another terrane, referred to collectively as the Hida Terrane.

3.2 Paleomagnetism

In the Mino Terrane, paleomagnetic investigations have been performed especially on bedded sedimentary rocks whose ages are paleontologically determined. Structural correction of the paleomagnetic directions which were obtained after thermal and alternating field demagnetization, gave shallow paleomagnetic inclinations, suggesting the low paleolatitudinal origin of the middle to upper Jurassic siliceous shales (Hattori, 1982). As shown in Fig. 3, even in taking measurement error into account, the paleolatitudes of the Jurassic and Triassic sediments differ widely from that of the Cretaceous granites of Southwest Japan summarized by Yaskawa and Nakajima (1974).

The paleomagnetic data presented by Hirooka et al. (1982) from the late Triassic to early Jurassic Funatsu granites in the Hida Terrane give another evidence of a paleolatitudinal difference between the Hida and the Mino Terranes. Six paleomagnetic results derived from the granites show that they are magnetized normally and that the paleomagnetic inclination cannot be distinguished from the present one, indicating no remarkable motion of the Hida Terrane after the Jurassic.

Paleobiogeographic studies dealing with the Mesozoic land floras in Japan (Kimura, 1980) demonstrate that the Jurassic and Cretaceous plant fossils in the Tetori Group of the Hida Terrane have much affinity with those of the Siberian paleofloristic province (temperate and moderately humid), and are quite different from those of the Outer Zone of Southwest Japan or the Indo-European province (subtropical and arid).

The paleogeographic position of the Hida Terrane contrasts strikingly with the

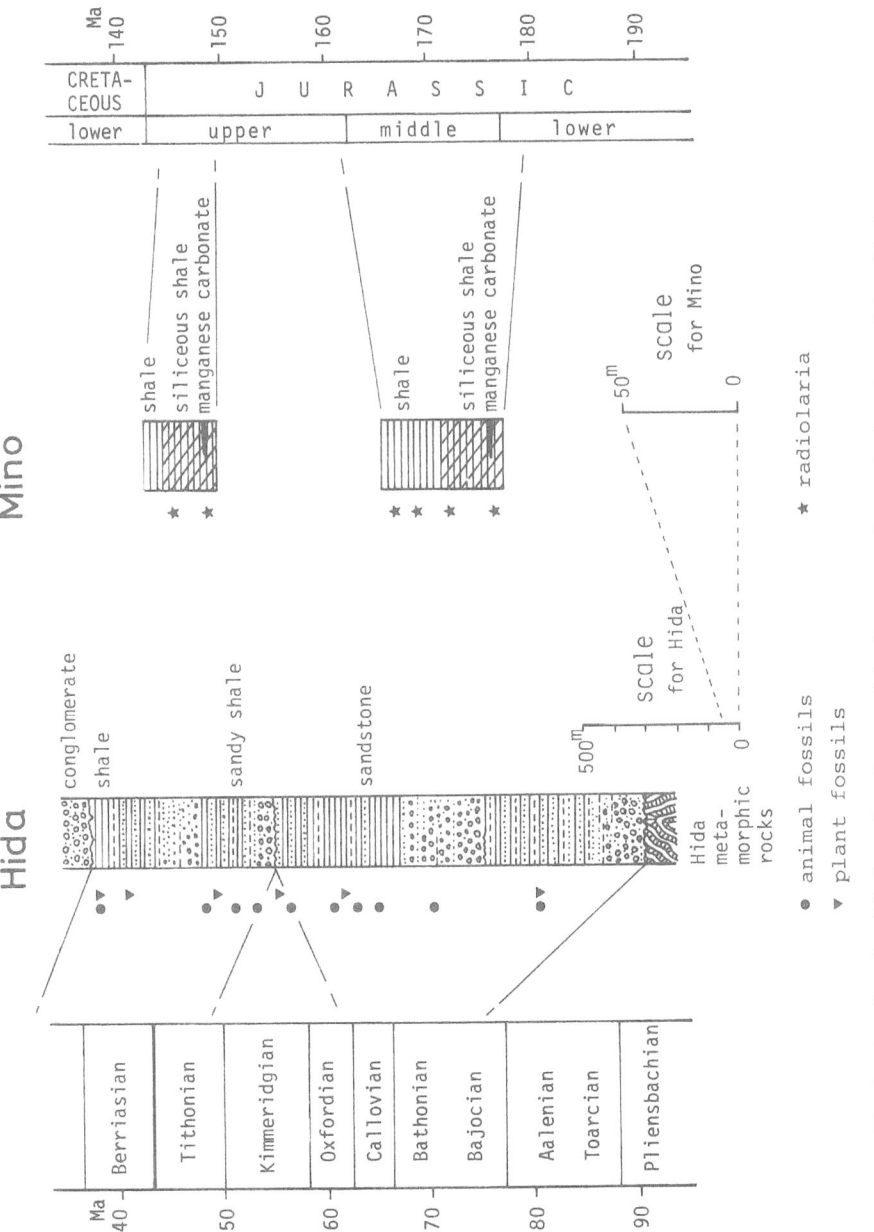

Fig. 2.   Stratigraphy of the Jurassic and a part of the Cretaceous Groups in the Mino and the Hida Terranes after KAWAI (1977), MIZUTANI (1981), and MIZUTANI et al. (1981b).

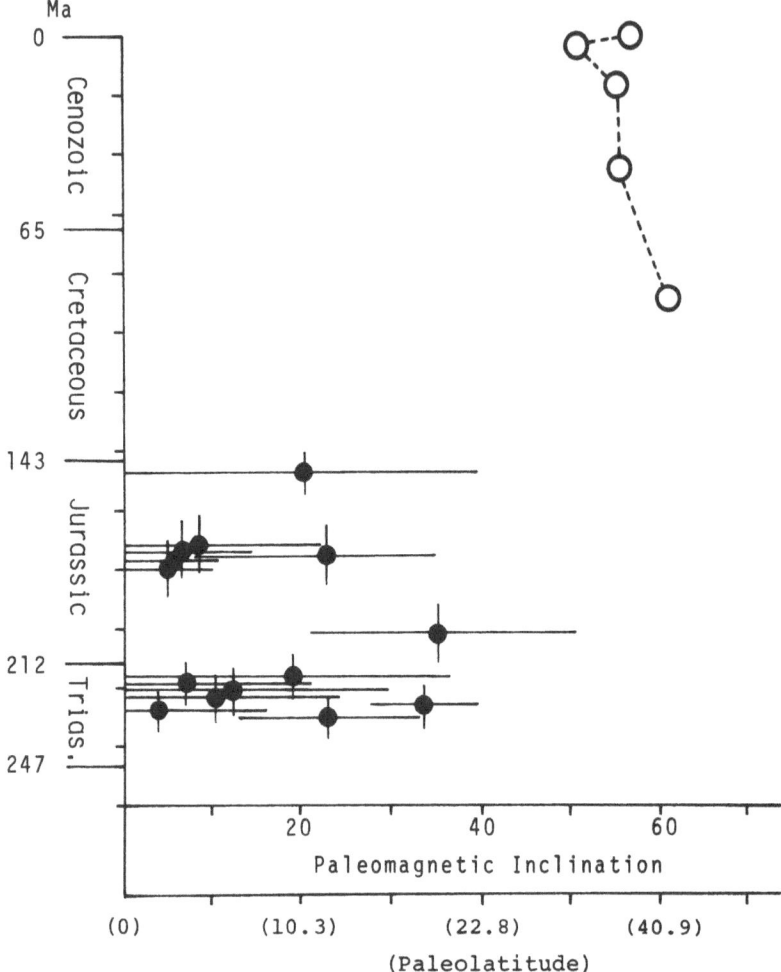

Fig. 3.   Change of paleolatitude for the Mesozoic rocks of central Japan. Open circles after YASKAWA and NAKAJIMA (1974), and closed circles after SHIBUYA and SASAJIMA (1980) and HATTORI (1982).

paleolatitude of the Mino Terrane inferred from the paleomagnetic and paleobiogeographic studies of the coeval sediments.

3.3   Interterrane relationship

All the geologic, paleontologic and paleomagnetic data described above, coupled with the additional paleomagnetic evidence reported by HATTORI and HIROOKA (1979), consistently show that the Jurassic siliceous shale, the Triassic chert and Permian greenstone in the sedimentary complex of the Mino Terrane had undoubtedly been drifted from the equatorial region, while the Hida Terrane did not move significantly during Triassic, Jurassic, and Cretaceous times.

Two models are possible in discussing the interterrane history in respect to the paleolatitudinal difference. A simple model is that the sedimentary complex of the Mino Terrane was entirely formed in the equatorial region and subsequently moved to the north, and collided with the Hida Terrane probably in early to middle Cretaceous time. The alternative is that the Triassic and Jurassic sea floor sediments had been conveyed from the south and accreted to the ancient continental margin where coarse clastics had been accumulating.

More data of paleomagnetism especially on the clastic sediments of the Mino Terrane are needed to account for the intraterrane features of the sedimentary complex of an olistostromal nature including Permian, Triassic, and Jurassic formations. It is obvious, however, that the Mino and the Hida Terranes came to be coalesced and amalgamated through the late Mesozoic time, as shown by the late Cretaceous Nohi rhyolite overlapping both terranes with marked unconformity.

4. Discussion

We have discussed the Mesozoic geology of central Japan, based on the Jurassic stratigraphy and paleomagnetic data, and recognized the Hida and the Mino Terranes, of which the latter drifted and came to the present position from the south. In the Japanese Islands, the similar tectonostratigraphic complexes are found along much of the length of Southwest Japan, where the middle to upper Jurassic siliceous shales participate commonly, Triassic chert formations are frequently intercalated in clastic facies of Jurassic age and the upper Paleozoic limestone-greenstone assemblages are enclosed in places as discussed by MIZUTANI (1981). It is particularly worth noting that red chert probably of Triassic age and Permian greenstone of the Tamba area give the shallow paleomagnetic inclination (KATSURA et al., 1980). All these areas closely resemble each other in all aspects, and it is highly probable that they constitute as a whole one single integrated terrane.

The geologic structure of the sedimentary complex of the Mino areas was studied by MIZUTANI (1964) and YOSHIDA (1972) who pointed out that the sedimentary formations are largely folded with a wavelength of 5–10 km. The fold axes sometimes plunge to the west and the axial traces of the fold structure in the Mino Terrane assume an arcuate form convex to the south or southeast. They turn their trends in a longitudinal direction in its western part, where Scharung or recess structure is recognized. In this part, the Permian limestone-greenstone assemblage (Fujiwara, Ryozen, and Ibuki) and bedded cupriferous pyrite ore deposites (Tsuchikura), very similar to those of Kieslager type, are distributed as shown in Fig. 1. We can speculate that an ancient ridge of a limestone-greenstone assemblage and cupriferous pyrite ore deposits lined almost perpendicularly to the convergent margin had impinged against a landmass at the occasion of accretion.

From the viewpoint of terrane analysis, the discussion can be extended further to the older history of the Hida Terrane, in which the following geologic units have been identified (HIROI, 1981); chronologically, the Hida Gneiss (600 Ma or older), the Circum-Hida Tectonic Zone including amphibolite, meta-gabbro (400 Ma), glaucophane schist (310–360 Ma), and lower to upper Paleozoic sedimentary rocks, the Unazuki Schist (210–250 Ma), and the Funatsu Granites (about 180 Ma). The spaciotemporal relation of these geologic units seems to show successive coalescing or welding of these terranes during late Paleozoic to early Mesozoic time (Fig. 4).

As expressed by HIROI (1981), the extensions of some geologic units in the Hida Terrane may be traceable far to the Korean region, although a fragmentary nature of the older units in the Hida Terrane makes it difficult to correlate accurately them with the other ones. In his discussion on the Unazuki Schist, Hiroi regarded this schist zone assignable to the Okcheon zone of South Korea. According to his correlation, the Okcheon zone includes a Precambrian massif, which is missing in the present position of the Hida Terrane. In spite of the result of a paleocurrent analysis that coarse clastic materials in the main part of the Mino area were derived from the present site of the Hida metamorphic complex (ADACHI and MIZUTANI, 1971), the source of orthoquartzite- and gneiss-clasts in the Jurassic Kamiaso conglomerate has not been found in that region so far. Probably, the Precambrian massif which had existed to the south of the ancient Hida area underwent intensive deformation and extensive denudation, and hence the record is recognized only as the Precambrian clasts in the

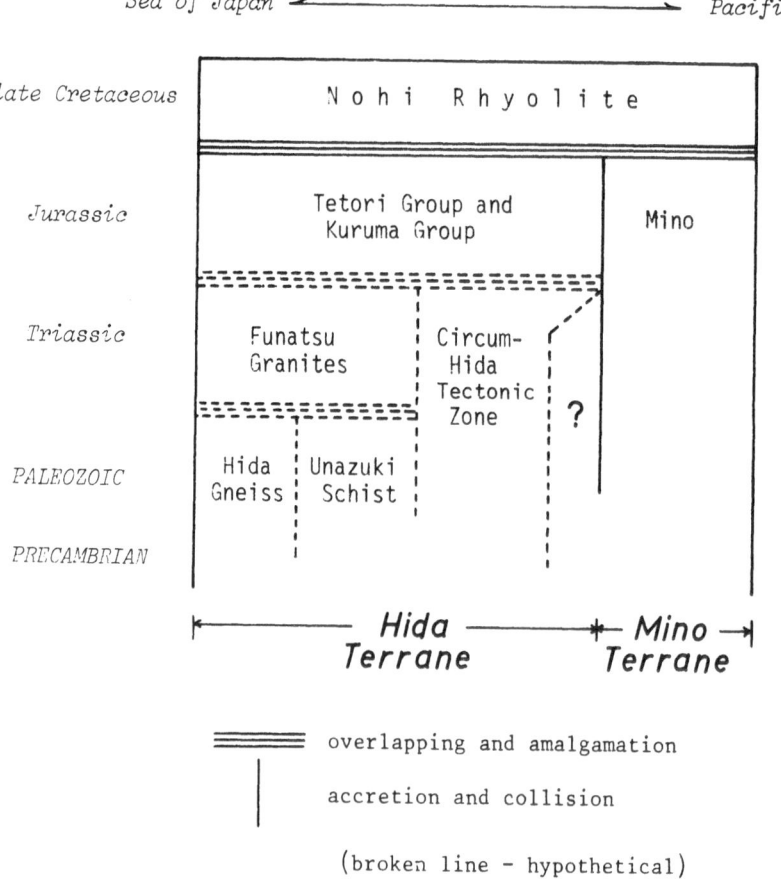

Fig. 4. Terranes in central Japan.

Kamiaso conglomerate. Alternatively, the Precambrian massif was disrupted and slivered along a strike-slip fault having lain on the southern border of an ancient landmass presumably equivalent to the present Hida Terrane.

As already pointed out by KOBAYASHI (1952), the eastern part of the Asian continent is composed of heterogeneous aggregate of fragments of continental crust and of orogenic belts which he named "Heterogen." Recent works by CONEY et al. (1980), BEN-AVRAHAM et al. (1981), and MCELHINNY et al. (1981) suggest that the Asian heterogen is essentially a collage of tectonostratigraphic terranes or accreted continental blocks, most of which were originally situated in lower paleolatitudinal regions occupying the Tethyan Ocean during late Paleozoic time. Systematic correlation work or reconstruction of several separated terranes of the older ages in Far East will be necessary for our better understanding of geologic history of East Asia.

We are much indebted to Dr. K. Hirooka of Toyama University and Dr. D. L. Jones of U.S.G.S., Menlo Park, for their helpful suggestions and discussions.

REFERENCES

ADACHI, M., Permian intraformational conglomerate at Kamiaso, Gifu Prefecture, central Japan, *J. Geol. Soc. Japan*, **77**, 471–482, 1971.

ADACHI, M. and S. MIZUTANI, Sole markings and paleocurrent system in the Paleozoic group of the Mino Terrain, Central Japan, *Mem. Geol. Soc. Japan*, **6**, 39–48, 1971. (in Japanese with English abstract).

BEN-AVRAHAM, Z., A. NUR, D. JONES, and A. COX, Continental accretion: From oceanic plateaus to allochthonous terranes: Science, **213**, 47–54. 1981.

CONEY, P. J., D. L. JONES, and J. W. H. MONGER, Cordilleran suspect terranes, *Nature*, **288**, 329–333, 1980.

HATTORI, I., The Mesozoic evolution of the Mino terrane, central Japan: A geologic and paleomagnetic synthesis, *Tectonophysics*, **85**, 313–340, 1982.

HATTORI, I. and K. HIROOKA, Paleomagnetic results from Permian greenstones in central Japan and their geologic significance, *Tectonophysics*, **57**, 211–235, 1979.

HIROI, Y., Subdivision of the Hida metamorphic complex, central Japan, and its bearing on the geology of the Far East in pre-Sea of Japan time, *Tectonophysics*, **76**, 317–333, 1981.

HIROOKA, K., T. NAKAJIMA, H. SAKAI, T. DATE, K. NITTAMACHI, and I. HATTORI, Accretion tectonics inferred from paleomagnetic measurements on Paleozoic and Mesozoic rocks in central Japan, this volume, pp. 179–194, 1982.

IGO, H., S. ADACHI, H. FURUTANI, and H. NISHIYAMA, Ordovician fossils first discovered in Japan, *Proc. Japan Academy*, **56**, *Ser. B*, 499–503, 1980.

KATSURA, I., J. NISHIDA, and S. SASAJIMA, Paleomagnetism of Permian red chert in the Tamba Belt, Southwest Japan, *Rock Magn. Paleogeophys.*, **7**, 130–135, 1980.

KAWAI, M., Jurassic System, in *Geology and Mineral Resources of Japan*, edited by K. Tanaka and T. Nozawa, pp. 166–181, Geological Survey of Japan, 1977.

KIMURA, T., The present status of the Mesozoic land floras of Japan, in *Prof. Saburo Kanno Memorial Volume*, pp. 379–413, 1980.

KOBAYASHI, T., Pre-Cambrian history of eastern Asia, *Jpn. J. Geol. Geography*, **22**, 39–54, 1952.

MCELHINNY, M. W., B. J. J. EMBLETON, X. H. MA, and Z. K. ZHANG, Fragmentation of Asia in the Permian, *Nature*, **293**, 212–216, 1981.

MIZUTANI, S., Superficial folding of the Paleozoic system of central Japan, *J. Earth Sci., Nagoya Univ.*, **12**, 17–83, 1964.

MIZUTANI, S., A Jurassic formation in the Hida-Kanayama area, central Japan, *Bull. Mizunami Fossil Museum*, **8**, 147–190, 1981. (in Japanese with English appendix of summary and paleontological description).

MIZUTANI, S., I. HATTORI, M. ADACHI, K. WAKITA, Y. OKAMURA, S. KIDO, I. KAWAGUCHI, and S. KOJIMA, Jurassic formations in the Mino area, central Japan, *Proc. Japan Acad.*, **57**, Ser. B, 194–199, 1981a.

MIZUTANI, S., N. IMOTO, A. YAO, K. ICHIKAWA, K. ISHIDA, K. NAKAZAWA, T. OTSUKA, D. SHIMIZU, and K. SUYARI, Triassic bedded chert and associated rocks in the Inuyama area, central Japan, in *The Second Int. Conf. Siliceous Deposits in the Pacific Region (IGCP Project 115)*, Guides to Excursions, edited by A. Iijima, pp. 156–210, 1981b.

SHIBATA, K., Geochronology of pre-Silurian basement rocks in the Japanese Islands, with special reference to age determinations on orthoquartzite clasts, in *The Basement of the Japanese Islands—Professor Hiroshi Kano Memorial Vol.*, pp. 625–639, 1979.

SHIBATA, K., M. ADACHI, and S. MIZUTANI, Precambrian rocks in Permian conglomerate from central Japan, *J. Geol. Soc. Japan*, **77**, 507–514, 1971.

SHIBATA, K. and M. ADACHI, Rb-Sr whole-rock ages of Precambrian metamorphic rocks in the Kamiaso conglomerate from central Japan, *Earth Planet. Sci. Lett.*, **21**, 277–287, 1974.

SHIBATA, K. and T. NOZAWA, Rb-Sr ages of the Hida metamorphic rocks from Kagasawa, Toyama Prefecture, *J. Jpn. Assoc. Mineral. Petrol. Econ. Geolog.*, **75**, 130, 1980 (in Japanese).

SHIBUYA, H. and S. SASAJIMA, A paleomagnetic study on Triassic-Jurassic system in Inuyama area, central Japan (Part I), *Rock Magn. Paleogeophys.*, **7**, 121–125, 1980.

YAO, A., T. MATSUDA, and Y. ISOZAKI, Triassic and Jurassic radiolarians from the Inuyama area, central Japan, *J. Geosci., Osaka City Univ.*, **23**, 135–154, 1980.

YASKAWA, K. and T. NAKAJIMA, Paleolatitudes of Southwest Japan, deducted from paleomagnetic results, *J. Geol. Soc. Japan*, **80**, 215–224, 1974 (in Japanese with English abstract).

YOSHIDA, S., Configuration of Yamaguchi zone—analytical study of a fold zone, *J. Fac. Sci. Univ. Tokyo, II*, **18**, 371–429, 1972.

Accretion Tectonics in the Circum-Pacific Regions, edited by M. Hashimoto and S. Uyeda, 179–194.
Copyright © 1983 by Terra Scientific Publishing Company (TERRAPUB), Tokyo.

Accretion Tectonics Inferred from Paleomagnetic Measurements of Paleozoic and Mesozoic Rocks in Central Japan

Kimio HIROOKA,* Tadashi NAKAJIMA,** Hideo SAKAI,*

Tetsuhiro DATE,* Kiyoshi NITTAMACHI,* and Isamu HATTORI**

*Department of Earth Sciences, Faculty of Science,
Toyama University, Toyama 930, Japan
**Geological Laboratory, Faculty of Education,
Fukui University, Fukui 910, Japan

Paleomagnetic results obtained for the Mesozoic and the Paleogene granitic rocks in the Hida Belt show a clear contrast with those of the Paleozoic to Mesozoic sediments in the Circum-Hida, the Mino, and the Shimanto Belts in the central part of Honshu Island, Japan. The sedimentary rocks have very shallow paleomagnetic inclinations, whereas the granitic rocks have the inclinations almost identical to that of the present geomagnetic field. The paleolatitudes computed from the inclinations indicate that the sedimentary rocks occurring in the Circum-Hida, the Mino, and the Shimanto Belts commenced in the Mesozoic to move northward after deposition in the equatorial region, and eventually collided with the Hida Belt which had been situated in the middle latitude since Jurassic age.

1. Introduction

Permo-Jurassic greenstones in the Mino Belt, central Japan, show very shallow paleomagnetic inclinations so that it can be considered that they were magnetized in the equatorial region and subsequently migrated toward the north (HATTORI and HIROOKA, 1979), while Cretaceous to Paleogene granitic rocks distributed in the southwest and the northeast part of Honshu Island give the inclination of the middle latitude (KAWAI et al., 1971). Kawai et al. (1961, 1971) pointed out the relative counterclockwise rotation of the northeast Honshu, on the basis of the big difference of declination between rocks from northeast and southwest Honshu. Several more paleomagnetic data of the pre-Neogene rocks in Japan have been reported on the Devonian to Cretaceous in Kitakami, northeast Honshu (MINATO and FUJIWARA, 1965; FUJIWARA, 1966), and on the Cretaceous to Paleogene in Chugoku, southwest Honshu (SASAJIMA et al., 1968). According to those results, Chugoku was in the latitude almost the same as the present latitude while Kitakami was situated in much lower latitude in the time of lower Permian to Triassic. These facts suggest that even the Honshu Island can not be described by a single tectonic history but be devided into several blocks which have suffered the different respective tectonism.

The late Paleozoic to Mesozoic paleomagnetism is, therefore, the most important in analyzing the tectonic history of Japanese Islands. The present authors attempted to make clear the history by the newly obtained results from a systematic paleomagnetic study in the central part of Honshu, where is observed a distinct zonal distribution of the typical geological provinces.

2. Paleomagnetic Sampling

In the central part of the Honshu Island there distribute seven major geological provinces, that is, the Hida, the Circum-Hida, the Mino, the Ryoke, the Sambagawa, the Chichibu, and the Shimanto Belts from the north to the south as shown in Fig. 1. Paleomagnetic samples were collected from granitic rocks of the Hida Belt and sedimentary rocks of the Circum-Hida, the Mino, and the Shimanto Belts.

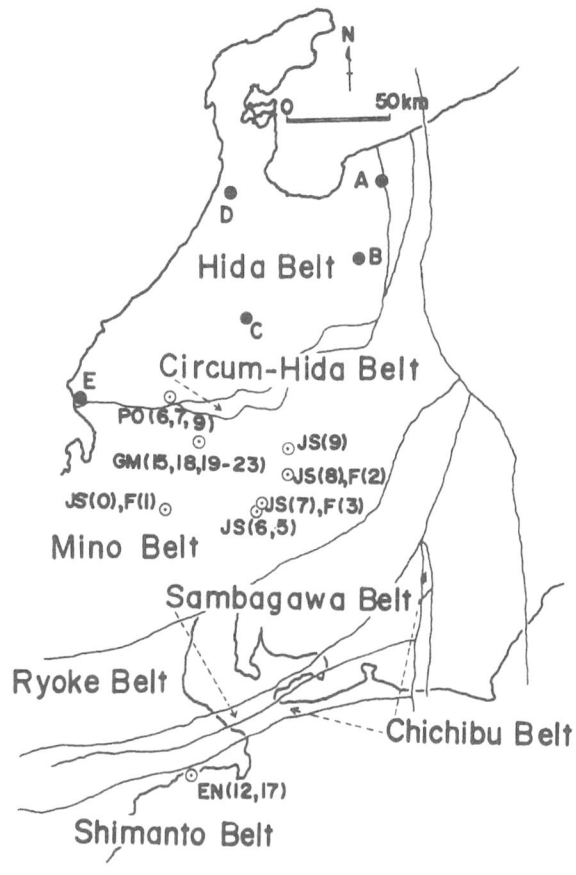

Fig. 1. Map showing the terrains and sampling localities in central Japan. ● ; sampling sites of granitic rocks, ◉ ; sampling sites of sedimentary rocks.

The Hida Belt is mainly composed of gneiss which is intruded by granite and partly covered by volcanics and sediments of Cretaceous, Miocene, and Quaternary ages. The granitic intrusion occurred twice in this area. The older intrusion was in the Jurassic period and the younger in the late Cretaceous to Paleogene time. The Circum-Hida Belt, a narrow tectonic zone, is fringing the Hida Belt. The geology of the main part of the former consists of complex olistostromal fragments of Ordovician to Permian sedimentary rocks. The Mino Belt, which is widely distributed to the south of the Circum-Hida Belt, is composed of Permian to Cretaceous pelagic sedimentary rocks such as chert, siliceous shale, mudstone, and greenstone. The Shimanto Belt is situating in the southernmost part of the Honshu Island and its elongated distribution extends westward to the Kyushu Island. Sedimentary rocks of Cretaceous and early Tertiary age are the main constituent members of this belt. Recently, precise micro-paleontological analyses of radiolarians and conodonts were carried out for the siliceous shales and the cherts in the Mino and Shimanto Belts, and detailed age-determinations have been done at several localities (YAO et al., 1980; YOSHIMURA et al., 1982).

We made paleomagnetic studies on rocks from 9 sites in five plutonic bodies of A, B, C, D, and E as is shown in Fig. 1. B and D are the Jurassic intrusives called the Funatsu Granites. Paleomagnetic samplings were made at two sites (GTY (11) and (12)) in B, and four sites (GK (0), (2), (3), and (5)) in D. The K-Ar ages of B and D are 174 and 163 Ma, respectively (NOZAWA, 1977). The remains are the late Cretaceous to Paleogene intrusives, the radiometric ages ranging from 63 to 88 Ma. Samples were collected at one site for each body.

From sandstone of the Motodo Formation which is disributed in the western part of the Circum-Hida Belt, we obtained paleomagnetic samples. The formation is considered to be the Triassic from the stratigraphical situations (OMURA, 1968). Sixty-two specimens taken from three sites (PO (6), (7), and (9)) were submitted to paleomagnetic measurements.

The lower Permian to upper Jurassic sedimentary rocks were collected from 16 sites (JS (0), (5), (6), (7), (8), (9), F (1), (2), (3), GM (15), (18), (19), (20), (21), (22), and (23)) in the Mino Belt. On the sedimentary rocks of sites GM (15), (18), (19), (20), (21), (22), (23), JS (7), (8), and (9), ages were determined paleontologically by the study of radiolarian fossils. The ages of GM (15), (18), (19), (20), (21), (22), and (23) are early to middle Permian. Site JS (7) belongs to the lower part of the middle Triassic, and JS (8) is of middle Jurassic age. Site JS (9) shows the youngest age of the uppermost Jurassic in the Mino Belt. The siliceous shale of site JS (9) were submitted to radiometric age determination of Rb-Sr method and the age obtained is 128 Ma (SHIBATA and MIZUTANI, 1980). This age indicates the time when Rb-Sr system in the rock was chemically closed. According to Shibata and Mizutani, the age of sedimentation estimated from radiolarian fossil analysis is 20 Ma older than the radiometric age. The Rb-Sr dating was carried out also on the rock of site JS (8), and the result shows 147 Ma (Shibata and Mizutani, personal communication) so that the age of sedimen-tation is assumed to have been 167 Ma in account with 20 Ma delay of closing the Rb-Sr system after the sedimentation. The other sites afford Jurassic sediments.

Paleomagnetic measurements were now on going on the samples from the

Shimanto Belt, and the results were so far obtained from two sites (EN (12) and (17)) of the upper Cretaceous. Radiolarian analyses show that both the sites are of Coniacian to Santonian stages (Mizutani, personal communication).

3. Measurements

Remanent magnetization was measured by means of a spinner magnetometer (Schonstedt SSM-1A). Five to 12 oriented samples were collected at each site and cored into specimens of 2.54 cm in the laboratory. One specimen is presented from each hand sample and was subjected to paleomagnetic measurements. After the measurements of natural remanent magnetization (NRM), all of the specimens were demagnetized stepwise by either of the two series of the alternating fields. One is the steps of 50, 100, 150, 200, 250, 300, 350, and 400 Oe and the other is 73, 145, 220, 293, 365, and 438 Oe. The higher demagnetizing steps were applied in some cases such as sites JS (5), (6), GM (15), (18), (19), (20), (21), (22), and (23). As for the each site, we decided the optimum demagnetizing step at which the smallest scattering of the remanent directions was obtained. The mean declination and inclination of the optimum step were adopted as the paleomagnetic data for the step.

In the cases that NRM has the smallest scattering, we regarded the step which gives the second smallest scattering as the optimum one in spite of NRM. In such a case the NRM direction is considered to be the composite vector of the original remanent and the overprinted soft secondary component.

Figure 2 shows the remanent direction changes caused by the stepwise demagnetization. Site JS (0) is an example of small scattering in magnetic direction (Fig. 2A) and site JS (9) is that of wide scattering (Fig. 2B). Generally, sites giving a small Fisher's angle of confidence in the original natural remanent magnetization do not change their direction significantly by the alternating field (a.f.) demagnetization.

The results of paleomagnetic measurements were tabulated in Table 1. Strike and dip of the bedding plane, geographic coordinates for the sampling sites were also listed in the table. Intensities for the remanent magnetization are in the order of 10^{-6} to 10^{-7} emu/fr for the ordinal sites in this study.

As for the most of the sites, about 30% of original NRM intensity remained after alternating field demagnetization of 400 Oe. The very hard remanent component, however, were observed in the cases of red chert such as samples of sites GM (15) and (23). No significant intensity decrease was detected up to 1,000 Oe in these cases. Examples of intensity decay curves are shown in Fig. 3.

Figure 4 represents Schmidt's stereographic projections of the mean magnetic directions and Fisher's circles of confidence for all the sites of the Hida, the Circum-Hida, the Mino, and the Shimanto Belts respectively. As is seen in B, C, and D of Fig. 4, paleomagnetic vectors show no systematic distribution on the sphere of the projections. Moreover, we can recognize no paleomagnetic vectors which have the similar direction to that of the present geomagnetic field nor that of the Cretaceous field estimated from granitic and volcanic rocks of both the inner side of Southwest Japan (SASAJIMA, 1981) and Northeast Japan (KAWAI et al., 1971; ITO et al., 1980). The sedimentary rocks of the Circum-Hida, the Mino, and the Shimanto Belts suffered

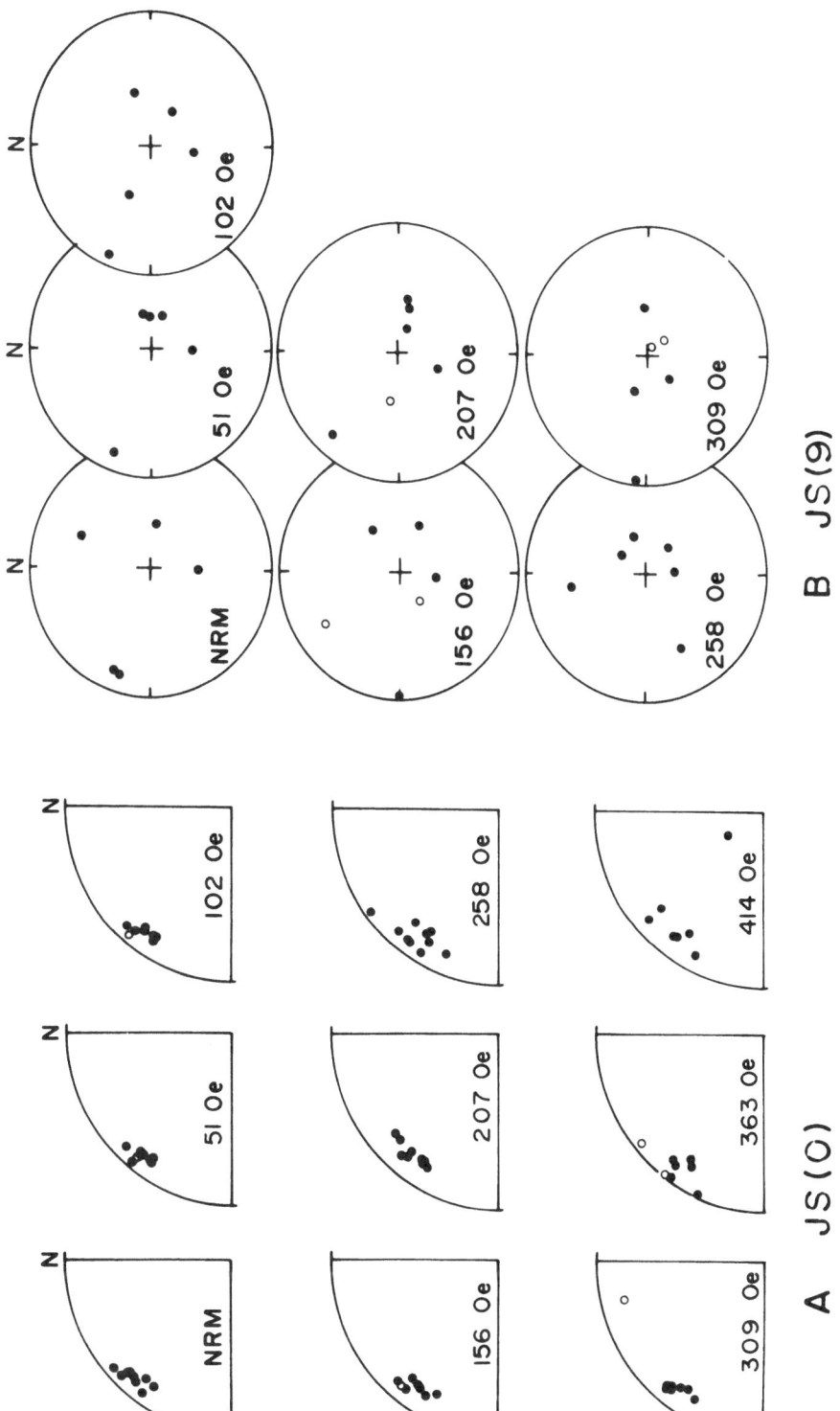

Fig. 2. Examples of directional changes of the remanence of sedimentary rocks through the a.f. demagnetization. Solid and open circles indicate the upper and lower hemisphere, respectively.

Table 1. Results of magnetic measurements, the coordinates of sampling sites and their bedding attitude.

Site	Locality		N	D(°)	I(°)	α_{95}(°)	k	J ($\times 10^{-6}$ emu/gr)	ODF (Oe)	Bedding**	
	Lat(N)	Lon(E)								Strike(°)	(Dip(°))
Hida Belt											
GTY(11)	36°32'	137°31'	9	−2.4	51.8	10.5	25.0	5.29	70		
GTY(12)	36°32'	137°31'	9	−6.4	54.2	6.2	70.2	13.7	141		
GK(0)	35°47'	136°48'	9	−0.8	64.2	11.2	22.0	9.12	50		
GK(2)	35°47'	136°48'	8	−28.6	40.4	18.2	10.3	2.06	150		
GK(3)	35°47'	136°48'	9	−11.8	53.2	17.6	9.5	4.24	100		
GK(5)	35°47'	136°48'	8	6.7	53.4	18.6	9.8	5.79	100		
GTY(8)	36°50'	137°40'	8	−24.3	52.6	37.8	3.1	0.89	100		
GTY(1)	36°14'	136°54'	8	−12.0	38.2	13.5	17.8	1.91	120		
GE(2)	35°57'	136°01'	10	−15.1	40.3	9.5	26.8	9.22	50		
Circum-Hida Belt											
PO(6)	35°51'	136°29'	17	85.4	60.3	7.9	21.3	2.81	150	N100E	72S
PO(7)	35°51'	136°29'	23	111.8	71.6	6.4	23.2	3.25	400	N100E	72S
PO(9)	35°51'	136°29'	22	105.4	52.4	3.5	81.3	6.52	200	N150E	64S

	Lat	Lon	N	D	I	α_{95}	k	J	ODF	Strike	
Mino Belt											
GM(15)	35°41'	136°40'	10	−106.8	−50.7	4.8	104.1	19.2	200	N30W	62N
GM(18)	35°40'	136°43'	6	−120.6	−22.8	19.4	12.8	202	200	N45E	50S
GM(19)	35°38'	136°43'	5	−54.5	71.2	15.0	26.8	6.50	100	N80W	80N
GM(20)	35°38'	136°43'	5	−35.8	62.2	22.8	12.2	117	200	N70W	67N
GM(21)	35°36'	136°45'	10	28.0	56.5	4.5	114.2	12.9	200	N80E	80N
GM(22)	35°41'	136°40'	6	207.4	−56.9	17.2	16.1	2.09	200	N43W	41N
GM(23)	35°41'	136°40'	10	262.5	−53.4	5.2	88.8	12.0	200	N25W	50N
JS(5)	35°24'	136°58'	5	216.0	67.1	24.7	10.6	3.39	220	N80E	90S
JS(7)	35°25'	136°59'	9	18.9	57.0	11.5	21.0	1.02	73	N50E	90N
JS(8)	35°32'	137°07'	10	44.2	75.4	8.2	36.0	2.56	207	N48E	75N
F(2)	35°32'	137°07'	9	51.6	47.9	25.7	5.0	1.0	300	N45E	70N
JS(0)	35°24'	136°28'	10	2.2	64.3	4.3	127.0	0.652	102	N17E	74W
F(1)*	35°24'	136°28'	8	24.3	55.2	15.6	13.6	1.0	300	N12E	75W
JS(6)	35°24'	136°58'	5	36.9	84.7	14.6	28.4	7.84	145	N80E	90S
F(3)*	35°24'	136°58'	9	15.2	86.0	17.2	10.0	4.0	100	N85E	85S
JS(9)	35°40'	137°09'	6	15.5	88.0	34.5	4.7	2.48	258	N70W	90S
Shimanto Belt											
EN(12)	34°16'	136°33'	7	−20.7	63.7	10.2	36.1	0.071	100	N52E	62W
EN(17)	34°16'	136°33'	9	177.7	−7.7	13.4	15.7	1.54	400	N50E	50W

Lat; latitude of sampling location, Lon; longitude of sampling location, N; number of samples, D; mean declination, I; mean inclination, α_{95}; Fisher's circles of confidence of 95 percent, k; Fisher's precision parameter, J; intensity of the remanence after optimum a.f. demagnetization (1 emu = 10^{-3} A/m), ODF; optimum demagnetization field (1 Oe = 0.1 mT).

* ; data are cited from Hattori (1982).

** ; strikes of bedding planes are the values referring to the geomagnetic north which ranges from 6.07° to 7.03°.

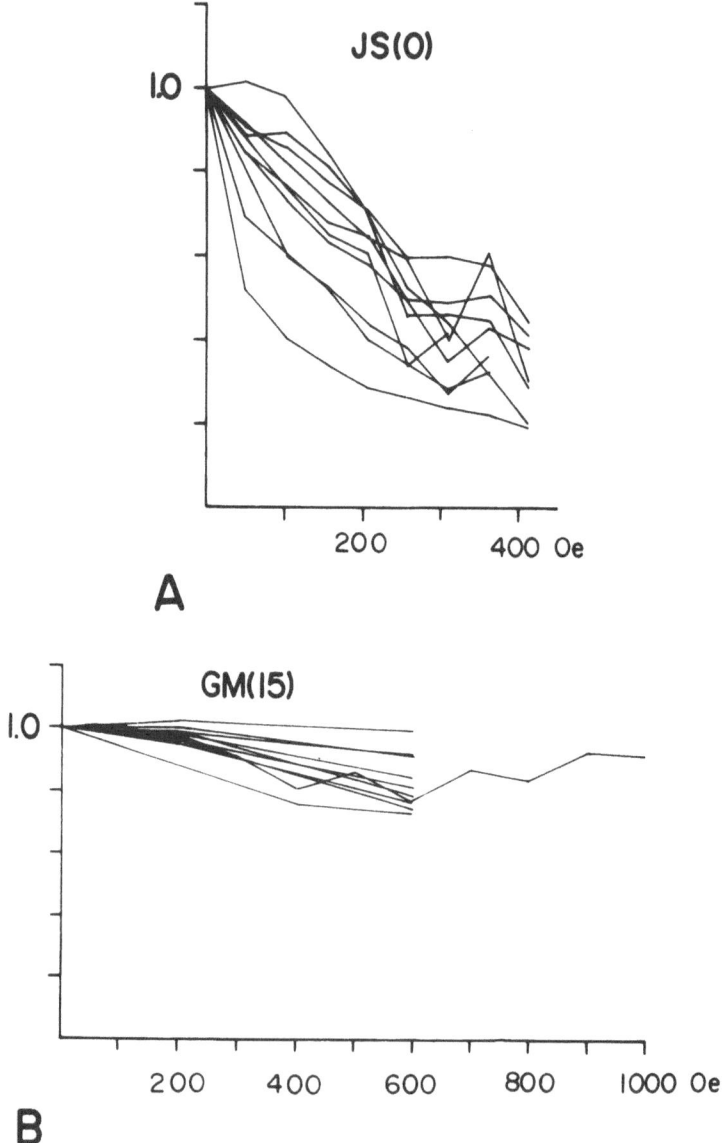

Fig. 3. Examples of normalized intensity decay curves with progressive a.f. demagnetization.

severe tectonism. The folded strata are steeply tilted so that we must restore the tilted bedding plane to the horizontal to know the paleomagnetic field of the time when the sedimentation took place. Paleomagnetic data such as mean declination and inclination, and paleolatitude after bedding correction are summarized in Table 2. Rock types and ages are also presented in the table. After the bedding corrections were made, the remarkably shallow inclinations of remanent direction are noticed for all the sites (Fig. 5).

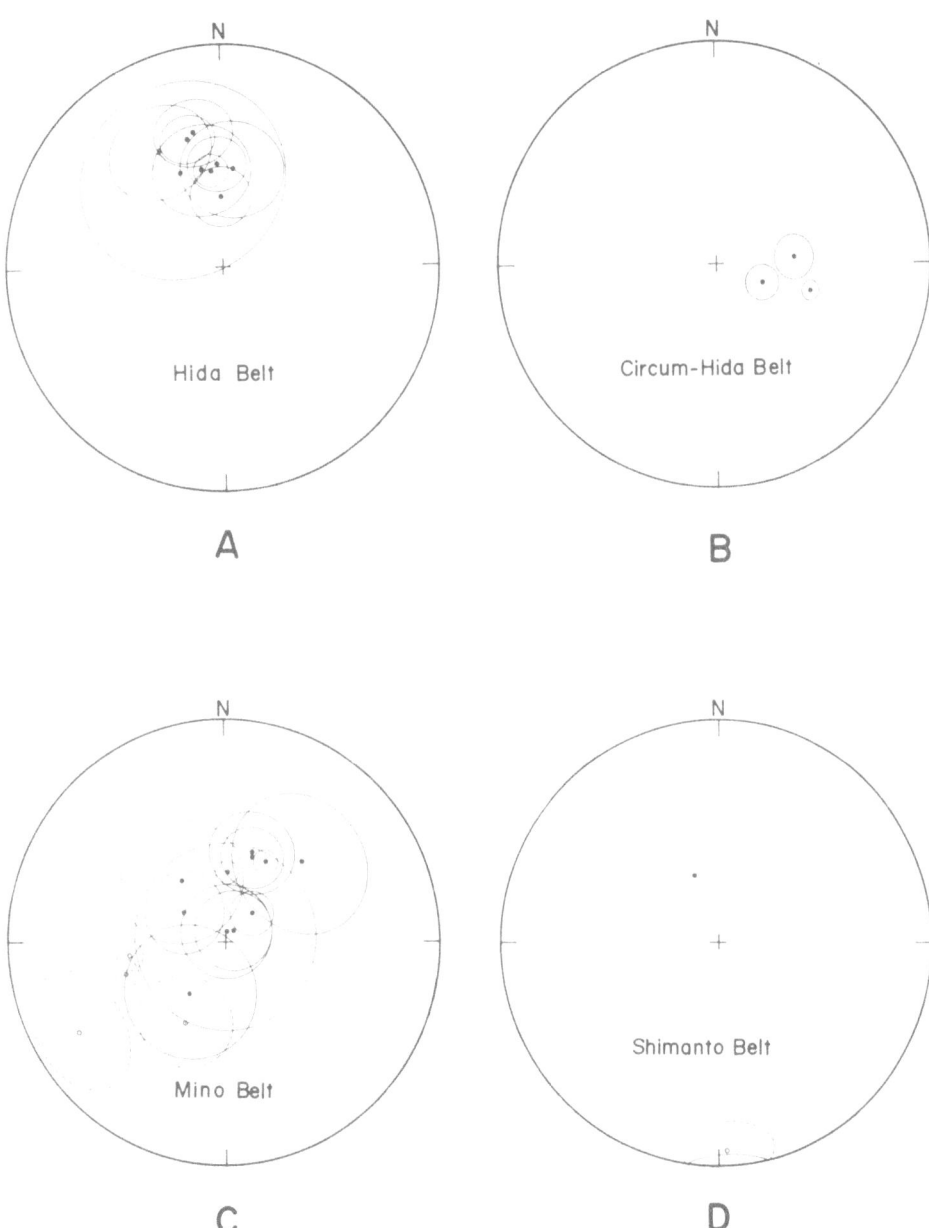

Fig. 4.   In-situ paleomagnetic directions of the Hida, the Circum-Hida, the Mino, and the Shimanto Belts plotted on Schmidt's equal area projections. Ovals represent the Fisher's circles of confidence of 95%. Open circles; upper hemisphere, solid circles; lower hemisphere.

Table 2.   Paleomagnetic results.

Site	Dc(°E)	Ic(°)	PL(°N)	Rock type	Age
Hida Belt					
GTY(11)			32.4	gr	174Ma
GTY(12)			34.7	gr	174Ma
GK(0)			46.0	gr	156Ma, 163Ma
GK(2)			23.1	gr	156Ma, 163Ma
GK(3)			33.8	gr	156Ma, 163Ma
GK(5)			34.0	gr	156Ma, 163Ma
GTY(8)			33.2	gr	69Ma, 88Ma
GTY(1)			21.5	gr	63Ma, 84Ma
GE(2)			23.0	gr	
Circum-Hida Belt					
PO(6)	151.8	19.4	−10.0	rss	Tr
PO(7)	165.4	11.4	−5.8	rss	Tr
PO(9)	191.9	43.1	−25.1	rss	Tr
Mino Belt					
GM(15)	248.8	9.0	−4.5	rc	L.–M. P
GM(18)	252.6	0.4	2.4	gs	L.–M. P
GM(19)	−12.4	−0.2	−0.1	gs	L.–M. P
GM(20)	−7.4	3.7	1.9	gs	L.–M. P
GM(21)	7.3	−15.0	−7.6	gs	L.–M. P
GM(22)	212.9	−16.4	8.4	rc	L.–M. P
GM(23)	254.1	−4.4	2.2	rc	L.–M. P
JS(5)	182.0	−13.7	6.9	rssh	U. Tr
JS(7)	−15.9	−13.1	−6.6	rc	U. Tr
JS(8)	−33.4	15.2	7.7	ssh	U. Tr
F(2)	−2.6	21.0	10.9	ssh	U. Tr
JS(0)	−53.6	11.0	5.6	ssh	M. Jr
F(1)*	−48.4	22.9	12.0	ssh	M. Jr
JS(6)	159.2	3.2	−1.6	rssh	M. Jr
F(3)*	166.9	8.6	−4.3	rssh	M. Jr
JS(9)	193.3	2.0	−1.0	ssh	U. Jr
Shimanto Belt					
EN(12)	−33.9	3.5	1.8	ssh	U. Cr
EN(17)	184.5	27.5	−14.6	ssh	U. Cr

Dc; mean declination after bedding correction, Ic; mean inclination after bedding correction, PL; paleolatitude, gr; granitic rock, rc; red chert, gs; greenstone, rssh; red siliceous shale, rss; red sandstone, ssh; siliceous shale, Tr; Triassic, L.-M. P; lower to middle Permian, U. Tr; upper Triassic, M. Tr; middle Triassic, U. Jr; upper Jurassic, U. Cr; upper Cretaceous.

*; data are cited from Hattori (1982).

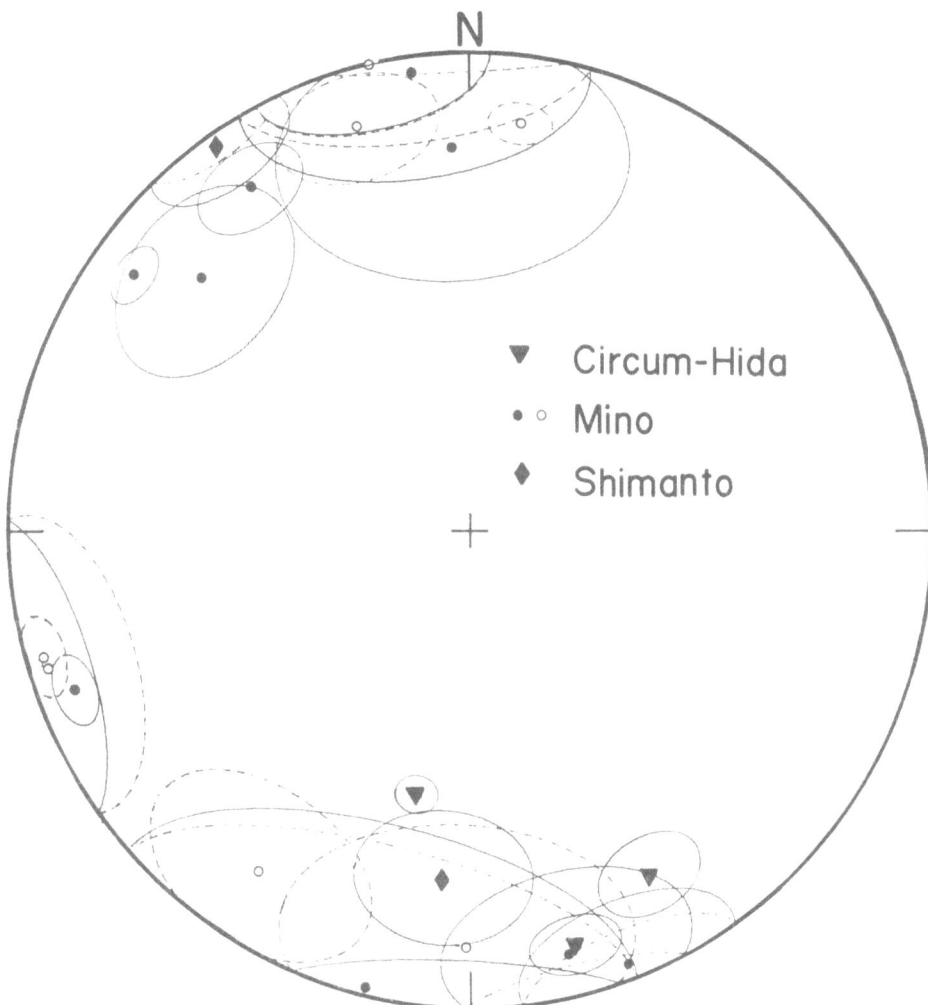

Fig. 5.   Paleomagnetic directions of sedimentary rocks after bedding correction. All the results obtained from the Circum-Hida, the Mino, and the Shimanto Belts are plotted altogether on a Schmidt's equal area projection. Open and solid symbols indicate on the upper and lower hemisphere of the projection, respectively.

On the contary, paleomagnetic declinations of granitic bodies in the Hida Belt do not deviate so much from the north and the inclinations vary within the range of 35° with the mean value of 50.3°. These directions are almost the same as that of the present-day geomagnetic field in Japan. We could not make bedding corrections on the granitic rocks in the belt, because of the difficulty to know the accurate horizontal plane at the time of the intrusion. We use declinations and inclinations listed in Table 1 as the paleomagnetic data in discussion.

4. Discussion

Paleomagnetic directions of the remanent magnetization of granitic rocks in the Hida Belt are summarized in Fig. 6. In spite of the wide distribution of the sampling sites and of wide range of the age from Jurassic to Paleogene, it is obvious that the remanent directions are so similar to each other that we get a small value of Fisher's angle of confidence of 7.1° by taking the mean direction of nine granitic sites in the belt.

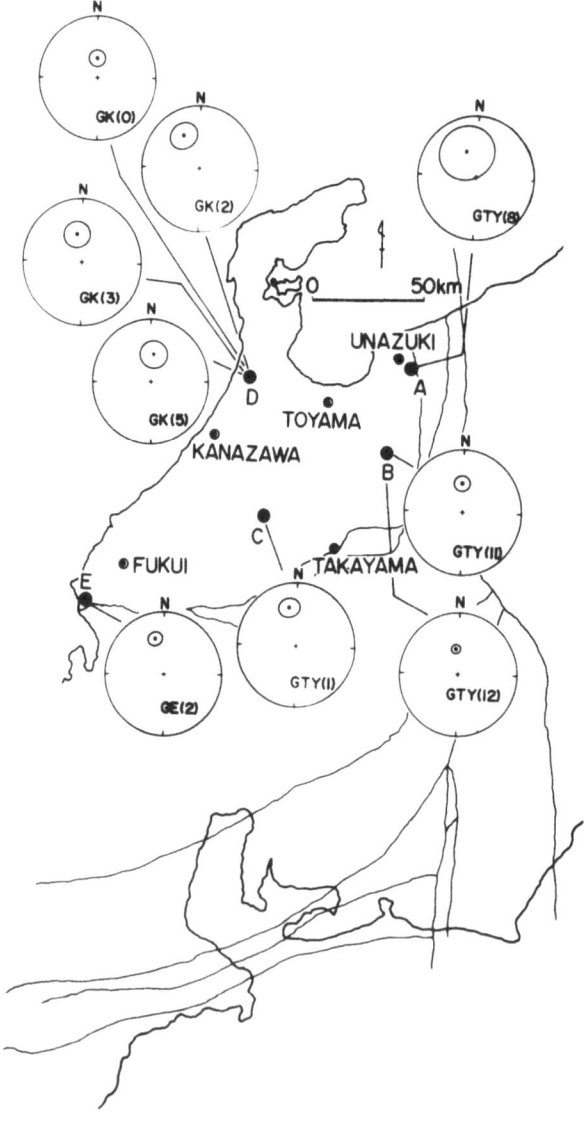

Fig. 6.   Paleomagnetic directions and the sampling sites in the Hida Belt. The mean directions of each site after the appropriate a.f. demagnetization are plotted on the lower hemisphere of Schmidt's net with the Fisher's circle of confidence of 95%.

Furthermore, the directions of paleomagnetic remanence are very close to the normal direction of the present geomagnetic field. This evidence suggests that the land block of the Hida Belt has made no detectable change of its relative position to the pole since Jurassic time.

On the other hand, remanent magnetizations of sedimentary rocks in the Circum-Hida, the Mino, and the Shimanto Belts show both normal and reversed polarities after making bedding correction. The conspicuous aspect of the results is that the inclinations are very shallow at all the sites as shown in Fig. 7. The inclinations range between 27.5° and − 16.4° except for one result from the Motodo Formation whose inclination is − 43.1°. This aspect indicates that the sediments were deposited in very

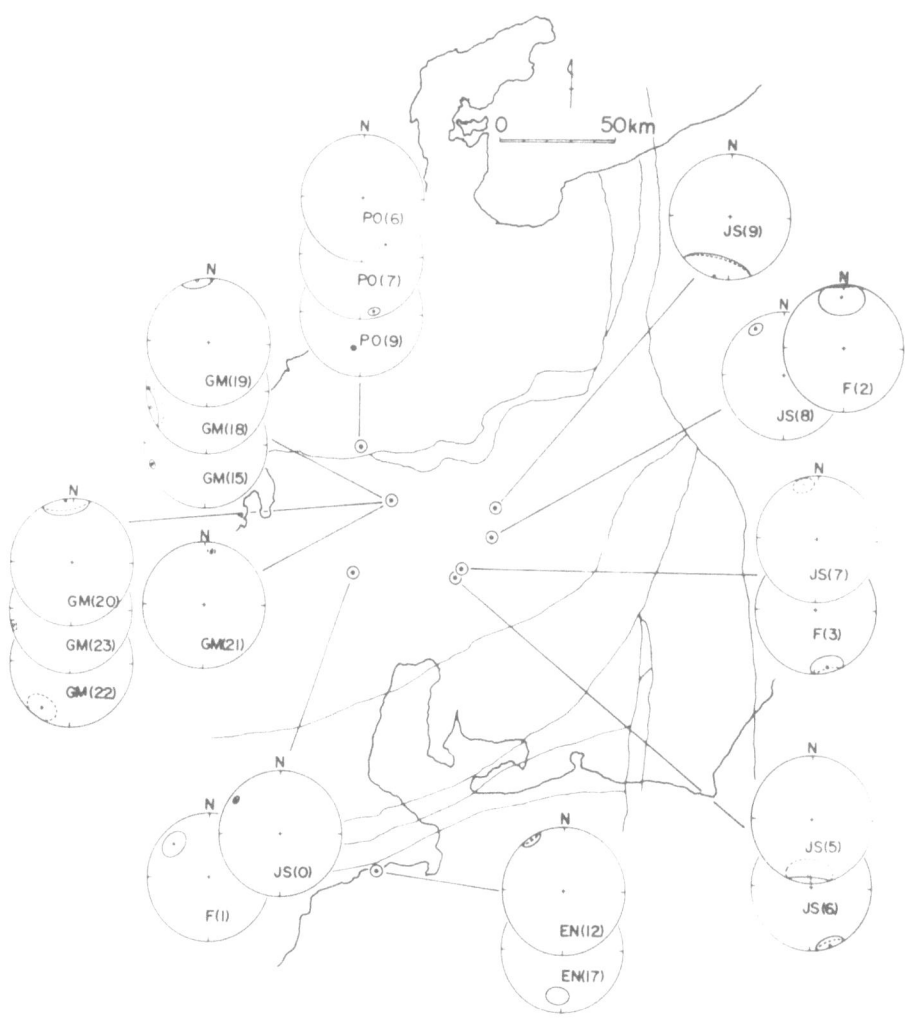

Fig. 7.   Paleomagnetic directions and sampling sites in the Circum-Hida, the Mino, and the Shimanto Belts.

low latitudes. The similar conclusion had been obtained by HATTORI and HIROOKA (1977, 1979) from the study of greenstones in the Mino Belt.

Regardless of the similarity of inclination values, sediments in the Circum-Hida, the Mino, and the Shimanto Belts show a big variety in declination as seen in Fig. 5. This is the consequence of the severe geotectonic movements which caused strong deformation of the strata. The folding axes of strata were plunged and distorted. It is difficult to put back the strata in their original state before folding. In the case that the folding axis is plunged or distored, it is almost impossible to restore the bedding plane to its original orientation by making a simple bedding correction. Only the inclination, therefore, can be used as the paleomagnetic information about the paleolatitude in reconstructing the paleogeographic map. The paleolatitudes computed from the inclinations indicate that all the sites of sedimentary rocks were formed in the equatorial region. The paleolatitudes for the four terrains of the Hida, the Circum-Hida, the Mino, and the Shimanto Belts are illustrated in Fig. 8.

Since we have no paleomagnetic data of the stable continent nearby the Japanese Islands to be referred to, we assume that the north pole of rotational axis of the earth has been fixed throughout the geological time. Relative movements of the land masses to the pole are discussed by using the paleomagnetic inclination.

As far as the geotectonic tilting of granitic bodies can be assumed small enough, the land mass of the Hida Belt seems to have keeping its position in the middle latitude since Jurassic period. The mean paleolatitude is 31.1°. The paleolatitude obtained from sedimentary rocks of the Circum-Hida, the Mino, and the Shimanto Belts are, on the

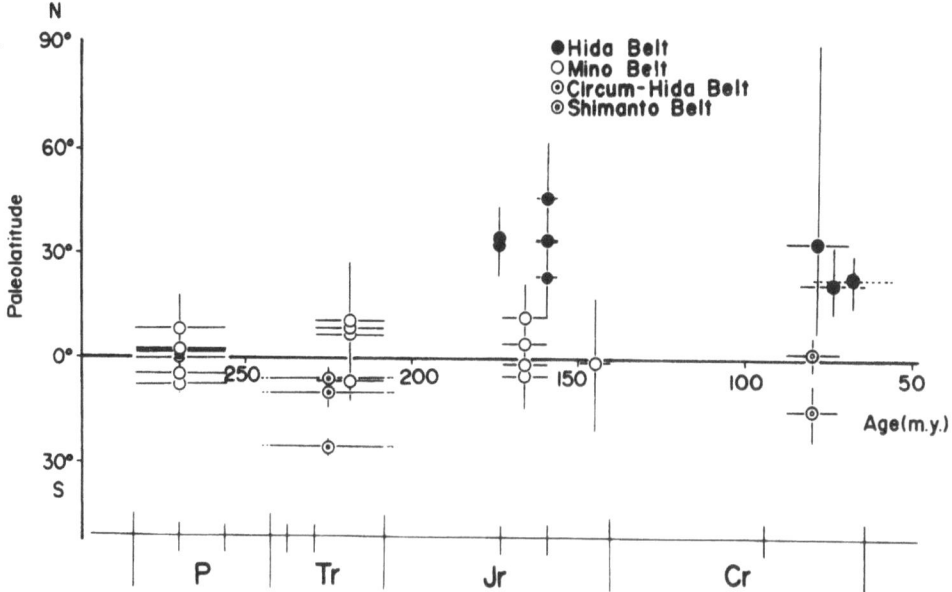

Fig. 8.   Change in the paleolatitude caluculated from the paleomagnetic inclinations obtained from some rocks in the Hida, the Circum-Hida, the Mino, and the Shimanto Belts. P; Permian, Tr; Triassic, Jr; Jurassic, Cr; Cretaceous.

contrary, very low of 10.9°N to 25.1°S. Especially, the results of the Mino Belt exhibit continuous low latitudes from the Permian to the upper Jurassic. Three sites in the Motodo Formation in the Circum-Hida Belt show the latitude of the southern hemisphere in the Triassic. The results obtained from two sites of silicious shale in the Shimanto Belt also give the low paleolatitude. The age of them estimated by radiolarian analyses is Coniacian-Santonian stage in the upper Cretaceous. It is noticeable that the sedimentary basin of the Shimanto Belt was in the equatorial region even in the late Cretaceous. From the geological evidences and paleomagnetic data of Valanginian stage, TAIRA (1981) remarked that the pelagic sediments of the Shimanto Belt were deposited in the low latitude in the time of the lower Cretaceous. He concluded that the sediments reached to a trench in Coniacian-Santonian stage by the evidence of the existence of terrigenous materials of this age. But our results show that the sediments were still in the equatorial region in that age.

The fact of the low values of paleolatitude of the Circum-Hida, the Mino and the Shimanto Belts indicates, therefore, that the sediments were deposited and magnetized in the place located near the equator. After the acquisition of remanent magnetization, the sediments migrated toward the north until they collided with and accreted to the land block of the Hida Belt which has been situated almost identical latitude of its present position since the Jurassic period. Although the ages of collision and accretion are not clear, the beginning of the northward migration of the pelagic sediments of the Circum-Hida Belt was not before 200 Ma ago and that of the Mino Belt must have been later than 140 Ma. The sediments of the Shimanto Belt did not commence its northward migration until 80 Ma.

Above-mentioned migration histories of the three terrains in central Japan indicate not only that all of the blocks consisting the Japanese islands moved together in one cluster from the south to the north as SASAJIMA (1981) pointed out, but that each terrain has the individual different tectonic history and is accreted one by one to the present position as the consequence of tectonism.

5. Conclusions

The paleomagnetic study of the granitic rocks in the Hida Belt, central Japan, revealed that the plutonic bodies were magnetized in the middle latitude not so different from their present latitude. The mean value of paleolatitude is 31.1°N.

Paleolatitude obtained from the sedimentary rocks in the Circum-Hida, the Mino, and the Shimanto Belts show a clear contrast with that of the granites in the Hida Belt. All the values of paleolatitude are within the range between 10.9°N and 25.1°S as is shown in Fig. 8 and Table 2. Consequently, the sedimentations were performed in the equatorial region.

The paleomagnetic data mentioned above leads us to the following tectonic history of the central part of Honshu. The pelagic sediments of the Circum-Hida, the Mino, and the Shimanto Belts were deposited in the equatorial region and magnetized there in the Mesozoic age. Then the sediments of those belts separately migrated toward the north until they collided with and accreted to the land block of the Hida Belt which has been situated in the middle latitude since Jurassic age.

REFERENCES

Fujiwara, Y., Paleomagnetic studies on some Mesozoic rocks in Japan, *J. Fac. Sci., Hokkaido Univ.*, **13**, 293–299, 1966.

Hattori, I. and K. Hirooka, Paleomagnetic study of the greenstone in the Mugi-Kamiaso area, Gifu Prefecture, central Japan, *J. Jpn. Assoc. Mineral. Petrol. Econ. Geol.*, **72**, 340–353, 1977.

Hattori, I. and K. Hirooka, Paleomagnetic results from Permian greenstones in central Japan and their geologic significance, *Tectonophysics*, **57**, 211–235, 1979.

Hattori, I., The Mesozoic evolution of the Mino terrane, central Japan: A geologic and paleomagnetic synthesis, *Tectonophysics*, in press, 1982.

Ito, H., K. Tokieda, and Y. Notsu, A paleomagnetic approach to possible tilting movements of Kitakami Mountains and Oshima Peninsula of Hokkaido, *Rock Magn. Paleogeophys.*, **7**, 113–120, 1980.

Kawai, N., H. Ito, and S. Kume, Deformation of the Japanese Islands as inferred from rock magnetism, *Geophys. J. Roy. Astron. Soc.*, **6**, 124–130, 1961.

Kawai, N., T. Nakajima, and K. Hirooka, The evolution of the island arc of Japan and the formation of granites in the Circum-Pacific Belt, *J. Geomagn. Geoelectr.*, **23**, 267–293, 1971.

Minato, M. and Y. Fujiwara, Magnetic pole positions determined by the Japanese Palaeozoic rocks, *Earth Sci. (Chikyu Kagaku)*, **78**, 21–22, 1965.

Nozawa, T., *Radiometric Age Map of Japan*, Geological Survey of Japan, 1977.

Omura, A., Sedimentological study of the Motodo Formation in the vicinity of Nishitani-mura, Ono-gun, Fukui Prefecture, central Japan, *J. Geol. Soc. Japan*, **74**, 217–231, 1968 (in Japanese with English abstract).

Sasajima, S., Pre-Neogene paleomagnetism of Japanese Islands (and vicinities), in *Paleoreconstruction of the Continents* edited by McElhinny and Valencio, *Geodynamics Series*, Vol. 2, pp. 115–128, 1981.

Sasajima, S., J. Nishida, and M. Shimada, Paleomagnetic evidence of a drift of the Japanese main island during the Paleogene period, *Earth Planet. Sci. Lett.*, **5**, 135–141, 1968.

Shibata, K. and S. Mizutani, Isotopic ages of siliceous shale from Hida-Kanayama, central Japan, *Geochem. J.*, **14**, 235–241, 1980.

Taira, A., Formative process of Shimanto Belt, *Kagaku*, **51**, 516–523, 1981 (in Japanese).

Yao, A., T. Matsuda, and Y. Isozaki, Triassic and Jurassic radiolarians from the Inuyama area, central Japan, *J. Geosci., Osaka City Univ.*, **23**, 135–155, 1980.

Yoshimura, M., S. Kido, and I. Hattori, Stylolitic cherts and radiolarian fossils in the Imajo area of the Nanjo Massif, Fukui Prefecture, central Japan, *Mem. Fac. Educ., Fukui Univ.*, **31** (II), 66–77, 1982 (in Japanese with English abstract).

Accretion Tectonics in the Circum-Pacific Regions, edited by M. Hashimoto and S. Uyeda, 195–206.

Accreted Oceanic Reef Complex in Southwest Japan

Kametoshi Kanmera and Hiroshi Nishi

Department of Geology, Faculty of Science,
Kyushu University, Fukuoka 812, Japan

Several large Early Carboniferous-Middle Permian limestones including the Akiyoshi, world famous for fusulinacean biostratigraphy, are isolatedly aligned in the Inner Zone of Southwest Japan. They rest on basaltic pedestals and are in fault contact with spicular siliceous shale and chert enclosing skeletal debris limestone and also with sandstone and mudstone, the latter containing many limestone blocks and breccias of various sizes. This complex disposition of heterogenous rock assemblages is best explained by accretion of a reef complex originated on an oceanic basaltic seamount and outskirted by coeval pelagic siliceous deposits, and trench-fill terrigenous sediments prior to deposition of the Upper Triassic rocks.

1. Introduction

Several large Upper Paleozoic limestone masses are isolatedly aligned in the inner side provinces of Southwest Japan, and the prominent is the Akiyoshi, Taishaku, Atetsu, Koyama, and Omi Limestones (Fig. 1). The first four are geogectonically situated in a non to weakly metamorphosed terrane between the high P/T Sangun Metamorphic Terrane on the north and south. The last lies at the northeastern end of the Hida Marginal Zone which forms a melange belt between the low P/T Hida Metamorphic Terrane on the north and the nonmetamorphic Paleozoic-Mesozoic Tamba Terrane on the south and comprises high P/T metamorphic rocks, serpentinites and associated ultramafic rocks, nonmetamorphic Paleozoic rocks, etc.

These limestones rest conformably on alkaline basalt flows and volcaniclastic rocks, which are exposed only 200–300 m thick though. Biostratigraphically they exhibit almost the same sequence ranging from Early Carboniferous to late Middle Permian. This volcanic-limestone sequence is best displayed in the Akiyoshi plateau. The Akiyoshi Limestone has been ascribed to a geosynclinal organic reef complex associated with submarine volcanic activity (Ota, 1968), and considered to grade laterally into terrigenous rocks and subordinately accompanied chert in the surrounding areas for the main reason that they contain many limestone lenses yielding the same fusulinacean faunas as in the Akiyoshi Limestone (Toriyama, 1954; Murata, 1961; Ota, 1968; Fujii and Mikami, 1970; Mikami, 1974; Schwan and Ota, 1977). Similar view has also been presented on the Taishaku and Koyama Limestones (Hase *et al.*, 1974; Okimura, 1975; Yokoyama *et al.*, 1979).

We examined stratigraphical and structural relationships between the Akiyoshi Limestone and surrounding siliceous and terrigenous rocks, and came to the

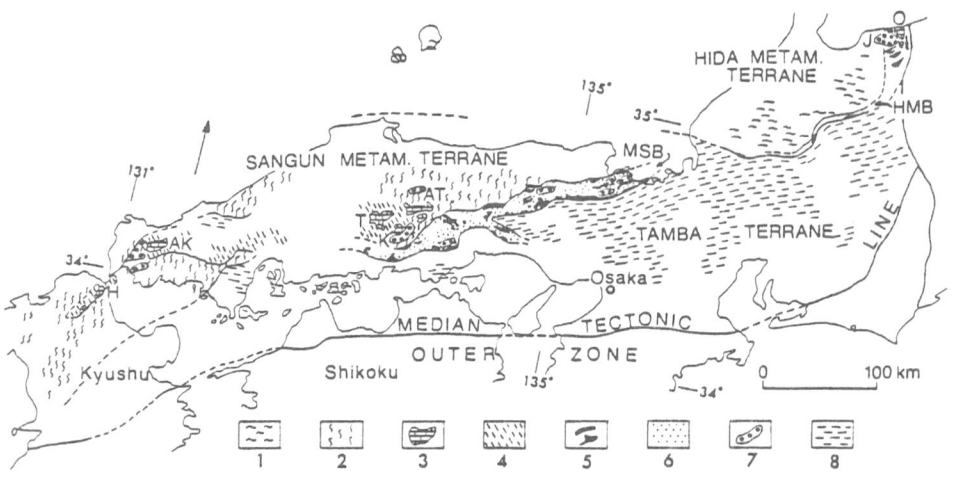

Fig. 1. Map showing the geotectonic position of Early Carboniferous-Middle Permian limestones in the Inner Zone of Southwest Japan (compiled from various data). 1, Hida metamorphic rocks and associated Triassic-Jurassic granites; 2, Sangun metamorphic rocks; 3, Early Carboniferous-Middle Permian limestone (AK, Akiyoshi; T, Taishaku; K. Koyama; AT, Atetsu; O, Omi; H. Hirao); 4, early Carboniferous-middle Permian siliceous rocks and late Permian-early Triassic (?) terrigenous rocks. 5, Basic-ultrabasic rocks (Ophiolites) in the Maizuru Structural Belt (MSB) and the Hida Marginal Belt (HMB); 6, Late Permian-Middle Triassic terrigenous rocks of the Maizuru Structural Belt; 7, Late Triassic and Early Jurassic (J) terrigenous rocks of a molasse facies; 8, Permian-Triassic-early Jurassic chert, siliceous shale and middle-late Jurassic terrigenous rocks of the Tamba Terrane.

conclusion that these rocks deposited primarily in different environments, that is on an isolated oceanic seamount and in its surrounding basin and a trench, respectively, and were subsequently accreted by subduction tectonics. The purpose of this paper is to describe the interrelationships among these heterogenous rock assemblages and to consider how they were complexly juxtaposed.

2. Isolated Oceanic Reef Complex Akiyoshi Limestone

The Akiyoshi Limestone is 16 km (NE-SW) by 7 km (NW-SE) wide and forms a low (300–400 m high) hummocky Karst. Its stratigraphy has been intensely studied by many paleontologists and is divided into 21 fusulinacean and brachiopod species-zones (OTA, 1977) that range from Early Carboniferous (Viséan) to early Middle Permian. The total thickness of these zones attains 770 m, although disconformities of short duration are known at several stratigraphic levels (OTA, 1977).

The following features would account for interpretation of an isolated oceanic reef complex of the Akiyoshi Limestone.

(1) The limestone rests conformably on a volcanic sequence consisting of alkaline basalt flows below and volcaniclastic sediments above, which are, however, totally exposed only 200 m thick. The lavas and volcaniclasts contain much amygdales, the

amount of which sometimes attains 30%, and are referable to the products of submarine eruption at shallow depths. The upper part of the volcaniclastic beds contains abundant crinoid oscicles and some brachiopods and rugosa, and grades upwards into bioclastic and oolitic limestones with biolithite in part (ETO, 1967).

(2) The limestone is white grey and lithologically belongs mainly to biosparrudite-biosparite and subordinately to biomicrudite-biomicrite and biolithite. Important rock-formers are fusulinaceans, crinoides, brachiopods, bryozoans, and dasycla-dacean and codiacean Green algae. Frame-builders of biolithite are colonial corals, bryozoans, solenoporacean Red algae, and stromatolites (Blue-green algae). Thus the rock-formers are all of shallow marine inhabitants. The biolithite is common in the lower part of the sequence, and the biomicrite facies becomes more common in the upper part (OTA, 1968; NAGAI, 1978; HASHIMOTO, 1979). Noteworthy is that the limestone is unstratified almost throughout the thickness except for the lowest part, and has no terrigenous intercalations.

(3) In addition oolitic limestone is commonly found at many stratigraphic levels. This implies that a part of the sedimentary area of the limestone was at a depth of several meters and under a strongly water-agitating condition in a tropical warm water environment.

Thus the Akiyoshi Limestone that lies on a basaltic pedestal forms a shallow marine sequence including reefal facies throughout the entire thickness. This volcanic-limestone sequence never grades laterally into terrigenous beds, but is induced to have primarily merged basinward with pelagic siliceous deposits as mentioned below.

3.   Non-Calcareous Rocks Surrounding the Akiyoshi Limestone

The Akiyoshi Limestone is at present surrounded by noncalcareous rocks, except for the northeastern area (see Fig. 2). They have been variously mapped and named by previous authors and much controversy exists on their stratigraphy, ages, geologic structures, and their interrelationships. On the basis of our investigation, however, they are grouped into the following three divisions.

3.1   Siliceous Pelagites Beppu Group

This is distributed in most areas surrounding the Akiyoshi Limestone, and comprises 2 to 10 cm thick bedded, pale green siliceous shale with very thin interbeds or partings of pale green shale in the lower (50–60 m thick), and 10 to 50 cm thick bedded to massive, pale green to milky chert in the upper (100–120 m). As has been described by FUJII (1972) in the Ota area, these are characterized by an abundance of sponge spicules and their fragments, although those contained in chert are often recrystallized. Spicules often show a preferred orientation of their long axes, and lamination is well exhibited by difference in size and amount of spicules. The matrix of siliceous shale is fine fragments of spicules and clay minerals such as chlorite and illite. The shale interbeds are much softer than siliceous shale, are made up largely of clay minerals with a small amount of spicules, and grade into siliceous shale. Relatively coarse-grained layers contain small fragments of plagioclase and decomposed relics of mafic minerals. This implies that the lower part

Fig. 2. Geological map of the Akiyoshi area (Akiyoshi Limestone area and the southeastern part of the Ota Group were adopted from OTA (1977) and KAWANO et al. (1963), respectively). Location of the mapped area is indicated as AK in Fig. 1. 1, Quaternary sediments; 2, Cretaceous sedimentary and igneous rocks; 3, Late Triassic Miné Group; 4, Sangun metamorphic rocks; 5, Ota Group (Late Permian-Early Triassic?); 6, Tsunemori Group (Late Permian?); 7, Beppu Group (late Early Carboniferous-late Middle Permian); 8, Akiyoshi Limestone Group (late Early Carboniferous-late Middle Permian) C, Carboniferous; P, Permian; dotted, recrystallized; 9, Basalt lava

of this group more or less contain volcanogenic material, but no terrigenous layers or grains are found at all in this group. This group is ascribed to pelagic deep-sea deposits.

Noteworthy is the sporadic occurrence of exogenetic skeletal limestones in this group. They are unstratified, irregular in form and mostly less than 3 m thick, with the thickest nearly 20 m and the largest extension about 100 m. They thicken and thin out abruptly within a short distance, and are discordantly enclosed in the spicular shale and chert. Many of them consist mainly of unsorted bioclasts of shallow benthic faunas such as fusulinaceans, brachiopods, bryozoans, crinoids, corals and calcareous algae, and are subordinately accompanied by ooids and limestone debris of a sand to a small pebble size. Sometimes basaltic fragments are also found. This type of limestone mostly occur in the lower and middle parts of this group, but some of the upper part are composed largely of limestone debris that are of a granule to pebble size and texturally various. At and near the levels of these limestones, isolated skeletal fragments are randomly scattered in siliceous rocks. The kinds of bioclasts and the lithology of limestone debris are all the same as those of the Akiyoshi Limestone, and these bioclasts, ooids, and limestone fragments are thought to have evidently been derived as submarine debris flows mainly of unconsolidated sediments with subordinate lithified limestone from the flank of the Akiyoshi reef complex.

A notable feature of these limestones is to merge into the surrounding siliceous shale and chert through a transitional zone, which is usually 10 to 30 cm thick and composed of a mixture of bioclasts, limestone debris, and siliceous mud containing many sponge spicules. In addition, the limestones, especially those comprising limestone debris, have a matrix that is characteristically filled with siliceous mud containing sponge spicules. These facts signify that the carbonate debris flowed onto and mixed with unconsolidated spiculous deposits.

It is evident that the bioclasts of limestones mentioned above are of the same age as the enclosing siliceous rocks. Those of the lower part of this group yield fusulinaceans of the *Millerella-Eostaffella* zone of late Visean age, and those in the middle part contain *Pseudofusulina* and *Triticites* of Early Permian. The upper part is defined as the *Neoschwagerina margaritae* zone of middle Middle Permian age. Thus this group is coeval with the almost whole thickness of the Akiyoshi Limestone, and its distribution suggests that it accumulated in the basin surrounding the volcanic seamount and overlying reef complex of Akiyoshi. Despite the great effort of finding microfossils such as conodonts and radiolaria from the siliceous rocks, no fossils favourable for age determination have been found (see Postscript).

The siliceous shale and chert have been considered to be conformably intercalated at many stratigraphic levels in terrigenous rocks (TORIYAMA, 1954; MURATA, 1961; KAWANO et al., 1963; OTA, 1968; FUJII, 1972; MIKAMI; 1974). However, the base of this group is everywhere truncated by a thrust fault. The upper limit is also often faulted, but there are some places where the chert is conformably followed by shale of the Tsunemori Group with an abrupt change of lithologies. In the Ota area the siliceous shale and chert of the Beppu Group are found either as thrust sheets or as exotic blocks in the clastic rocks.

This group is in fault contact with the Akiyoshi Group. Therefore the original relationship between the two groups cannot be known, but common occurrence of siliceous sponge spicules in the lowest part of the Akiyoshi Limestone and some intercalations of spicular chert in its basal volcaniclastic beds suggest interfingering lithofacies change of the two groups at the flank of the Akiyoshi Seamount. It is noteworthy in this context that the marginal facies of the Lower and lower Middle Carboniferous sections in the Atetsu (KOIKE, 1967), the Taishaku (HASE *et al.*, 1974), and the Koyama Limestone (YOKOYAMA *et al.*, 1979) are mainly composed of bioclastic and limestone debris, and have many thin interbeds of calcareous spicular chert and some beds of volcaniclastic rocks.

3.2   Tsunemori Group

This group is distributed on the north, west and southwest of the Akiyoshi Limestone. Its main part consists of massive mudstone with subordinate thinly alternating fine-grained sandstone and silty shale. Mudstones are intensely scaly cleaved. Sandstones are usually 1–5 cm, rarely up to 15 cm thick. Towards the upper part of this group 10 to 30 cm thick bedded alternation of muddy sandstone and mudstone becomes predominant, and in the uppermost part there are many slump beds of sandstone. The thickness of this group is difficult to estimate because of strong deformation, but may probably be less than 500 m.

The alternating beds of the main and the upper part often show graded bedding, but stratal as well as intrastratal soft-sediment deformation of various kinds are common. They irregularly and abruptly thin and thicken to form uneven hummocks, discordant injection lenses and wedges of sand layers, and disharmonic folds of the upper bedding surfaces. Internally deeply buried load casts, and isolated pods and stringers of sand foundered into underlying mudstone are common. In addition, minor asymmetrical folds and nearly lying folds with a very thinly pulled off over-turned limb, and laminations displaced by a planeless fault are characteristically recognized. These folds and planeless faults generally show a southerly vergence. It is evident that these deformations resulted mainly from intrastratal flow of sandy beds before lithification.

The sandstone slumps in the uppermost part vary greatly in thickness and abruptly terminate. They show an extremely disordered structure exhibiting an irregularly contorted or a chaotically disconnected form with many glide planes or isoclinal folds. Associated with them there occur some pebbly mudstone containing angular slabs of sandstone along with rounded pebbles of igneous rocks. The sandstones are well bedded in the thickness less than 30 cm, and well sorted with some thin laminae containing much carbonaceous matter. These features suggest their primary deposition in a shallow environment.

An important feature of the Tsunemori Group is to contain many limestone blocks and breccias in the upper part. These are particularly abundant in areas adjacent to the Akiyoshi Limestone, but some are found even in areas a few kilometers away from it. They are randomly scattered in mudstone and alternating muddy sandstone and mudstone. The limestone blocks are various in shape and in size up to 30 m across, and have a sharp discordant contact with the surrounding mudstone. The limestone breccias occur as lenticular bodies and abruptly thicken and thin out. Their thickness ranges from less than 1 m to several tens of meters and their extension is up to a few hundred meters.

The limestone breccias are unstratified and unsorted, and composed of heterogeneous, angular to subangular fragments of limestones of various ages ranging from Middle Carboniferous to late Middle Permian. They have narrow interstices filled with secondary calcite in the inner part, but have varied interstices that were filled with a mixture of minute limestone fragments and terrigenous mud in the marginal part. They abruptly change into surrounding mudstone, in which isolated limestone clasts are scattered.

A striking feature of these limestone breccias is the large size of some of the included clasts, which commonly attain a few meters across. An interesting observation on the internal structure is that several neighbouring clasts can be perfectly restored into one and the same clast. This sort of disposition of the included clasts can be produced only by mechanical brecciation of consolidated rocks. The included clasts are light-coloured, and the contained fusulinaceans and lithologic features clearly show their derivation from the Akiyoshi Limestone. Abrupt termination of rock bodies, lack of sorting and stratification, coarse texture, angularity of composing clasts, and other lines of evidence indicate that these blocks and breccias are submarine slides mainly of lithified limestone.

In addition to limestone blocks and breccias, there are some beds of calclithite, detrital carbonate sandstone, in some places. These occur adjacent to and at nearly the same levels as the limestone breccias, and are well sorted and interbedded with fine-grained sandstone and siltstone. They consist of water-worn, subrounded limestone debris and bioclasts of a sand to granule size with scattered grains of terrigenous quartz and feldspars, and are thought to have been produced by reworking and transportation of limestone slide materials. The interbedded sandstones also contain some limestone fragments and bioclasts of a sand size.

The limestone blocks and clasts of the breccias and calclithite beds are of various ages ranging from Middle Carboniferous to late Middle Permian, among which the most abundant are of the *Yabeina-Lepidolina* zone, the youngest in the Akiyoshi Limestone.

The limestones in question have been considered to be in situ lenses in the Tsunemori Group by many workers and to represent the same age as the surrounding terrigenous rocks. However, as explained above, they are all of exotic origin and are ascribed to the collapse of the Akiyoshi Limestone. Thus at least the upper part of the Tsunemori Group is younger than late Middle Permian, although no fossils have been found from the terrigenous rocks. It is overlain with a distinct unconformity by the Upper Triassic (Carnian-Norian) Miné Group on the west.

The Tsunemori Group continues eastward underneath at least the western half tract of the Akiyoshi Limestone. It has long been thought that the Akiyoshi Limestone merges laterally into and is also conformably overlain by the Tsunemori Group, although both the groups of this area are overturned as a whole (e.g. TORIYAMA, 1954; OTA, 1968; FUJII and MIKAMI, 1970; SCHWAN and OTA, 1978). However, the Akiyoshi rests discordantly on the Tsunemori in part with a glide plane and in other part with a tight contact. The glide plane is represented by a sheared, weakly slickensided plane. It is undulatory, but as a whole is nearly horizontal. The tight contact is shown by a sharp boundary without any transitional lithologies. Noteworthy is that the Akiyoshi is in contact with the Tsunemori with rocks of different ages and with different attitudes at every exposure in such a way as at one exposure with a steeply dipping Middle Carboniferous rocks, at another with a flat-lying Early Permian rocks and so on. This discordant relation implies an extraneous slide contact of the Akiyoshi Limestone on the Tsunemori mudstone, and it is clearly shown by the oblique distribution of the fossil zones of the Akiyoshi to the boundary line between the two groups, as seen even in the simplified map of Fig. 2.

3.3   Ota Group

The Ota Group is distributed on the south to southeast of the Akiyoshi Limestone. Its lower part consists of mudstone and thin bedded alternation of fine-grained sandstone and shale. The mudstone is fairly intensely cleaved and contorted. It contains many slump beds of sandstone and is similar in lithologic features to the upper part of the Tsunemori Group. This lower member seems to continue laterally into the upper part of the Tsunemori Group.

The upper part of the Ota Group is composed of fine- to medium-grained, massive sandstone with some coarse-grained, in part pebbly sandstone which characteristically contains subangular fragments of black shale, limestone, siliceous shale and chert, and skeletal debris of various kinds. The skeletal debris are much abraded and fragmentary, and are of detrital origin. The contained fusulinacean debris belong to many species and genera of Carboniferous and Permian ages, of which the youngest is *Lepidolina multiseptata*. Therefore, it is clear that the upper part of the Ota Group is younger than Middle Permian, but no exact age cannot be known on account of lack of endogenetic fossils.

The Ota Group is considered to have primarily underlain by the Beppu Group. It attains about 700 m in thickness.

4.   Geologic Structure

4.1   Akiyoshi Limestone

As the Akiyoshi Limestone is unstratified and has no distinct difference in colour and gross textures throughout the sequence, its geologic structure can be ascertained only by means of biostratigraphic zonation on fossils and mapping of fossil-zones. Great efforts by many paleontologists have resulted in fossil-zone maps and large-scale structural ones which are, however, still very much diverse among authors. Recently OTA (1977) and SCHWAN and OTA (1977) presented a compiled map of the Akiyoshi Limestone and interpret that the southern tract forms a normal sequence gently dipping north, but the northern tract represents an inverse limb of a flatlying overturned anticline which thrust on the southern normal sequence.

As seen in a simplified map of the Akiyoshi Limestone in Fig. 2, the major disruption structure is characterized by the prevalence of thrust faults trending NEE-SWW and verging SE, which imply the compressional movement from NWW to SEE. Detailed mapping by MORINAGA and OTA (1971), NAGAI (1978), and NAGAI and OTA (1980) have revealed that the thrust sheets are intersected by NE-SW and NW-SE cross-faults and steeply dipping parallel faults to form a complex block-mosaic structure which is well shown by interruptions of fossil-zones within a short distance as exemplified in Fig. 3. The block-mosaic structure is manifest particularly in the western tract of the Akiyoshi Plateau where the Akiyoshi Limestone lies on the stratigraphically younger Tsunemori Group and also locally on the coeval Beppu Group, which are exposed as inliers at some places, with a tectonic or displaced fall-in contact. It should be also noted that the block-mosaics are accompanied by remarkable fracturing or brecciation of the limestone in their marginal part, especially in areas adjoining the Tsunemori and the Beppu Group. These structures suggest that the Akiyoshi Limestone of the western half tract was a large gravitational slide associated with the thrust faulting.

4.2   Beppu, Tsunemori, and Ota Groups

The Ota Group is bounded on the northwest mostly by the Beppu Group and in part by the basal volcanic rocks of the Akiyoshi Group with a nearly vertical fault zone. Along this zone a narrow tectonic melange comprising basaltic volcaniclastic rocks, chert, sandstone, and shale of those groups are discontinuously found. As seen in Fig. 2, the siliceous rocks of the Beppu Group and the sandstone and shale of the Ota Group are juxtaposed in many belts and have been considered to be a conformable alternating sequence (TORIYAMA, 1954; MURATA, 1963; KAWANO et al., 1963; FUJII, 1972; SCHWAN and OTA, 1977). However, as already mentioned, the Beppu Group is older in age than the Ota Group, but at present it tectonically lies on the latter in most exposures and its base is everywhere truncated by a southerly verging thrust fault. Its upper limit is also often faulted.

Both the Ota and the Beppu Group are participated together in a slightly overturned and more or less isoclinal fold system with the NEE-SWW trending and northerly dipping axis. This fold system is apparently subsequent to the imbricate thrusting of the Beppu Group over the Ota Group. The subparallel distribution of the two groups indicates that the Beppu Group tectonically emplaced on the Ota Group in a system of very low-angle thrusting to form a flat-lying imbricate structure before the upright folding. In this connection it should be noted that small isolated chert blocks up to about 20 m (not indicated in Fig. 2) are sporadically found in sandstones and shales of the Ota Group near the chert beds. These are probably slump masses detached from the front of proceeding main chert masses in the process of its overthrusting onto the Ota Group.

Almost the same stratigraphical and structural relationships are recognized between the Beppu and the Tsunemori Group on the north and northwest of the Akiyoshi Limestone. As stated before, the Beppu Group stratigraphically underlies the Tsunemori Group, but in many cases it tectonically rests on the latter with a gently undulating thrust. Thus the two groups form an imbricate structure of a southerly vergence. To the west of Beppu there is a narrow tectonic inlier of the Tsunemori on one hand, and are some outliers of the Beppu resting tectonically on the Tsunemori on the other. As in the Ota Group there are some chert blocks of various sizes in the Tsunemori.

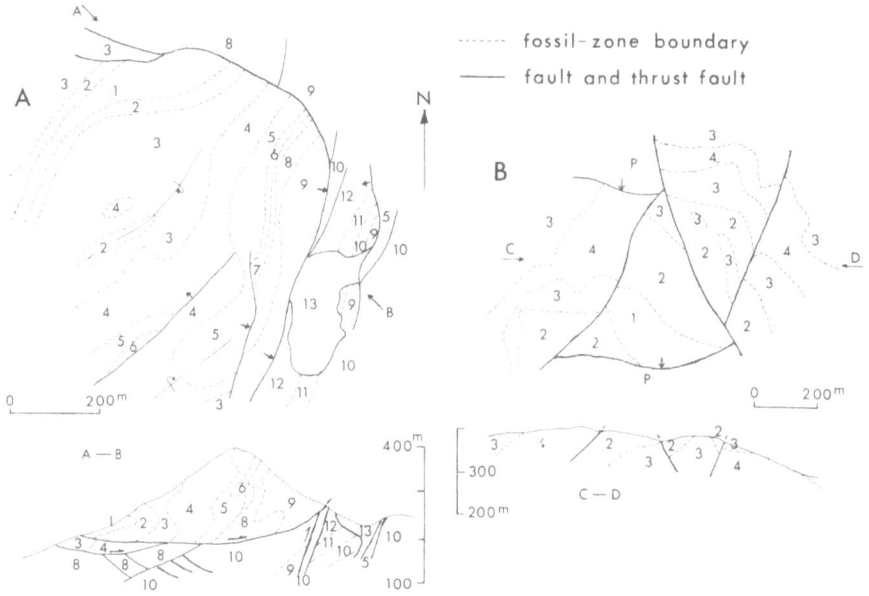

Fig. 3. Simplified fossil-zones and structural maps of the Sumitomo Quarry (A) (Adopted from MORINAGA and OTA, 1971) and the Ryugoho (B) (Adopted from NAGAI, 1978). Location of the mapped area is indicated in Fig. 2. 1–6, Early Carboniferous-Late Carboniferous (1, *Nagatophyllum* zone; 2, *Millerella-Eostaffella* zone; 3, *Profusulinella* zone; 4, *Fusulinella* zone; 5, *Beedina* zone; 6, *Triticites* zone); 7–12, Early Permian-late Middle Permian (7, *Triticites simplex* zone; 8, 9, *Pseudofusulina* zone; 10, *Parafusulina* zone; 11, 12, *Neocshwagerina* zone); 13, Tsunemori Group.

In short, the Akiyoshi Limestone and the surrounding rocks are characterized by the prevalence of a southerly verging imbricate structure, although modification by subsequent deformation is fairly remarkable.

5. Summary

The Early Carboniferous to late Middle Permian Akiyoshi organic reef complex developed on the basaltic pedestal of an isolated oceanic seamount that reached shallow depths and was widely encircled by coeval deep-sea pelagites of the Beppu Group consisting of spicular chert and siliceous shale with volcanogenic clay partings. From the reef flanks bioclastic and reef debris were occasionally flowed into the area of deposition of siliceous spiculite. When these deposits moved to and arrived at an arc trench, the siliceous pelagites were conformably overlain by trench-fill sediments of the Tsunemori Group comprising mudstone and turbidites. At the inner margin of the trench the two groups were scraped to form an accretionary prism in such an imbricated array as has been well demonstrated by multichannel reflection profiles in Nankai Trough (KARIG *et al.*, 1975; NASU *et al.*, 1982). Common occurrence of stratal and intrastratal soft-sediment deformation including asymmetrical folds of a southerly vergence in the Tsunamori Group suggests its accretion in the unconsolidated state.

Following the accretion of the Beppu and the Tsunemori Group in the northern area, the Akiyoshi seamount collided with the landward wall of the trench, and only the uppermost part of the volcanic pedestal and the overlying limestone were detached by southerly verging thrusts and were incorporated in the accretionary prism. At the same time large-scale gravitational collapse took place in the western side of the seamount and shed a large amount of limestone blocks and breccias into the trench-fill terrigenous sediments of the Tsunemori Group. Subsequently the Ota Group accumulated in the trench that removed to the south of the consumed Akiyoshi seamount and was juxtaposed with the thrust slices of the underlying Beppu Group.

The accretionary emplacement of the Akiyoshi Limestone was taken place probably in late Permian to early Triassic times, because the Tsunemori Group containing blocks and breccias of the Akiyoshi Limestone and the thrust sheets of the Tsunemori and the Beppu Group are overlain by the Carnian-Norian Miné Group with a clinounconformity.

Reviewed by M. Hashimoto. We thank him for helpful comments on the manuscript.

REFERENCES

ETO, J., A lithofacies analysis of the lower portion of the Akiyoshi Limestone Group, *Bull. Akiyoshi-Dai Sci. Museum*, **4**, 7–42, 1967.

FUJII, A., Ota Formation of the Yamaguchi Group in the Akiyoshi district, *J. Geol. Soc. Japan*, **79**, 309–321, 1972.

FUJII, A. and T. MIKAMI, Tsunemori Formation—its relation with the Akiyoshi Limestone, *J. Geol. Soc. Japan*, **76**, 545–557, 1970.

HASE, A., Y. OKIMURA, and T. YOKOYAMA, The Upper Paleozoic formations in and around Taishaku-dai, Chugoku massif, Southwest Japan, with special reference to the sedimentary facies of limestones, *Rep. Geol. Sci. Hiroshima Univ.*, **19**, 1–39, 1974.

HASHIMOTO, K., Bio- and litho-facies of the Akiyoshi Limestone Group in the southern area of the Akiyoshi Plateau, *Bull. Akiyoshi-Dai Sci. Museum*, **14**, 1–26, 1979.

KARIG, D. E., J. C. INGLE *et al.*, Site 298, in *Initial Reports of Deep Sea Drilling Project*, **31**, 317–350, U.S. Govt. Printing Office, Washington, D. C., 1975.

KAWANO, M., E. TAKAHASHI, H. ISHIKAWA, S. MATSUGAKI, and S. HARADA, Ota Group—on the geologic structure and age—, *Bull. Akiyoshi-Dai Sci. Museum*, **12**, 1–33, 1963.

KOIKE, T., A Carboniferous succession of conodont faunas from the Atetsu Limestone in Southwest Japan, *Sci. Rep. Tokyo Kyoiku Daigaku, Sec. C*, **93**, 279–318, 1967.

MIKAMI, T., Geologic study of the Beppu Formation, in *Earth, Human and Education, Prof. H. Kusumi Memorial volume*, 165–173, 1974.

MORINAGA, Y. and M. OTA, Subsurface geology of the Akiyoshi Limestone Group in the Maki and Kyoei area, Shuho Town, Southwest Japan, *Bull. Akiyoshi-Dai Sci. Museum*, **7**, 25–56, 1971.

MURATA, M., On the geological structure of the Akiyoshi Plateau, *Contrib. Inst. Geol. Paleontol., Tohoku Univ.*, **53**, 1–46, 1961.

NAGAI, K., Litho- and bio-facies of reef limestones in the Ryugoho area of the Akiyoshi Limestone Plateau, *Bull. Akiyoshi-Dai Sci. Museum*, **13**, 15–34, 1978.

NAGAI, K., and M. OTA, On the geology of the Minami-Dai area of the Akiyoshi Limestone Plateau, Yamaguchi Prefecture, *Rep. Earth Sci., General Educ., Kyushu Univ.*, **21**, 7–15, 1980.

NASU, N., H. KAGAMI *et al.*, Multichannel seismic reflection data across Nankai Trough, *IPOD-JAPAN Data Ser.*, **4**, 1–34, 1982.

OKIMURA, Y., Geosynclinal development of the northern part of the Chugoku Belt based on facies analysis of the limestone groups, *Assoc. Geol. Collab. Japan, Monogr.*, **19**, 49–56, 1975.

OTA, M., The Akiyoshi Limestone Group; A geosynclinal organic reef complex, *Bull. Akiyoshi-Dai Sci. Museum*, **5**, 1–44, 1968.

OTA, M., Geological studies of Akiyoshi, Part I, General Geology of the Akiyoshi Limestone Group, *Bull. Akiyoshi-Dai Sci. Museum*, **12**, 1–33, 1977.

SCHWAN, W. and M. OTA, Geological studies of Akiyoshi, Part II, Structural tectonics of the Akiyoshi Limestone Group and its surroundings, *Bull. Akiyoshi-Dai Sci. Museum*, **12**, 35–110, 1977.

TORIYAMA, R., Geology of Akiyoshi, Part II. Stratigraphy of the non-calcareous groups developed around the Akiyoshi Limestone Group, *Mem. Fac. Sci., Kyushu Univ., Ser. D*, **5**, 1–46, 1954.

YOKOYAMA, T., A. HASE, and Y. OKIMURA, Sedimentary facies of Koyama Limestone, *J. Geol. Soc. Japan*, **85**, 11–25, 1979.

Note added in proof

After the manuscript was recieved by the editor, radiolarian fossils of Early to Middle Permian ages were found from several horizons of the middle part of the Beppu Group by Mr. Uchiyama who has been engaged in research for the graduation thesis under a supervision of the senior author.

Accretion Tectonics in the Circum-Pacific Regions, edited by M. Hashimoto and S. Uyeda, 207–218.
Copyright © 1983 by Terra Scientific Publishing Company (TERRAPUB), Tokyo.

Tectonic Environments and Crustal Section of the Outer Zone of Southwest Japan

S. HADA and T. SUZUKI

Department of Geology, Kochi University,
Kochi 780, Japan

The Outer Zone of Southwest Japan can be divided into two terranes by the Kurosegawa Tectonic Zone, according to the tectono-stratigraphic and geophysical features. The Northern Terrane which is the area to the north of the Kurosegawa Tectonic Zone mainly consists of nappes with recumbent folds and thrust sheets derived from the orderly stratigraphic sequences. This is considered to be the terrane dominated over the marginal sea-type tectonic environment. The Southern Terrane which is situated to the south of the Kurosegawa Tectonic Zone is characterized by the existence of the accretionary complex, and the Pacific-type tectonic environment is postulated for this terrane. The crustal structure has been changed from continental to oceanic across the Kurosegawa Tectonic Zone. Consequently, the Kurosegawa Tectonic Zone is indicated as the upper slope discontinuity which is characterized by the serpentinite melange that the various kinds of blocks different in lithology and age were gathered from the varied spaces and depths.

1. Introduction

Southwest Japan faces to the Nankai Trough which forms part of the Philippine Sea–Asian plate boundary, and is divided into the Inner Zone on the Japan Sea side and the Outer Zone on the Pacific side by the major fault called the Median Tectonic Line. The Outer Zone is characterized by the distinct zonal distribution of the pre-Neogene rocks and has been divided into the following three belts arranged from north to south (Fig. 1):

(1) Sanbagawa Belt (High-pressure metamorphic rocks).

(2) Chichibu Belt (Non- or weakly metamorphic upper Paleozoic to middle Mesozoic strata).

(3) Shimanto Belt (Non- or weakly metamorphic upper Mesozoic to upper Tertiary strata).

However, the Outer Zone of Southwest Japan can be divided into two terranes by the Kurosegawa Tectonic Zone, if we consider its tectono-stratigraphic characteristics (SUZUKI *et al.*, 1979; HADA *et al.*, 1979). Accordingly, the following three-fold division of the Outer Zone of Southwest Japan is accepted in this paper: the Northern Terrane (the Sanbagawa Belt and the northern subbelt of the Chichibu Belt), the Kurosegawa Tectonic Zone, and the Southern Terrane (the middle and southern subbelts of the Chichibu Belt and the Shimanto Belt).

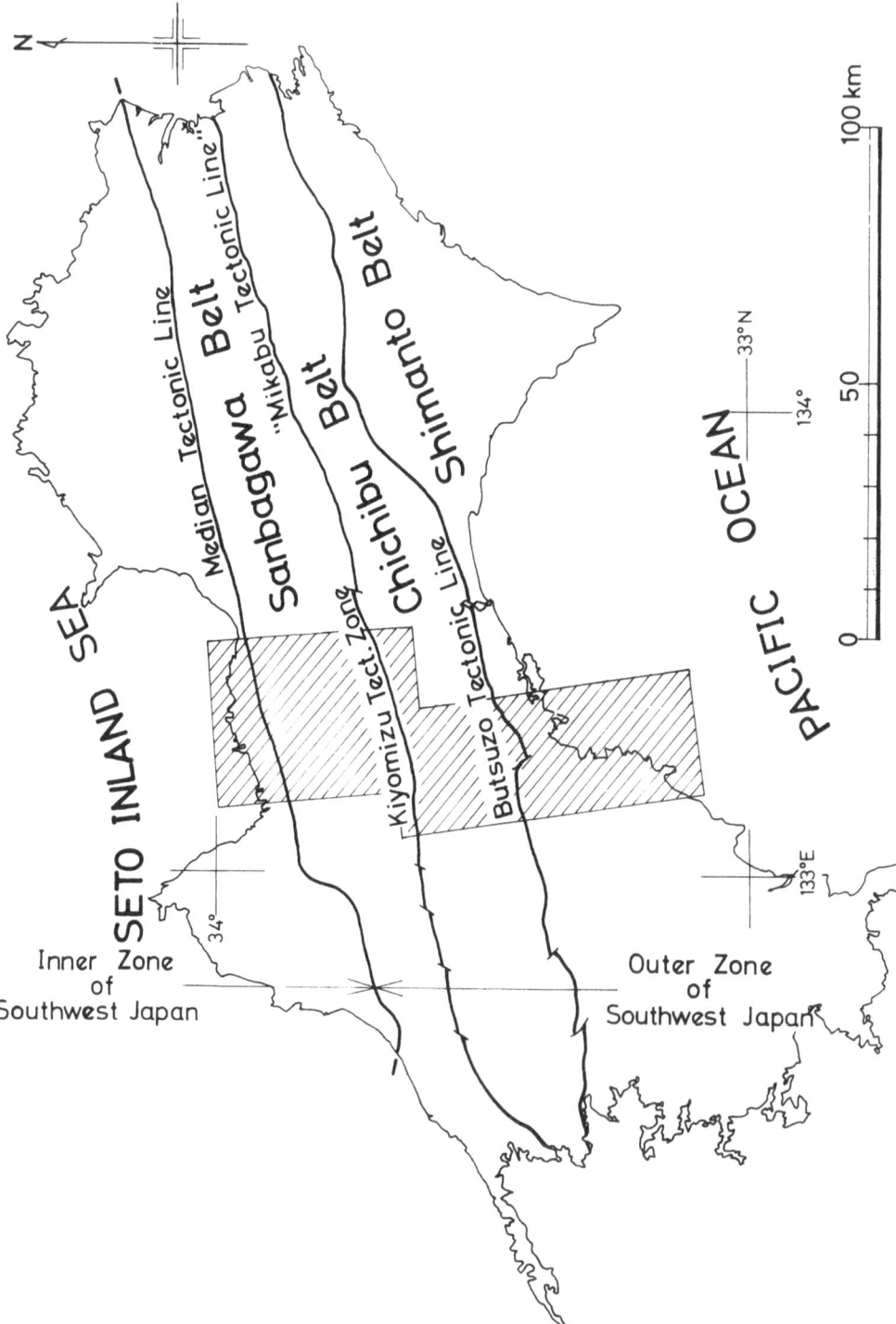

Fig. 1. Pre-Neogene tectonic divisions in Shikoku. The study area is contained within the box.

The distinct geologic unit in the Southern Terrane is the tectonic melange of a chaotically and pervasively sheared mudstone which contains various sizes of tectonic blocks of heterogeneous lithology. The melange unit separates stratigraphically coherent unit throughout its extent. On the contrary, the Northern Terrane is occupied by the geometrically rather simpler thrust sheets derived from orderly stratigraphic sequences. The Kurosegawa Tectonic Zone was originally defined as a fault zone characterized by a kind of the tectonic uplift area which is composed of numerous lenticular bodies derived from the basement rocks (ICHIKAWA et al., 1956) and has been interpreted as a suture zone characterized by the serpentinite melange (SUZUKI et al., 1976; SUZUKI, 1977; HADA et al., 1979; MARUYAMA, 1981).

We describe here the tectonic environments and geophysical characteristics of the Outer Zone of Southwest Japan, and give its crustal section. The geologic significance of the Kurosegawa Tectonic Zone will be discussed in particular.

2. Tectono-Stratigraphic Framework

Figure 2 is the generalized map showing the lithic belts of the Outer Zone in the middle part of Shikoku and its composite cross section. These are originally compiled in the scale of 1:200,000 from the results of our study, HARA et al. (1977) and TSUKUDA (1980). The striking contrast of the tectono-stratigraphic characteristics in the both sides of the Kurosegawa Tectonic Zone is clearly depicted in the figure.

The Southern Terrane which is the area to the south of the Kurosegawa Tectonic Zone is occupied by the two tectono-stratigraphic units (HADA et al., 1979; SUZUKI and HADA, 1979):

(1) Chaotically and pervasively sheared mudstone which includes various sizes of slabs and blocks composed of sandstone, coherently bedded alternation of sandstone and mudstone, greenstone (mostly composed of pillow lava), chert, and red shale. The age of the argillite "matrix" is consistently younger than that of the blocks (chert and red pelagic shale). The penetrative foliation is developed in the matrix, which shows, in places, the schistose appearance. This unit is best described as a tectonic melange.

(2) Flysch sediments of weakly deformed sandstone and mudstone alternating in various thickness. The alternation keeps sufficient continuity of layers except in the part of slumping masses, and deformed into open to close large-scale folds.

Noteworthy feature in this terrane is the existence of the rock unit having exotic ages, lithologies, metamorphic assemblages or deformational histories. Greenstones are very fragmentary as isolated lenticular bodies and are immersed only in the argillite matrix of the tectonic melange. The tectonic melange forms a narrow strip averaging 1.5 km wide that separates the weakly deformed, but stratigraphically coherent flysch sediments with the northerly dipping high-angle faults which gave rise to the imbrication with the polarity proceeding to the south.

The Northern Terrane which is the area to the north of the Kurosegawa Tectonic Zone consists of the Sanbagawa Belt and the northern subbelt of the Chichibu Belt. The Sanbagawa Belt is composed of high-pressure metamorphic rocks chiefly derived from the Mesozoic sediments (SUYARI et al., 1980), pyroclastics and volcanics. Recently, HARA et al., (1977) distinguished the nappes with recumbent folds and also

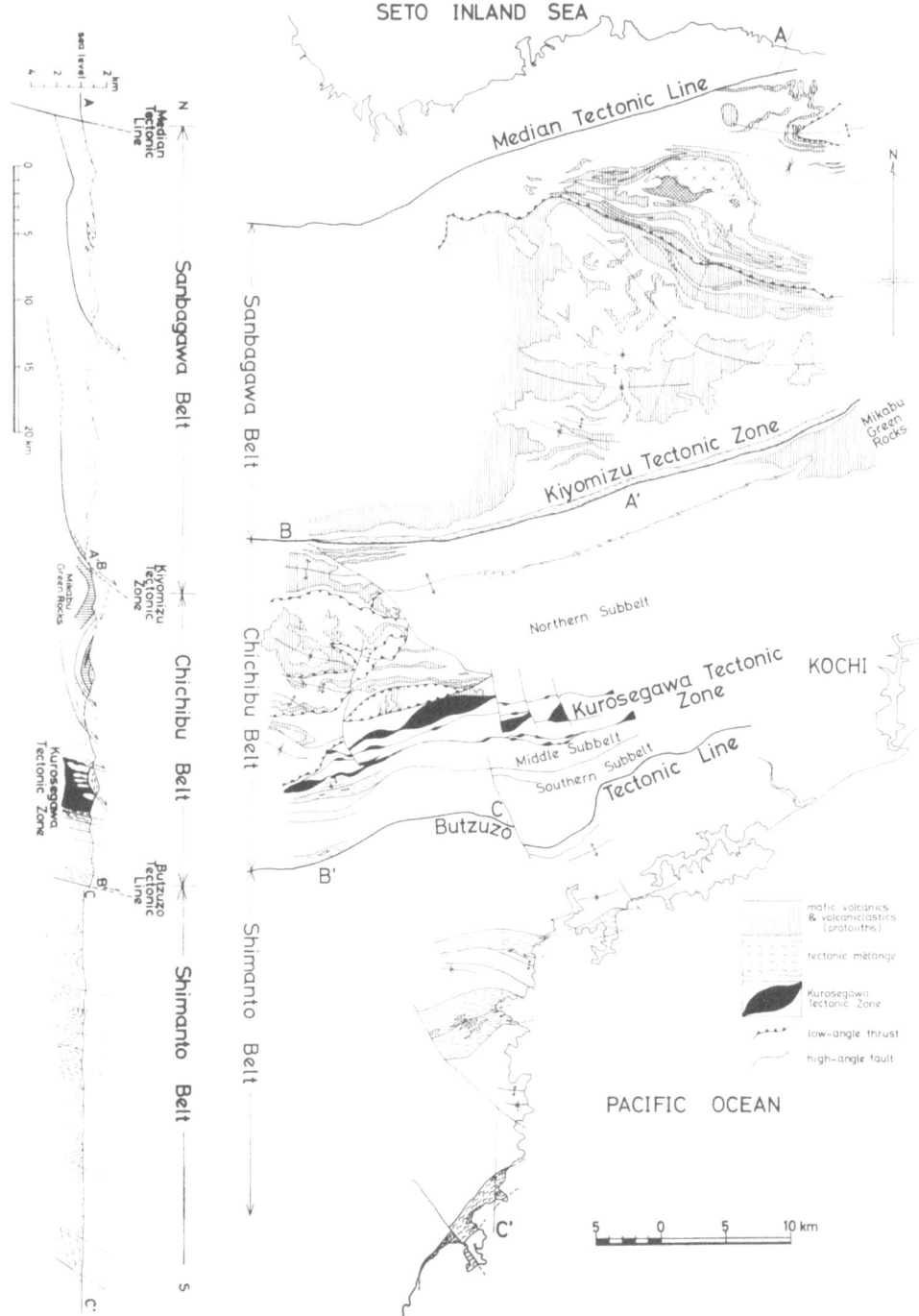

Fig. 2.   Generalized map showing the lithic belts of the Outer Zone of Southwest Japan in the middle part of Shikoku and its composite cross section.

the two main phases of the tectonic movements in the Sangabawa Belt. The older one is that of the Nagahama-Ozu phase in Middle to Late Jurassic time and is characterized by the formation of nappes with large-scale recumbent folds. The recumbent folds show the vergence towards the south owing to the upthrusting emplacement of the Sanbagawa crystalline schists from north to south. Their axial trends have a tendency to be oriented parallel or subparallel to the general trend of the Sanbagawa Belt. The younger one is that of the Hijikawa phase during latest Jurassic and earliest Cretaceous time, and is characterized by the formation of large-scale upright folds oriented in en echelon fashion.

TSUKUDA (1980) and HADA (1981) made clear that the northern subbelt of the Chichibu Belt has been also affected by the same tectonic movements as in the Sanbagawa Belt. Figure 3 summarizes the tectono-stratigraphic characteristics of the both sides of the Kurosegawa Tectonic Zone. It schematically indicates the lithofacies, ages and the structural relationships of the formations. Recent enormous progress in the micropaleontology of conodonts and radiolarians made possible to discuss the accurate age of the formation. The Upper Triassic-Jurassic clastic formations in the northern subbelt of the Chichibu Belt are interpreted to be rather autochthonous (parautochthonous). These formations contain large boulders as granitic and rhyolitic rocks comparable to rocks of the Kurosegawa Tectonic Zone (TSUKUDA, 1980) and their sandstones characteristically contain rock-fragments derived from granitic rocks. These formations are structurally overlain by the older allochthonous formations (Upper Carboniferous and Permo?-Triassic) mainly consisting of chert, limestone, greenstone, and black shale. The development of the thrust sheets including nappes (TSUKUDA et al., 1981) corresponds to the tectonic movement of the Nagahama-Ozu phase in the Sanbagawa Belt. Successively, the parautochthonous and allochthonous formations have been affected by the tectonic movement of the Hijikawa phase which is characterized by the formation of the upright folds. Greenstone in this terrane contains not only pillowed basalt but also the abundant reworked sediments of basaltic composition and generally occurs as continuous layers or large bodies intimately coexisting even with psammitic rocks (Fig. 2).

Accordingly, the two terranes having the contrasting tectono-stratigraphic characteristics are juxtaposed across the Kurosegawa Tectonic Zone. The tectonic zone extends over 650 km along the middle part of the Outer Zone from Kyushu to the Kii Peninsula. The lenticular bodies characterizing the tectonic zone are generally more or less elongated and parallel to the general trend of the surrounding rocks of the Chichibu Belt, but are always separated from them by faults, thus appearing as exotic elements in the Chichibu Belt. The constituent rocks of the Kurosegawa Tectonic Zone are as follows: (1) unmetamorphosed Siluro-Devonian formations, (2) granitic and high-grade metamorphic rocks (about 400 m.y.), (3) medium-grade metamorphic rocks (about 400 m.y.), (4) jadeite-glaucophane schists of 240 m.y., and (5) pumpellyite-glaucophane schists of 310–380 m.y. (MARUYAMA, 1981; BANNO et al., 1981). These are wholly or partly enclosed by serpentinites suggesting that the deep-seated shear zone attaining to the upper mantle was present at least in certain geologic age and that the Kurosegawa Tectonic Zone is a serpentinite melange zone (SUZUKI et al., 1976; SUZUKI, 1977; HADA et al., 1979; MARUYAMA, 1981).

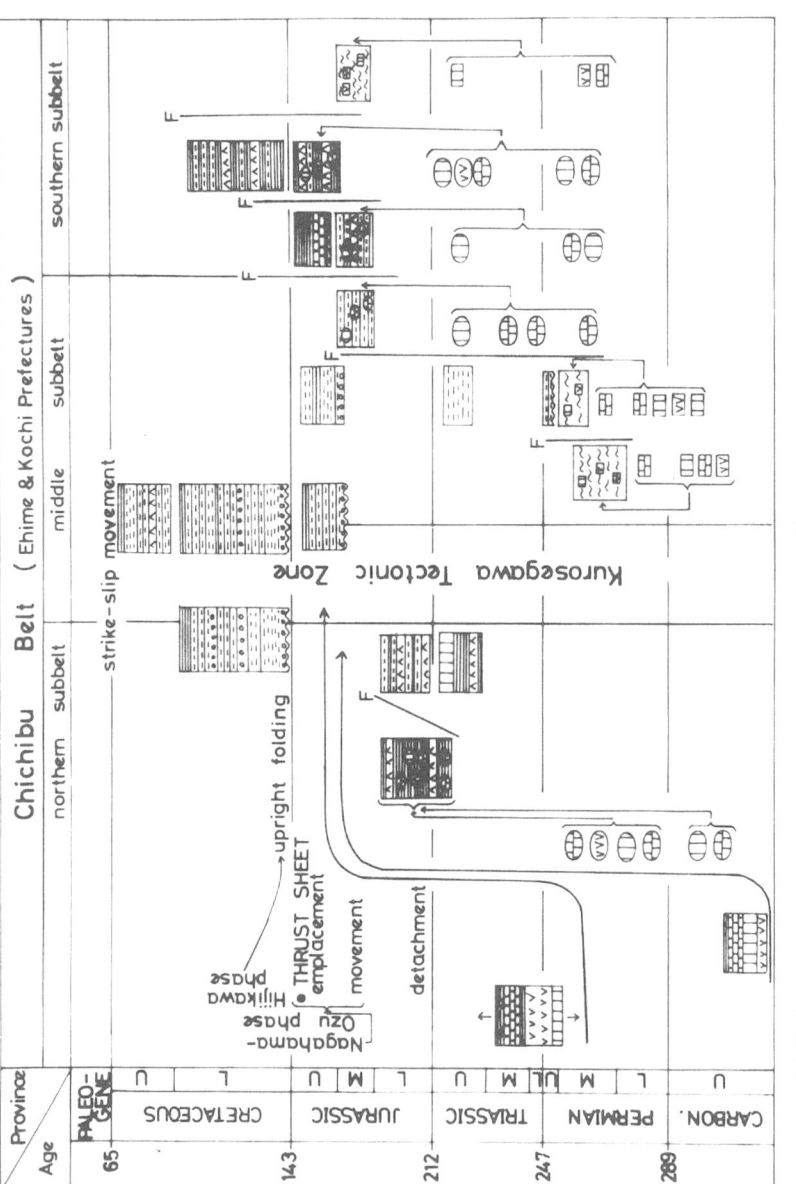

Fig. 3. Tectono-stratigraphic correlation chart of the Chichibu Belt in Kochi and Ehime Prefectures. 1. chert, 2. limestone, 3. mafic volcanics and volcaniclastics, 4. acidic tuff, 5. mudstone, 6. sandstone, 7. olistolith, 8. tectonic block, 9. tectonic melange. Northern subbelt; Sources: ISOZAKI and MATSUDA (1980), ISOZAKI et al. (1981), HADA (1981). Middle and Southern subbelts; Sources: HADA (1974), TAIRA et al. (1979), NAKATANI and YAO (1980), SATO and MATSUDA (1981), MATSUOKA and YAO (1981).

3. Geophysical Data

Recently, KIMURA and OKANO (1980) proposed a model of the crust and the upper mantle in Shikoku, which is based on apparent velocities of P wave from shallow earthquakes. The structure section of the continental part in this model is successfully joined to that of the ocean part which was determined by YOSHII *et al.* (1973). As a matter of fact, the correspondence of the geologic structure and the crustal model was made clear by this study. Figure 4 is a revised model of the crustal structure by Okano and Kimura. This is the model for a vertical cross section in the N22°W-S22°E direction approximately perpendicular to the geologic structures. The focal depth distribution of earthquakes projected on the same cross section is also overprinted on the figure.

The M-discontinuity is deeper than 35 km in the Sanbagawa Belt, but the crustal layers are gradually thinner towards the south of the Kurosegawa Tectonic Zone. Consequently, the crust in the area to the north of the Kurosegawa Tectonic Zone seems to be composed of a continental crust involving both the granitic and basaltic layers. Although the Conrad-discontinuity can not be determined accurately, a model in which the velocities of P wave gradually increase in proportion to the depth from just beneath the seismically active zone in the crust to the Conrad (6.3 to 6.7 km/sec) is proposed. The M-discontinuity abruptly rises up to the depth level of about 22–23 km in the southern side of the Kurosegawa Tectonic Zone with disappearance of the

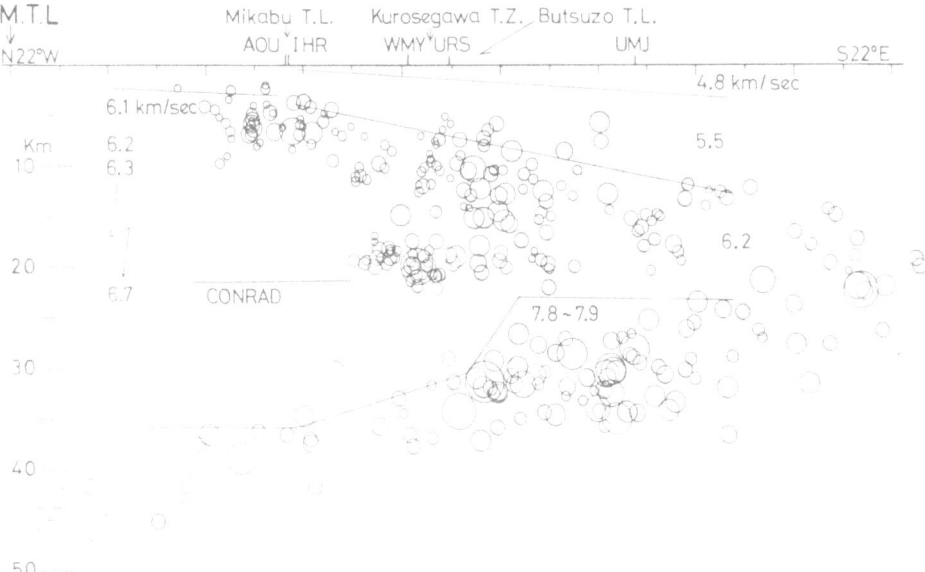

Fig. 4. A proposed model of the crust and the uppermost mantle in a vertical cross section in the direction of N22°W-S22°E. Focal depth distribution of earthquakes is projected on the same cross section. Larger and smaller circles indicate the magnitude of earthquakes. AOU, IHR, WMY, URS, and UMJ indicate positions of observation stations.

basaltic layer in the ocean side. Consequently, the 6.2 km/sec layer which extends oceanward under the Tosa-kaidan (forearc basin) is directly underlain by the layer with P wave velocity of 7.8–7.9 km/sec (mantle layer). It means that the boundary of the crustal structure between the typical continent and ocean is situated on the southern side of the Kurosegawa Tectonic Zone (HADA *et al.*, 1982a).

The abrupt change in crustal structure across the Kurosegawa Tectonic Zone is also suggested from the distribution of foci of earthquakes (Fig. 4). It is really remarkable that a blank on the focal distribution of earthquakes (aseismic zone) is existent under the Kurosegawa Tectonic Zone and it is observed to dip north steeply. The upper boundary of the focal distribution of crustal earthquakes is very shallow in the Sanbagawa Belt and inclined downward to the south in the Shimanto Belt. It approximately corresponds to the upper boundary of the 6.1–6.2 km/sec layer. On the other hand, the lowermost boundary of the focal distribution of the crustal earthquakes is remarkably discontinuous across the aseismic zone under the Kurosegawa Tectonic Zone. Namely, its depth is about 12 km in the area to the north of the Kurosegawa Tectonic Zone and is, in turn, 20–22 km in the area to the south of the tectonic zone. HADA *et al.* (1982a) suggested that the deeper part of the Kurosegawa Tectonic Zone extends to this aseismic zone.

4. Summary and Discussion

On the basis of the lithologic, structural, and geophysical data summarized above, it is considered that the tectonic environments of the Outer Zone of Southwest Japan are as follows: (1) The rock formations in the Northern Terrane to the north of the Kurosegawa Tectonic Zone were formed in the marginal sea-type tectonic setting (HADA *et al.*, 1979). The hypothesized marginal sea was existed in the northern part of the present Kurosegawa Tectonic Zone in pre-Middle Jurassic time and was closed completely accompanying the emplacement of nappes and thrust sheets over the older arc from the north during Middle and Late Jurassic time. (2) The tectonic melange in the Southern Terrane to the south of the Kurosegawa Tectonic Zone is recognized as the accretionary complex. It includes tectonically off-scraped blocks from the subducting ocean lithosphere (pelagic sediments and greenstones) as well as the trench-fill and slope sediments that have been tectonically deformed and ultimately kneaded into the material added to the continental or island-arc margin. The flysch sediments can be best documented as the forearc basin sediments deposited on the "basement" of the accretionary complex. Accordingly, the Pacific-type tectonic environment is postulated for the Southern Terrane.

Figure 5 is a proposed schematic crustal section in the Outer Zone of Southwest Japan. The 6.1–6.6 km/sec layer in the area to the north of the Kurosegawa Tectonic Zone indicates a granitic layer involving the older arc basement. Across the Kurosegawa Tectonic Zone, the crustal structure has been changed from the continental to oceanic one. This boundary is situated at by far the more landward side as compared with that in Northeast Japan which has been considered to be a typical island-arc. We consider that this difference is directly attributable to the geologic characteristics of the Outer Zone of Southwest Japan. In the area to the south of the

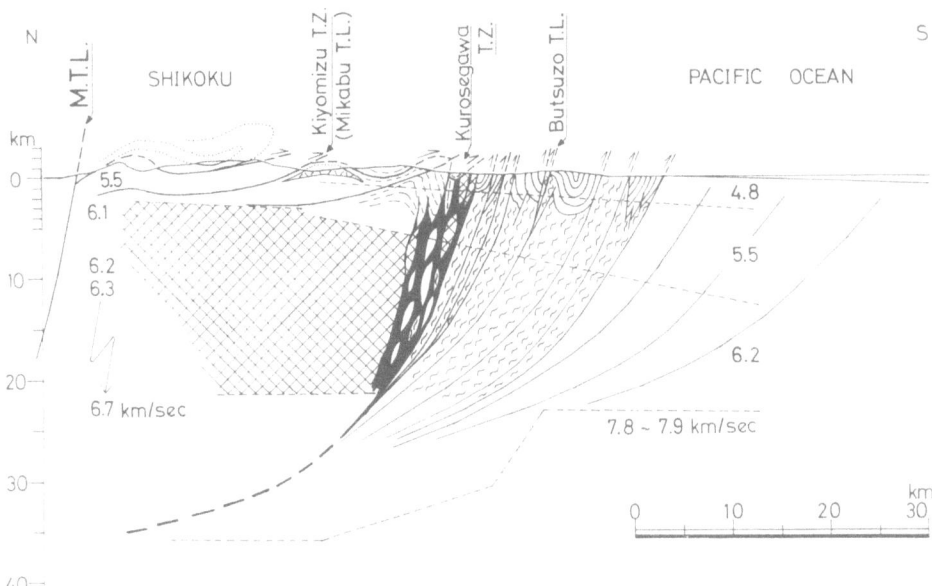

Fig. 5.   A proposed schematic crustal section in the Outer Zone of Southwest Japan. Reticulate pattern indicates the granitic layer involving the older arc basement.

Kurosegawa Tectonic Zone, the existence of much larger volumes of the accretionary complex is expected comparing to the case of Northeast Japan. Although there has been controversy for many years concerning the nature of the crust beneath the forearc basin (CURRAY *et al.*, 1977; HAMILTON, 1977; DICKINSON and SEELY, 1979; KARIG *et al.*, 1980; KIECKHEFER *et al.*, 1980), we suggest that the basin off Shikoku is underlain by accretionary complex and probably is not directly underlain by any upperplate continental and oceanic crusts. The 6.2 km/sec layer which is directly underlain by 7.8–7.9 km/sec layer is interpreted as being composed of highly metamorphosed part of the accretionary complex and probably includes plutons of granitic rocks which were produced by the partial melting of deeper parts of the accretionary pile.

The present Kurosegawa Tectonic Zone corresponds to the suture zone which marks the boundary between the Northern and Southern Terranes. Across the tectonic zone, the accretionary complex being constructed under the Pacific-type tectonic environment in the ocean side adjoins to the continental crust in the landward side. On this point of view, the Kurosegawa Tectonic Zone is identified as being the upper slope discontinuity defined by KARIG and SHARMAN (1975). It is very difficult to restore the relevant tectonic setting of the constituent rocks of the Kurosegawa Tectonic Zone which are enclosed by serpentinites. During the emplacement of serpentinites derived from the mantle wedge, various kinds of rocks different in lithology and age were gathered and squeezed upward from the various spaces and depths. HADA *et al.* (1979) interpreted that granitic rocks and high- and medium-grade metamorphic rocks in the tectonic zone correspond to the constituent rocks of various depths of the continental crust involving the older arc basement. The granitic rocks and the calc-alkaline rocks of the Siluro-Devonian formations have been considered to have formed a volcano-plutonic formation (KANO, 1971; MAEJIMA and YOSHIKURA, 1976) and may be generated in the active continental margin under the tensionally tectonic conditions (HADA *et al.*, 1979). High *P* and low *T* schists such as jadeite-glaucophane schists are interpreted as being the metamorphic rocks derived from the deeper parts of the accretionary prism adjoining to the Kurosegawa Tectonic Zone in the ocean side. The timing of the emplacement of serpentinites would be mainly during the tectonic movements of the Nagahama-Ozu and Hijikawa phases in Middle Jurassic to earliest Cretaceous time.

Figure 6 illustrates the proposed crustal section after the tectonic movement of the Hijikawa phase in earliest Cretaceous time. At this time, the hypothesized marginal sea was closed completely and the entire terrane to the north of the Kurosegawa Tectonic Zone in the Outer Zone had become a part of the frontal arc. In the mean time, the

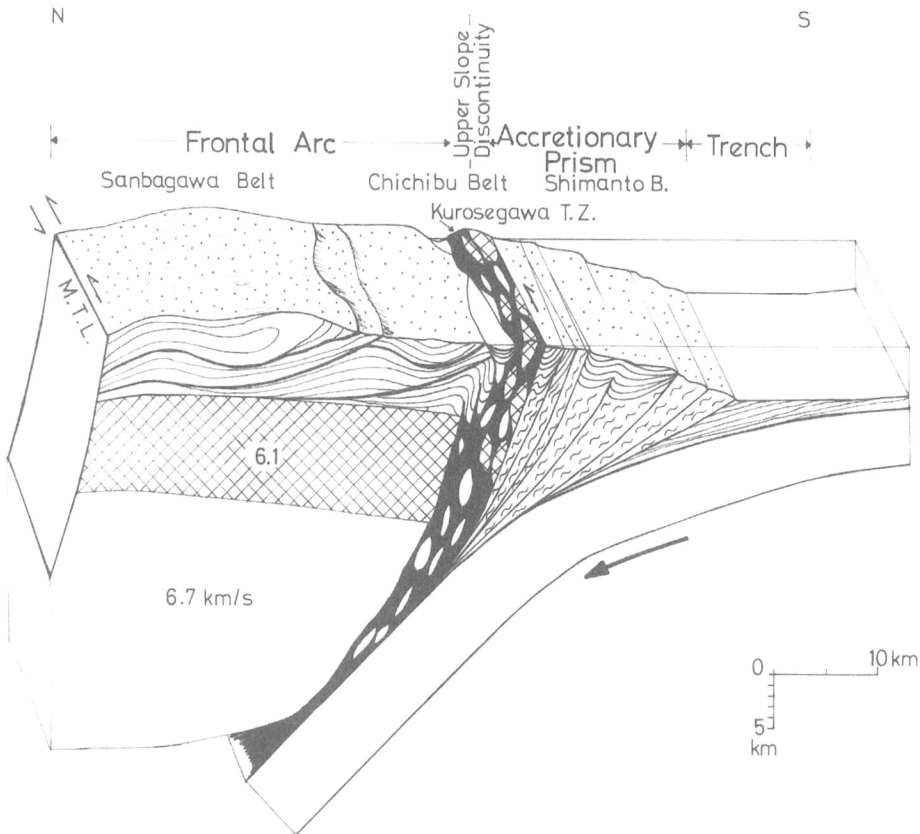

Fig. 6. Restored diagram illustrating the proposed crustal section after the tectonic movement in earliest Cretaceous time. Reticulate pattern indicate the granitic layer involving the older arc basement.

subducting process had been proceeded at the present site of the Shimanto Belt to the south of the Kurosegawa Tectonic Zone.

In post-Cretaceous time, the strike-slip movement has been known in the area of the Kurosegawa Tectonic Zone inferred from the fracture analysis of the granitic rocks in the tectonic zone (HADA, 1972; 1974) and the structural analysis of the Cretaceous strata distributing along the tectonic zone (ICHIKAWA and IKUMA, 1977; HADA *et al.*, 1982b). Because of this movement, the Kurosegawa Tectonic Zone became a zone of much complicated geology.

The authors would like to express their appreciation to Prof. J. F. Dewey of the State University of New York and Mr. S. Yoshikura of Kochi University for the valuable discussions and suggestions. Thanks are extended to Dr. M. Hashimoto of National Science Museum in Tokyo who read this manuscript critically.

REFERENCES

BANNO, S., S. MARUYAMA, T. NAKAJIMA, and T. MATSUDA, Kurosegawa Tectonic Zone in Southwest Japan, in *Oji International Seminar on Accretion Tectonics, Abstract*, p. 15, 1981.

CURRAY, J. R., G. G. SHOR, Jr., R. W. RAITT, and M. HENRY, Seismic refraction and reflection studies of crustal structure of the eastern Sunda and western Banda arcs, *J. Geophys. Res.*, **82**, 2479–2489, 1977.

DICKINSON, W. R. and D. R. SEELY, Structure and stratigraphy of forearc region, *Bull. Am. Assoc. Petrol. Geol.*, **63**, 2–31, 1979.

HADA, S., Fracture system developed in the Mitaki Igneous Rocks of Mt. Torigata, Kochi Prefecture, Japan, *Res. Rep. Kochi Univ.*, **21**, 63–85, 1972(in Japanese with English abstract).

HADA, S., Construction and evolution of the intrageosynclinal tectonic lands in the Chichibu Belt of western Shikoku, Japan, *J. Geosci., Osaka City Univ.*, **17**, 1–52, 1974.

HADA, S., Structure of the Chichibu Belt in the Agawa-Niyodo area, Kochi Prefecture, *Studies on Late Mesozoic Tectonism in Japan*, **3**, 39–47, 1981(in Japanese).

HADA, S., T. SUZUKI, S. YOSHIKURA, and N. TSUCHIYA, The Kurosegawa Tectonic Zone in Shikoku and tectonic environment of the Outer Zone of Southwest Japan, in *The basement of the Japanese Islands— Professor Hiroshi Kano Memorial Volume*, pp. 341–368, Akita Univ., 1979(in Japanese with English abstract).

HADA, S., T. SUZUKI, K. OKANO, and S. KIMURA, Crustal section based on the geological and geophysical features in the Outer Zone of Southwest Japan, *Geol. Soc. Japan, Mem.*, **21**, in press, 1982a.

HADA, S., M. TASHIRO, and T. IKUMA, Geological structure of the Upper Cretaceous strata in the Monobe area, Shikoku, *Paleont. Soc. Japan, Spec. Pap.*, **26**, in press, 1982b.

HAMILTON, W., Subduction in the Indonesian region, in *Island Arcs, Deep Sea Trenches, and Back-Arc Basins*, edited by M. Talwani and W. C. Pitman, pp. 15–31, AGU, Washington, D.C., 1977.

HARA, I., K. HIDE, K. TAKEDA, E. TSUKUDA, M. TOKUDA, and T. SHIOTA, Tectonic movement in the Sambagawa Belt, in *The Sambagawa Belt*, edited by K. Hide, pp. 307–387, Hiroshima Univ., 1977(in Japanese with English abstract).

ICHIKAWA, K. and T. IKUMA, Extension of the Kaminirogawa Tectonic Line (Chichibu Belt in Shikoku), *Rep. Geol. Soc. Japan, Kansai Branch*, **81**, 3–4, 1977 (in Japanese).

ICHIKAWA, K., K. ISHII, C. NAKAGAWA, K. SUYARI, and N. YAMASHITA, Die Kurosegawa-Zone, *J. Geol. Soc. Japan*, **62**, 82–103, 1954(in Japanese with German abstract).

ISOZAKI, I. and T. MATSUDA, Carboniferous dolomite confirmed by conodonts in the northern subbelt of the Chichibu Belt, *Rep. Geol. Soc. Japan, Kansai Branch*, **86**, 6–7, 1980 (in Japanese).

ISOZAKI, I., T. MATSUDA, and H. SANO, Conodont biostratigraphy of the dolomites distributed in the Odo area of the northern subbelt of the Chichibu Belt, Kochi Prefecture, *Rep. Geol. Soc. Japan, Kansai Branch*, **88**, 1–2, 1981 (in Japanese).

KARIG, D. E. and G. F. SHARMAN, Accretion and subduction in trenches, *Geol. Soc. Am. Bull.*, **86**, 377–389, 1975.

KARIG, D. E., M. B. LAWRENCE, G. F. MOORE, and J. R. CURRAY, Structural framework of the fore-arc basin, NW Sumatra, *J. Geol. Soc. London*, **137**, 77–91, 1980.

KANO, H., Studies on the Usugimu conglomerates in the Kitakami Mountains—Studies on the granite-bearing conglomerates in Japan, No. 22, *J. Geol. Soc. Japan*, **77**, 415–440, 1971(in Japanese with English abstract).

KIECKHEFER, R. M., G. G. SHOR, Jr., J. R. CURRAY, W. SUGIARTA, and F. HEHUWAT, Seismic refraction studies of the Sunda Trench and forearc basin, *J. Geophys. Res.*, **85**, 863–889, 1980.

KIMURA, S. and K. OKANO, Structure of the lower crust and the uppermost mantle in Shikoku, Japan, *J. Seismol. Soc. Japan, Ser. 2*, **33**, 157–168, 1980(in Japanese with English abstract).

MAEJIMA, W. and S. YOSHIKURA, Conglomerate of Permian Ukiishi Formation in the north of Yuasa, Wakayama Prefecture, *J. Geol. Soc. Japan*, **82**, 643–654, 1976(in Japanese with English abstract).

MARUYAMA, S., The Kurosegawa melange zone in the Ino district to the north of Kochi City, central Shikoku, *J. Geol. Soc. Japan*, **87**, 659–683, 1981.

MATSUOKA, S. and A. YAO, The Jurassic radiolarian assemblages in the Sakawa region, Kochi Prefecture, *Rep. Geol. Soc. Japan, Kansai Branch*, **89**, 4–5, 1981(in Japanese).

NAKATANI, T. and A. YAO, The radiolarian assemblage of the correlatives of the Torinosu Group in the western part of Shikoku, *Rep. Geol. Soc. Japan, Kansai Branch*, **86**, 5–6. 1980(in Japanese).

SATO, K. and T. MATSUDA, Re-examination of ages of the "Permian" in the middle subbelt of the Chichibu Belt by conodonts and radiolarians at the Sakawa area, Kochi Prefecture, *Rep. Geol. Soc. Japan, Kansai Branch*, **88**, 2–3, 1981(in Japanese).

SUYARI, K., Y. KUWANO, and K. ISHIDA, Discovery of the Late Triassic conodonts from the Sambagawa Metamorphic Belt proper in western Shikoku, *J. Geol. Soc. Japan*, **86**, 827–828, 1980(in Japanese).

SUZUKI, T., The Kurosegawa Tectonic Zone and the Chichibu Belt in Shikoku, in *The Sambagawa Belt*, edited by K. Hide, pp. 153–164, Hiroshima Univ., 1977(in Japanese with English abstract).

SUZUKI, T. and S. HADA, Cretaceous tectonic melange of the Shimanto Belt in Shikoku, Japan, *J. Geol. Soc. Japan*, **85**, 467–479, 1979.

SUZUKI, T., S. HADA, and S. YOSHIKURA, The history of development of the Kurosegawa Tectonic Zone with special reference to the Kurosegawa Tectonic Zone in Kochi Prefecture, *Toko-kiban (Basement of the Island Arc)*, **3**, 57–58, 1976 (in Japanese).

SUZUKI, T., S. HADA, and S. YOSHIKURA, Tectonic environment of the outer side of Southwest Japan, *Chikyu (The Earth Monthly)*, **1**, 57–62, 1979 (in Japanese).

TAIRA, A., K. NAKASEKO, J. KATTO, M. TASHIRO, and S. SAITO, New observations on the "Sanbosan Group" in western Shikoku, *Geologic News*, **302**, 22–35, 1979(in Japanese).

TSUKUDA, E., Allochthonous bodies of the Kurosegawa Tectonic Zone origin found in the "Chichibu Paleozoic Formation" in the northern part of Yokokura-yama, Kochi Prefecture, *Res. Struct. Geol. Assoc.*, **25**, 37–43, 1980 (in Japanese with English abstract).

TSUKUDA, E., I. HARA, R. TOMINAGA, M. TOKUDA, and T. MIYAMOTO, Geologic structure of the Chichibu Belt in the middle and western part of Shikoku, *Studies on Late Mesozoic Tectonism in Japan*, **3**, 49–59, 1981(in Japanese with English abstract).

YOSHII, T., W. J. LUDWIG, N. DEN, S. MURAUCHI, M. EWING, H. HOTTA, P. BUHL, T. ASAMA, and N. SAKAJIRI, Structure of Southwest Japan margin off Shikoku, *J. Geophys. Res.*, **78**, 2517–2525, 1973.

Accretion Tectonics in the Circum-Pacific Regions, edited by M. Hashimoto and S. Uyeda, 219–230.
Copyright © 1983 by Terra Scientific Publishing Company (TERRAPUB), Tokyo.

Accretionary Melange of Cretaceous Age in the Shimanto Belt in Japan

T. Suzuki and S. Hada

Department of Geology, Kochi University,
Kochi 780, Japan

From the geological, lithological, structural, and paleontological data of the Shimanto Belt, it was made clear that Lower Cretaceous melange blocks including a part of oceanic crust are scattered in the Upper Cretaceous matrix of the melange zone. Moreover, Upper Cretaceous turbidite sequence is bounded with the melange zone by northerly dipping high-angle thrusts. Accretionary melange of Cretaceous age may have been tectonically imbricated with the turbidite sequence during underthrusting in plate tectonic process. These assemblages in the Belt may represent a typical product of accretionary prism in a tectonic framework of the Pacific-type continental margin.

1. Introduction

Concerning to the tectonic environment of the Outer Zone of Southwest Japan in Late Mesozoic time, areas on both sides of the Kurosegawa Tectonic Zone (Fig. 1), which is considered to be a suture zone characterized by serpentinite melanges at present, make a remarkable contrast with each other (Suzuki *et al.*, 1979; Hada *et al.*, 1982). The southern side area, a main part of which is occupied by the Shimanto Belt, was in a "Pacific-type" tectonic environment characterized by imbricated complex-structures. On the contrary, the northern area in a "marginal sea-type" which is accompanied by large-scale nappes with low angle thrusts and recumbent folds.

The Shimanto Belt in Kyushu, Shikoku, and Kii Peninsula is separated from the Chichibu Belt to the north by the Butsuzo Tectonic Line and faces to the Pacific Ocean to the south (Fig. 1). The northern part of the Belt is occupied mainly by the Cretaceous sediments and the southern part by the Paleogene. However, their detailed succession has been not yet established in many places, because of complicated geologic structure and poverty of megafossils.

The purpose of this paper is to discuss certain aspects concerning the tectonic framework of the Shimanto Belt and especially to elucidate developing process of accretionary melange* in this Belt. Conclusively, we emphasize that the Shimanto Belt

---

*"Accretionary melange" bears a meaning of rock groups which have been accreted tectonically after they fell in a trench by gravity-sliding and deposited as sedimentary melange including a part of oceanic crust. In other words, it means that originally sedimentary melange formed by submarine sliding near a trench is transformed into tectonic melange which is scraped off and accreted to continental or island-arc margin by plate tectonic process.

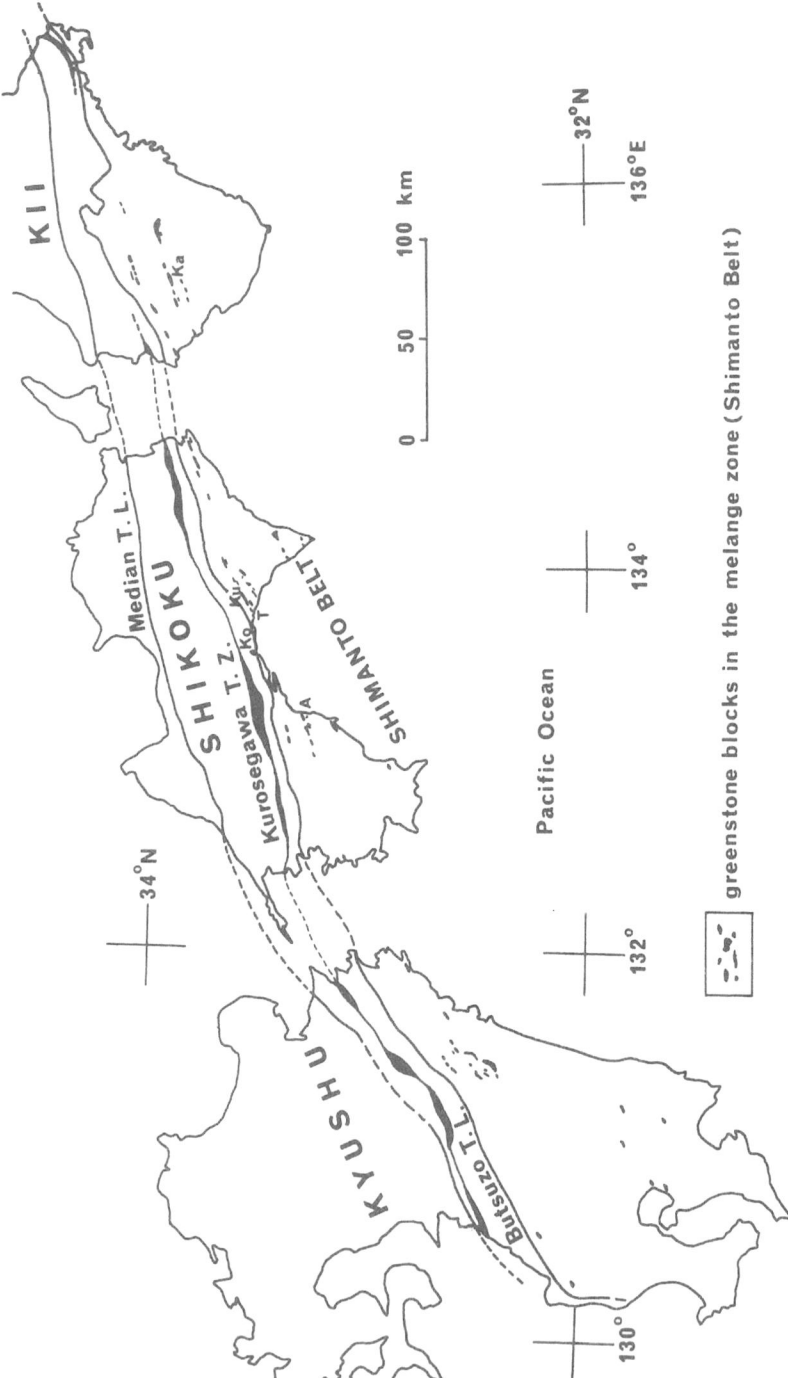

Fig. 1. Major structural elements in Outer Zone of Southwest Japan and greenstone blocks in the Shimanto Belt. A : Awa, Ka : Karuigawa, Ko : Kochi City, Ku : Kue, T : Tei.

consists of two assemblages characterized respectively by accretionary melange and turbidite sequence. The one of these two assemblages is considerably different from the other in lithologic facies, geologic structure, and environment of formation (SUZUKI and HADA, 1979). Moreover, the fundamental geologic structures of these assemblages are characterized by thrusting which gave rise to imbrication dipping to the north.

2. Geologic, Lithologic, and Tectonic Setting

Strata of the northernmost part of the Shimanto Belt (so-called "Morotsuka Group" in Kyushu, "Susaki Formation" in Shikoku, and "Hidakagawa Group" in Kii Peninsula) were thought to be mainly Neocomian to Santonian in age. They are assigned to two main assemblages which have a contrasting lithology and style of deformation. One assemblage consists mainly of intensively sheared and steeply dipping strata of mudstone. It includes mappable rock units composed of greenstones mainly of pillowed basalt and pelagic cherts as relatively isolated lenticular bodies in more abundant argillite matrix partly alternating with thin layers of acidic tuff. Cliff-scale observation indicates that the argillite matrix includes various sizes of slabs and blocks composed of dismembered sandstones, greenstones, limestones, pelagic red-shales, and cherts (Fig. 2). No large-scale fold is observed, and all the upward directions determined on pillow lavas face to the north. Penetrative foliation is developed in the argillite matrix trending consistently east-west and dipping steeply north in general. It shows intensively schistose appearance in some places. Judging from these structural style, this assemblage is best described as "melange zone"

Fig. 2. Sandstone blocks in the highly deformed argillite matrix of the melange zone of the Susaki Formation in the southern part of Awa region.

(Suzuki et al., 1978; Suzuki and Hada, 1979; Taira et al., 1979). On the contrary, the other assemblage consists mainly of sandstone and mudstone alternating in various thickness and of monotonous lithology, and its degree of deformation is very low (Fig. 3). It keeps sufficient continuity of layers except in the part of slumping masses and is deformed into open or close large-scale folds. Their axis-trends are nearly east-west. Sedimentary structures such as graded bedding and sole marks are sometimes recognizable and the assemblage here is designated as "turbidite sequence" (Suzuki and Hada, 1979) or "flysch unit" (Hada et al., 1982).

Both assemblages of contrasting lithology and tectonic style are generally bounded each other by conspicuous strike faults. Judging from characteristics in structural style and rock facies of both assemblages, it is reasonable to consider that the melange zone is a stratigraphically lower member than the turbidite sequence and originally the latter unconformably overlaid the former. This thought is supported by paleontological data, as described later. We suppose that this relationship may comparable to a tectonic unconformity characteristic of an arc-trench gap environment.

The mappable rock units composed of greenstones and pelagic cherts as relatively lenticular bodies which are generally 20–300 m thick, are scattered *only* in the melange zone. In general, greenstones contain massive basaltic lava, pillowed basalt, pillow breccia, hyaloclastite, and their reworked sediments. Commonly several different types

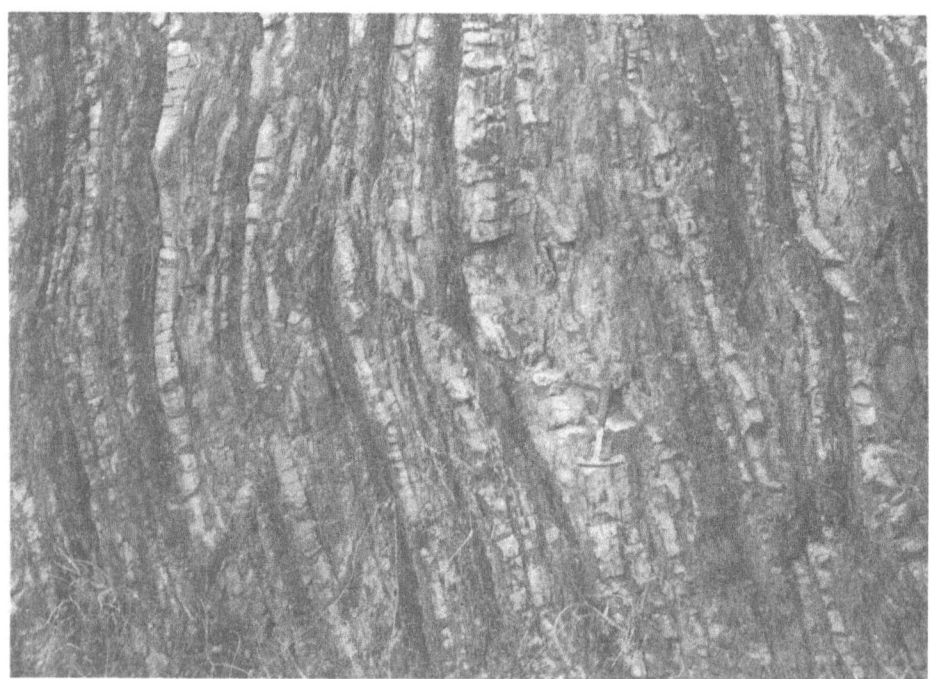

Fig. 3.   Monotonous layers of sandstones and mudstones in the turbidite sequence of the Susaki Formation in the southern part of Awa region.

of pillows such as trapdoor, flattened, elongate, cylinder, and hollow bulbous pillows observed at the ocean spreading ridge (BALLARD and MOORE, 1977), are recognized in individual flow units of the Shimanto Belt. Moreover, characteristics of non-vesicularity, variolitic structure and quench texture recognized in almost all of pillowed basalts in the Belt, suggest extrusion in relatively deep water environment such as ocean-floor. This consideration is not inconsistent with conclusion drawn from their chemical compositions. Namely, bulk chemical compositions of the basaltic rocks of the Shimanto Belt coincide with nearly that of the ocean-floor basalts (SUGISAKI et al., 1979). Chemical data of calcium rich clinopyroxenes show a tendency of change of magma-type from tholeiitic (lower part) toward alkali-basalt type (upper part) in a volcanic sequence at Kue, eastern Shikoku (Fig. 1), interesting in correspondence to that of volcanic islands or seamounts, although more detailed chemical information is necessary (SUZUKI and HADA, 1979). By the contents of Ti, Ni, and Cr, it is available to distinguish the origin of tholeiitic rocks (e.g. JAKES and WHITE, 1972; ISHIZUKA, 1981). On the Karuigawa region of the Hidakagawa Group, central part of Kii Peninsula (Fig. 1), as described later again, Ti (5,900–13,500 ppm), Ni (39–112 ppm) and Cr (71–380 ppm) contents of greenstones are not similar to that of island-arc tholeiite or oceanic island tholeiite, but very nearly the same as abyssal tholeiite (SUZUKI and YAMAGUCHI, 1981). Conclusively, the basaltic rocks in the Shimanto Belt are considered to be ocean-floor basalts formed originally under deep water environment, including partly that of volcanic islands or seamounts.

Paleomagnetism of melange blocks including pillow lava, inter-pillow limestone and red-shale was studied by KODAMA (1981) in the Tei region in eastern Shikoku (Fig. 1). Directions of natural remnant magnitization of the melange blocks, both before and after bedding correction within each blocks, were significantly different not only from the present field's direction but within blocks. He concluded that the shallow inclination (0 to 20) relative to bedding planes suggests that the melange blocks were formed further south than the present latitude and they are allochthonous bodies removed to the present position by plate tectonic process.

There is rare occurrence of coarse-grained basic rocks and ultramafics in the Shimanto Belt. They are recognized in only four regions through the Belt. Figure 4 shows an occurrence of basic and ultramafic rock complex in the melange zone of the Hidakagawa Group of Kii Peninsula at Karuigawa (Fig. 1). They consist of six lenticular bodies surrounded by serpentinites or altered rocks (magnesite + serpentine + spinel + quartz) in the sheared pelitic matrix of the melange zone. The serpentinites include various sizes of slabs and blocks composed of serpentinized harzburgite, rodingite, gabbro, diabase, plagiogranite(?), and pillow lavas. These basic and ultramafic rocks are considered to be a kind of dismembered ophiolitic rocks. Hitherto metamorphic grade of the Shimanto Belt has generally been considered to belong to the prehnite-pumpellyite facies (HASHIMOTO et al., 1970; IMAI et al., 1971; SUZUKI, 1977), although IMAI et al. (1971) discovered regional metamorphic rocks belonging to the actinolite zone in the northeastern part of this Belt in Kyushu. However, it is important that most of gabbro and diabase in the melange zone of the Hidakagawa Group mentioned above, are metamorphosed under the prehnite-pumpellyite facies after they have suffered metamorphism of the greenschist or amphibolite conditions facies,

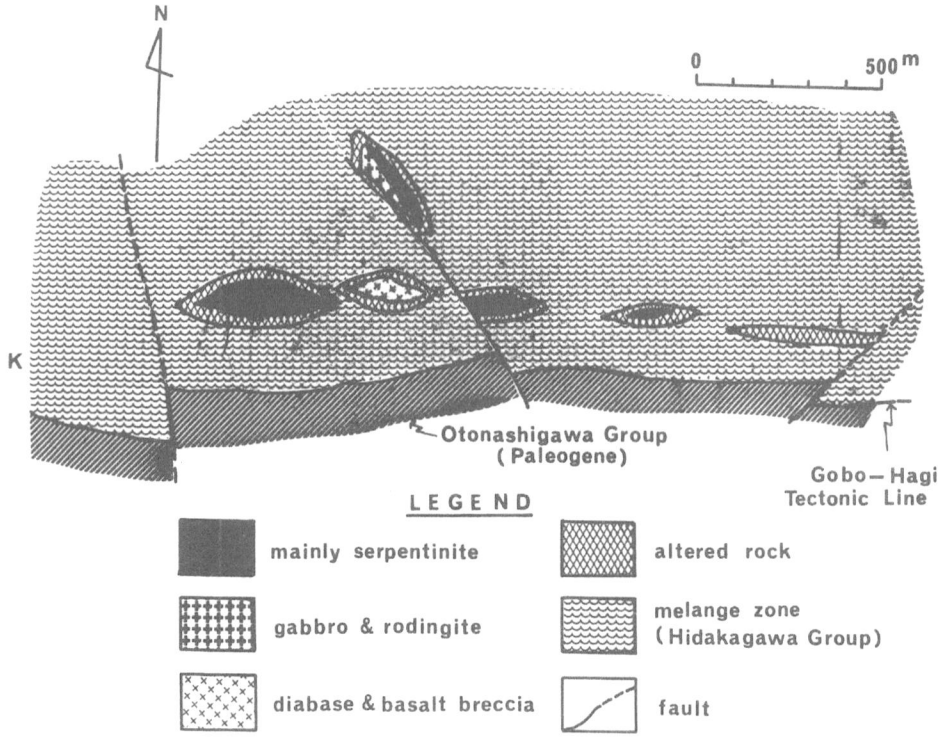

Fig. 4.   Geologic map of the Cretaceous Hidakagawa Group of the central part of Kii Peninsula.   K :
Karuigawa.

although ultramafics have not been dated. On the contrary, basaltic rocks belong only
to the prehnite-pumpellyite facies through whole the Shimanto Belt. We recognized by
X-ray method that illites in the sheared pelitic matrix surrounding melange blocks in
this region are phengitic, which may be higher pressure-type than muscovite common
in mudstones of the turbidite sequence there (SUZUKI and YAMAGUCHI, 1981).
Therefore, it is possible that these metamorphisms are comparable to ocean-floor
metamorphism and/or subduction metamorphism during active underthrusting.

    Sandstones in the melange zone occur always as lenticular bodies in sheared pelitic
matrix (Fig. 2). We can easily distinguish them from those of turbidite sequence with
the naked eye and microscopic features. A standard quartz-feldspar-rock fragment
diagram from the sandstones of the Susaki Formation and of the Hidakagawa Group is
shown in Figure 5. Sandstone blocks in the melange zone are fine-grained and fairly
well sorted with comparatively better rounded grains. Amounts of clay matrix are over
15%. Rock fragments consist of chert, polycrystalline quartz and basaltic rock in thin
sections. Therefore, the most sandstones of the melange zone are grouped into lithic-
wacke type according to OKADA's (1968) definition (Fig. 5). Unusual chert-rich lithic
sandstones were also reported from the Alaskan melange zone (CONNELLY, 1978).
DICKINSON (1978) and CONNELLY (1978) considered that recycling of uplifted

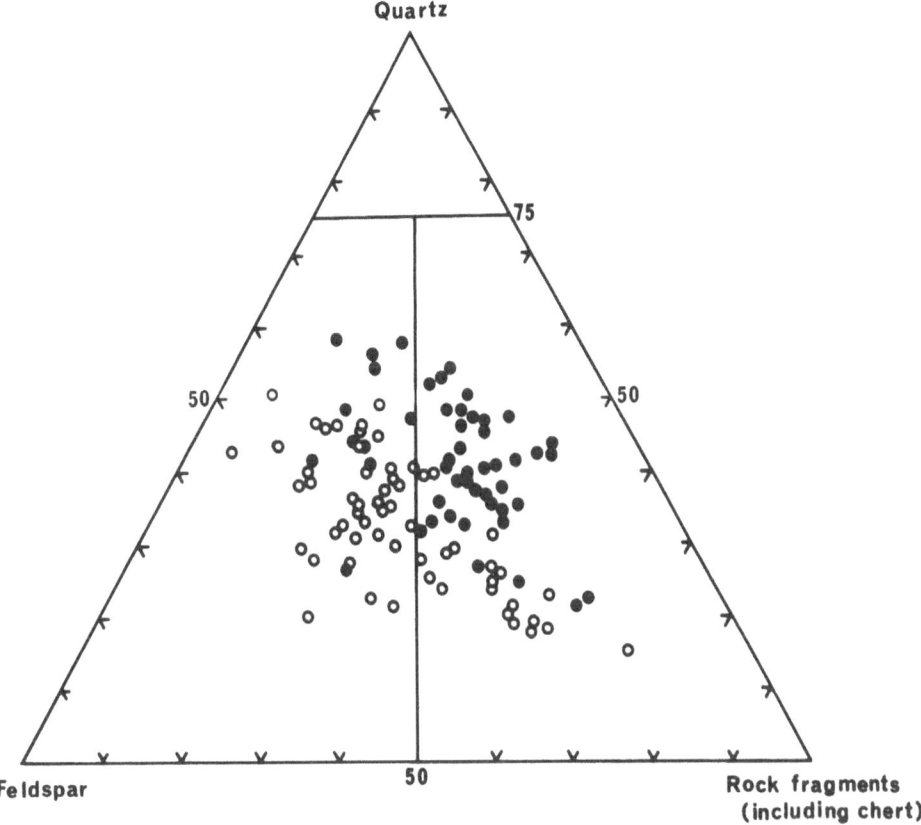

Fig. 5. Q-F-R diagram for the sandstones of the Susaki Formation and the Hidakagawa Group. Open circle: sandstone of the turbidite sequence; Solid circle: sandstone of the melange zone.

subduction complexes forms the important sources for these sandstones.

On the contrary, almost all of sandstones in the turbidite sequence are characterized by more smaller amounts of clay matrix, medium and subangular grains, and relatively bad sorting. Rock fragments are characterized by considerably terrigenous materials such as volcanic, granitic, metamorphic, and pelitic rocks and a small amount of chert in thin sections (SUZUKI *et al.*, 1978; SUZUKI and HADA, 1979). We recognized high average amounts of potassium feldspar by dyeing thin sections. That is, the average amounts of potassium feldspar and the average ratio of potassium feldspar to total feldspar (K/F) are 10.24% and 0.41 respectively; those of the melange blocks are 2.58% and 0.17 (SUZUKI and HADA, 1979). They are grouped into feldspathic arenite or feldspathic wacke of the Okada's definition (Fig. 5).

There is a distinctive difference between the sandstones of the melange zone and the turbidite sequence in Fig. 5, although the compositional field is continuous. It is suggested that the sandstone blocks of the melange zone have nature of more distal facies as compared with those of the turbidite sequence, because the former is characterized by the higher contents of stable grains and the lower contents of terrigenous rock fragments. Therefore, these sandstone blocks are considered to be products of recycling of uplifted subduction complex or distal facies originally deposited near the trench, as discussed later.

3. Geologic Profile Based on Geological, Structural, and Paleontological Data

Recent geophysical and geological studies on a continental margin indicate that an oceanic trench is a place of plate convergence. Moore and Karig (1976) have proposed that the type of sedimentary assemblage discussed in this paper is associated with the Shikoku subduction zone along the Nankai Trough where the Philippine Sea plate is underthrusted beneath the Asian plate. Kanmera (1976a, b) has revealed that the site of the development of the Shimanto Belt corresponds to that formed under the tectonic environment of modern arc-trench gap.

A typical example of geologic profile of the Susaki Formation in the Shimanto Belt at Awa (Fig. 1), southwestern part of Kochi City, is shown in Fig. 6. Chaotically and pervasively sheared melange zone including greenstones and pelagic cherts and weakly deformed turbidite sequence alternating sandstones and mudstones in good order, are tectonically imbricated by many thrusts dipping to the north, forming a sandwitch-like structure.

The assemblage in the turbidite sequence always shows Upper Cretaceous age from fossil evidences. The age of radiolarian siltstones is Coniacian (*Archaeodictyomitra* cf. *rhadina* Assemblage) (Hada *et al.*, 1982). From the rock facies and sedimentary structure, the assemblage may correspond to comparatively shallower water sediment of the midslope or slope basin. On the other hand, the age of matrix of the melange zone here ranges from Cenomanian to Coniacian (*Dictyomitra duodecimcostata* Assemblage), while that of blocks is Valanginian to Aptian (*Archaeodictyomitra* cf. *conica* Assemblage) (Hada *et al.*, 1982). Although the age of the greenstones is uncertain, the Valanginian red chert is intimately associated with the greenstones in this region (Fig. 6). Taira (1981) also recognized similar relationships in some places of the Shimanto Belt in Shikoku. The fact that melange blocks of various ages are older than the surrounding pelitic sediments supports the inference that these melange blocks are exotic. Pelitic rocks surrounding exotic blocks may have been deposited, mixing with the blocks, under comparatively deeper water environment such as near-trench. Therefore, the assemblage in the melange zone and turbidite sequence in the Shimanto Belt may represent the typical product of accretionary prism in the Pacific-type continental margin.

4. Conclusion

Figure 7 shows a possible development picture of the Shimanto Belt during the Late Cretaceous time. Original rocks of the melange zone have been brought from both continental and oceanic crusts. Products by submarine sliding may have derived from normally faulted outer- and inner-slope of the trench and/or from inner-slope by thrust faults. Deep-sea sedimentary melange sequences including trench-fill sediments may have deformed to tectonic melange under the process of tectonic compression such as

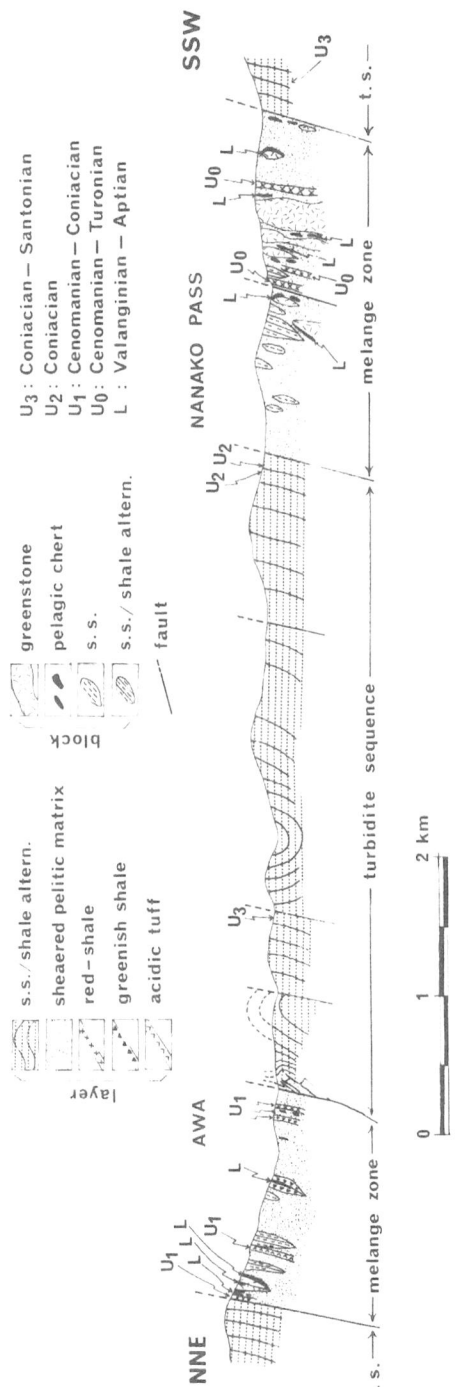

Fig. 6.   Simplified geologic profile of the Susaki Formation at Awa.

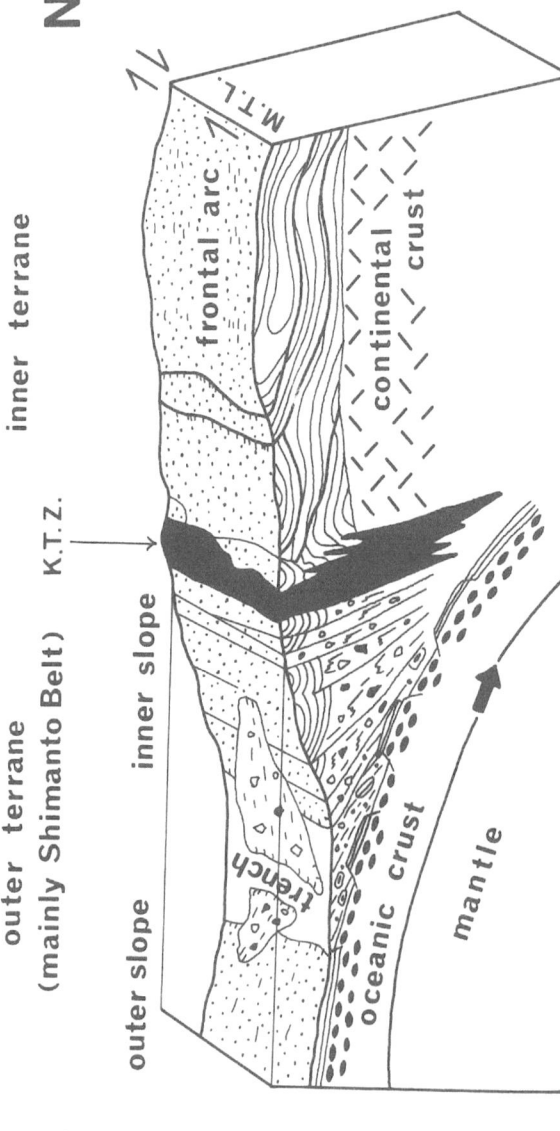

Fig. 7. A possible development picture of the Shimanto accretionary melanges in the Late Cretaceous stage (no vertical exaggeration). M.T.L.: Median Tectonic Line; K.T.Z.: Kurosegawa Tectonic Zone.

subduction or accretion. During underthrusting, brittle-type rocks were further broken into blocks of various size and suspended in the less competent matrix of mudstone. The highly sheared accretionary melange has been tectonically imbricated with the folded turbidite sequence. Many thrust faults which are steeply dipping to the north, separated the former from the latter and gave rise to imbrication with a southward polarity. In conclusion, the tectogenesis in the Shimanto Belt corresponds to the site of tectonic environment of modern arc-trench gap. There is no evidence of disruption by large-scale strike-slip fault in the Shimanto Belt.

The authors wish to express their sincere gratitude to Prof. J. F. Dewey of the State University of New York for valuable discussion and suggestion, and to Dr. M. Hashimoto of National Science Museum for critical reading of the manuscript. We also thank Mr. Y. Yamaguchi of the Tosa Senior High School for field works and chemical analyses.

REFERENCES

BALLARD, R. D. and J. G. MOORE, *Photographic Atlas of the Mid-Atlantic Ridge Rift Valley*, 114 pp., Springer-Verlag, New York, Heidelberg, Berlin, 1977.

CONNELLY, W., Uyak complex, Kodiak islands, Alaska: A Cretaceous subduction complex, *Geol. Soc. Am., Bull.*, **89**, 755–769, 1978.

DICKINSON, W. R., Plate tectonic influences on sandstone compositions, *Abstracts with programs, Geol. Soc. Canada, Min. Assoc. Canada and Geol. Soc. Am.*, **3**, 389, 1978.

HADA, S., T. SUZUKI, S. YOSHIKURA, and N. TSUCHIYA, The Kurosegawa Tectonic Zone in Shikoku and tectonic environment of the Outer Zone of Southwest Japan, in *The Basement of the Japanese Islands— Professor Hiroshi Kano Memorial Volume*, Akita Univ., 341–368, 1979 (in Japanese with English abstract).

HADA, S., T. SUZUKI, K. OKANO, and S. KIMURA, Crustal section based on the geological and geophysical features in the Outer Zone of Southwest Japan, *Geol. Soc. Japan, Mem.*, **21**, in press, 1982.

HASHIMOTO, M., S. IGI, Y. SEKI, S. BANNO, and G. KOJIMA, *Metamorphic Facies Map of Japan*, Geol. Survey of Japan, 1970.

IMAI, I., Y. TERAOKA, and K. OKUMURA, Geologic structure and metamorphic zonation of the northeastern part of the Shimanto terrane in Kyushu, Japan, *J. Geol. Soc. Japan*, **77**, 207–220, 1971(in Japanese with English abstract).

ISHIZUKA, H., Geochemistry of the Horokanai ophiolite in the Kamuikotan tectonic belt, Hokkaido, Japan, *J. Geol. Soc. Japan*, **87**, 17–34, 1981.

JAKES, P. and A. J. R. WHITE, Major and trace element abundances in volcanic rocks of orogenic areas, *Geol. Soc. Am., Bull.*, **83**, 29–40, 1972.

KANMERA, K., Comparison between past and present geosynclinal sedimentary bodies I, *Kagaku (Science)*, **46**, 284–291, 1976a(in Japanese).

KANMERA, K., Comparison between past and present geosynclinal sedimentary bodies II, *Kagaku (Science)*, **46**, 371–378, 1976b(in Japanese).

KODAMA, K., Paleomagnetism of the Shimanto Belt in Shikoku, *Abstracts Oji Int. Seminar on Accretion Tectonics*, 22, 1981.

MOORE, J. C. and D. E. KARIG, Sedimentology, structural geology, and tectonics of the Shikoku subduction, Southwestern Japan, *Geol. Soc. Am., Bull.*, **87**, 1259–1268, 1976.

OKADA, H., Classification and nomenclature of sandstones: *J. Geol. Soc. Japan*, **74**, 371–384, 1968(in Japanese with English abstract).

SUGISAKI, R., T. SUZUKI, K. KANMERA, T. SAKAI, and H. SANO, Chemical compositions of green rocks in the Shimanto Belt, southwest Japan, *J. Geol. Soc. Japan*, **85**, 455–466, 1979.

Suzuki, T., Paleo-volcanism and metamorphism in the Outer Zone of Southwest Japan, *Nauka, Moscow, USSR*, 148–163, 1977 (in Russian with English abstract).

Suzuki, T. and S. Hada, Cretaceous tectonic melange of the Shimanto Belt in Shikoku, Japan, *J. Geol. Soc. Japan*, **85**, 467–479, 1979.

Suzuki, T. and Y. Yamaguchi, Genetic consideration of ultramafic and basic rocks in the southern part of the Hidakagawa Group, Wakayama Prefecture, in *Abstracts 88th Annual Meet., Geol. Soc. Japan*, 368, 1981 (in Japanese).

Suzuki, T., S. Hada, H. Umemura, K. Kado, Y. Sakamoto, and H. Nakagawa, Genetic consideration of the green rocks in the Shimanto Belt—with special reference to mode of occurrence—, *Chikyu kagaku (Earth Sci.)*, **32**, 321–330, 1978(in Japanese with English abstract).

Suzuki, T., S. Hada, and S. Yoshikura, Tectonic environment of the outer side of Southwest Japan: *Chikyu (The Earth Monthly)*, **1**, 57–62, 1979(in Japanese).

Taira, A., The process of formation of the Shimanto Belt, *Kagaku (Science)*, **51**, 516–523, 1981(in Japanese).

Taira, A., K. Nakaseko, J. Katto, M. Tashiro, and S. Saito, New observations on the "Sanbosan Group" in western Shikoku, *Geologic News*, **302**, 22–35, 1979(in Japanese).

Accretion Tectonics in the Circum-Pacific Regions, edited by M. Hashimoto and S. Uyeda, 231–241.

Paleomagnetism of the Shimanto Belt in Shikoku, Southwest Japan

K. KODAMA,* A. TAIRA,* M. OKAMURA,* and Y. SAITO**

*Department of Geology, Kochi University, Kochi 780, Japan
**Department of Geology, National Science Museum, Tokyo 160, Japan

Within the Shimanto terrain in Shikoku, various blocks and slivers of pillowed basaltic rocks, radiolarian chert and pelagic carbonate rocks are incorporated in fine-grained sediment matrix. They form a so-called melange facies. A paleomagnetic study was carried out on the melange complexes and surrounding rock associations. Reliable results were obtained from three areas: Tei and Okitsu areas of Cretaceous melange and turbidite associations, and Cape Muroto area of gabbroic intrusives of Miocene age. Samples in Tei area were from pillow lavas, inter-pillow carbonate rocks and reddish shale, together with the terrigenous sediments which tectonically overlie the Tei melange. In Okitsu area, mudstones and intercalated thin tuffaceous sediments were collected. Directions of natural remanent magnetization of the melange blocks, both before and after tilting correction, are significantly different not only from the present geomagnetic field but between blocks. The directions after bedding correction have low inclination (0 to −20°) and deflected declination (20° to 90° westwards). Although the declinations are unreliable due to the chaotic nature of melange blocks, the shallow inclinations suggest that the melange blocks were formed in an equatorial region. Mudstones overlying the Tei melange, on the other hand, show bedding corrected remanence directions not far from the present field both in inclination and declination. From the distal turbidites at Okitsu and gabbroic dikes at Cape Muroto we obtained, after tilting correction, remanence directions clearly identified as reverse magnetization near the present latitude. These results suggest that the terrigenous deposits and some intrusive rocks consisting of a major portion of the Shimanto terrain are autochthonous, while the melange blocks are allochthonous or exotic complexes emplaced by some plate tectonic process.

1. Introduction

Recent progress in stratigraphic and paleontologic studies of the Shimanto belt, a thick Cretaceous to Tertiary highly deformed sedimentary belt (subduction complex) lying along the outermost part of Southwest Japan, has revealed that several zones of chaotic mixture of various masses, such as pelagic sediments and metabasalts, occur as allochthonous blocks in tectonic contact with terrigenous sediments of flysch facies (e.g., TAIRA et al., 1980). Based upon extensive radiolarian age analysis and lithologic examination of both the melange rocks and the surrounding clastic sediments, it has been suspected that the blocks in melange facies were parts of the oceanic plate, which were mixed with terrigenous sediments in a deep-sea environment like trench by subduction process (TAIRA *et al.*, 1980). In order to obtain another independent data on the origin of the melange complex, we have undertaken paleomagnetic studies of the Shimanto belt in Shikoku province since 1980. We report here the results of the preliminary study carried out on the Cretaceous melange and turbidite associations and a Miocene gabbroic dike.

2. Geologic Setting of the Shimanto Belt

Separated from the Chichibu belt (Paleozoic to Early Mesozoic subduction complex) to the north by the Butsuzo Tectonic Line, the Shimanto belt extends continuously from the south Nansei Islands to Boso Peninsula (Fig. 1). The belt is underlain mainly by thick sequences of graywacke and shale representing a variety of turbidite facies. The layers of the Shimanto belt are dipping generally northward at high angles showing a clear zonal distribution along their strike. Younging southward from the uppermost Jurassic to the Lower Miocene, the layers are regionally highly folded into isoclinal folds, and reveal imbricate structures bounded by major thrusting. The Lower Miocene rocks in the southernmost Shimanto belt are locally overlain, with angular unconformity, by Middle Miocene and later sediments accompanied, in some places, with contemporaneous acidic to basic volcanics and granites (KANMERA and HASHIMOTO, 1980).

The Shimanto belt in Shikoku province, as well as in the other regions, consists generally of alternated strata of sandstones and argillaceous rocks of various thicknesses. These rocks are commonly accompanied with slumps and olistostromes.

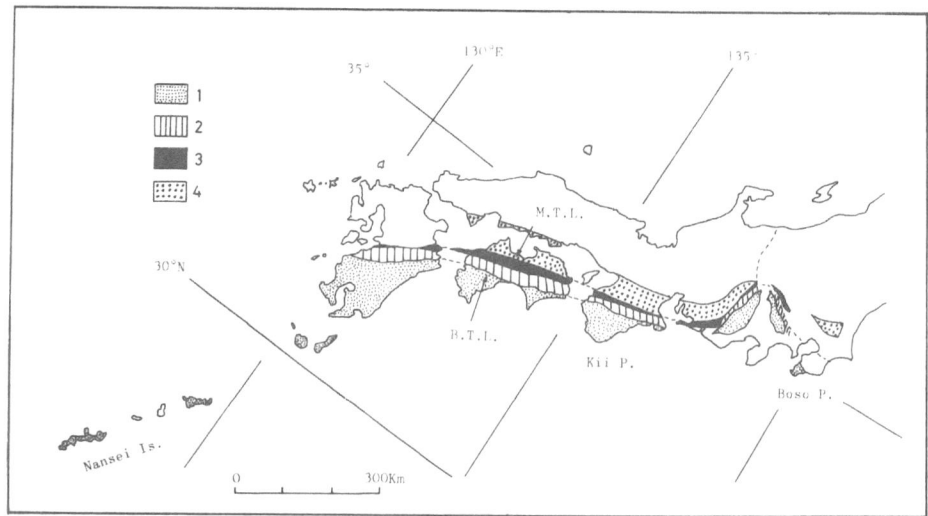

Fig. 1. Geologic outline map of the outer zone of Southwest Japan. 1: Shimanto belt, 2: Chichibu belt, 3: Sanbagawa metamorphic belt, 4: Ryoke metamorphic belt, M.T.L.: Median Tectonic Line, B.T.L: Butsuzo Tectonic Line.

According to the recent paleontologic studies of the region (OKAMURA, 1980), the turbidite facies as a whole tends to be younging southward from the northernmost belt of Albian age. In contrast, each turbidite belt, a few kilometers in thickness, indicates a northward younging trend. The belts of turbidite facies are separated in some places by a zone of so-called melange facies that is defined as a chaotic mixture of various blocks of sandstone, chert, basic volcanics and hyaloclastites in highly sheared shale matrix (TAIRA et al., 1980). The size of the blocks of basic volcanics is varied from several centimeters to over several hundreds of meters. Frequently larger blocks exhibit well-preserved pillow structures. Generally they are unmetamorphosed or only slightly metamorphosed to zeolite to pumpellyte-prehnite facies. Some pillow lavas are closely accompanied with pelagic sediments bearing radiolaria which shows the age significantly older (Valanginian to Santonian) than that of the melange matrix (Campanian) and the surrounding turbidite sequences (Coniacian to Campanian) (OKAMURA, 1980; TAIRA et al., 1980). For example, one of the northern melange zones yields radiolarian chert of Valanginian to Hauterivian age that is embedded in the shale matrix bearing Campanian radiolaria. In another melange zone, on the other hand, Albian radiolarian chert is overlain by Coniacian to Santonian turbidite. Although no absolute age of the associated basic lavas has been reported, the systematic relationship so far obtained by radiolarian age analysis suggests that blocks within the melange zones were not deposited in-situ but are likely to be older exotic blocks.

3. Paleomagnetic Sampling

We visited seven localities for paleomagnetic sampling as shown in Fig. 2. Four localities (AW, KU, OK, TI) are in the Cretaceous Shimanto belt and three (MT, OB, SK) in the Paleogene Shimanto. Except for the gabbroic dike at site MT, basaltic pillow lavas in melange zones were collected from all the sites, and their natural remanent

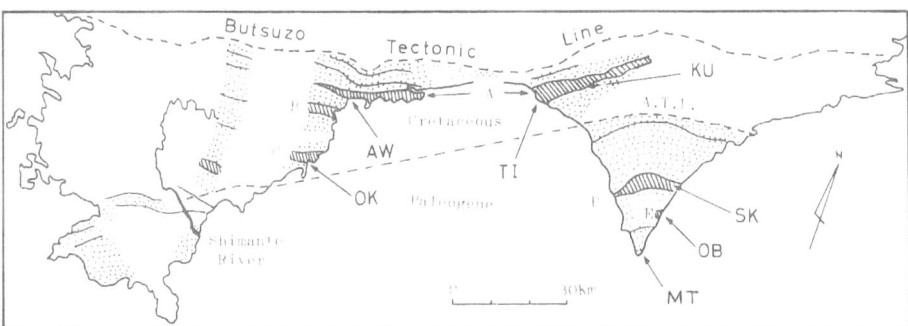

Fig. 2.   Generalized geologic map of the Shimanto belt in Shikoku showing site locations for paleomagnetic study. Dotted areas are alternations of sandstones and shales. TI, OK: Cretaceous melange zones and terrigenous sediments, AW, KU: Cretaceous melange zones, OB, SK: Paleogene melange zones, MT: Miocene gabbroic dike complex at Cape Muroto. Shaded areas are melange zones of Cretaceous (A, B, C) and Paleogene (D, E) ages. Cretaceous and Paleogene Shimanto belts are divided by Aki Tectonic Line (A.T.L.).

magnetization (NRM) was measured. After the stability test by progressive alternating field and thermal demagnetization of the NRM, the melange rocks at KU and SK were precluded for further investigation because of their unstable magnetic behaviors. Two other melange rocks at OK and OB were also rejected because no primary horizontal plane could be inferred due to severe internal folding or flexure. We will report hereafter the results from the following four localities: Tei melange and nearby terrigenous sediments (TI), Awa melange (AW) that is presumably a westward continuation of the Tei melange, the turbidite associations near Okitsu melange (OK), and the Miocene gabbroic dike complex at Cape Muroto (MT).

In the Tei melange, samples were collected from blocks of basaltic pillow lavas, inter-pillow limestone and red-coloured shale. For each of these samples, the initial bedding plane was estimated. In the case that bedding of pillow blocks could not be inferred with sufficient certainty by the observation of their downwarping shapes, we used the bedding of the associated materials such as inter-pillow sediments or surrounding hyaloclastite layers. Especially the inter-pillow carbonate deposits at two localities (TI and AW), each about 50 to 70 cm thick between basaltic pillows, showed clear stratification or lamination as a good indicator of the paleo-horizontal plane. Combining the bedding of these inter-pillow materials and the direction of protrusion resulting from molding of the pillows before solidification, we estimated the primary horizontal planes for the tilting correction of in-situ remanence directions. Interestingly, recent paleontologic study (OKAMURA *et al.*, 1980) disclosed that these laminated carbonate rocks contain abundant recrystallized nanno-fossils of Valanginian to Barremian age (Lower Cretaceous). It has been inferred that the origin of these carbonate rocks was nanno-ooze deposited on the basaltic lava. Taking into account that the Valanginian to Hauterivian chert blocks are also interbedded with basaltic tuff in the melange zone, it may be reasonable to assume that the eruption of the basaltic lavas and deposition of the pelagic sediments were more or less contemporaneous (TAIRA *et al.*, 1980). In Awa melange, we collected samples from pillow lavas and the associated sediments, but stable magnetization was found only for the inter-pillow carbonate rocks. Besides these melange blocks, samples were collected from mudstones within the distal turbidite which tectonically overlies the Tei melange. According to the paleontologic study (OKAMURA, 1980), these terrigenous rocks contain radiolarians of Coniacian to Santonian (Upper Cretaceous) age, that is much younger than the melange rocks. This age-lithology relationship is similar to those found in the other Cretaceous melange zones in the Shimanto belt. Similar terrigenous sediments were also sampled in the Okitsu area (site OK in Fig. 2) from mudstones and intercalated thin tuffaceous sediments, both representing turbidite facies. Although no reliable age-lithology relationship has been confirmed in this area, the age of the detrital sequences is suspected to be Upper Cretaceous, probably Coniacian to Campanian, based on several radiolarian ages so far reported in the nearby areas and the lithological similarities to the other Cretaceous Shimanto sequences. The last paleomagnetic result we report here is on a gabbroic intrusive complex outcropping locally at Cape Muroto. The complex, over 100 m thick, develops for about 500 m along the strike of the surrounding sediments showing clear evidence of thermal contacts. The age is believed to be Lower Miocene based on the radiolarian dating of the nearby sediments (A. Taira, personal communication, 1982) and preliminary absolute dating of the intrusion itself (Y. Takigami, personal communication, 1982). Samples were collected from two localities in the core of the complex, each about 10 m apart.

4. Paleomagnetic Results

A total of 202 specimens were thus obtained for magnetic measurement from 15 sites in the four areas introduced earlier (TI, AW, OK, and MT in Fig. 2). Among these 15 sites, seven sites are within the melange zones and other seven are in the area of detrital sediments and one in an intrusive complex. Most of the measurements were made with a spinner magnetometer of the Geophysical Institute, University of Tokyo, and some very weakly magnetized specimens (argillaceous rocks in TI and OK) were measured with a super-conducting magnetometer of Kobe University. Before summarizing the paleomagnetic results, let us briefly introduce rock-magnetic behaviors of representative specimens and the result of field test to pillow breccias. Figure 3 illustrates the behaviors during progressive alternating field (AF) and thermal demagnetization for several pilot specimens of melange rocks. Because both basaltic pillow lavas and inter-pillow limestones showed stable behaviors during AF and thermal treatments, the remaining samples were cleaned magnetically by the optimum demagnetization field with peak intensity of 200 to 400 Oe that was defined as the one yielding the best grouping. Reddish shale in the Tei melange was effectively demagnetized by thermal treatment up to 500°C. The terrigenous sediments and gabbroic rocks were also demagnetized by AF treatment, and the optimum field intensities were determined by the same criterion as in the case of melange rocks.

In order to supplement the laboratory tests for the stability of remanent magnetization, we applied a conglomerate test in Graham's sense (GRAHAM, 1949) to small fragments of pillow basalts occurring in various sizes within the hyaloclastite layers in the Tei melange. We collected seven pebbles of several to several tens of centimeters in size, and measured their NRM directions. As shown in Fig. 4 they seem to carry more or less secondary components which were erased easily by AF demagnetization. The cleaned directions are scattered away from the direction of the present geomagnetic field at the locality. To estimate the statistical significance of the dispersion, we applied a test of randomness derived by Watson (WATSON, 1956); namely, on the assumption of N randomly directed vectors, the length of resultant vector R_0 which will be exceeded by R with any stated probability P was calculated. If an observed value of R is less than R_0, it can be concluded that the hypothesis of randomness is true on the significance level of P. If not, we agree to reject the hypothesis of randomness. In our case the length of the seven cleaned vectors became 4.45 that was less than the value of 4.89 in Watson's table at the corresponding sample size and probability of 0.01. Therefore, the cleaned remanence directions of pillow lavas can be regarded as the primary directions.

All of the paleomagnetic data thus obtained are listed in Table 1, where data from the melange rocks and the other rocks are shown separately for the sake of convenience. It is evident from the table that the magnetic directions of the melange

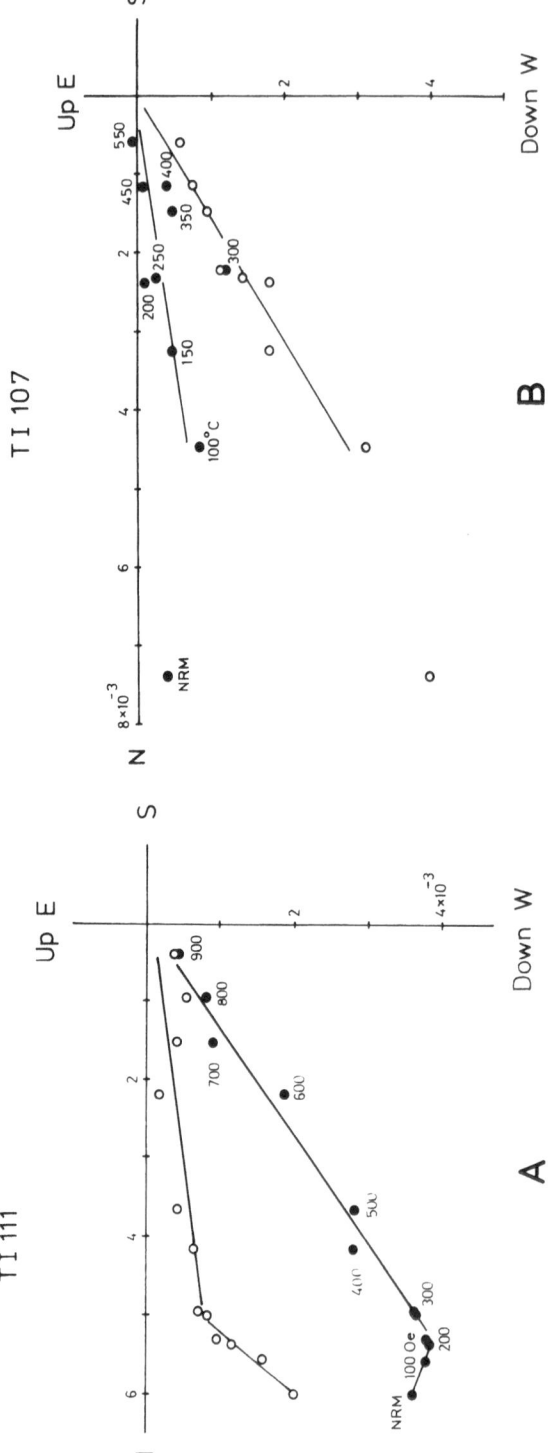

Fig. 3.   Orthogonal plots of successive endpoints of magnetization vector during progressive AF or thermal demagnetization for basaltic pillow lava from site TI (A, B), inter-pillow limestone from site AW (C) and TI (D), and red-coloured shale from site TI (E). Numbers are intensities of peak fields in Oersted (A, C) or demagnetization temperatures in centigrade degrees (B, D, E). Solid dots represent projections onto horizontal plane, and open circles onto north-south vertical plane. Units indicated for the axis are in emu.

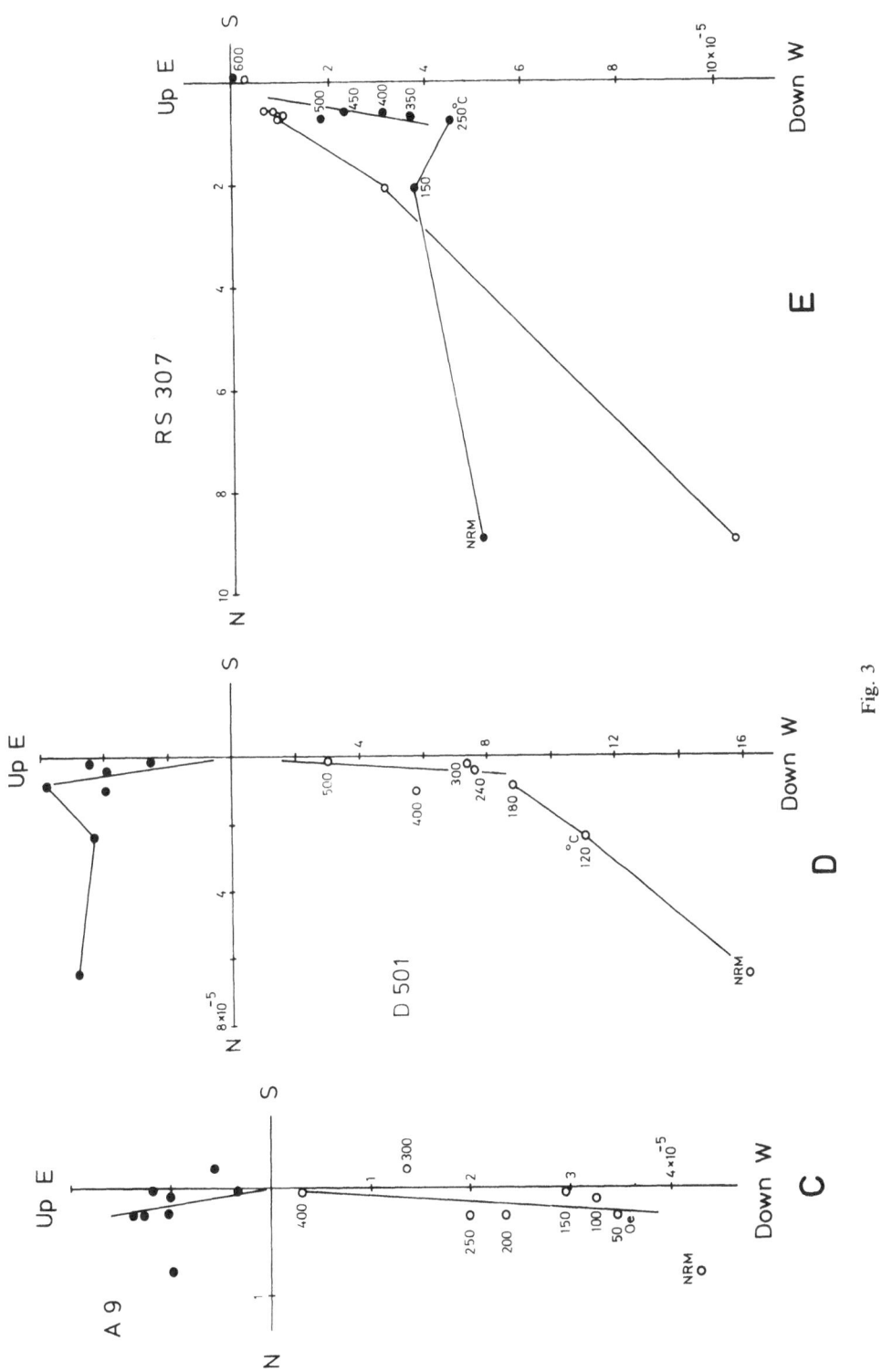

Fig. 3

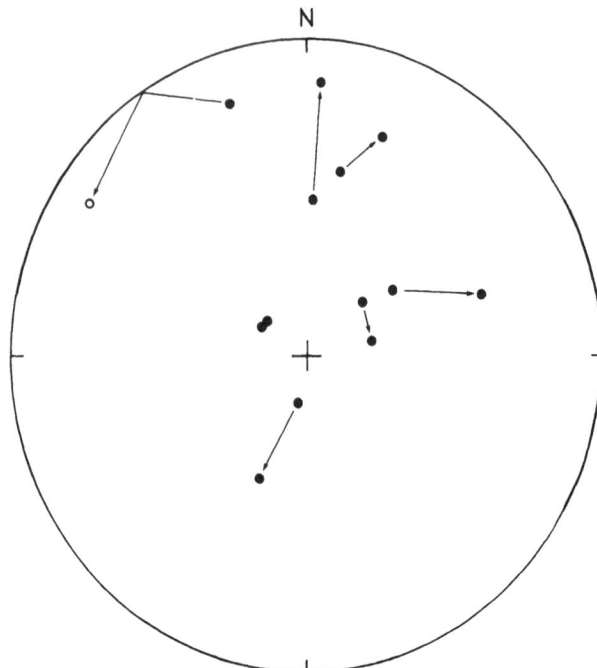

Fig. 4. Results of the conglomerate test for pillow breccias in site TI. Directional changes from initial measurement to AF demagnetization step of 200 Oe are indicated by arrows. Solid (open) circles are plotted on the lower (upper) hemisphere by equal-area projection. Statistical analysis is discussed in the text.

rocks scatter widely block by block both before and after bedding correction, while corrected directions of the detrital rocks tend to cluster well around the direction of the normal or reverse polarity near the present latitude (32°N). Further notable is that the directionality of melange rocks can be characterized by their scattered declinations and shallow inclinations. The different declinations can be best interpreted in the way that the blocks in the melange zone were more or less rotated during the initial emplacement. The consistent shallow inclinations, on the other hand, may mean that the rocks had been formed at nearly the same latitude. The absolute paleolatitude calculated on the assumption of a geocentric axial dipole field ranges 0° to 12°. If the polarity of melange rocks can be assigned to the normal one, the melange rocks may have been magnetized in southern hemisphere between 0° and about 12°. Although the assumption of normal polarity is not warranted, it is significant that the paleolatitude of melange rocks is different both from the latitude of the present localities and the paleolatitude of terrigenous rocks. The virtual geomagnetic pole positions corresponding to the data in the table are plotted in Fig. 5. The figure, where reverse polarities of some rocks were converted to normal ones, demonstrates that the apparent poles of melange rocks, in contrast with those of the other detrital rocks, distribute dispersedly and quite far from the north pole.

Table 1. Paleomagnetic data from the Shimanto terrain.

Site	Rock type	N	I	D	I*	D*	k	α_{95}	λ	$\Delta\lambda$	θ	ϕ
TI 100	PB	15	33°	342	−25	341	13	11°	13°	6°	40°N	338°E
TI 116	PB	10	30	254	−4	273	56	7	2	4	1	44
TI 128	PB	9	60	284	−5	318	35	9	3	5	37	10
TI D	PB	19	54	314	−22	324	57	5	11	3	34	357
TI E	IL	11	67	299	−8	325	39	8	4	4	40	2
TI F	RS	5	15	265	−20	261	39	13	10	7	−14	40
AW A	IL	19	61	78	3	29	29	6	2	3	48	267
TI A1	MS	12	80	173	30	354	36	7	16	4	71	355
TI A2	MS	11	72	172	38	2	16	12	21	8		
TI A3	MS	26	70	182	29	336	17	7	15	4		
OK A	TS	10	−21	220	−49	164	24	10	30	9	71	40
OK B	MS	23	−48	223	−40	147	10	10	23	7	54	36
OK C	VS	8	−34	245	−51	191	21	13	32	12	86	248
MT 1	GB	11	−45	276	−49	146	59	8	30	7	61	44
MT 2	GB	13	−44	267	−45	154	23	9	27	7	66	35

Rock type: PB = pillow basalt, IL = inter-pillow limestone, RS = red-coloured shale, MS = mudstone, TS = tuffaceous sandstone, VS = volcanic sandstone, GB = gabbro. N: number of specimens. I, D: in-situ inclination and declination. $I*$, $D*$: inclination and declination after correction for bedding tilt. k: Fisher's precision parameter. α_{95}: semi-angle of 95% confidence cone. λ: absolute paleolatitude. $\Delta\lambda$: error of paleolatitude. θ, ϕ: latitude and longitude of virtual geomagnetic pole (VGP for mudstone from site TI was obtained by averaging the mean directions of site TI A1 to A3 because the samples were collected from the same horizon).

5. Discussions

The significant point in interpreting the present paleomagnetic data will be the difference of paleontologic ages between melange rocks and other detrital rocks. The age contrast leads us to speculate that the melange complex may be exotic blocks which had been incorporated into the younger terrigenous sediments, and therefore the time of mixing can be regarded as more or less contemporaneous with the time when detrital rocks deposited. In addition to this age constraint, our paleomagnetic result suggests that the Upper Cretaceous terrigenous sediments were formed not so far from the present localities. In other words, the place of the emplacement of melange rocks may be near the present positions. The embedded melange rocks, in contrast, may have been formed in an equatorial region, possibly between 0° and 12° in southern or northern hemisphere. Thus the paleomagnetic and paleontologic results enable us to estimate the transport velocity of the Cretaceous melange complex, that is, the northward component of the motion of the plate on which the melange constituents were placed. If we assume that metabasalts and pelagic sediments in the Cretaceous melange zones had been formed at the equatorial region in the Valanginian time and afterwards they were incorporated near their present localities into detrital sediments during the Campanian, the duration of transport would be roughly 50 m.y. which gives the northward component of transport velocity about 10 cm/yr. Considering the plate

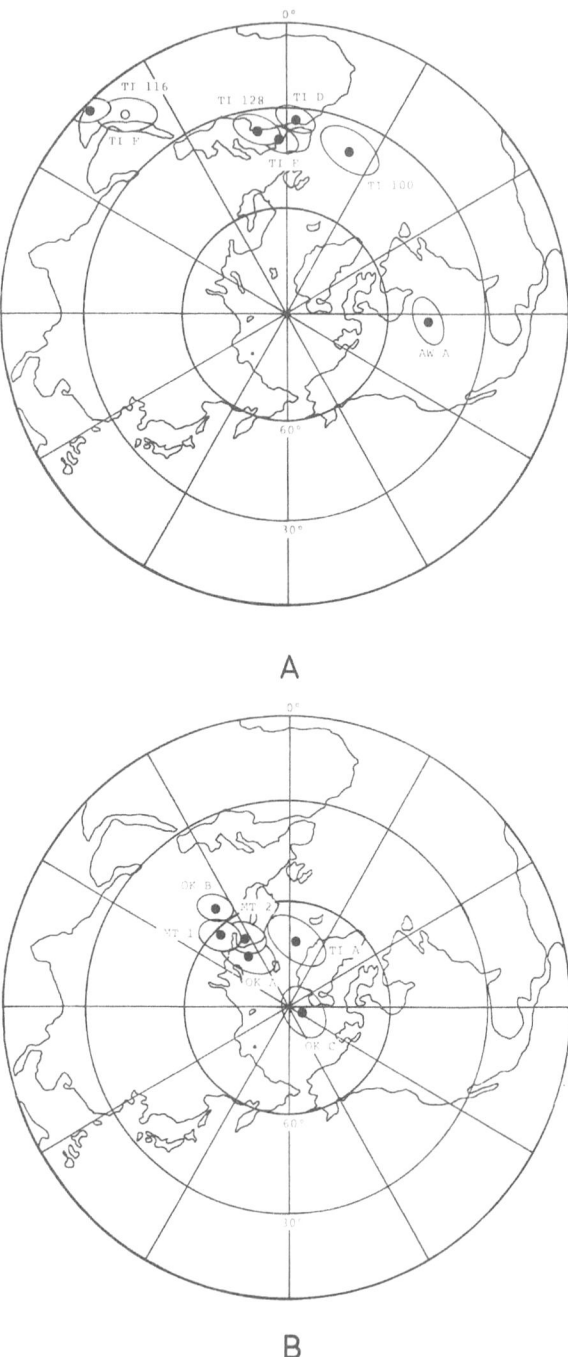

Fig. 5. Virtual geomagnetic poles for the mean directions of the melange rocks (A) and terrigenous rocks (B) from the Shimanto belt in Shikoku. Equal-area projection.

tectonic reconstruction by HILDE *et al.* (1977), this tentative value may correspond to the motion of the Kula plate that had subducted beneath the Asian plate until Upper Cretaceous. Since the motion of the Kula plate presumably had some latitudial component, the total velocity may be over 10 cm/yr that is, in any case, comparable with that of present plate movements.

We thank Prof. N. Niitsuma (Shizuoka University) for many valuable comments and suggestions on testing paleomagnetic reliability. We are also grateful to Mr. K. Shimamura (Tohoku Univ.) and many students of Kochi Univ. for kindly helping our field work. Messrs. H. Nishimura and Y. Murata of Kobe Univ. assisted the measurement by cryogenic magnetometer. The manuscript was reviewed by Prof. S. Uyeda (Univ. of Tokyo) whom we thank for encouraging us throughout the work.

REFERENCES

GRAHAM, J. W., The stability and significance of magnetism in sedimentary rocks, *J. Geophys. Res.*, **54**, 131–167, 1949.

HILDE, T. W. C., S. UYEDA, and L. KROENKE, Evolution of the western Pacific and its margin, *Tectonophysics*, **38**, 145–165, 1977.

KANMERA, K. and M. HASHIMOTO, The Shimanto Belt, in *The Geology of Japan, Earth Science, Vol. 15* edited by K. Kanmera, M. Hashimoto, and T. Matsuda, pp. 54–58, Iwanami-shoten, Tokyo, 1980(in Japanese).

OKAMURA, M., Radiolarian fossils from the northern Shimanto Belt (Cretaceous) in Kochi prefecture, Shikoku, in *Geology and Paleontology of the Shimanto Belt*, edited by A. Taira and M. Tashiro, pp. 153–178, Rinyakosaikai Press, Kochi, Japan, 1980 (in Japanese with English abstract).

OKAMURA, M., A. TAIRA, M. TASHIRO, and J. KATTO, Inter-pillow nannolimestone from the melange zones of the Shimanto Belt, in *Geology and Paleontology of the Shimanto Belt*, edited by A. Taira and M. Tashiro, pp. 215–216, Rinyakosaikai Press, Kochi, Japan, 1980(in Japanese with English abstract).

TAIRA, A., M. OKAMURA, J. KATTO, M. TASHIRO, Y. SAITO, K. KODAMA, M. HASHIMOTO, and T. CHIBA, Lithofacies and geologic age relationship within melange zones of Northern Shimanto Belt (Cretaceous), Kochi prefecture, Japan, in *Geology and Paleontology of the Shimanto Belt*, edited by A. Taira and M. Tashiro, pp. 179–214, Rinyakosaikai Press, Kochi, Japan, 1980 (in Japanese with English abstract).

WATSON, G. S., A test of randomness of directions, *M.N., Geophys., Suppl.*, **7**, 160–161, 1956.

Japanese Cenozoic Ophiolites

Accretion Tectonics in the Circum-Pacific Regions, edited by M. Hashimoto and S. Uyeda, 245–260.
Copyright © 1983 by Terra Scientific Publishing Company (TERRAPUB), Tokyo.

Mineoka Ophiolite Belt in the Izu Forearc Area—Neogene Accretion of Oceanic and Island Arc Assemblages on the Northeastern Corner of the Philippine Sea Plate

Yujiro OGAWA[1]

Department of Geology, Kyushu University, Hakozaki, Fukuoka 812, Japan

A singular type of accretion of ophiolitic rocks and forearc sedimentary rocks occurred during the Miocene and Pliocene on the northeastern corner of the Philippine Sea plate. The area is sandwiched between the Honshu (continental and volcanic) and Izu (volcanic) forearcs. The Mineoka ophiolite belt exhibits an intense cataclastic deformation and zeolite facies metamorphism. Ophiolite protruded to make olistostromal sediments, a chaotic mixture containing both continental and volcanic arc-derived clastics. The Miocene to Pliocene Izu forearc sediments were also trapped and accreted to the Honshu arc-side to make the Miura fold belt. Fault and fold systems in the two belts represent a right-handed en echelon arrangement, and they may have been generated by thrust shearing in the right-lateral transpressive stress field around the obliquely subducting Philippine Sea plate. Dismembered ophiolite was thus emplaced after the earliest stage of obduction of the Philippine Sea plate.

1. Introduction

Dismembered ophiolite usually appears in a tectonic zone, forming an ophiolite melange. It represents a suture, an ancient zone of joining of oceanic lithosphere (DEWEY, 1977). Ophiolite melange is characterized by brecciated and sheared serpentinite which contains blocks of ophiolitic rocks, such as peridotite, dunite, basalt, chert, limestone, and other pelagic sedimentary rocks and often metamorphic rocks. Some ophiolite melanges are characterized by cataclastic deformation and rather high geothermal metamorphism in rocks from amphibolite to zeolite facies. SALEEBY (1978, 1979) described the Kings-Kaweah ophiolite along the Western Foothills of the Sierra Nevada batholith, and suggested the process of emplacement of the oceanic lithosphere along the transpressive continental margin (SALEEBY, 1981). In some cases ophiolite olistostrome becomes the diagnostic assemblage in the ophiolite melange such as in the Lichi melange in East Taiwan documented by PAGE and SUPPE (1981) and SUPPE et al. (1981). Such modes of occurrence of ophiolite melange suggest that they may have been derived through several stages including ophiolite protrusion along a transform fault and the mixing of arc- or continent-derived clastics on a continental margin.

Similar ophiolite melange occurs in a young tectonic belt, in the Mineoka ophiolite belt in the Miura and Boso Peninsulas, central Japan, which is a particular

---

[1]Present address: Department of Geology, Imperial College, London SW7, U.K.

zone of the frontal part of an arc-arc crossing area (Fig. 1, OGAWA, 1982). Ophiolitic rocks appear to be highly dismembered along a right-handed en echelon fault zone and also appear as clastic rocks of olistostromal origin within forearc sedimentary sections. Most of the ophiolitic rocks are dated as late Eocene, and the covering and mixing sedimentary rocks are of lower Miocene to Pliocene in age. TONOUCHI and KOBAYASHI (1980, 1982) suggest from geophysical data that heavy materials, probably a thin slice of oceanic lithosphere, exist under the peninsulas dipping southwest, and its northern edge appears to the surface along the Mineoka ophiolite belt. They also verify that the pillow lavas were emplaced at first at latitude of 24° by means of paleomagnetic study, and explain that the ophiolite was subsequently emplaced by the northeastward obduction of the oldest part of the Philippine Sea plate after about 1,000 km northward movement. The tectonics is roughly explained as such but the structural and paleogeographical significance is much more complicated. Ophiolite and continent- and island arc-derived sediments were mixed and sheared and further trapped against the Honshu arc side during Miocene and Pliocene.

In this paper I describe the general occurrence of the ophiolitic rocks in the Mineoka ophiolite belt, especially their structural and stratigraphical features. I conclude that the ophiolite was emplaced at several stages, through obduction followed by thrust shearing and related protrusion in the oblique subduction area of the Philippine Sea plate under the Eurasian plate. These emplacements were connected with northward collision of the Izu volcanic arc to the Honshu arc (OGAWA, 1982).

2. Geologic Setting

The Mineoka ophiolite belt is situated in the Miura and Boso Peninsulas, on the convex side of the Honshu arc, occupying the structural high area on the Honshu arc side behind the Sagami trough (Figs. 1 and 2). The Philippine Sea plate on its northeastern corner is now subducting beneath the Eurasian plate along the Sagami and Suruga troughs. The recent tectonics around the Sagami trough are analyzed by ANDO (1974), SENO, (1977), MATSUDA et al. (1978), NAKAMURA and SHIMAZAKI (1981) and others. They showed that during at least the last 5 m.y., a right-lateral strike-slip component accompanied with predominant north or northwestward subduction has occurred along the trough. The plate movement now is nearly parallel to the plate boundary, hence very greatly oblique subduction occurs. The slip-vector analysis of earthquakes along the northern boundary of the Philippine Sea plate deduces relatively northwestward movement at about 3 to 5 cm/year in average around the triple junction area (SENO, 1977; MINSTER and JORDAN, 1979).

Around the Sagami trough area, characteristic features of subduction topography are observed. Trench, trench slope, structural highs, and forearc and back arc basins are distinct (KIMURA, 1976; KIMURA and NASU, 1979), though active volcanism does not occur on the arc side of the Sagami trough subduction. The Okinoyama bank belt and the Hayama-Mineoka uplift belt (Fig. 2) correspond to the structural high areas of the subduction system.

A characteristic features is that the subducting plate, the Philippine Sea plate, has a land area, the Izu volcanic arc. The Izu arc has been colliding against the Honshu arc

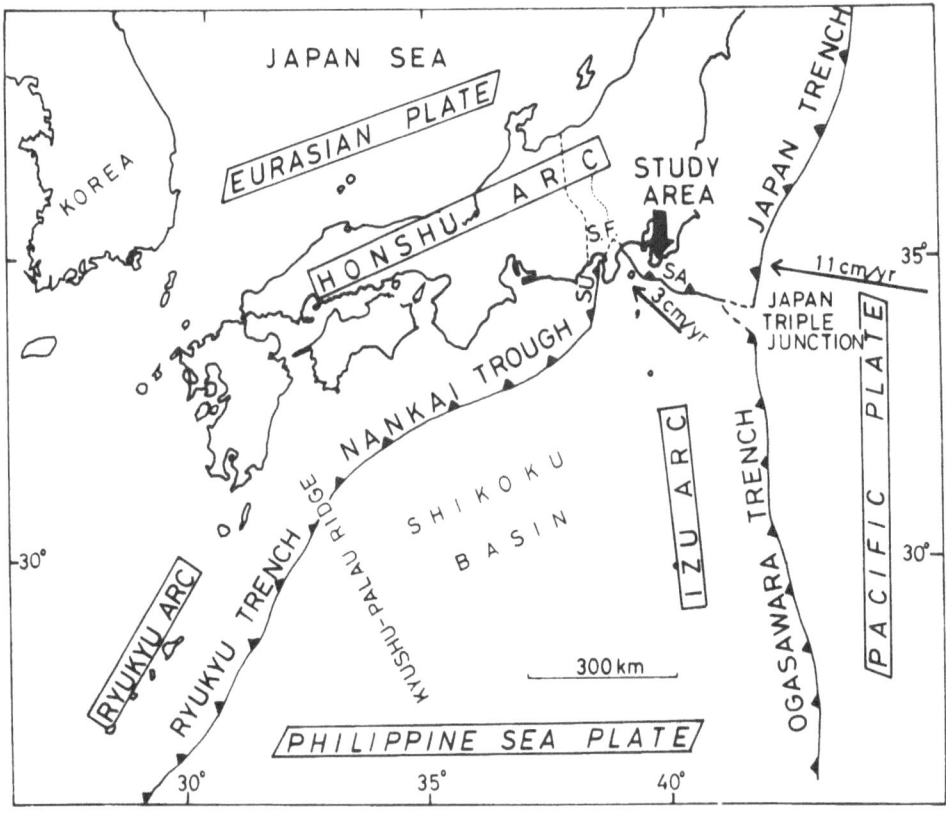

Fig. 1. Index map of the study area. Plate motion relative to the Eurasian plate is adopted from SENO
(1977) for the Philippine Sea plate and from MINSTER and JORDAN (1979) for the Pacific plate. S.F.: South
Fossa Magna, SU: Suruga trough, SA: Sagami trough. Read 130°, 135°, 140° for 30°, 35°, 40° in longitude.

and consequently Miocene voluminous island arc type or back arc type volcanic and
volcaniclastic rocks piled up in and around the South Fossa Magna area together with
continent derived clastics, and eastward or southward verging thrusting and
recumbent folding occurred to make the Tanzawa orogenic belt (MATSUDA, 1962,
1978; SEKI et al., 1969). The northward convex bending of the structure in the Honshu
arc to the north of the Izu arc may have been caused by the colliding of the Izu arc
(MATSUDA, 1978; OGAWA and HORIUCHI, 1978; OGAWA, 1982). Bending occurred
earlier than the Miocene, probably around the Cretaceous according to these studies,
while Niitsuma (1982) proposed much younger bending and colliding in Plio-
Pleistocene. Left and right-lateral strike-slip tectonics also occurred in the western and
eastern side of the Fossa Magna area respectively (KIMURA, 1965; OGAWA and
HORIUCHI, 1978). Above all to the east of the Izu arc, in the Miura and Boso Peninsulas,
Izu forearc sedimentary rocks chiefly made of Miocene to Pliocene or Pleistocene
volcaniclastic rocks are distributed around the dismembered ophiolite belt forming the
Miura fold belt (OGAWA and HORIUCHI, 1978). Particular tectonic belt and fold belt,
both the Mineoka ophiolite belt and Miura fold belt, show en echelon arrangement

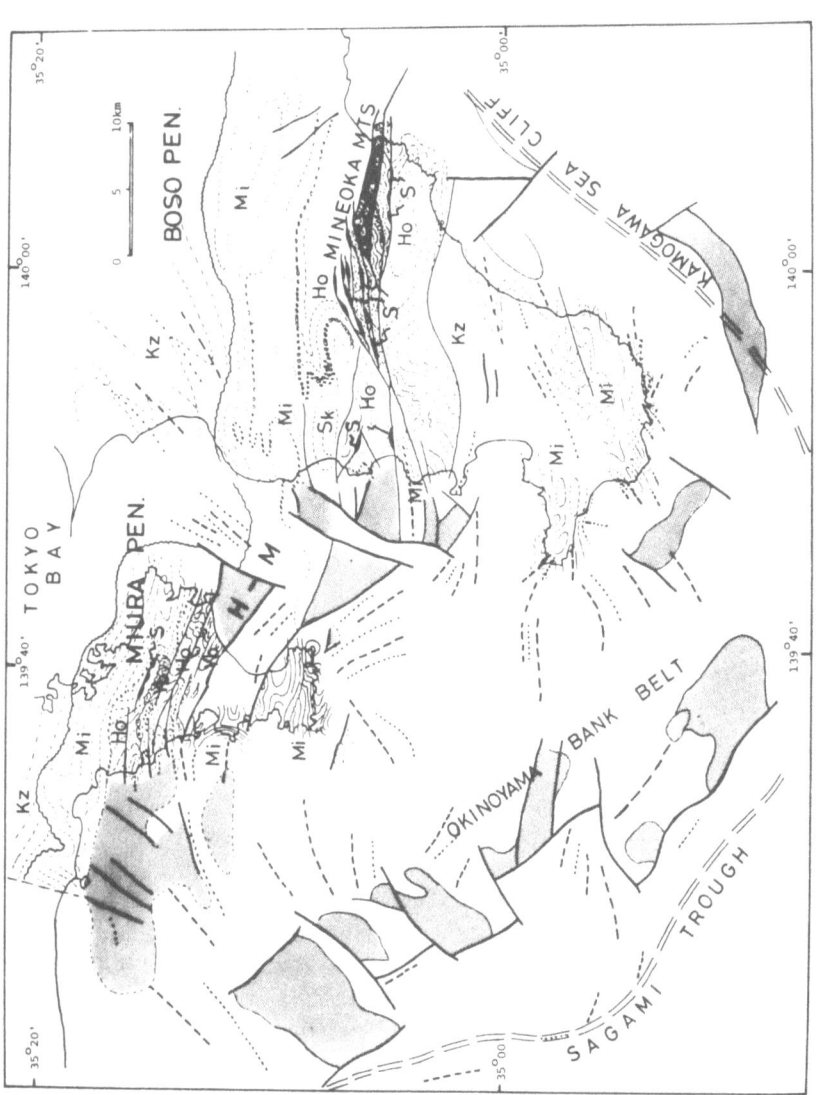

Fig. 2.   Simplified structural map of the Miura-Boso Peninsulas. S : serepentinite, Ho : Hota Group, Yb : Yabe Group, Sk : Sakuma Group, Mi : Miura Group, Kz : Kazusa Group. H-M indicates the genral trend of the Hayama-Mineoka uplift belt. Shaded area showing the submarine outcrop of the pre-Miura Group, after KIMURA (1976).

roughly parallel to the present plate boundary (Fig. 2). Structural and stratigraphic features also reveal a very particular type of sedimentation and deformation (OGAWA, 1982). In a word, the geology is characterized by mixing of three components of clastics; the provenances of which are the Honshu continental arc, the Izu volcanic arc and the Mineoka ophiolite. These geological units are lithologically as follows: The oldest rock assemblage structurally appears along the Okinoyama bank belt (KIMURA, 1976) and the Hayama-Mineoka uplift belt (KOIKE, 1957; Fig. 2). They are the Mineoka, Hota, and Yabe Groups, and the covering younger assemblage is distributed around the belt. They are the Miura and Kazusa Groups (Fig. 6).

The Mineoka Group is composed of two distinct assemblages; one is the Mineoka ophiolite and the other is the pelagic and terrigenous clastic rocks. The ophiolite is described in the next chapter. The pelagic and terrigenous clastics are composed of the lower siliceous or calcareous claystone or mudstone with thin intercalations of acidic tuff beds, and of the upper arkosic turbidite. They are much more strongly compacted and deformed than the later sedimentary rocks. The Mineoka ophiolite and clastics are correlated to the southern half of the Shimanto Supergroup, especially to the Setogawa Group to the west of the Fossa Magna (KOIKE, 1957). The sedimentation age might be early Tertiary, from Paleogene to lower Miocene in the Setogawa (IIJIMA et al., 1981), but the precise age of the Mineoka is not well known. *The Hota Group* is composed of lower Miocene tuffaceous siltstone, sandstone, and thin layers of acidic tuff. Conglomerates comprising pebbles both of ophiolitic rocks, andesitic or rhyolitic rocks and other sedimentary rocks which were derived from both volcanic arc and continent are intercalated. Strata are more or less deformed by penecontemporaneous faulting and folding, even largely slumped. *The Yabe Group* unconformably overlies the Hota Group with basalt boulders (KIMURA et al., 1976), and is composed characteristically of greenish altered pumice fall deposits and their derivatives. They are called the Kojima Formation in the Mineoka Mountains locally (Figs. 2, 3 and 4). Pumiceous rocks are altered locally into montmorillonite and clinoptilolite, but no strong deformation is observed in a microscopic scale. The Sakuma Group to the west of the Mineoka Mountains in Figs. 2 and 3 is lithologically slightly different in the extent of tuffaceous silstone. It includes middle Miocene foraminiferas, and may be the equivalent to the Yabe Group.

The Miura and Kazusa Groups are widely distributed and unconformably overlie the Hota or Yabe Group. The Miura and Kazusa Groups form gentle but complicated fold structures, the axes of which arrange en echelon or in curl (Fig. 2). The groups range from upper Miocene to Pliocene and Pleistocene in age, and are composed mainly of volcanic-arc derived clastics with considerable amounts of tuff fall deposits. Sequences are largely tuffaceous. Scoriaceous and pumiceous clastics of sandstone and conglomerate are intercalated within tuffaceous siltstone in the lower part of the Miura and Kazusa Groups. Clastics have textures either of turbidite or debris flow, and often show cross bedding (Fig. 5). Large scale submarine canyon and channel fill sediments are developed in several horizons, especially in the southwestern parts of the peninsulas. Judging from the paleocurrent analysis, most of the sediments were derived from south or southwest in the southern part of the peninsulas, and from west or northwest in the northern part (OGAWA, 1982). The clastic grains are coarse in the southern and western parts, and gradually become finer toward north and east. Thus the clastics were dominantly derived from the volcanic arc lying to the west, i.e. the Izu volcanic arc (KOIKE, 1957; MATSUDA, 1962; OGAWA, 1982).

These volcanic forearc type formations, later than the Mineoka Group, never suffered from intense compaction nor deformation, but only deformed into gentle undulation in a large scale, and nearly penecontemporaneous faulting and folding were developed unde very shallow depths (OGAWA and HORIUCHI, 1978; OGAWA, 1982).

3.   Tectonics of the Mineoka Ophiolite Belt

3.1   Lithology

The ophiolitic rocks are now strongly dismembered, and blocks or layers of much harder rocks appear as knolls and remarkable knockers surrounded by sheared serpentinite and clastic rocks (Fig. 4). The original ophiolite succession is presumed as follows; the lowest are ultramafics mostly of harzburgite and rarely of dunite (UCHIDA and ARAI, 1978), now largely serpentinized, and followed upward by gabbro and dolerite, basalt with dolerite dykes, chert, and limestone (Fig. 6). The basalts are mostly pillow basalt of tholeiitic composition and rarely alkali-olivine basalt and picrite (KANEHIRA, 1976; TAZAKI and INOMATA, 1980). Chert is radiolarian and limestone is micritic, and they are often interclatated with acidic tuff beds. These sequences are attributed to the upper part of the oceanic lithosphere and its covering pelagic sediments around an island arc.

Basaltic rocks are not schistose but weakly metamorphosed. Chlorite and pumpellyite are the common metamorphic minerals and actinolite occurs in some rocks. Blocks of epidote-hornblende schist on the order of 10 m are distributed in the easternmost part of the Mineoka Mountains (Figs. 3 and 4; KANEHIRA et al., 1968). In the blocks several layers of quartz schist and psammitic schist are intercalated parallel to the general foliation of the epidote-hornblende schist. Tectonic conditions for these metamorphic rocks are not well known, but obduction of oceanic lithosphere accompanied by oceanic sediments may be preferable. Pillow basalt in the Kamogawa harbor contains nanno-ooze limestone. Umber is interbedded between the pillow basalt and volcaniclastic mudstone near Mineoka-Sengen (TAZAKI et al., 1980).

3.2   Age

The pillow lava is dated as 29.9–39.9 Ma by K-Ar method (TAKIGAMI et al., 1980). Muscovite in the psammitic schist associated with the epidote-hornblende schist gives a K-Ar age of 38.0 Ma (YOSHIDA, 1974). Nanno-ooze in the interpillow limestone mentioned above indicates ages ranging from late Eocene to Miocene (M. Okamura, personal communication, 1981), and the radiolarian assemblage of the chert near Mineoka-Sengen indicates the lower Tertiary (K. Nakaseko, personal communication, 1980). Thus the radiometric and fossil ages of the ophiolitic rocks and their surroundings are all of early Tertiary, mostly of late Eocene age. The age around 40 Ma is coincident to the time of the change of the movement direction of the Pacific plate (JACKSON et al., 1972), and possibly the ophioltic rocks and their covering sedimentary rocks were of the oldest oceanic crust and pelagic sediments of the juvenile

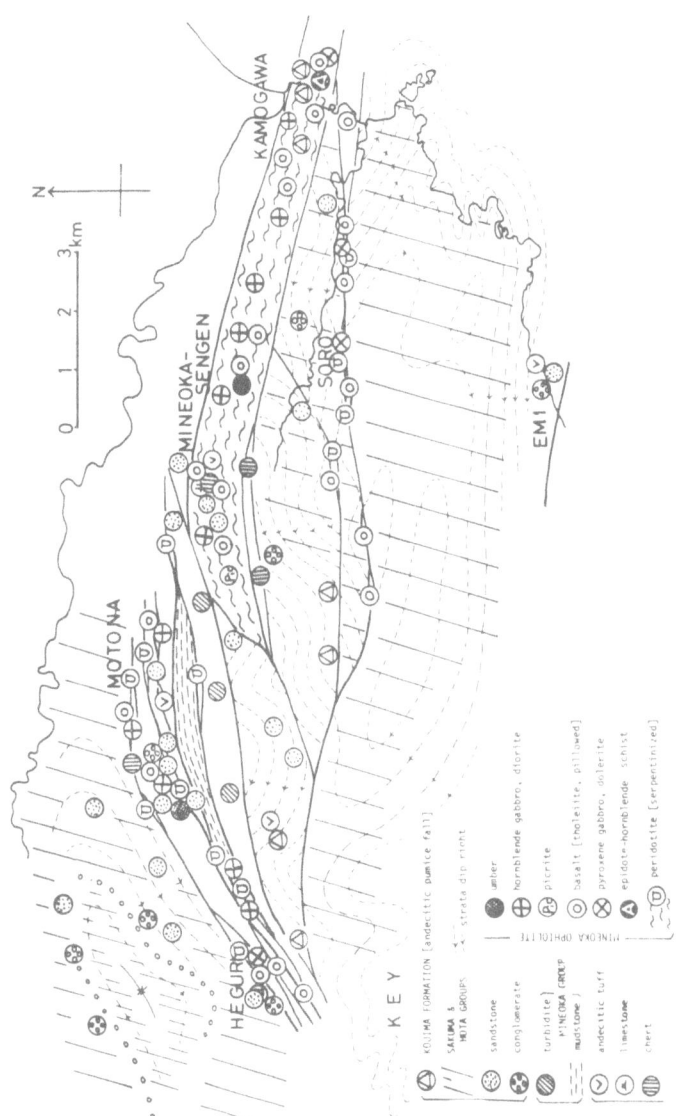

Fig. 3. Locality map of the ophiolitic and other rocks in the Mineoka Mountains.

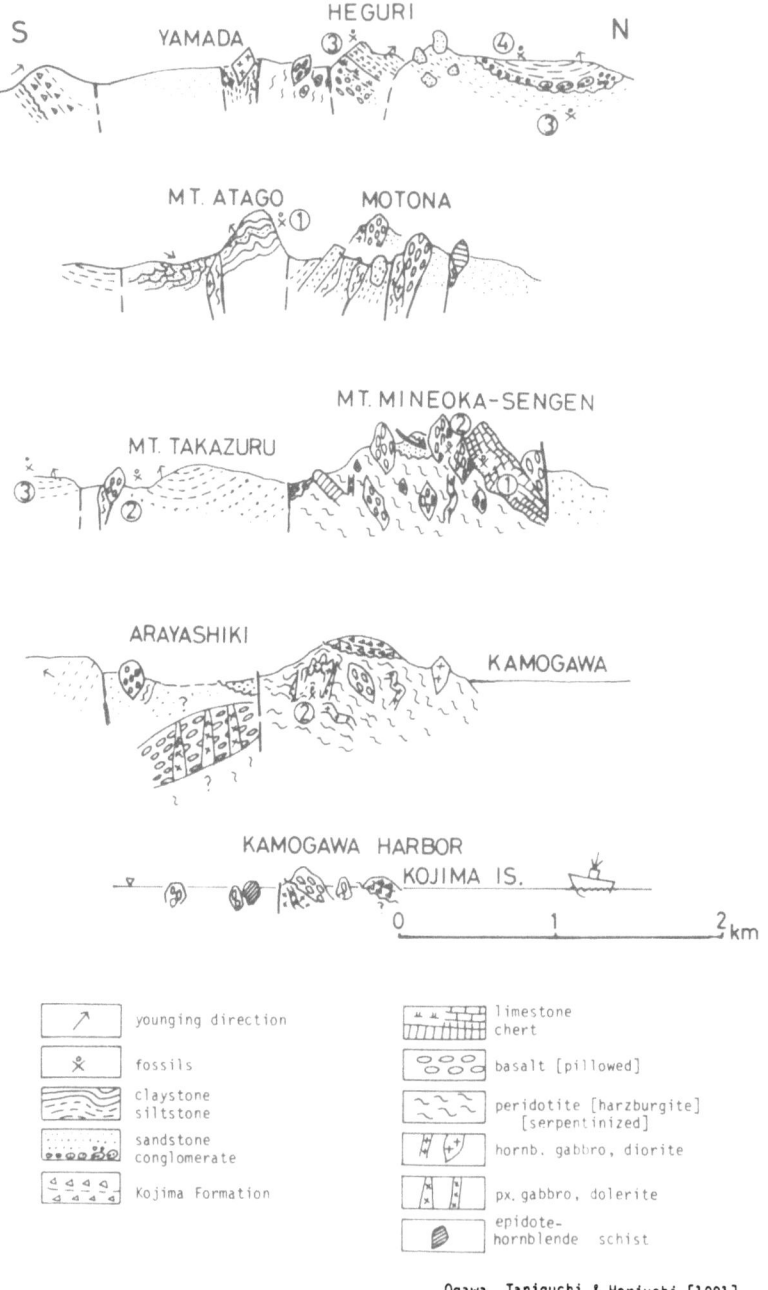

Fig. 4. Schematic (vertically exaggerated) profiles of the Mineoka ophiolite belt in the Mineoka Mountains. Number at the fossil mark means; 1: early Tertiary (including presumed), 2: early Miocene, 3: early to middle Miocene, 4: middle Miocene.

stage of the Philippine Sea plate (TONOUCHI and KOBAYASHI, 1982).

On the other hand, fossils in mudstone around the ophiolitic rocks, which is apparently interbedded with basaltic lava or unconformably overlies the basaltic conglomerate, show slightly younger ages of early Miocene or early to middle Miocene (OGAWA, 1982). Mudstone which is interbedded apparently with basalt near Mineoka-Sengen yields early Miocene radiolarians, *Minetosphaera* sp. (K. Nakaseko, personal communication, 1980). Mudstone intercalated structurally with sheared serpentinite at Kamogawa yields foraminiferas such as *Globigerinita unicava* and *Globigerinoides* sp. cf. *trilobus* (N 6–7 of Blow's zones) (YOSHIDA, 1974). Mudstone overlying glauconitic sandstone above basaltic conglomerate and galuconitic limestone at Heguri yields an early to middle Miocene radiolarian assemblage which includes *Stichocorys delmontensis* and *Cyrtocapsella tetrapera*. These two species are the characteristic ones in the most sections of the Hota Group (OGAWA, 1982).

To summarize, the ages of the ophiolitic rocks are mostly of late Eocene and those of the sedimentary rocks around the blocks of ophiolitic rocks, some of which are in structural contact, are of early Miocene or early to middle Miocene.

3.3 Deformation and mineralization

The ophiolitic rocks characteristically suffered from cataclastic deformation and zeolite mineralization. Cataclastic deformation is remarkable in gabbro, dolerite, and epidote-hornblende schist in the Kamogawa harbor and in serpentinite everywhere (Fig. 5–1). Deformation is weak or rare in basalt. Stylolitic texture is common in chert and limestone. The cataclastic texture in the epidote-hornblende schist is striking; porphyroclasts are surrounded by crushed matrix of the same rocks. Original metamorphic foliation within large blocks is traceable from block to block over ten meters in outcrop, and is parallel to the foliation of the psammitic and quartz schist. The mode of deformation is classified as the microbreccia of HIGGINS (1971). Hornblende gabbro within the serpentinite is often deformed into mylonite in part. These textures of the cataclastic rocks resemble very much the certain sedimentary textures of debris flow type conglomerate or sandstone. But the textures of the cataclastic rocks (e.g. Fig. 5–1) are distinct from sedimentary ones from the following criteria; (1) monolithologic, (2) boundaries between the blocks and matrix are usually straight and distinct, (3) cataclastic texture, porphyroclast and matrix, and rotation or tail structure are common, (4) distinct sedimentary conglomerate and sandstone in the region are polylithologic, cemented with calcite, and much softer (e.g. Fig. 5–2), and the two can easily be separated (refer Fig. 5–1 and 2).

Deformation in the deeper level may have turned the ophiolitic rocks into the mylonite or microbreccia as seen in gabbro or epidote-hornblende schist, and that in the middle level led to the brecciation and crushing as seen in gabbro and dolerite, and the deformation in upper level involved the shearing only as seen in some serpentinite. Some serpentinite is sheared intensely and during the shearing the thin pegmatitic gabbro layers became isoclinally folded or rotated. Such styles of structures indicate that the cataclastic deformation, shearing or folding have occurred in various depths successively or progressively while the protrusion of the ophiolite occurred to make it into the ophiolite melange (KARSON and DEWEY, 1978) (Fig. 6).

Another characteristic feature of the dismembered ophiolite is zeolite mineralization. Zeolite is commonly developed in veins in microgabbro, dolerite, basalt, and tuffaceous sediments within bedded chert, but never in epidote-hornblende schist nor in other formations later than the Mineoka ophiolite, such as the Hota, Yabe, and Miura Groups. In some rocks prehnite and laumontite are common, and in other rocks, analcite, natrolite or clinoptilolite is common. Greenschist and pumpellyite-bearing basalt are often recognized in the pebbles or grains of the ophiolitic rocks-derived conglomerate and sandstone (ARAI, 1981; Ogawa, unpublished data). These rocks as well as the epidote-hornblende schist blocks are included within serpentinite melange. There is the possible metamorphism in or around a transform fault or oceanic ridge itself as reported by Fox et al. (1976), but the exact relations for the systematic metamorphism and tectonics are not well known yet.

Thus the cataclastic deformation and zeolite mineralization are intense in the ophiolitic rocks. But the strata unconformably overlying the ophiolitic rocks, the Hota and Yabe Groups never suffered from any cataclastic deformation and zeolite mineralization except for the sheared blocks within the fault zones. Furthermore, zeolite veins occurs only in the ophiolitic rocks with rare exceptions in the Yabe Group. The Yabe Group is only gently undulated in a wide scale and does not show any microscopic deformation. Sandstone and conglomerates of middle early Miocene in and around the ophiolite belt contain deformed and metamorphosed grains and pebbles as detritus. Therefore the chief deformation and mineralization took place only in the Mineoka ophiolitic rocks before the sedimentation of the Hota Group; before the middle early Miocene.

3.4   Ophiolite clastics

Particular types of coarse sandstone and conglomerate which contain some ophiolitic clasts are distributed in and around the ophiolite belt (Fig. 3). Such clasts include gabbro, basalt, serpentinite, chromspinel, metamorphic rocks, chert, and limestone or ophicalcite, and are usually cemented by calcite matrix (Fig. 5–2). The stratigraphic position is correlated to the lowest horizon of the Hota Group. Clastic rocks rich in serpentinite fragment are characteristically developed within the ophiolite belt around Mineoka-Sengen. Around it other kinds of clastics, which also contain abundant fragments of granite-origin (minerals such as quartz, alkali feldspar, and plagioclase) and clasts of acidic volcanic rocks beside the ophiolitic rocks are in contact with ophiolitic rocks bodies. Some sandstone and conglomerate contain metamorphosed grains or pebbles of the greenschist facies (ARAI, 1981) and the pumpellyite facies. Acidic volcanic rocks occur abundantly in the Hota Group as pebbles and sand grains, and they may have been derived from the volcanic arc lying to the west, the Izu arc.

These particular types of clastics which contain either or both of ophiolitic rock fragments, acidic volcanic fragments or granitic rock fragments, are widely distributed in and around the Mineoka ophiolite belt. Clasts are unsorted but rather well rounded. In Emi, several kilometers south to the Mineoka Mountains (Fig. 3), coarse sandstone and conglomerate rich in volcanic fragments contain huge blocks of acidic tuff which alternate with siliceous radiolarian sedimentary rocks. The blocks are very similar to

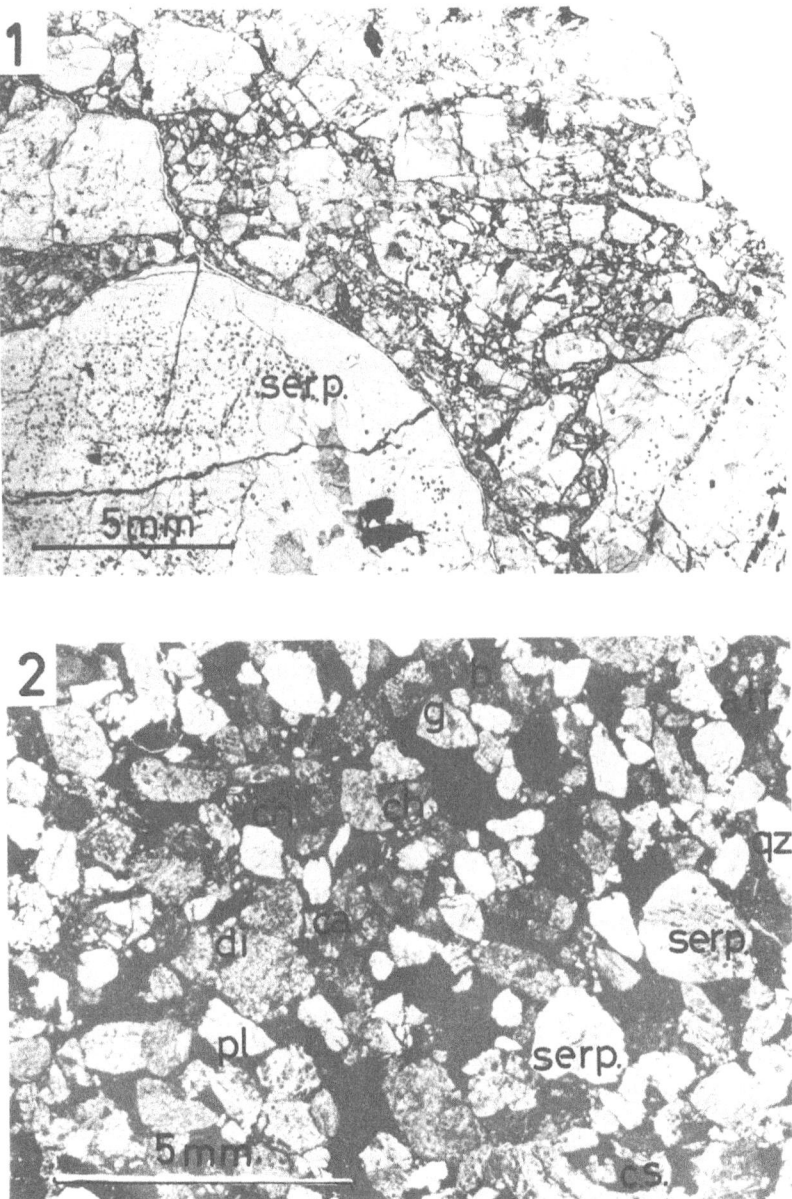

Fig. 5. Photomicrographs of serpentinite microbreccia at the west of the Kinugasa Station (1), and serpentine sandstone at Mineoka-Sengen (2). Note the straight grain boundary and breeciated chips in (1), and rounded grains and good sorting in (2). Abbriviations in (2), serp: serpentinite or serpentine, pl: plagioclase, di. diorite, ca: calcite, ch: radiolarian chert, g: micrographic texture of quartz and plagioclase, b: altered basalt, qz: quartz, a.tf.: acidic tuff, and c.s.: chromespinel.

the tuff beds within the interlayered chert and limestone in the Mineoka ophiolite belt at Mineoka-Sengen (Fig. 4). The acidic tuff blocks in Emi contain zeolite veins of analcite and clinoptilolite, and chlorite. The coarse clastics in Emi are correlated to the late early Miocene by silicoflagellate biostratigraphy by SAWAMURA and NAKAJIMA (1980). Most of the formations containing clasts of serpentinite or ophiolite are the lower member of the Hota Group, middle early Miocene.

The ophiolitic rock fragments or blocks are also contained within the other stratigraphic horizons as follows. Basal conglomerate of the Yabe and Sakuma Groups, which unconformably overlie the Hota Group, also contain abundant pebbles or boulders of ophiolitic rocks, especially of pillow basalt (KIMURA et al., 1976). The conglomerates also contain rocks common in the Mesozoic or Paleozoic terrane in the Honshu arc, such as chert, limestone, and sandstone or slate. Thus the clastic rocks around the Mineoka ophiolite belt are composed of detritus which were supplied from three terranes: the Honshu arc, Izu arc and the ophiolite itself.

Pebbles and boulders within the middle horizon of the Miura Group also contain abundant ophiolitic rocks such as serpentinite, gabbro, and altered basalt besides quartzose mylonite. They may have been reworked from the other groups around the ophiolite belt.

3.5   Structure and tectonics

The Mineoka ophiolite belt is revealed as right-handed en echelon fault zones, which are lined by sheared serpentinite. The surrounding strata out of the ophiolite belt show a gentle but complicated fold structure, the axes of which are not always parallel to the fault zone, but are arranged either en echelon or highly obliquely, and sometimes the axes rest in a rotated fashion (Figs. 2 and 3).

The Miocene Hota, Yabe and Miura Groups and Plio-Pleistocene Kazusa Group all show such a fold structure (OGAWA and HORIUCHI, 1978; OGAWA, 1982). Though each group overlies the substrata by an angular unconformity, each group shows the similar en echelon structure with each other. These strata are only slightly compacted in spite of the complicated fault and fold structures. They indicate that the deformation succeeded step by step during the forearc sedimentation from the early Miocene to Pliocene or even Pleistocene. The weak compaction and complicated structures indicate the structures were formed in the shallower part just after or slightly after the sedimentation in a shearing stress field dominated by strike-slip. Sedimentation and deformation were simultaneously occurred as documented by HOWELL et al. (1980).

The shearing sense around the ophiolite belt is probably right-lateral and transpressive, because most of the faults and folds arrange in right-handed en echelon as mentioned previously. Right-lateral shearing brings the surrounding strata into either or both thrust shearing and Riedel shearing (TCHALENKO, 1968). But the strata around the Mineoka ophiolite belt also have en echelon folds at the same time. This suggests that the structures were developed in a right-lateral transpressive stress field (HARLAND, 1971). Most of the structures have been successively formed during the sedimentation from the lower Miocene to Pliocene when not only the ophiolitic rocks but also the overlying clastics are mixed together to make the ophiolite melange and the fold belt. Serpentinite melange was formed by protrusion along the thrust shears.

Figure 2 also provides the submarine geology which shows the fold axes of Pliocene and Pleistocene formations after KIMURA (1976). The folds show en echelon and a rotational arrangement. This area is said to suffer from right-lateral transpressive shearing by the recent tectonics around the northeastern corner of the Philippine Sea plate. In some parts, normal faults with a component of "eduction" (NAKAMURA and SHIMAZAKI, 1981) is presumed, but in a wide sense the plate boundary is dominated by transpression due to oblique subduction of the Philippine Sea plate. The region in the Miocene time had also a similar tectonic condition as in the present.

4.   Speculation on the Tectonic Development

Up to the late Eocene (about 40 Ma) the Kula or Pacific plate moved northward, but when the movement of the Pacific plate changed to westward at about 40 Ma ago (JACKSON *et al.*, 1972), the Philippine Sea plate was trapped and the Pacific plate began to subduct along the present Japan-Bonin (Ogasawara) trenches (HILDE *et al.*, 1977; MATSUDA, 1978). The Izu volcanic arc occurred parallel to the subduction along the new trenches, then the Japan Triple Junction off Kanto happened to occur as a TTT type triple junction (Fig. 1). The Philippine Sea plate itself rifted along the north-south trending oceanic ridges now recognized as a fossil ridge in the Shikoku basin (KOBAYASHI and NAKADA, 1978). The Philippine Sea plate may have independently subducted northward during the Miocene (KINOSHITA, 1980). On account of the northward movement of the Philippine Sea plate, the Izu arc collided to the Honshu arc to make the plate boundary between the Izu arc, and the triple junction changed from the trench type to the highly obliquely transpressive type. Lateral shearing accompanying slight subduction may have prevailed along the boundary since the early Miocene about 20 Ma till Recent. Sedimentation around the boundary had two aspects. One is that the sedimentation was affected by simultaneous deformation along the transpressive boundary, and deformation successively occurred during the forearc sedimentation; the other is that the area was supplied by the three components of clastics from the Honshu continental arc, the Izu volcanic arc, and the ophiolitic rocks. The ophiolitic rocks may have belonged to the earliest stage of the Philippine Sea plate, because no sea floor as young as about 40 Ma exists around the Pacific plate side, and the area corresponds to the easternmost part of the Philippine Sea plate which is the earliest part of the plate, the age being consistent.

As described before, the area treated in this paper has rested as the right-lateral transpressive field since the Philippine Sea was trapped and the Izu arc began to collide to the Honshu arc. The age was at least the middle early Miocene, because the first appearence of the detrital ophiolitic rocks is in the lower Hota Group of that age. Probably this stage is correlated to the obduction of the Philippine Sea plate which was followed by the dismembered ophiolite protrusion. Before this stage the chief metamorphism of the greenschist facies and the epidote-hornblende schist facies had already took place. The zeolite metamorphism may have occurred during the protrusion, and after the deposition of the Yabe Group. Cataclastic deformation took place during the ophiolite protrusion. Ophiolite clasts were supplied at several stages during the forearc sedimentation in Miocene, but some of them having been reworked,

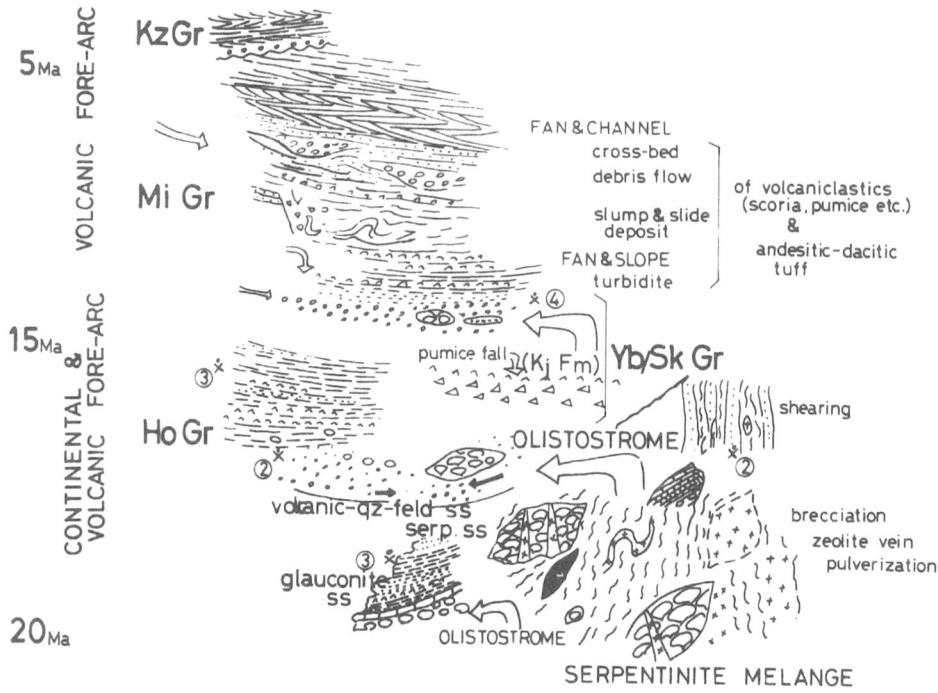

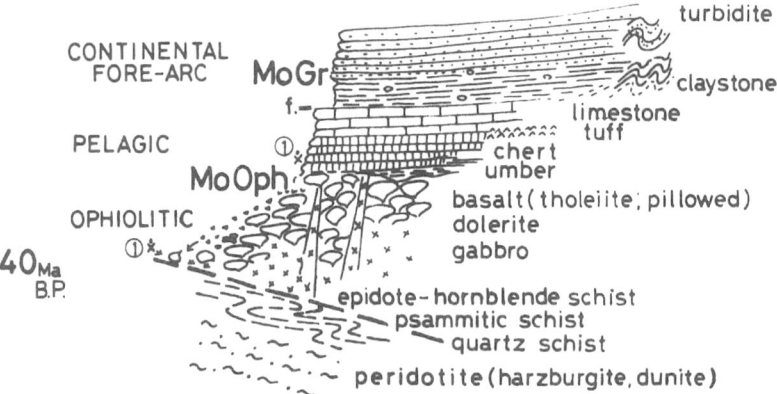

Fig. 6. Schemantic profile showing development of the ophiolite (serpentinite) melange and covering forearc sediments. Mo means Mineoka. Other abbriviations of the groups are the same in Fig. 2. Numbers at the fossil marks are the same in Fig. 4. Not to scale.

from the ophiolite belt.

Such complicated sedimentation and deformation took place in the northeastern corner of the Philippine Sea plate. The chief cause is that the area rests just in the sandwiched area between the Honshu and Izu arcs and the ophiolite itself. Such an area much resembles the present Sagami trough (Ogawa, 1982).

I thank K. Kanmera, N. Nasu, K. Kobayashi, and H. Okada for discussion and encouragement throughout the study. Thanks are extended to K. Fujioka, S. Tonouchi, K. Horiuchi, and H. Taniguchi for discussion in the field and laboratories. M. Hashimoto and D. S. Cowan are appreciated for review and improvement of the manuscript. This work was supported in part by the Grant-in-Aid from the Ministry of Education, Japan (Sogo A-43401) and by the funds from the Cooperative Program (No. 81103) provided by the Ocean Research Institute, University of Tokyo.

REFERENCES

ANDO, M., Seismo-tectonics of the 1923 Kanto Earthquake, *J. Phys. Earth*, **22**, 263–277, 1974.

ARAI, S., Igneous and ultramafic rocks in the Mineoka belt, Boso Peninsula, in *Excursion guide book of Geological Society of Japan*, pp. 59–72, 1981.

DEWEY, J. F., Suture zone complexities: A review, *Tectonophysics*, **40**, 53–67, 1977.

FOX, P. J., E. SCHREBER, H. ROLETT, and K. McCAMY, The geology of the Oceanographer fracture zone: A model for fracture zone, *J. Geophys. Res.*, **81**, 4117–4128, 1976.

HARLAND, W. R., Tectonic transpression in Caledonian Spitsbergen, *Geol. Mag.*, **108**, 27–42, 1971.

HIGGINS, M. W., Cataclastic rocks, *U. S. Geol. Surv. Prof. Paper*, **687**, 97 pp., 1971.

HILDE, T. W. C., Uyeda, S. and Kroenke, L. Evolution of the Western Pacific and its margins, *Tectonophysics*, **38**, 145–165, 1977.

HOWELL, D. G., J. K. CROUCH, H. G. GREENE, D. S. McCULLOCH, and J. G. VEDDER, Basin development along the late Mesozoic and Cenozoic margin: a plate tectonic margin of subduction, oblique subduction, and transform tectonics, in *Sedimentation in oblique-slip mobile zones*, edited by P. F. Ballance and H. G. Reading, *Spec. Publ. Int. Assoc. Sediment.*, **4**, pp. 43–62, 1980.

IIJIMA, A., M. UTADA, R. MATSUMOTO, K. KIMIYA, and Y. WATANABE, Field guide to Tertiary siliceous deposits in the southern Setogawa terrane in central Shizuoka Prefecture, Japan, in *Second Int. Conf. IGCP Project #115, Siliceous deposits in the Pacific region*, pp. 86–155, 1981.

JACKSON, E. D., E. A. SILVER, and G. B. DALRYMPLE, Hawaiian-Emperor chain and its relation to Cenozoic circumpacific tectonics, *Geol. Soc. Am. Bull.*, **83**, 601–617, 1972.

KANEHIRA, K., Mode of occurrence of serpentinite and basalt in the Mineoka district, southern Boso Peninsula, *Mem. Geol. Soc. Japan*, **13**, 43–50, 1976.

KANEHIRA, K., Y. OKI, S. SANADA, M. YABE, and F. ISHIKAWA, Tectonic blocks of metamorphic rocks at Kamogawa, southern Boso Peninsula, *J. Geol. Soc. Japan*, **74**, 529–534, 1968.

KARSON, J. and DEWEY, J. F., Coastal complex, western Newfoundland: An Early Ordovician oceanic fracture zone. *Geol. Soc. Am. Bull.*, **89**, 1037–1049, 1978.

KIMURA, M., Marine Geology in the Sagami-Nada Sea and its Vicinity, in *Marine Geology Map Series, 3, and Its Explanatory Text*, 9 pp., Geological Survey of Japan, 1976.

KIMURA, M., and N. NASU, Submarine tectonics around the Japanese Islands—with special reference to tectonics of the Sagami trough, *Bull. Coll. Sci., Univ. Ryukyus*, **28**, 131–141, 1979.

KIMURA, M., M. YUASA, Y. MASAKI, and Y. KANIE, Finding of pillow lava from the Oligocene-Miocene formation of the Miura Peninsula, *Bull. Geol. Surv. Japan*, **27**, 451–457, 1976.

KIMURA, T., Tectonic movement in the Southern Fossa Magna, central Japan, analyzed by the minor structures in its southwestern area, *Jpn. J. Geol. Geography*, **36**, 63–85, 1965.

KINOSHITA, H., Paleomagnetism of sediment cores from Deap Sea Drilling Project Leg 58, Philippine Sea, in *Initial Reports of the Deep Sea Drilling Project, Vol. 58*, edited by G. de Vries Klein, K. Kobayashi, *et al.*, pp. 765–768, U. S. Government Printing Office, Washington, D.C., 1980.

KOBAYASHI, K. and M. NAKADA, Magnetic anomalies and tectonic evolution of the Shikoku inter-arc basin, *J. Phys. Earth*, Suppl., **26**, S391–S402, 1978.

KOIKE, K., Structural development of the South Kanto district, *Chikyu-Kagaku (Earth Science)*, **34**, 1–17, 1957.

MATSUDA, T., Crustal deformation and igneous activity in the South Fossa Magna, Japan, *Geol. Mon. Am. Geophys. Union*, **6**, 140–150, 1962.

MATSUDA, T., Collision of the Izu Bonin arc with the central Honshu—Cenozoic tectonics of the Fossa Magna, Japan, *J. Phys. Earth*, Suppl., **26**, S409–S421, 1978.

MATSUDA, T., Y. OTA, M. ANDO, and N. YONEKURA, Fault mechanism and recurrence time of major earthquakes in southern Kanto district, Japan, as deduced from coastal terrace data, *Geol. Soc. Am. Bull.*, **89**, 1610–1618, 1978.

MINSTER, J. B. and T. H. JORDAN, Rotation vectors for the Philippine and Rivera plates, *EOS*, **60**, 958, 1979.

NAKAMURA, K. and K. SHIMAZAKI, Sagami-Suruga troughs and plate subduction, *Kagaku (Science)*, **51**, 490–498, 1981.

NIITSUMA, N., Magnetic stratigraphy of the Japanese Neogene and the development of the island arc of Japan, *J. Phys. Earth*, Suppl., **26**, S367–S378, 1978.

NIITSUMA, N., Touchstone for plate tectonics—Southern Fossa Magna, *Chikyu (The Earth Monthly)*, **4**, 326–333, 1982.

OGAWA, Y., Tectonics of some forearc fold belts in and around the arc-arc crossing area in central Japan : in *Trench-forearc Geology*, edited by J. K. Leggett, *Geol. Soc. London, Spec. Publ.*, **10**, 49–61, 1982.

OGAWA, Y., and K. HORIUCHI, Two types of accretionary fold belts in central Japan, *J. Phys. Earth*, Suppl., **26**, S321–336, 1978.

PAGE, B. M. and J. SUPPE, The Pliocene Lichi melange of Taiwan : Its platetectonic and olistostromal origin, *Am. J. Sci.*, **281**, 193–227, 1981.

SALEEBY, J., Kings River ophiolite southwest Sierra Nevada foothills, California, *Geol. Soc. Am. Bull.*, **89**, 617–636, 1978.

SALEEBY, J., Kaweah serpentinite melange, southwest Sierra Nevada foothills, California, *Geol. Soc. Am. Bull.*, **90**, 29–46, 1979.

SALEEBY, J., Ocean floor accretion and volcanoplutonic arc evolution of the Mesozoic Sierra Nevada : in *The Geotectonic Development of California, Rubey Volume 1*, edited by W. G. Ernst, pp. 132–181, 1981.

SAWAMURA, K. and T. NAKAJIMA, Miocene silicoflagellate zones in the Boso Peninsula, *Bull. Geol. Surv. Japan*, **31**, 333–345, 1980.

SEKI, Y., Y. OKI, T. MATSUDA, K. MIKAMI, and K. OKUMURA, Metamorphism in the Tanzawa Mountains, central Honshu, *J. Jpn. Assoc. Mineral. Petrol. Econ. Geol.*, **61**, 1–24, 50–75, 1969.

SENO, T., The instantaneous rotation vector of the Philippine Sea plate to the Eurasian plate, *Tectonophysics*, **42**, 209–226, 1977.

SUPPE, J., J. G. LIOU, and W. G. ERNST, Paleogeographic origins of the Miocene East Taiwan ophiolite, *Am. J. Sci.*, **281**, 228–246, 1981.

TAKIGAMI, Y., I. KANEOKA, and M. HIRANO, K-Ar and ^{40}Ar-^{39}Ar age determination of the Mineoka ophiolite, *Bull. Volcanol. Soc. Japan*, **23**, 308, 1980.

TAZAKI, K. and M. INOMATA, Picrite basalts and tholeiitic basalts from Mineoka tectonic belt, Central Japan, *J. Geol. Soc. Japan*, **86**, 653–671, 1980.

TAZAKI, K. and M. INOMATA, and K. TAZAKI, Umbers in pillow lava from the Mineoka tectonic belt, Boso Peninsula, *J. Geol. Soc. Japan*, **86**, 413–416, 1980.

TCHALENKO, J. S., The evolution of kink-bands and the development of compression textures in sheared clays, *Tectonophysics*, **6**, 159–174, 1968.

TONOUCHI, S. and K. KOBAYASHI, Magnetic anomaly of ophiolite and paleomagnetics, *Chikyu (The Earth Monthly)*, **2**, 375–379, 1980.

TONOUCHI, S. and K. KOBAYASHI, Paleomagnetic and geotectonic investigation of ophiolite suites and surrounding rocks in south-central Honshu, Japan, this volume, pp. 261–288, 1982.

UCHIDA, T. and S. ARAI, Petrology of ultramafic rocks from the Boso Peninsula and the Miura Peninsula, *J. Geol. Soc. Japan*, **84**, 561–570, 1978.

YOSHIDA, Y., Finding of foraminiferas from the Mineoka Mountains, Chiba Prefecture, *Chishitsu News*, *Geol. Surv. Japan*, **233**, 30–36, 1974.

Accretion Tectonics in the Circum-Pacific Regions, edited by M. Hashimoto and S. Uyeda, 261–288.

Paleomagnetic and Geotectonic Investigation of Ophiolite Suites and Surrounding Rocks in South-Central Honshu, Japan

Shyoji TONOUCHI and Kazuo KOBAYASHI

Ocean Research Institute, University of Tokyo, Nakano, Tokyo 164, Japan

Paleomagnetic investigation of basaltic rocks in the Hayama-Mineoka and Setogawa ophiolite belts, south-central Japan has indicated that both belts are most possibly composed of a tectonically obducted ocean floor. Existence of a tilted subterranean layer of the high density oceanic rocks in the southern Boso Peninsula has also been inferred from gravity and magnetic total force anomalies. Systematic deviations of paleomagnetic directions from that of the present field have been found with island-arc type rocks older than 11 Ma in Fujigawa and Tanzawa districts adjacent to these ophiolite belts. These results seem to suggest that the collision and obduction started at a time older than 11 MaBP along the Hayama-Mineoka and Setogawa ophiolite belts. This zone of collision and obduction is geographically different from the present collision zone between Izu Peninsula and central Honshu arc as well as the present subduction at Sagami and Suruga Troughs. Jumps of a subduction zone in a geological period may be suspected.

1. Introduction

It is now widely accepted that "ophiolite" is a ready-to-hand sample of the ancient ocean floor composed of deep-sea sediments, pillow and sheeted basalts, gabbros, and ultramafic rocks. A mechanism of obduction was postulated to explain emplacement of the ocean floor toward the continental margins (COLEMAN, 1971; DEWEY and BIRD, 1971; DEWEY, 1976).

However, the origin and mode of emplacement of ophiolite suites are still much debated particularly in Japan. In the traditional manner the ophiolite is defined by its lithological assemblage alone ("Steinmann's trinity") and not directly relevant to the ocean floor materials. A difficulty with most of the Japanese ophiolite complexes lies in their occurrence that their outcrops are generally very limited in size. Stratigraphic correlation between individual outcrops is in most cases difficult. Dip and strike of each exposed stratum can hardly be determined. Each unit of ophiolite assemblage is often dismembered or even lacking. Some of the units such as serpentinites were sometimes assumed to have been emplaced by a sedimentary process or land-slide (LOCKWOOD, 1971).

Three possibilities may remain as a plausible explanation of the origin of such presumably "ophiolitic" complexes;

(1) They consist of tectonically obducted ancient ocean floor accreted along the continental margin.

(2) They are parts of mélange derived from the ancient ocean floor but subject to erosion and secondary deposition by land-slide or similar mechanism of transportation.

(3) They are not at all related to the ancient ocean floor but originating from a magmatic intrusion in the ancient geosynclinal zone.

The first explanation implies that the zone of the ophiolite occurrence is a geosuture at which two plates converged at one time in a geological period.

The second possibility is also indicative of the existence of paleotrench along the ophiolite belt. However, an alternative interpretation is valid, i.e. the oceanic blocks composing the ophiolitic sequence may have been transported from a distant area and have been deposited in a tectonic depression unrelated to the ancient trench. If the latter was the case, the ancient plate boundary can not be identified by the ophiolite occurrence alone.

The third is a classical interpretation of ophiolite suite and still supported by a number of investigators. Some geophysical evidences are needed to judge its validity in addition to geological information.

The authors have attempted to make use of paleomagnetic techniques in the study of the emplacement processes of ophiolite complexes and their environs, since the stable natural remanent magnetization (NRM) of rocks (Sections 2 and 4) is a powerful indicator of subsequent tectonic movements including both tilting and horizontal drift as well as the paleolatitude of the original site of formation.

If a unit of ophiolite suite was emplaced by a sedimentary process such as a land-slide, stable component of its NRM should be scattered among blocks since the depositional alignment of magnetic moments can not proceed in such large blocks. If measured directions of NRM of rocks collected at several geographically separated outcrops are well concentrated or change systematically, it can be concluded that the unit moved together in this area by one tectonic process as a whole. In this way the above possibilities (1) and (2) can be distinguished.

If the original horizontal plane can be identified and stable directions of NRM are corrected with its dip and strike (bedding correction), the inclination of the original NRM can be reproduced. Paleolatitude of the site of formation of the rock can thus be determined, since the magnetic inclination is a unique function of the magnetic latitude. Conversely when the original bedding plane can not be recognized, deviation of the concentrated direction of measured NRM from the ancient field direction may provide information on the dip and strike (bedding angle) of the tectonic tilting, if the paleolatitude of its formation is assumed. By this procedure a possible general feature of the subterranean ophiolite body can be reconstructed.

Occurrences of rock assemblages corresponding to ophiolite have been reported from various districts in Japan. Horokanai ophiolite in the Kamuikotan tectonic belt, Hokkaido (ISHIZUKA, 1980) and Yakuno ophiolite in the Tamba-Maizuru zone, inner southwestern Japan (ISHIWATARI, 1978) have been investigated in detail and their oceanic crust origin has most convincingly been demonstrated. Green rocks occurring in Mikabu zone at Toba district (NAKAMURA, 1971) have recently been attributed to the oceanic crust origin (NAKAMURA, 1981). Existence of a paleotrench associated with a lost ocean or an ancient marginal basin has been postulated with the Kurosegawa zone in the southwestern Japan (MARUYAMA, 1978; SUZUKI *et al.*, 1979; HORIKOSHI, 1979). Hayama-Mineoka tectonic belt in Boso and Miura Peninsulas (OGAWA and HORIUCHI, 1978; OGAWA, 1980; TAZAKI and INOMATA, 1980b) and Setogawa belt in central Japan (ARAI and UCHIDA, 1979) are considered to be composed of ophiolite by some of the investigators (OGAWA, 1981; OHASHI and SHIRAKI, 1981). Ages of rocks comprising the ophiolites in Hokkaido and the southwestern Japan are Cretaceous and Permian, respectively, while rocks of Hayama-Mineoka and Setogawa belts are Neogene.

Our rock-magnetic studies (TONOUCHI *et al.*, 1981; TONOUCHI and KOBAYASHI, 1982a) have indicated that these Neogene ophiolites are much less altered than Cretaceous ones so as to preserve more stable and presumably primary directions of NRM particularly in basaltic rocks. We have thus investigated paleomagnetic properties of the Hayama-Mineoka and the Setogawa ophiolite belts much in detail. In order to confirm the paleomagnetic results, observed anomalies of gravity and magnetic field on the Hayama-Mineoka ophiolite belt and its surroundings have also been analyzed. Alkali basalts in Takakusayama and Ryuso districts adjacent to the Setogawa belts have been studied to examine their possible oceanic seamount origin but the results were reported elsewhere (TONOUCHI and KOBAYASHI, 1982b). Then for the purpose of testing the possible regional tectonic movement coupled with the ophiolite obduction as well as of identifying age of obduction, paleomagnetic directions of island arc type igneous rocks occurring in the surrounding areas such as Fujigawa and Tanzawa districts have been studied.

2. Paleomagnetism of the Hayama-Mineoka Ophiolites

2.1 Outline of geology and paleomagnetic samples

The Hayama-Mineoka belt trends in a roughly ESE-WNW direction in the central portion of Boso and Miura peninsulas about 50 km south of Tokyo (Fig. 1) (NAKAJIMA *et al.*, 1981). In the midst of Boso Peninsula it is topographically characterized by an elongated Mineoka Hill with faulted cliffs on its north and more gentle slope on its south (ARAI, 1981). The strata are exposed at Kamogawa pier on the east coast of Boso Peninsula as well as at the crest of Mineoka Hill and appear to extend to the shelf and slope east off Kamogawa (KIMURA, 1976). Outcrops of igneous rocks are very poor in Miura Peninsula and western portion of Boso Peninsula in which the Hayama-Mineoka belt is overlain with unconformity by sedimentary layers of Hota and Miura groups. The Hota Group consists of Miocene siltstone and sandstone intercalated by tuff and is generally dipping southward.

Ultramafic rocks (harzburgite and dunite) occur at the crest of Mineoka Hill and at Kinugasa Town near Yokosuka but most of their outcrops are small in size. Paleomagnetic samples were collected at two sites at Mineoka Hill (sites 8 and 19 in Fig. 2) and one site at Kinugasa. Gabbros occur at the crest of Mineoka Hill. Seven oriented samples of hornblend gabbro were collected at site 10 and three oriented samples of gabbro pegmatite were collected at site 14 (Fig. 2).

Pillow and dyke basalts are found throughout the Hayama-Mineoka belt (KANEHIRA, 1976). One hundred and one oriented samples of pillow basalts were

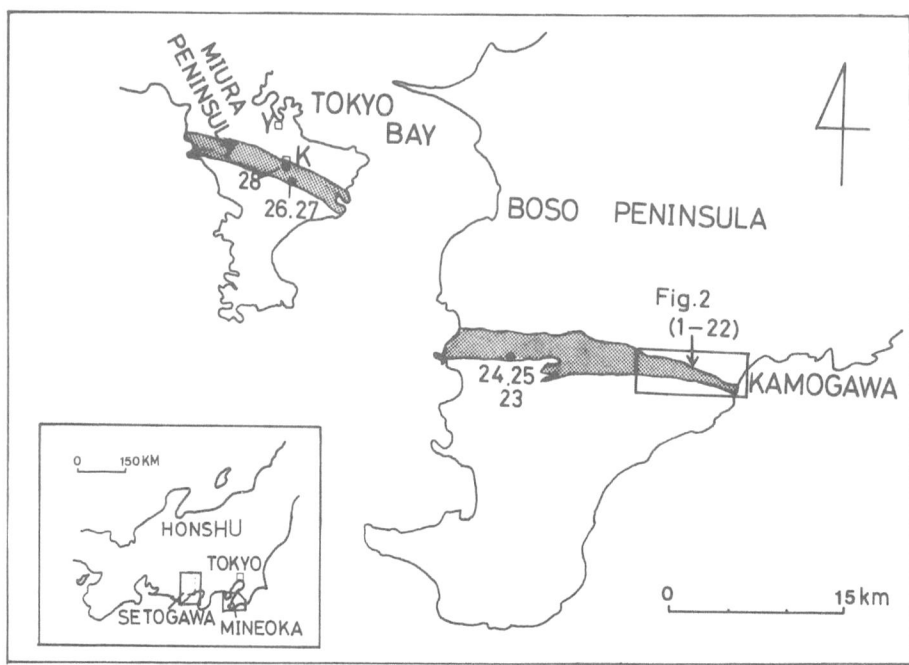

Fig. 1.   Location of the Hayama-Mineoka ophiolite belt (shaded zones) and sampling sites, Y; Yokosuka City, K; Kinugasa JNR Station.

collected for paleomagnetic measurement at 12 sites and 85 oriented samples of dyke basalts were collected at six sites (Fig. 2 and Table 1). Chemical compositions of three dyke basalts from Boso (sites 1, 5, and 29) and one dyke basalt from Miura Peninsula (site 27) analyzed by wet chemical method with the aid of H. Haramura indicate that these rocks are similar to abyssal tholeiites in compositions of both major and minor elements. Tazaki and Inomata (1980b) have reported similar results with seven basalts from the same belt. Alkali basalt and picrite basalt are also found but their amount seems to be much smaller than that of tholeiite. Nine picrite basalts were samples for paleomagnetism at the western part of Mineoka Hill (site 22).

Epidote-hornblende schist and quartz schist are exposed at Bentenjima and Byobujima islands located near Kamogawa coast (sites 6 and 7) but reconaissance measurement showed that these rocks are too weakly magnetized (10^{-6} emu/cc) to be measured with our spinner magnetometer. Siliceous shales and cherts are exposed on the northern part of the Mineoka Hill (e.g. site 18) but their magnetization is too weak to be measured. Umber (chocolate-colored sediment composed of very fine particles of geothite $\alpha FeO(OH)$ and small amount of pyrolusite MnO_2) occurs immediately above tholeiitic pillow lava on Mineoka Hill (Tazaki and Inomata, 1980a) but it is too fragile to be paleomagnetically measured. Its occurrence is very similar to the Cyprus umber overlying the Troodos Massif (Gass, 1963) and also to some DSDP cores often found immediately above the ocean floor pillow basalts (Klein et al., 1978; 1980). It is,

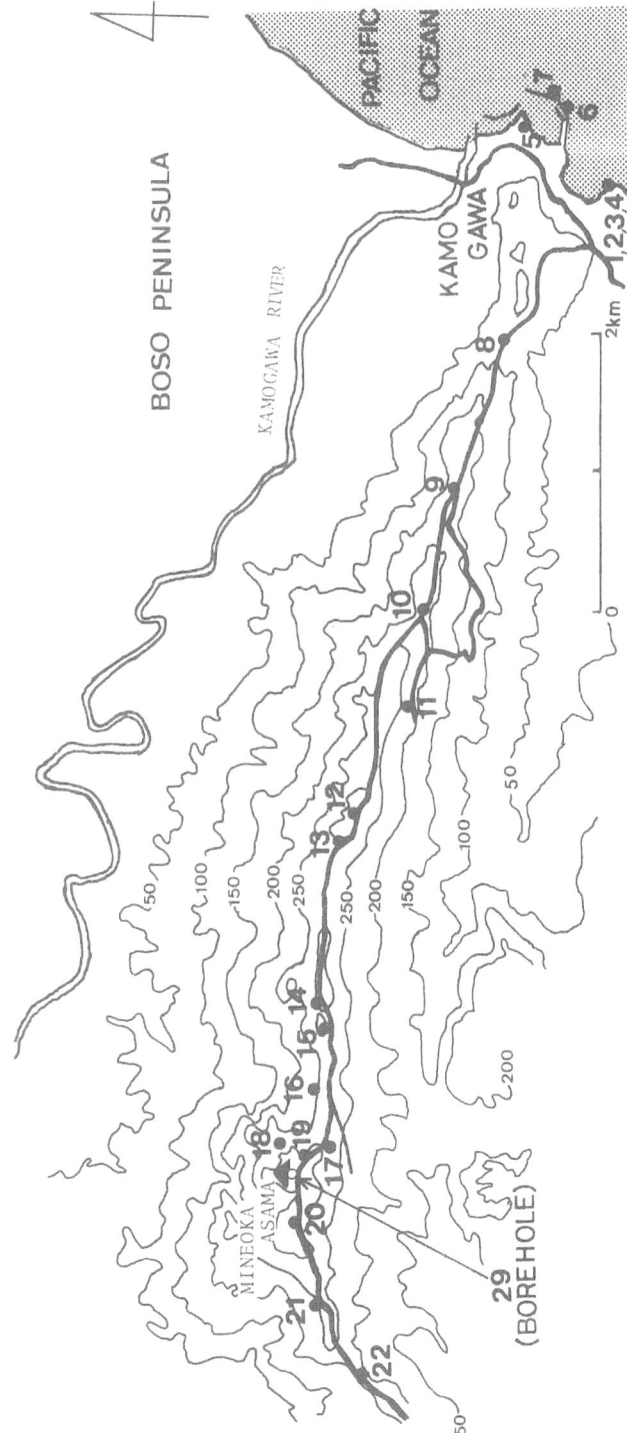

Fig. 2. Topography (contours in m) and location of sampling sites (numerical figures denote site number) in the southeastern Boso Peninsula (denoted by a rectangle in Fig. 1). Solid lines denote roads.

Table 1. Paleomagnetic properties of the Hayama-

Site No.	Locality	Rock Type	No. of oriented sample	J_n ($\times 10^{-4}$ emu/cc)
1	Shinyashiki, Kamogawa coast	Dyke Basalt	14	14.47
2	Shinyashiki, Kamogawa coast	Dyke Basalt	7	13.42
3	Shinyashiki, Kamogawa coast	Pillow Basalt	14	26.62
4	Shinyashiki, Kamogawa coast	Pillow Basalt	16	13.56
5	Fishing pier, Kamogawa	Dyke Basalt	12	13.53
6	Byobujima, Kamogawa	Epidote-hornblende Schist	—	0.03
7	Bentenjima, Kamogawa	Dyke Basalt	10	24.17
8	Mineoka Hill, crest	Ultramafic Rock	3	9.85
9	Mineoka Hill, crest	Pillow Basalt	6	9.79
10	Mineoka Hill, north of crest	Hornblende Gabbro	7	0.69
11	Mineoka Hill, crest	Ultramafic Rock	8	2.78
12	Mineoka Hill, crest	Pillow Basalt	7	8.06
13	Mineoka Hill, crest	Pillow Basalt	2	13.78
14	Mineoka Hill, crest	Gabbro pegmatite	3	0.88
15	Mineoka Hill, southern slope	Pillow Basalt	4	33.99
16	Mineoka Hill	Pillow Basalt	7	14.66
17	Mineoka Hill	Pillow Basalt	4	9.92
18	Mineoka Hill, north of crest	Chert	—	0.03
19	Mineoka Hill, crest	Ultramafic Rock	5	4.48
20	Mineoka Hill, crest	Pillow Basalt	3	9.63
21	Mineoka Hill, southern slope	Pillow Basalt	12	17.35
22	Western part of Mineoka Hill	Picrite	9	43.67
23	Uenodai, western Boso	Dyke Basalt	12	22.14
24	Uenodai, western Boso	Pillow Basalt	9	3.03
25	Uenodai, western Boso	Pillow Basalt	—	n.d.
26	Hirasaku, Miura Penin.	Pillow (weathered)	17	15.21
27	Hirasaku, Miura Penin.	Dyke Basalt	30	11.03
28	near Kinugasa JNR Stn.	Ultramafic Rock	16	3.28
29	Drilled core, Mineoka Hill crest	Pillow and Dyke Basalt	37	35.17

therefore, an evidence of submarine eruption.

K-Ar ages of basalts occurring on Mineoka Hill and at the Kamogawa coast are 29.3 to 39.9 Ma (TAKIGAMI et al., 1980). Ar^{39}-Ar^{40} ages of the Kamogawa basalts range between 40 and 50 Ma (KANEOKA et al., 1981). Epidote-hornblende schist from Bentenjima Island off Kamogawa was dated to be 38 Ma (YOSHIDA, 1974). It must be noted here that these Eocene to Oligocene ages are those of eruption or metamorphism of the rocks but not of their tectonic emplacement. The ages of microfossils contained in shales and cherts on Mineoka Hill have not been identified.

2.2 Paleomagnetic results

Intensity and direction of NRM of rocks have been measured using a 5 Hz spinner magnetometer. More than two cylindrical specimens were cut from each sample and

Mineoka ophiolites in Boso and Miura Peninsulas.

Q_n	MDF	D	I	k	α_{95}	T_{c1} (°C)	Thermo-magnetic curve	X-value
0.74	35–55	16.0	7.8	15.01	5.50	315–413	I	0.76
0.99	150	6.0	27.1	101.76	4.00	430–440	I	0.76
5.98	160–170	144.8	−20.9	23.01	5.50	440–460	I	0.82
0.90	45	36.7	5.3	65.08	3.17	n.d.	n.d.	n.d.
0.57	90	−85.1	21.6	1.82	31.71	n.d.	n.d.	n.d.
n.d.	n.d.	n.d.	n.d.	—	—	n.d.	n.d.	n.d.
1.74	n.d.	2.2	14.8	21.50	7.63	n.d.	n.d.	n.d.
1.44	220–270	−28.7	49.7	44.76	18.60	n.d.	n.d.	0.00
2.64	90	12.8	32.0	108.61	4.96	n.d.	n.d.	n.d.
1.15	370–470	162.9	46.9	18.21	14.37	n.d.	n.d.	n.d.
0.70	100–270	21.8	40.4	10.79	12.70	575	n.d.	0.00
0.75	90	56.1	55.4	65.03	5.20	n.d.	n.d.	n.d.
3.76	n.d.	29.7	32.9	40.93	40.10	460	I	0.83
0.69	130	158.1	−18.1	8.52	33.40	575	R	n.d.
4.80	n.d.	9.9	−8.7	38.89	12.41	n.d.	n.d.	n.d.
1.31	n.d.	170.3	−6.1	1.12	54.05	475–485	I	n.d.
2.74	45	−16.4	−40.6	192.77	4.40	560	R	n.d.
n.d.	n.d.	n.d.	n.d.	—	—	n.d.	n.d.	n.d.
1.82	210	−23.4	62.9	155.13	5.40	n.d.	n.d.	n.d.
1.10	n.d.	3.6	49.6	151.49	7.49	n.d.	n.d.	n.d.
2.44	45	134.8	−8.8	12.59	10.14	500	I	n.d.
30.36	230	4.6	39.8	105.17	3.61	440–450	I	0.20
1.28	160–180	19.5	28.5	59.70	4.11	n.d.	n.d.	n.d.
1.07	170	−16.0	19.3	34.08	7.48	n.d.	n.d.	n.d.
n.d.	n.d.	n.d.	n.d.	—	—	n.d.	n.d.	n.d.
3.67	270–360	−112.2	−39.2	1.43	48.43	475	I	n.d.
6.89	120–230	−47.8	−15.4	8.89	8.62	380–455	I	n.d.
0.49	40–120	−28.4	−40.2	2.21	23.30	555–575	R	n.d.
6.78	70–140	n.d.	$\begin{cases} +29.6 \\ -26.6 \end{cases}$	—	—	395–490	I	0.75

the results were averaged for each sample. Susceptibility K was measured with an AC bridge so that Königsberger ratio $Q_n = J_n/KH$ was calculated where $H = 0.45$ Oe. The results are listed in Table 1.

Intensity of NRM of pillow basalts ranges from 3×10^{-4} to 88×10^{-4} emu/cc, except for that of highly weathered pillow basalts from Miura Peninsula with much lower values. Dyke basalts have NRM intensity ranging from 2×10^{-4} to 41×10^{-4} emu/cc but their majority has intensity smaller than that of pillow basalts apparently due to coarser grain size of ferromagnetic minerals. Ultramafic rocks have a range of NRM intensity from 0.8×10^{-4} to 14×10^{-4} emu/cc.

The NRM intensity of pillow basalts of the Mineoka ophiolite is roughly comparable to that of the oceanic basalts recovered by DSDP and dredge hauls (LOWRIE, 1977; STEINER et al., 1978) as well as other ophiolitic basalts reported from various sites in the world (BUTLER *et al.*, 1975, 1976; LEVI and BANERJEE, 1977; LEVI *et al.*, 1978). Königsberger ratio Q_n is 3.5 in average for the Mineoka pillow basalts, indicating that the NRM dominates induced magnetization in magnetic anomalies.

Forty selected specimens were progressively demagnetized in an alternating magnetic field (a.f.) with maximum peak amplitude up to 600 Oe. Median destructive field (MDF) of NRM of pillow basalts is 90 to 350 Oe, which is larger than that of dyke basalts (30–150 Oe). Microscopic observation of polished sections of these specimens indicates that higher stability of pillow basalts is owing to the finer grain sizes of constituent ferromagnetic minerals. A.f. demagnetization of anhysteretic remament magnetization (ARM) performed in 800 Oe a.f. parallel to a steady field of 0.45 Oe also shows that dyke basalts are magnetically less stable than pillow basalts but are probably still useful for paleomagnetic studies.

NRM of ultramafic rocks (sites 8, 11, 19, and 28) have sufficienty high stability to record the ancient field direction. However, microscopic observation revealed occurrence of fine particles of magnetite surrounding olivine crystals. It seems likely that the magnetite was recrystallized when the rocks were serpentinized so as to acquire chemical remament magnetization after the formation of the ocean floor. These rocks are, therefore, not used for tectonic considerations.

Thermomagnetic analysis, analysis of opaque minerals by means of an electron probe microanalyzer (EPMA), and microscopic observation of opaque minerals were carefully made to infer the origin of NRM of the paleomagnetic samples. Details of the results were published elsewhere (TONOUCHI and KOBAYASHI, 1982a) and only the initial Curie temperature T_{c1}, thermomagnetic reversibility (I denotes irreversible and R reversible) and X-value of titanomagnetite ($X\mathrm{Fe_2TiO_4} - (1 - X)\mathrm{Fe_3O_4}$) are listed in Table 1. It was concluded from these rock-magnetic tests that basaltic rocks in the Hayama-Mineoka ophiolite belt are similar to the ocean floor basalts which can record the direction of the ancient geomagnetic field.

Progressive alternating field demagnetization of NRM of some pilot specimens indicated that a.f. of about 100 Oe was optimum to obtain the least scatter in the direction of NRM among samples at each site. All the samples thereafter were demagnetized by an a.f. of 100 Oe and the mean values of declination, inclination, Fisher's precision parameter k, and circle of confidence at 95% probability after demagnetization α_{95} were obtained as listed in Table 1. Sites 5, 13, 14, 16, and 26 with large scatter in the NRM direction ($\alpha_{95} > 30$) are excluded from the tectonic considerations, since block movement within site is suspected.

Both normal and reversed polarities of magnetization are found with basalts and gabbros in the Mineoka belt. A similar and more detailed record of sequences of normal and reversed magnetic polarities have been revealed from a vertical 92 m long core of pillow and dyke basalts drilled by the Kokudobosai Co. Ltd. on the south side of a small mount Mineoka Asama (Fig. 2) on the crest of Mineoka Hill (INOUE *et al.*, 1980). Intensity and direction of NRM of 38 horizons were measured with 115 cylindrical specimens having a length of 2.7 cm and 1.25 cm in diameter. Declination is only arbitrary, while inclination of NRM shows a pattern of normal-reverse sequence similar to that often found in the DSDP basement cores (JOHNSON and HALL, 1978; RYALL *et al.*, 1977; FALLER *et al.*, 1978; JOHNSON, 1979; KINOSHITA, 1980). Absolute values of inclination are roughly constant throughout the cores (30° for normal and 27° for reverse polarity in average) except for transitional zones between normal and reverse (Fig. 3). This result seems to indicate that the 92 m thick basaltic layer was a part of the ancient ocean floor as a whole. Formation of this basaltic layer by accumulation of landslide products is, therefore, impossible.

Beneath the basaltic layer, about 25 m of serpentinized ultramafic rock was drilled. Contact plane of the two kinds of rocks could not be identified. A fault movement may have caused this superposition of basalts immediately overlying ultramafic rocks. As the recovered pieces of ultramafic rocks were too fragmented, their NRM could not be measured. Neither geological nor paleomagnetic consideration would, therefore, be possible for the origin of such an occurrence of ultramafic rock.

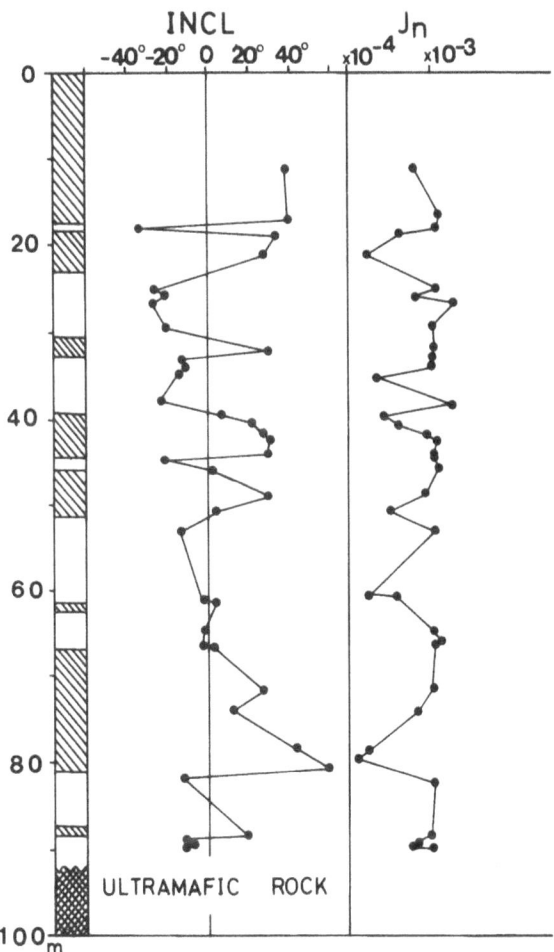

Fig. 3. Paleomagnetic results of basaltic cores from a borehole drilled at the crest of Mineoka Hill.

2.3   Tectonic implications and paleolatitude

Paleomagnetic measurement described in the preceding sections has shown that at the majority of the sites the directions of stable NRM within each site are concentrated; α_{95} is smaller than $15°$ except for sites 5, 13, 14, 16, and 26 at which rocks are either very fragmented or highly weathered (Table 1). Moreover, within-site averages of NRM directions are generally quite similar to one another among sites in one area. For example, NRM directions of five sites in Kamogawa coast (sites 1, 2, 3, 4, and 7) are nearly equal to one another, showing low inclinations and easterly declinations. Some of the sites (9, 15, and 23) on Mineoka Hill and in the western Boso Peninsula and Miura Peninsula also have systematically low inclination, although the declinations in one western Boso site (24) and Miura site (27) are westerly. Consistency of NRM directions among such separated sites in one belt is very much in favor of the tectonic obduction of ancient ocean floor as the origin of the Hayama-Mineoka ophiolite.

The tectonic movement can also be revealed by surface features of some eruptive lava flows. Surface of a ropy lava located about 20 m away from site 1 (dyke basalt at Shinyashiki, Kamogawa coast) has strike of N78°W and dip of 27°S. Pillow lava overlying the ropy lava and dyke shows nearly the same orientation. Although oriented paleomagnetic samples were not collected from these ropy and pillow lavas because they are too fragile and fragmented, it seems reasonable to assume that dykes were tilted together with the ropy and pillow lavas, as no fault is recognized between lavas and dykes. NRM directions of dyke basalts have been corrected on the assumption that the ropy and pillow lavas were primarily formed in a horizontal plane and were later tilted together with dykes by a tectonic process.

The average direction of NRM of dyke rocks from site 1 is $I = 34°$, $D = 12°$ after the above-mentioned bedding correction (Fig. 4). Its inclination is still significantly low compared to that expected for the present position of the site ($I = 55°$ at $35°N$). This difference in inclination (~ 20) can be explained as a result of either a secular variation of the geomagnetic field or a northward drift of the rocks.

If an axial dipole field is assumed, the paleolatitude of the site is calculated to be $23°N$, which is $12°$ south of the present latitude. Although the effect of secular variation can not be disregarded, it seems possible that these rocks were a part of the floor of the Philippine basin which drifted northward by about 2,000 km after its formation (roughly 40 to 50 Ma BP). Such a northward drift of the Philippine basin has already been demonstrated by shape analysis of magnetic anomalies (LOUDEN, 1977) and NRM of DSDP sediments (LOUDEN, 1977; KINOSHITA, 1980; KLEIN and KOBAYASHI, 1980).

No bedding correction has been successful at the other sites in the Hayama-Mineoka belt because identification of the horizontal plane using pillow structure is difficult. Therefore, the presumable tectonic tilting was calculated with each site on the assumption that the rocks were originally magnetized in a direction parallel to the axial dipole field at a latitude of $23°N$ (Table 2). Such an interpretation of paleomagnetic data is essentially the same as the deduction of rotation of a landmass from a deflected declination of NRM and would be valid as long as the assumption on the original location of these rocks is correct.

Uncorrected directions of NRM at a few sites on the crest of Mineoka Hill (sites 9,

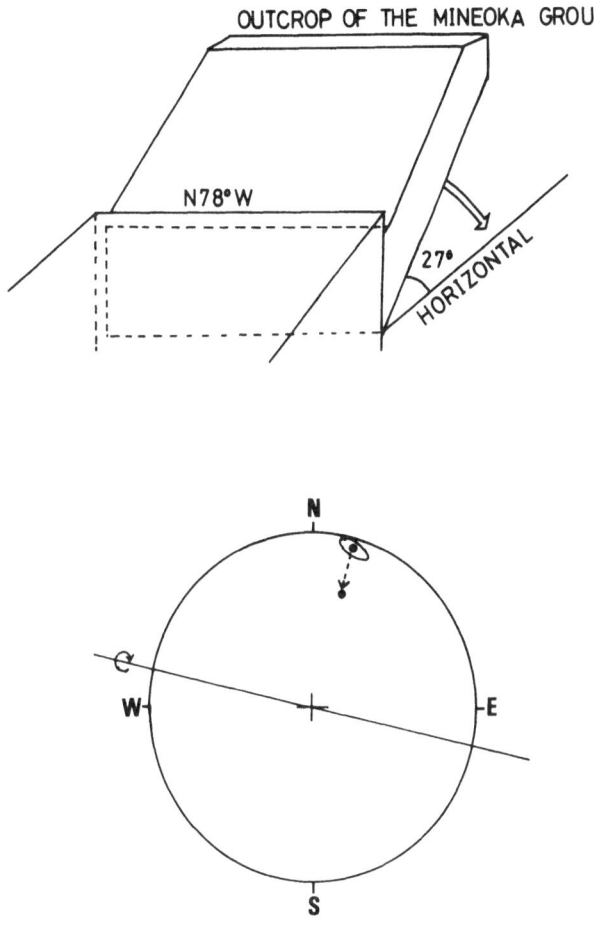

Fig. 4.   Averaged direction of NRM with its oval of 95°ₒ confidence and its bedding correction (a broken arrow in the lower figure). Solid circles in the Schmidt projection net indicate downward inclinations.

Table 2.   Paleomagnetically deduced bedding correction of each site (see text).

Site No.	Rock type	D	I	Bedding	
				Strike	Dip
2	Dyke Basalt	6.0	27.1	N42°W	11°S
3	Pillow Basalt	144.8	−20.9	N20°E	36°S
4	Pillow Basalt	36.7	5.3	N48°W	42°S
7	Dyke Basalt	2.2	14.8	N79°W	22°S
15	Pillow Basalt	9.9	−8.7	N80°W	22°S
16	Pillow Basalt	170.3	−6.1	N79°E	30°S
21	Pillow Basalt	134.8	−8.8	N26°E	48°S
23	Dyke Basalt	19.5	28.5	N13°W	19°S
24	Pillow Basalt	−16.0	19.3	N74°E	21°S

20, 22, and 29) are roughly parallel to the field direction at 23°N. This fact may be explained as indicating that relatively small blocks of ophiolitic complexes without appreciable tilting exist on the crest of the Hill.

2.4 Gravity and magnetic anomalies caused by the ophiolite belt in Boso and Miura Peninsulas

A belt of positive Bouguer gravity anomaly of about 60 mgal (Fig. 5) along the Hayama-Mineoka belt in Boso Peninsula (MATSUDA and SUDA, 1964) indicates existence of an appreciable amount of high density materials under this belt and definitely excludes possibility of landslides or other sedimentary processes for emplacement of the high density rocks. Simple model calculation of gravity has indicated that a southward dipping subterrenean high density layer inferred from the present paleomagnetic results may cause the major portion of the observed gravity anomaly.

An aeromagnetic survey was recently conducted around this area by Hydrographic Department, Maritime Safety Agency of Japan. Geological Survey of Japan (HORIKAWA et al., 1979; OGAWA et al., 1979) published an aeromagnetic map in some part of Boso Peninsula and sea east off the Boso coast. It is evident that a belt of negative magnetic anomaly greater than 200γ in amplitude lies on the Hayama-Mineoka ophiolite belt (Fig. 6).

The observed negative magnetic anomaly belt can be explained by a southward tilting layer the same as assumed for gravity interpretation, if its magnetization is northward and nearly horizontal or slightly upward, just as shown by paleomagnetic measurement.

3. Paleomagnetism of the Setogawa Ophiolite

3.1 Outline of geology of the Setogawa Group

The Setogawa Group occurs in a belt located in Shizuoka Prefecture, west of Mt. Fuji in the south-central Honshu, Japan and trends roughly N-S (Fig. 7). The strata steeply dipping westward are recognized by sedimentary structures such as sole marks (WADA, 1976). Fossils found in the Setogawa Group indicate Oligocene age of sedimentation. The western margin of the Group adjoins to the Cretaceous Shimanto Group along the Sasayama Tectonic Line.

Serpentinized ultramafic rocks have been found as small outcrops in the western margin of this belt. Pillow and dyke basalts widely occur in the belt (ARAI et al., 1978). Gabbro occurs in the eastern portion of this belt. Cherts with thickness of about 250 m are exposed in the upper reaches of Abegawa River. Major element analysis of basaltic rocks in this Group (sites 1 and 2) by H. Haramura indicates that they are most probably abyssal tholeiite. Similar chemical compositions have been reported for six samples of basalts from the same belt by OHASHI and SHIRAKI (1981), who also found a small amount of high-magnesian andesite and alkali basalt in the central part of the belt. Similarity of basalts to the deep ocean floor as well as coexistence of chert, gabbro, and ultramafic rocks seems to indicate that these complexes are ophiolites as postulated by OHASHI and SHIRAKI (1981).

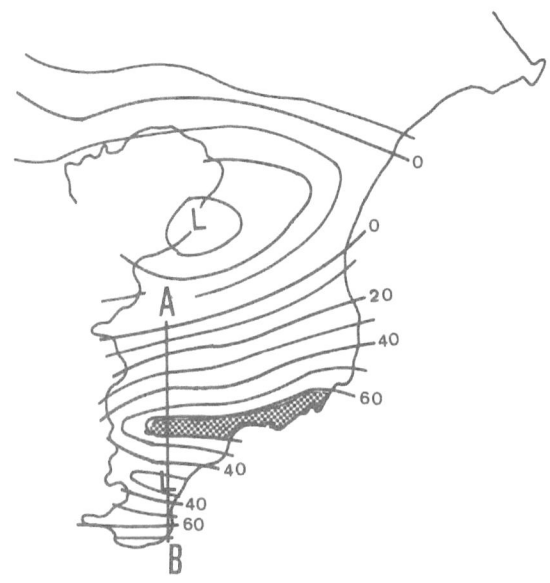

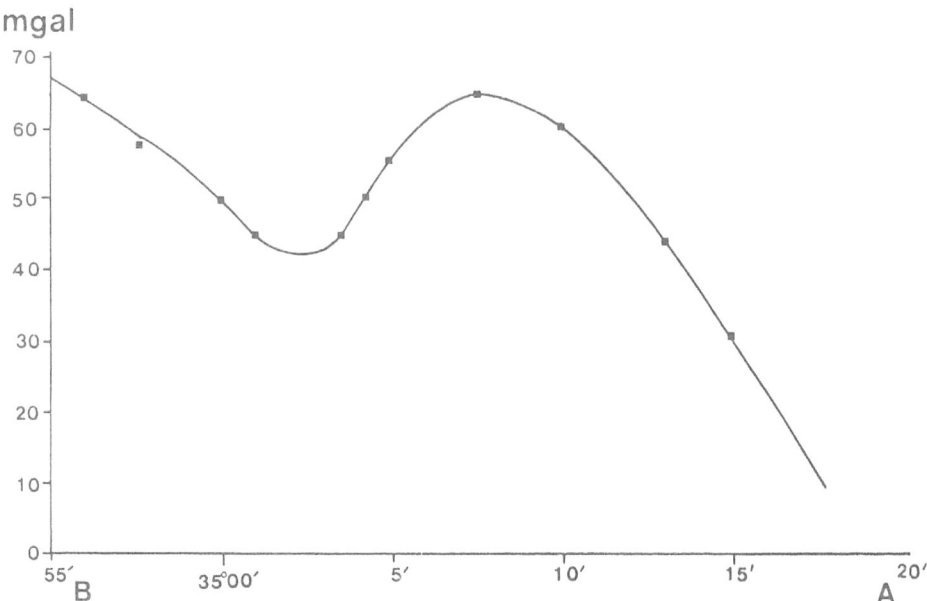

Fig. 5.   Distribution of the Bouguer gravity anomaly in Boso Peninsula (numerical figures beside contour in units of mgal) and its NS profile along a line AB.

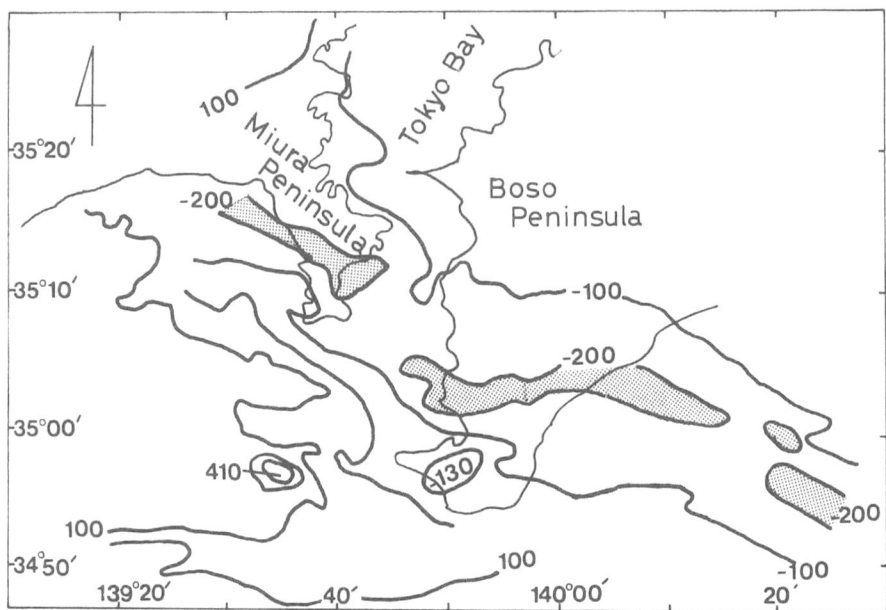

Fig. 6.   Geomagnetic total force anomaly observed in Boso and Miura Peninsulas and surrounding seas,
compiled from data by Hydrographic Department, Maritime Safety Agency of Japan (unpublished), and
Geological Survey of Japan (HORIKAWA *et al.*, 1979; OGAWA *et al.*, 1979). Contours in γ (nT).

3.2   Paleomagnetic results and their tectonic implications

Ninety-eight oriented samples of pillow and dyke basalts were collected for
paleomagnetic measurement at 12 sites in the Setogawa belt (Fig. 7, Table 3). Intensity
of NRM of 24 samples of dyke and pillow basalts from sites 1 and 2 in the southern part
of the belt ranges from 3 to 37×10^{-4} emu/cc. Q_n of these rocks is much larger than
unity. Their MDF is 80 to 130 Oe. Scatter in NRM directions within site becomes the
smallest after a.f. demagnetization by a peak field of 100 Oe. Circles of 95% confidence
are smaller than 10 degrees. Thermomagnetic measurement as well as microscopic and
EPMA analyses has revealed that these samples contain titanomaghemites with x
$= 0.74 - 0.84$ and $T_c = 440 - 480°C$, which are quite similar to those of ocean floor
basalts. These results indicate that NRM directions of these samples are paleomagnetically very reliable and can be used as magnetic fossils.

The averaged directions of NRM thus obtained with samples from sites 1 and 2 are
both reversed and deflected westward (Table 3). These directions can be explained if we
assume that these rocks were originally magnetized parallel to the geomagnetic field
with reverse polarity and then tilted westward.

Stratigraphic investigation of this belt (WADA, 1976) indicated that general tilting
of the sedimentary layers of this Group is about 60° westward with a strike of roughly
N-S. If the paleomagnetic directions are corrected by this tilting, the original
declinations become 175.9 at site 1 and 153.3 at site 2, while the inclinations remain
unchanged (-29.1 and -25.5). The corrected paleomagnetic directions can be

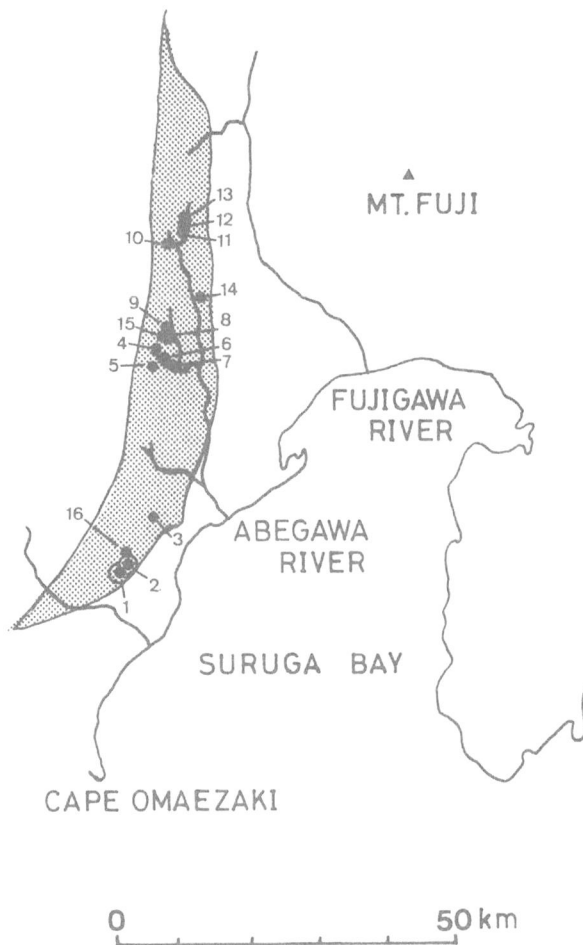

Fig. 7. Distribution of the Setogawa Group (shaded zone) and sampling sites of ophiolitic rocks. Double circles denote sites at which paleomagnetically applicable samples were collected. At the other sites titanomagnetites contained in basalts are changed to pyrrhotite and pyrite and can not be used for records of the original magnetic directions. Numerical figures denote site number. As to location of this area refer to an index map in Fig. 1.

explained if these basalts were formed at a latitude of about 20°N.

Dyke and pillow basalts from sites 6 through 13 in the northern part of the Setogawa belt have intensity of NRM by more than one order of magnitude weaker than that of basalts from sites 1 and 2. Scatter in NRM of basalts represented by α_{95} in Table 3 is appreciably large at sites 6 to 13. The scatter is not reduced by a.f. demagnetization. The deviations of the averaged directions of NRM of these samples from the present field direction are within their circles of 95% confidence α_{95} and, therefore, statistically insignificant. It may, therefore, be suspected that these rocks acquired a predominating secondary magnetization (possibly chemical remanent

Table 3. Paleomagnetic properties of the Setogawa ophiolites (Mt; magnetite, Po; pyrrhotite, Py; pyrite).

Site No.	Locality	Rock type	No. of sample	J_n ($\times 10^{-4}$ emu/cc)	Q_n	MDF	D	I	k	α_{95}	T_{ct} (°C)	Thermo mag.	Opaque
1	Southern part of Setogawa belt	Dyke Basalt	11	20.71	1.74	80–100	−124.1	−29.1	38.8	8.4	440	I	TiMt $X = 0.74$
2	Southern part of Setogawa belt	Pillow Basalt	13	19.77	5.47	100–110	−142.7	−25.5	23.74	9.1	480	I	TiMt $X = 0.84$
3	North of Takakusayama	Dyke Basalt	12	0.36	0.30	150–580	88.9	54.8	9.2	15.9	575	R	Mt ($X = 0.00$)
4	Nakakochi River	Dyke Basalt	6	10.95	1.03	90–130	−2.7	10.5	18.2	16.1	575	R	Mt
5	In a branch of Nakakochi River	Ultramafic Rock	10	48.73	1.42	45–55	−1.3	58.0	4.8	15.9	575	R	Mt
6	Nakakochi River	Dyke Basalt	9	0.03	0.08	600	132.8	−3.4	1.1	180.0	n.d.	n.d.	Mt, P_o
7	Nakakochi River	Dyke Basalt	5	0.44	0.53	45	166.7	32.4	1.4	114.9	325	I	Mt, P_o
8	Upper branch of Senmata River	Dyke Basalt	10	0.63	0.55	110	−119.2	−19.1	3.2	50.6	330	I	Mt, P_o, P_y
9	Upper branch of Senmata River	Pillow Basalt	4	0.08	0.08	100	−6.2	−18.1	1.8	71.9	n.d.	n.d.	Mt, P_o
10	Upper branch of Senmata River	Dyke Basalt	8	0.77	1.62	50–100	4.3	37.4	8.6	20.0	355	I	Mt, P_o
11	Upper branch of Abegawa	Dyke Basalt	6	0.47	2.14	600	8.1	18.8	17.2	16.6	n.d.	n.d.	Mt, P_o, P_y
12	Upper branch of Abegawa	Dyke Basalt	6	0.41	0.80	600	18.8	8.1	36.8	11.2	320	I	Mt, P_o
13	Upper branch of Abegawa	Dyke Basalt	8	0.14	0.52	90	−0.9	18.2	5.1	27.2	310	I	Mt, P_o, P_y
14	Eastern edge of belt	Gabbro	—	<0.03	—	—	n.d.	n.d.	—	—	n.d.	—	P_o, P_y
15	Upper branch of Senmata River	Chert	—	<0.03	—	—	n.d.	n.d.	—	—	n.d.	—	—
16	Southern part of Setogawa belt	Chert	—	<0.03	—	—	n.d.	n.d.	—	—	n.d.	—	—

magnetization) parallel to the present geomagnetic field after the tectonic movement.

Thermomagnetic analysis and microscopic observation of polished sections have revealed occurrence of pyrrhotite (ferromagnetic iron sulfide) and pyrite in these rocks showing a low intensity and a large scatter in directions of NRM. Existence of pyrrhotite can be easily recognized by heating, because it is thermally unstable and often decomposes to either hematite or magnetite by heating at temperatures above 300°C or so (KOBAYASHI and NOMURA, 1972). Occurrence of iron sulfides suggests the prevalence of reducing circumstance at relatively low temperatures after the formation of these rocks. Hydrothermal activity in the ocean floor may have caused such a condition. A weak chemical remanent magnetization was acquired by pyrrhotite formed during the hydrothermal activity.

Dyke basalts from sites 4 and 5 contain nearly stoichiometric magnetite. NRM of samples from site 3 is weak, while that from site 4 is as strong as that from sites 1 and 2. Scatter in NRM directions is moderately large. It may be inferred from these results that dyke basalts at sites 3 and 4 have undergone an alteration under an oxidizing condition at a moderately high temperature (T $>$ 350°C). NRM of these rocks may not be the original thermoremanent magnetization acquired at the time of solidification of rocks from molten magma.

NRM of ultramafic rocks (site 5) is also due to stoichiometric magnetite precipitated along the margin of hydrated olivine crystals (TONOUCHI and KOBAYASHI, 1982a). Its direction is parallel to that of the present geomagnetic field and may possibly have been acquired after the tectonic movement. Some gabbros and cherts were sampled from sites 14, 15, and 16, but their magnetization is too weak (less than 10^{-6} emu/cc) to be accurately measured by our magnetometer.

The present paleomagnetic investigation mentioned above has indicated that only 24 among 98 samples or only two sites among 13 sites of sampling may be useful for tectonic applications of paleomagnetic directions. Nevertheless, the paleomagnetic results with both dyke and pillow basalts from sites 1 and 2 have indicated the lower paleolatitude ($\sim 20°$N) of sites of formation of these rocks in a good agreement with the paleomagnetically inferred paleolatitude of the Hayama-Mineoka basalts. This paleomagnetic implication together with proximity of chemical composition, petrography and mode of occurrence of some basaltic rocks to the abyssal tholeiite suggests that cherts, pillow and dyke basalts, gabbros, and ultramafic rocks found in the Setogawa belt were parts of the ancient ocean floor located roughly at the same latitude as that of the Hayama-Mineoka ophiolites and that they drifted northward and were accreted to the southern margin of Honshu.

4. Paleomagnetism of Some Adjacent Rocks

Location of the western extension of the Hamaya-Mineoka belt is yet uncertain. A postulated zone of extension is generally covered by a thick pile of younger extrusive and intrusive rocks such as Tanzawa and Mt. Fuji. For the purpose of identifying location of the western extension of the belt and determining the age of the tectonic event forming the Hayama-Mineoka and Setogawa ophiolitic belts, paleomagnetic directions of some rocks occurring in this postulated zone have been investigated.

4.1   Paleomagnetism of alkali basalts and basaltic andesites in Fujigawa district

One hundred and thirty-three oriented samples were collected at 11 sites located in the upper part of Fujigawa River situated east of the Setogawa belt. Among them 56 samples of basalts at five sites (sites 3, 4, 9, 10, and 11) were collected from Kushigatayama Formation exposed south and west of Kofu City. K-Ar age of a basalt (34 Ma according to Nishiyama and Ueda, 1976) and foraminifera zones (N3 to N6) indicate that this formation is of Oligocene to early Miocene.

Twenty-eight oriented samples of alkali basalts and ten oriented dacite samples were collected at sites 1, 2, and 5 in Takahagi Formation exposed northeast of Kofu City. K-Ar age of 10.6 Ma (Nishiyama and Ueda, 1976) and zone N10 of foraminifera (Shimazu et al., 1971) indicate that Takahagi Formation is of middle Miocene. Thirty-nine basaltic andesites at sites 6, 7, and 8 were from Onuma Formation younger than Takahagi Formation. Direction and intensity of NRM of these samples have been measured and the usual tests of paleomagnetic reliability have been made. It has been revealed by thermomagnetic and microscopic analyses that samples from site 7 contain pyrrhotite which may have been formed by hydrothermal alteration. Samples from site 3 may be unsuitable to the paleomagnetic applications, because their outcrops are too fragmented and their NRM directions are, probably as a result of fragmentation, too much scattered.

There exists a remarkable tendency in the NRM directions of samples from this districts except for sites 3 and 7; directions of NRM of samples from Takahagi Formation (sites 1, 2, and 5) and Onuma Formation (sites 6 and 8) are roughly parallel to that of the present field (samples from site 1 are reversedly magnetized but still parallel to the present direction), while the directions of Kushigatayama Formation (sites 4, 9, 10, and 11) are significantly different from the present field direction, implying existence of a tectonic movement (Figs. 8, 9, and Table 4).

4.2   Paleomagnetism of quartz diorites and basaltic rocks in Tanzawa district

A quartz diorite intrusive mass of middle to late Miocene in age, 20 km long in EW direction and 5 km wide occurs in Tanzawa districts. Eighty-nine oriented samples of quartz diorites were collected at sites 4 to 14. Intrusion of the quartz diorite is supposed to have occurred in late Miocene. Gabbroic rocks (sites 1–3, south of quartz diorite and sites 15–18 north of the mass) and basalts (sites 19–21 in Togatake south of quartz diorite) surrounding the intrusive mass were also sampled for paleomagnetic measurement. Eleven oriented samples of basaltic andesites were collected from site 22 on the north coast of Lake Sagami, about 20 km northeast of the quartz diorite mass. Eighteen samples of andesites in the Misaka Group were collected at site 23 located about 15 km northwest of the mass. Ages of rocks surrounding the quartz diorite are unknown but certainly older than intrusion of quartz diorite, because some of the surrounding rocks are partly metamorphosed to crystalline schist under the thermal and mechanical influence of the intrusion.

Results of paleomagnetic measurement are summarized in Figs. 8 and 9 and Table 5. Samples from sites 2, 3, 10, 12, 19, and 20 were found to be highly altered by microscopic and X-ray analyses and are excluded from the paleomagnetic consideration. It is thus evident from these figures and table that most of unaltered rocks

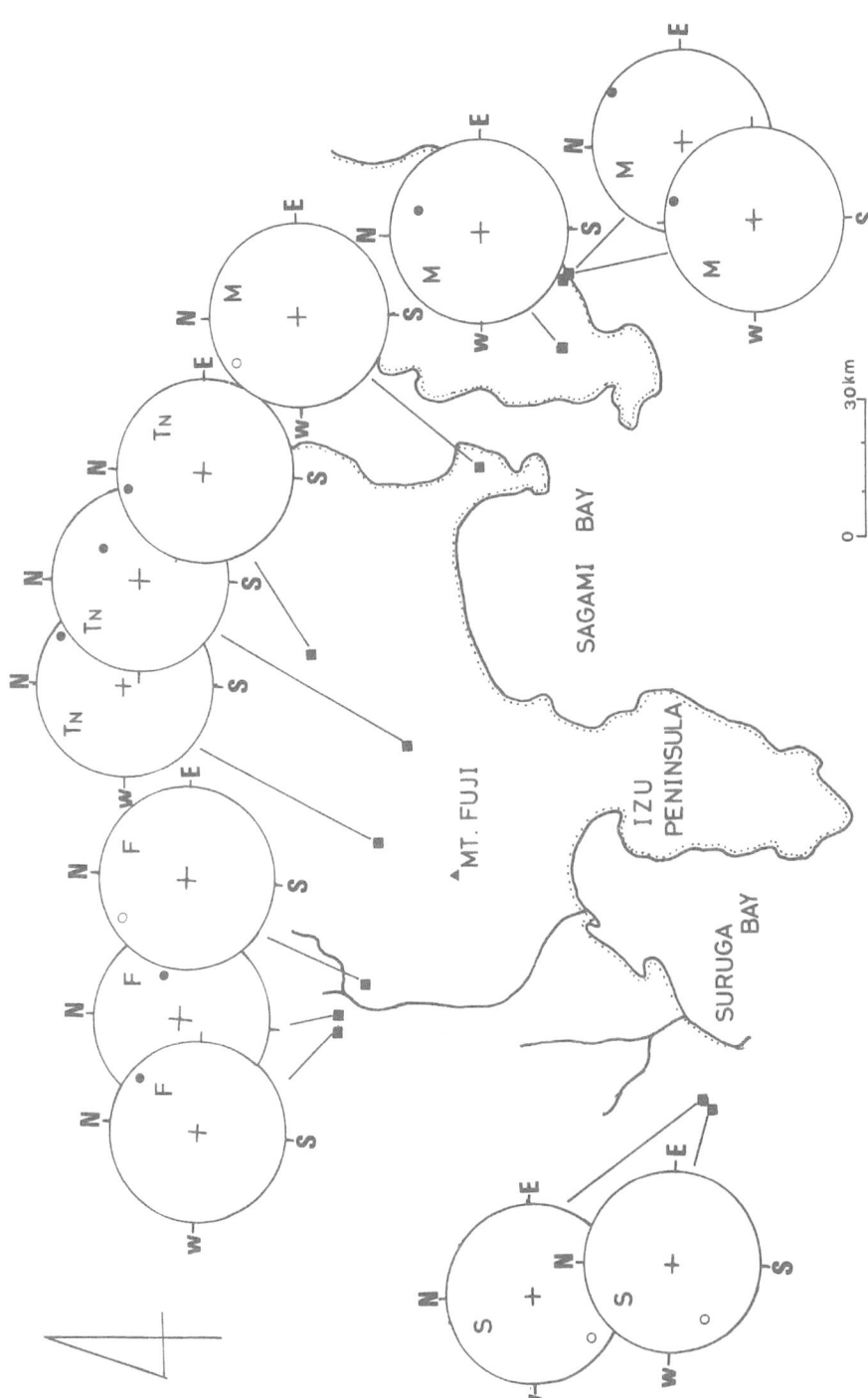

Fig. 8. Averaged directions of NRM at some representative sites of ophiolite belts and surrounding rocks with older ages. Solid and hollow small circles in the Schmidt projection nets denote lower and upper inclination. M; Hayama-Mineoka belt, T_N; Tanzawa district (its periphery), F; Fujigawa district, S; Setogawa.

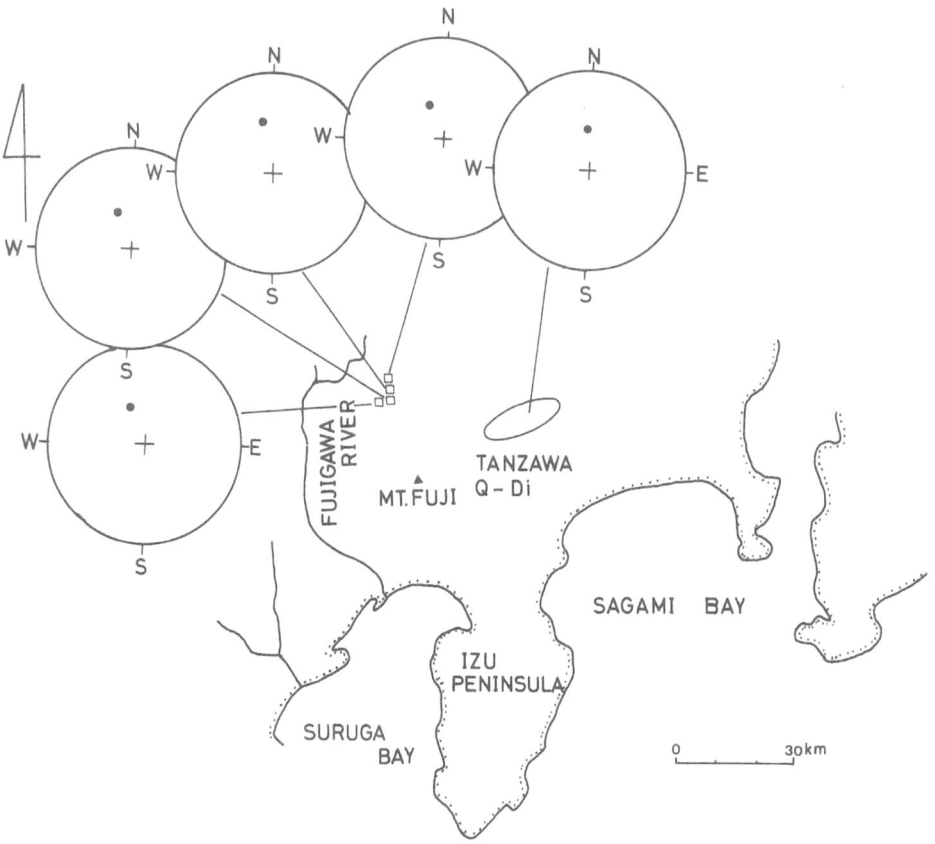

Fig. 9.   Averaged directions of NRM of quartz diorite from Tanzawa and rocks younger than about 11 Ma collected from Fujigawa districts. Note that NRM directions are roughly parallel to that of the present geomagnetic field.

surrounding the quartz diorite (sites 1, 21, 22, and 23) have NRM directions appreciably deflected from the direction of the present field. Their inclination is generally low in good agreement with that of majority of basaltic rocks from Hayama-Mineoka belt.

In contrast to the directions of NRM of older rocks, those of quartz diorite are roughly parallel to that of the present field, although scatter in directions represented by radius of circle of confidence α_{95} in Table 5 is relatively large probably due to viscous components in NRM. If the NRM directions of all the quartz diorite samples from Tanzawa are statistically treated together, their average is $D = -1.0$, $I = 54.2$, $\alpha_{95} = 6.7$, which is quite agreeable with that of the present field. Gabbroic rocks from north of the quartz diorite intrusion (sites 15–18) are also magnetized roughly parallel to the present field.

It may be inferred from the present paleomagnetic results that a tectonic

Table 4. Summary of magnetic properties of rocks from Fujigawa district.

Site No.	Locality	Rock type	No. of oriented sample	J_n ($\times 10^{-4}$ emu/cc)	Q_n	MDF	D	I	k	α_{95}	T_{cl} (°C)	Thermomagnetic curve	X-value
1	Takahagi northeast of Kofu	Basalt (Takahagi F.)	12	6.43	1.52	330–400	177.7	38.1	15.1	12.1	575	R	0.24
2	Takahagi northeast of Kofu	Basalt (Takahagi F.)	16	16.67	2.50	60	−37.6	70.1	139.0	2.7	535	I	0.26
3	Middle part of Tokiwa River	Dyke basalt (Kushigatayama F.)	4	23.73	2.85	65	147.1	42.1	471.6	4.2	—	—	n.d.
4	Middle part of Tokiwa River	Dyke basalt (Kushigatayama F.)	11	2.71	0.43	370	−129.5	−35.6	5.2	20.0	555	R	0.22
5	North of Lake Motosu	Dacite (Takahagi F.)	10	0.75	0.71	90	−27.6	61.9	24.3	8.9	565	R	0.27
6	North of Lake Motosu	Basaltic andesite (Onuma F.)	16	1.15	0.62	90–170	−13.7	48.9	34.1	6.4	510	I	0.27
7	West of Lake Motosu	Basaltic andesite (Onuma F.)	14	1.56	0.65	100–240	−17.0	56.9	33.7	6.6	470	I	n.d. (+Py)
8	West of Lake Motosu	Basaltic andesite (Onuma F.)	9	2.53	0.55	90–180	−19.5	59.0	56.1	6.5	570	R	n.d.
9	Netsuko	Basalt (Kushigatayama F.)	16	1.07	0.78	580	−29.9	−22.2	21.7	7.3	530	I	0.22
10	West of Kofu	Basalt (Kushigatayama F.)	10	0.23	0.20	70–500	67.2	44.7	13.6	14.4	—	—	n.d.
11	West of Kofu	Basalt (Kushigatayama F.)	15	1.48	0.72	600	35.1	18.1	21.6	8.4	—	—	0.31

Table 5. Summary of magnetic properties of rocks from Tanzawa district.

Site no.	Locality	Rock type	No. of oriented sample	J_n (×10⁻⁴ emu/cc)	Q_n	MDF	D	I	k	α_{95}	T_{c1} (°C)	Thermo-magnetic curve	X-value
1	Tanzawa mountain	Gabbroic rock	3	6.60	0.83	50–90	8.9	19.6	52.3	12.3	n.d.	—	n.d.
2	Tanzawa mountain	Gabbroic rock	4	0.71	0.16	75	109.5	−34.5	18.7	21.7	n.d.	—	n.d.
3	Tanzawa mountain	Gabbroic rock	7	0.85	1.13	120	124.3	−38.4	1.0	107.0	575	R	Py
4	Tanzawa mountain	Quartz diorite	5	20.33	1.71	45–70	−13.6	73.5	11.9	23.1	n.d.	—	n.d.
5	Tanzawa mountain	Quartz diorite	12	13.84	0.98	40–75	−16.4	34.8	5.1	21.3	575	R	0.005 + II
6	Tanzawa mountain	Quartz diorite	3	1.71	0.33	55	43.0	58.8	68.5	9.3	n.d.	—	n.d.
7	Tanzawa mountain	Quartz diorite	4	1.04	0.29	45	−85.8	67.1	35.6	15.6	n.d.	—	n.d.
8	Tanzawa mountain	Quartz diorite	3	1.77	0.39	240	−9.6	76.0	94.5	7.0	n.d.	—	n.d.
9	Tanzawa mountain	Quartz diorite	13	2.56	0.30	30	−8.2	52.0	11.3	15.9	n.d.	—	n.d.
10	Tanzawa mountain	Quartz diorite	16	0.04	0.19	350–420	20.0	32.6	49.1	5.5	n.d.	—	n.d.
11	Tanzawa mountain	Quartz diorite	6	0.40	0.13	35	−82.9	37.4	30.6	14.0	n.d.	—	n.d.
12	Tanzawa mountain	Quartz diorite	11	0.52	0.13	40	105.9	73.4	1.1	73.4	n.d.	—	n.d.
13	Tanzawa mountain	Quartz diorite	8	0.41	0.11	50	25.4	48.4	22.4	11.9	575	R	n.d.
14	Tanzawa mountain	Quartz diorite	11	0.86	0.12	n.d.	−12.7	55.2	2.5	32.1	565	R	n.d.
15	Tanzawa mountain	Gabbroic rock	4	1.52	0.19	35–45	−10.2	59.1	12.7	26.7	n.d.	—	n.d.
16	Tanzawa mountain	Gabbroic rock	10	2.78	0.24	35	−47.7	51.7	6.5	28.3	575	R	0.07
17	Tanzawa mountain	Gabbroic rock	13	2.56	0.16	45	26.8	52.1	4.0	20.3	575	R	n.d.
18	Tanzawa mountain	Gabbroic rock	6	5.58	0.34	n.d.	−9.2	54.2	43.9	11.7	575	R	n.d.
19	Togatake	Basalt	17	0.31	0.14	50–105	−9.8	27.5	3.2	23.9	n.d.	—	n.d.
20	Togatake	Basalt	4	0.28	0.06	80	32.0	46.2	272.6	5.5	n.d.	—	n.d.
21	Togatake	Basalt	12	2.92	1.31	n.d.	0.4	0.5	3.2	19.8	575	R	0.32
22	North of Lake Sagami	Basaltic andesite	11	3.46	0.62	45–75	−12.7	17.0	20.9	10.2	500	R	0.40
23	Misaka mountain	Andesite	18	1.88	0.51	90–170	36.4	14.8	9.4	11.8	545	R	0.22
24	East of Tsuru	Sediment	9	0.42	0.28	1–40	8.5	39.3	109.7	4.9	510	R	n.d.

movement occurred before the intrusion of quartz diorite mass and possibly before that of northern gabbroic rocks but after eruptions of Lake Sagami basaltic andesite and Misaka andesite. Ages of the latter two have not been precisely determined yet, but are probably of middle Miocene. If so, time of the tectonic event is implied to be between middle to late Miocene.

5.   Summary and General Tectonic Considerations

The following implications have been derived from the present paleomagnetic investigation of the Hayama-Mineoka and Setogawa ophiolite belts together with some adjacent rocks;

(1)   Most of basaltic rocks from Hayama-Mineoka and Setogawa belts except for sulfurized basalts in the northern Setogawa belt are reliable as paleomagnetic tools, indicating their paleolatitude if their bedding is properly corrected, or providing information on their tectonic movement if their paleolatitude is given.

(2)   Directions of NRM of the paleomagnetically reliable basaltic rocks mentioned above are so well concentrated among sites in each district as to indicate that the mode of emplacement of these basaltic rocks is not a detrital one but a tectonic process such as obduction.

(3)   A vertical core of 92 m thick basalts (pillow and dykes) drilled on the crest of Mineoka Hill has a sequence of normal and reversed polarities of NRM similar to that of the oceanic layer 2 revealed by DSDP. Such paleomagnetic and structural patterns of a basaltic layer as well as chemical composition of basalts equivalent to abyssal tholeiite show that basaltic rocks occurring in Hayama-Mineoka and Setogawa belts were formed in the deep ocean floor from the mid-oceanic ridge.

(4)   Paleomagnetic directions of the eastern Mineoka basalts after tilt correction as well as paleomagnetic inclination of supposedly flat basaltic layer on the Mineoka Hill crest imply possibility of their origin at a latitude ($\sim 23°N$) significantly lower than the present position ($35°N$), although the effect of the secular variation of the geomagnetic field can not be completely ruled out.

(5)   Shallow inclination ($I \fallingdotseq 10°$) of observed NRM of basaltic rocks at majority of sites in the Hayama-Mineoka belt can be best explained if they have been tilted southward by about 30 after their magnetization and northward drift.

(6)   Existence of a tilted subterrenean layer of rocks with high density and stable remanent magnetization with shallow inclination is consistent with observed gravity and magnetic anomalies in the southern part of Boso and Miura Peninsulas.

(7)   Directions of NRM of island-arc type rocks occurring in the adjacent areas (Fujigawa and Tanzawa) are nearly parallel to that of the present field if their ages are younger than about 11 Ma, while some older rocks have NRM directions significantly and systematically deflected from the present field direction. Although no oceanic tholeiite has been found in these areas, the Hayama-Mineoka tectonic belt seems to extend westward to these areas.

These paleomagnetic results imply that a microcontinental block including the areas of southern part of Boso and Miura Peninsulas as well as the present Suruga Bay coast collided with the Honshu arc at a time between about 30 Ma and 11 MaBP. A part of the oceanic crust existing between the two landmasses before the collision was tectonically obducted and is now exposed as the ophiolite complexes. The line of collision was located along a zone connecting the ESE-WNW trending Hayama-Mineoka belt to the roughly NS trending Setogawa belt (Fig. 10), although the area of the western extension of Hayama-Mineoka belt is covered by later intrusive and volcanic materials.

Existence of paleosubduction before the start of collision of the two blocks can be suspected along the Hayama-Mineoka and Setogawa belts. Direction of the relative motion of the two converging plates was probably NNW as suggested by the paleostress axis prior to 12 MaBP derived from the configuration of intrusive bodies in the central Honshu (TAKEUCHI *et al.*, 1979). OGAWA and HORIUCHI (1978) and OGAWA (1981) described some evidences of right-lateral slip motion along the trend of the Hayama-Mineoka belt. Such a strike slip motion may have been caused by a slightly oblique motion between the two colliding blocks or initiated after two plates changed their relative motion to WNW direction.

If the relative motion of convergence is NNW, the paleosubduction along the Setogawa belt should have been quite oblique. Such an oblique subduction may have caused an extensional stress field immediately behind the trench. Widespread hy-

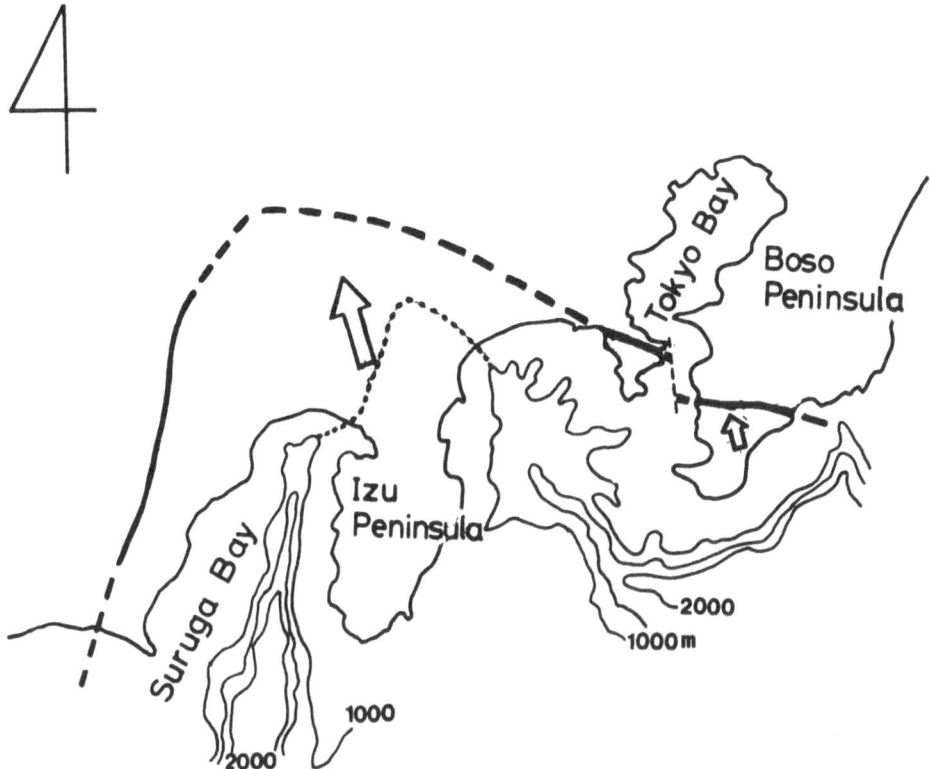

Fig. 10. Schematic illustration of collision in middle Miocene inferred from the present investigation (Dotted line indicates the present collision boundary).

drothermal activity in obducted basaltic rocks as indicated in this paper by the widespread occurrence of pyrrhotite and pyrite in place of magnetite as well as coexistence of high magnesian andesites in the Setogawa belt (OHASHI and SHIRAKI, 1981) are consistent with prevalence of extensional stress, since it has been suggested that high magnesian andesite such as boninite occurs only when the primitive hydrous magma erupted very rapidly under the extensional field (KUSHIRO, 1974; 1982; KOBAYASHI, 1983). As the high magnesian andesite occurs at the fore-arc zone which is much closer to the trench axis than usual arc volcanic front as seen with the present Bonin Islands, it could have been obducted together with the oceanic crust when the two landmasses sandwich the zone. Frequent occurrences of high magnesian andesite and its derivatives together with ophiolite may possibly be explained in the same way.

Position of the paleosubduction mentioned above was quite different from the present trenches, i.e. Sagami and Suruga Troughs. Direction of the present motion of the Philippine-Sea plate relative to Honshu (Asia plate) in WNW since at least 3 MaBP as indicated by the tectonic stress and focal mechanism of earthquakes. The subduction zone jumped perhaps simultaneously with the change in direction of relative motion about 5 MaBP. At present, Izu Peninsula is colliding with the central Honshu, but ISHIBASHI (1978) suggested from seismicity data that an embryo of subduction is being created in an area south of Izu Peninsula. If such a zone of subduction developes in future, the present Sagami and Suruga Troughs will be a suture zone accompanied by an ophiolite belt. Nankai Trough trending ENE will then be more straightly connected with Japan and Izu-Bonin Trenches.

The authors wish to express their sincere thanks to Dr. S. Arai, Tsukuba University, Dr. T. Tiba, National Science Museum and many others for their kind guidance in the field work. We owe Dr. M. Nomura for his computer program of calculation of gravity and magnetic anomalies. We are grateful to Mr. H. Haramura for chemical analyses of some rocks. We acknowledge Hydrographic Department, Maritime Safety Agency of Japan to put their unpublished data of magnetic anomalies at our disposal.

REFERENCES

ARAI, S., Igneous and ultrabasic rocks of the Mineoka belt in Boso Peninsula, in *Guidebook of Field Trip*, pp. 59–86, *Geol. Soc. Japan*, 1981(in Japanese).

ARAI, S. and T. UCHIDA, Highly Magnesian Dunite from the Mineoka Belt, Central Japan, *J. Jpn. Assoc. Mineral. Petorol. Econ. Geol.*, **73**, 176–179, 1978.

ARAI, S. and T. UCHIDA, Estimation of the equilibrium condition of the ultramafic rocks in the Setogawa belt, Shizuoka Prefecture, *Sci. Rep. Shizuoka Univ. Earth Sci.*, **4**, 19–24, 1979.

ARAI, S., K. SHIMOGAWA, and T. TAKAHASHI, On the mode of emplacement of ultramafic-mafic rocks in the Setogawa belt, Shizuoka Prefecture, *J. Geol. Soc. Japan*, **84**, 691–693, 1978.

BUTLER, R. F., S. K. BANERJEE, and J. H. STOUT, Magnetic properties of oceanic pillow basalts from Macquarie Island, *Nature*, **257**, 302–303, 1975.

BUTLER, R. F., S. K. BANERJEE, and J. H. STOUT, Magnetic properties of oceanic pillow basalts: evidence from Macqurie Island, *Geophys. J. Roy. Astron. Soc.*, **47**, 179–196, 1976.

COLEMAN, R. G., Plate tectonic emplacement of upper mantle peridotites along continental edge, *J. Geophys. Res.*, **76**, 1212–1222, 1971.

DEWEY, J. F. and J. M. BIRD, Origin and emplacement of the ophiolite suite: Appalachian ophiolites in Newfoundland. *J. Geophys. Res.*, **76**, 3179–3206, 1971.

Dewey, J. F., Ophiolite obduction, *Tectonophysics*, **31**, 93–120, 1976.

Faller, A. M., M. Steiner, and K. Kobayashi, Paleomagnetism of basalts and interlayered sediments drilled during DSDP Leg 49 (N-S transect of the northern mid-Atlantic Ridge) in *Initial Reports Deep Sea Drilling Project, Vol. 49*, edited by B. P. Luyendyk, J. R. Cann *et al.*, pp. 769–780, U. S. Government Printing Office, Washington, D. C., 1978.

Gass, I. G., Is the Troodos Massif of Cyprus a fragment of Mesozoic ocean floor? *Nature* (London), **220**, 39–42, 1963.

Horikawa, Y., K. Tsu, and K. Ogawa, Aeromagnetic anomalies and geological features in the Hitachi (Ibaraki Prefecture)-Kamogawa (Chiba Prefecture) area, Part I, Data acquisition, processing and interpretation, *Bull. Geol. Surv. Japan*, **30**, 487–511, 1979.

Horikoshi, Ei, Downward extension of the Kurosegawa tectonic zone based on the distribution of Quaternary volcanoes, *J. Geol. Soc. Japan*, **85**, 427–434, 1979.

Inoue, I., S. Tonouchi, and H. Maeyama, A consideration on topographic features and geological structure at the Mineoka uplifted zone in Boso Peninsula—with special emphasis on the correlation between landslide topography and geology, *Karuishigaku zasshi (Zeit. Bimssteinskunde)*, **6**, 123–135, 1980(in Japanese).

Ishibashi, K., Plate convergence around the Izu collision zone, central Japan: development of a new subduction boundary with a temporary transform belt, in *Abstracts of Papers, International Geodynamics Conference, Western Pacific and Magma Genesis*, Tokyo, 66–67, 1978.

Ishiwatari, A., A preliminary report on the Yakuno ophiolite in the Maizuru zone, inner southwest Japan, *Chikyu Kagaku*, **32**, 301–310, 1978.

Ishizuka, H., Geology of the Horokanai ophiolite in the Kamuikotan tectonic belt, Hokkaido, Japan, *J. Geol. Soc. Japan*, **86**, 119–134, 1980.

Johnson, H. P., Magnetization of the oceanic crust, *Rev. Geophys. Space Phys.* **17**, 215–226, 1979.

Johnson, H. P. and J. M. Hall, A detailed rock magnetic and opaque mineralogy study of the basalts from the Nazca Plate, *Geophys. J. Roy. Astron. Soc.*, **52**, 45–64, 1978.

Kanehira, K., Modes of occurrence of serpentinite and basalt in the Mineoka Districts, southern Boso Peninsula, *Mem. Geol. Soc. Japan*, **13**, 43–50, 1976.

Kaneoka, I., Y. Takigami, S. Tonouchi, T. Furuta, Y. Nakamura, and M. Hirano, Pre-Neogene volcanism in the central Japan based on K-Ar and Ar-Ar analyses, in *Abstracts 1981 IAVCEI Symposium-Arc volcanism*, Tokyo and Hakone, 166, Vol. Soc. Jpn., 1981.

Kimura, M., Geological structure of south Kanto, in *Marine Geology*, edited by N. Nasu, pp. 155–181, Univ. Tokyo Press, 1976 (in Japanese).

Kinoshita, H., Paleomagnetism of sediment cores from Deep Sea Drilling Project Leg 58, Philippine Sea, in *Initial Reports of Deep Sea Drilling Project, Vol. 58*, edited by G. Klein, K. Kobayashi *et al.*, pp. 765–768, U.S. Government Printing Office, Washington, D.C., 1980.

Klein, G. and K. Kobayashi, Geological summary of the north Philippine Sea, based on Deep Sea Drilling Project Leg 58 results, in *Initial Reports of Deep Sea Drilling Project, Vol. 58*, edited by G. Klein, K. Kobayashi *et al.*, pp. 951–962, U.S. Government Printing Office, Washington, D.C., 1980.

Klein, G., K. Kobayashi, H. Chamley, D. H. Curtis, H. J. B. Dick, E. J. Echols, D. M. Fountain, H. Kinoshita, N. G. Marsh, A. Mizuno, G. V. Nisterenko, H. Okada, J. R. Sloan, D. M. Waples, and S. M. White, Off-ridge volcanism and seafloor spreading in the Shikoku Basin, *Nature*, **273**, 746–748, 1978.

Klein, G., K. Kobayashi *et al.*, *Initial Reports of Deep Sea Drilling Project, Vol. 58*, 1021 pp., U.S. Government Printing Office, Washington, D.C., 1980.

Kobayashi, K., Cycles of subduction and Cenozoic arc activity in the northwestern Pacific margin, in *Geodynamics of Western Pacific*, edited by T. W. C. Hilde and S. Uyeda *AGU-GSA*, **10**, in press, 1983.

Kobayashi, K. and M. Nomura, Iron sulfides in the sediment cores from the Sea of Japan and their geophysical implications, *Earth Planet. Sci. Lett.*, **16**, 200–208, 1972.

Kushiro, I., Melting of hydrous upper mantle and possible generation of andesitic magma: an approach from synthetic systems, *Earth Planet. Sci. Lett.*, **22**, 294–299, 1974.

Kushiro, I., Petrology of high-MgO bronzite andesite resembling boninite from site 458 near the Mariana trench, in *Initial Reports of Deep Sea Drilling Project, Vol. 60*, edited by D. Hussong, S. Uyeda *et al.*, pp. 731–734, 1982.

LEVI, S. and S. K. BANERJEE, The effects of alterations on the natural remanent magnetization of three ophiolite complexes: possible implications for the oceanic crust, *J. Geomagn. Geoelect.*, **29**, 421–439, 1977.

LEVI, S., S. K. BANERJEE, S. BESKE-DIEHL, and B. MOSKOWITZ, Limitations of ophiolite complexes as models for the magnetic layer of the oceanic lithosphere, *Geophys. Res. Lett.*, **5**, 473–476, 1978.

LOCKWOOD, J. P., Sedimentary and gravity-slide emplacement of serpentine, *Geol. Soc. Am. Bull.*, **82**, 919–936, 1971.

LOUDEN, K. R., Paleomagnetism of DSDP sediment, phase shifting of magnetic anomalies, and rotations of the west Philippine basin, *J. Geophys. Res.*, **82**, 2989–3002, 1977.

LOWRIE, W., Intensity and direction of magnetization in oceanic basalts, *J. Geol. Soc. London*, **133**, 61–82, 1977.

MARUYAMA, S., Dismembered ophiolite in the Kurosegawa tectonic belt, *Kaiyokagaku*, **10**, 287–296, 1978(in Japanese).

MATSUDA, T. and Y. SUDA, Gravity Anomaly Map in Japan *Rep. Geol. Surv. Japan*, **209**, 1–8, 1964(in Japanese).

NAKAJIMA, T., H. MAKIMOTO, J. HIRAYAMA, and S. TOKUHASHI, Geology of the Kamogawa District, Quadrangle ser., *Tokyo (8)*, No. 95, Geol. Sur. Japan, 1981.

NAKAMURA, Y., Petrology of the Toba ultrabasic complex, Mie Prefecture Central Japan, *J. Fac. Sci. Univ. Tokyo, Ser 2*, **18**, 1–51, 1971.

NAKAMURA, Y., On the Mikabu green rocks in Toba district, *Abstr. Ann. meet., Geol. Soc. Japan*, p. 367, 1981(in Japanese).

NISHIYAMA, K. and Y. UEDA, The Neogene of Yamanashi Prefecture—Research on the geochronolgy and the Neogene stratigraphy of the southern Fossa Magna, central Japan, *Mem. Geol. Soc. Japan*, **13**, 349–366, 1976(in Japanese).

OHASHI, F. and K. SHIRAKI, High-magnesia and High-silica Volcanic Rock in the Setogawa Ophiolite, *J. Jpn. Assoc. Mineral Petrol Econ. Geol.*, **76**, 69–79, 1981(in Japanese).

OGAWA, Y., Beard-like veinlet structure as fracture eleavage in the Neogene siltstone in the Miura and Boso Peninsulas, central Japan. *J. Fac. Sci. Univ. Kyushu*, **13**, 321–327 1980(in Japanese).

OGAWA, Y., Tertiary tectonics in Boso and Miura Peninsulas—ophiolites and Izu fore-arc basin sediments trapped in the Honshu arc, *Chikyu*, **3**, 411–420, 1981(in Japanese).

OGAWA, Y. and K. HORIUCHI, Two Types of Accretionary Fold Belts in Central Japan, *J. Phys. Earth*, **26**, Suppl., S321–S336, 1978.

OGAWA, K., Y. HORIKAWA, and K. TSU, Aeromagnetic anomalies and geological features in the Hitachi (Ibaraki Prefecture)-Kamogawa (Chiba Prefecture) Area, Part II, magnetic and geological provinces, *Bull. Geol. Surv. Japan*, **30**, 549–569, 1979.

RYALL, P. J. C., J. M. HALL, J. CLARK, and T. MILLIGAN, Magnetization of oceanic crustal layer 2—results and thoughts after DSDP Leg 37, *Can. J. Earth Sci.*, **14**, 684–706, 1977.

SHIMAZU, M., A. TABUCHI, and T. KUSUDA, Geological structure of the northeastern part of the Tanzawa mountainland *J. Geol. Soc. Japan*, **77**, 77–89, 1971 (in Japanese).

STEINER, M., R. DAY, K. KOBAYASHI, and A. FALLER, Summary of magnetic observations of Leg 49, in *Initial Reports of Deep Sea Drilling Project*, *Vol. 49*, edited by B. P. Luyendyk, J. R. Cann et al., 807–812, U.S. Government Printing Office, Washington, D.C., 1978.

SUZUKI, T., S. HADA, and S. YOSHIKURA, Geotectonic evolutionary model of the outer zone of southwestern Japan, *Chikyu*, **1**, 57–62, 1979(in Japanese).

TAKEUCHI, A., K. NAKAMURA, Y. KOBAYASHI, and K. HORI, Neogene stress field in central Honshu viewed from dyke swarms, *Chikyu*, **1**, 447–452, 1979(in Japanese).

TAKIGAMI, Y., I. KANEOKA, and M. HIRANO, K-Ar and ^{40}Ar-^{39}Ar dating on ophiolites from Mineoka, Abstracts Fall meeting of Vol. Soc. Japan, *Bull. Vol. Soc. Jpn.*, **25**, 308, 1980.

TAZAKI, K. and M. INOMATA, Umbers in pillow lava from the Mineoka tectonic belt, Boso Peninsula, *J. Geol. Soc. Japan*, **86**, 413–416, 1980a.

TAZAKI, K. and M. INOMATA, Picrite basalts and tholeiitic basalts from Mineoka tectonic belt, central Japan, *J. Geol. Soc. Japan*, **86**, 653–671, 1980b.

TONOUCHI, S., T. FURUTA, and K. KOBAYASHI, Magnetic properties of the ophiolite suites; Boso-Miura Peninsulas (Cenozoic) and Maizuru District (Paleozoic), *J. Geogr.* **90**, 14–24, 1981(in Japanese).

Tonouchi, S. and K. Kobayashi, Magnetic properties of Cenozoic ophiolites in the Hamaya-Mineoka and Setogawa belts, south-central Honshu, Japan, *J. Geomagn. Geoelectr.*, in press, 1982a.

Tonouchi, S. and K. Kobayashi, Paleomagnetism of alkali basalts occurring in Takakusayama and Ryuso districts, south-central Honshu, Japan, *J. Geomagn. Geoelectr.*, in press, 1982b.

Wada, N., The geology of the Abe-gawa District, Shizuoka Prefecture, Central Japan, *J. Geol. Soc. Japan.*, **82**, 581–593 1976(in Japanese).

Yoshida, Y., Discovery of foraminifers in the Mineoka Hills, Chiba Prefecture, *Chishitsu News* (Geology Monthly), *Geol. Surv. Japan*, **233**, 30–36 1974(in Japanese).

Southwest to South Pacific

Accretion Tectonics in the Circum-Pacific Regions, edited by M. Hashimoto and S. Uyeda, 291–306.
Copyright © 1983 by Terra Scientific Publishing Company (TERRAPUB), Tokyo.

A Possible Mechanism of Episodic Spreading of the Philippine Sea

Seiya Uyeda* and Robert McCabe**

*Earthquake Research Institute,
University of Tokyo, Tokyo 113, Japan
**Department of Geology and Mineralogy,
Kyoto University, Kyoto 606, Japan

Philippine Sea is composed of three parts with ages younging eastwards. They are the West Philippine Basin of late Mesozoic to early Tertiary age, the Shikoku and Parece Vela Basins of ca 30–17 m.y. age and the Mariana Trough of ca 6–0 m.y. age. There is a variety of hypotheses on the origin of the West Philippine Sea, but the younger basins are now believed to have been formed by a series of episodic separations between the over-riding plate and the trench line. If we take the view that the position of the trench line tends to be fixed to the mantle frame of reference, due possibly to the anchoring effect of long subducted slab, back-arc spreading is possible when the over-riding plate retreats landward. Episodic spreading, thus, calls for episodic retreating motion of the over-riding plate. Thus, in turn, in the case of the Philippine Sea, would call for episodic subduction at its western margin. At present, the subduction at the Ryukyu-Luzon-Mindanao trench system is causing the active spreading of the Mariana Trough. It is proposed that intermittent collisions of buoyant features at subduction zones on the western margin were the cause of episodicity of subduction. This proposal was examined by looking at the possible collision tectonics in the Philippines during the late Cenozoic period. It was found that large scale collisions might have taken place at times when back-arc spreading ceased, i.e. at the early Oligocene time and middle Miocene time.

1. Introduction

There are various types of subduction zones (Uyeda and Kanamori, 1979; Uyeda, 1979; Dewey, 1980; Uyeda, 1981, 1982). Classifications of subduction zones by these authors are essentially similar, although there are minor differences. Two endmembers are the Mariana-type with active back-arc spreading and the South American type without it. The latter is called the Chilean-type by Uyeda and Kanamori, and the Pervian-type by Dewey. Many other subduction zones are classified as intermediate of the two endmembers. The differences, most apparant in the existence or non-existence of back-arc spreading, are probably due to the differences in the degree of mechanical coupling between the over-riding and underthrusting plates: the Mariana-type represents a low coupling or low stress regime while the Chilean or Peruvian-type represents a high coupling or high stress regime. As to the origin of such differences in the degee of coupling or the state of stress, it has been recognized that the convergence rate itself has little relevance (CHASE, 1978; UYEDA and KANAMORI, 1979; DEWEY, 1980). The possible origin appears to lie, rather, in the differences of the relative motion between the over-riding plate and the trench line. Let us denote the velocity vectors of the over-riding plate and of the trench line in the absolute frame of reference as V_U and V_T. As to the frame of reference, it may be possible to adopt the hot-spot frame for convenience as representing the mantle frame of reference. When V_U and V_T have a divergent component, the Mariana-type or low stress subduction will result, whereas when they have a convergent component, the Chilean-Peruvian-type or high stress subduction will occur. The intermediate type corresponds to neither cases. If we assume that the trench line velocity is generally small because the deep subducted oceanic slab is considered to be anchored to the mantle (HYNDMAN, 1972; TULLIS, 1972), i.e. if $V_T \doteqdot 0$, the only important factor becomes V_U. CHASE (1978) and UYEDA and KANAMORI (1979) demonstrated that the present absolute motions of plates support the above premise. On the other hand, MOLNAR and ATWATER (1978) suggested that the age of the subducting oceanic plate controls V_T, i.e. older, and heavier, plates have greater seaward component of V_T, resulting in the separation of the over-riding plate and the trench line, that in turn gives rise to the Mariana-type subduction and younger, and lighter, plates tend to produce the other endmembers. DEWEY (1980) treating the cases in a more general fashion, called this retreating motion of trench lines "roll-back."

The Philippine Sea has three major sub-units. They are the West Philippine Basin of the late Mesozoic to early Tertiary age (LOUDEN, 1976; SHIH, 1980), the Shikoku-Parece Vela Basin of the Oligocene-Miocene age (KOBAYASHI and ISEZAKI, 1976; MROZOWSKI and HAYES, 1979), and the Mariana Trough of the late Miocene to the present age (HUSSONG and UYEDA, 1981). KARIG (1971) suggested that this series of basins, younging eastward, has been formed through a process of episodic back-arc spreading during the Cenozoic time. Such a rapid episodicity is difficult to be attributed to the changes in the 'roll-back' vectors that depends on the age of the sea floor. In this paper, applying the above explained basic ideas on subduction, we intend to explain the mechanism of the proposed episodic spreading of the Philippine Sea in terms of episodic retreat of the over-riding plate. Episodic retreat of the over-riding plate of the Philippine Sea would require, from geometrical reasons, episodic subduction along its western margin. We suggest that such an episodic subduction could originate from a stop-go process controlled by intermittent major collisions at the western margin, i.e. along the Philippines.

2.   Evolution of the Philippine Sea

As explained in the previous section, the Philippine Sea consists of three major units with different ages (Fig. 1d). KARIG (1971) contended that all three basins have been formed through an episodic extensional back-arc spreading process, the present Mariana arc having originally been attached to the eastern margin of the Asian plate and migrated eastward to the present position. Since the orientations of the magnetic lineations and the Central Basin Fault of the West Philippine Basin are almost at right angles to that of the Mariana arc-trench system, it's origin does not appear to be the same as younger basins to the east where the magnetic lineations are more or less

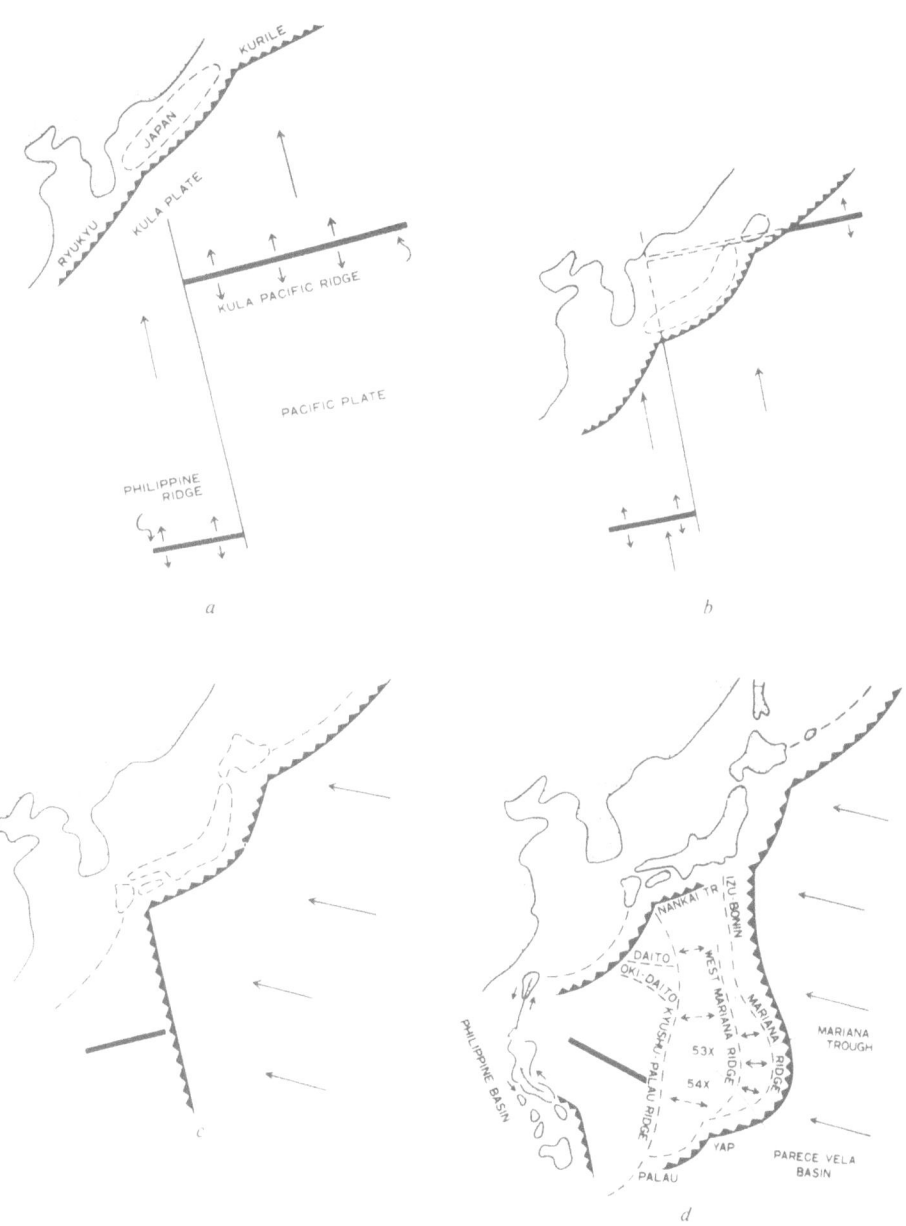

Fig. 1. Possible model of the evolution of the Philippine Sea (UYEDA and BEN-AVRAHAM, 1972).

parallel to the arc-trench system. Uyeda and Ben-Avraham (1972) proposed a modified tectonic model for the Philippine Sea, as shown in Fig. 1, a–d. In their model, the West Philippine Basin is a trapped portion of a pre-existed ocean, possibly the Pacific Ocean. This pre-existed ocean consisted of the Kula and Pacific plates and a spreading center that now is the Central Basin Fault. Along the eastern margin of that sea was a long transform fault which was "transformed" to a subduction boundary when the Pacific plate changed its direction of motion at about 40 m.y. ago from NNW to WNW. An island arc (proto-Mariana arc) was formed along this subduction boundary and trapped the West Philippine Basin. Then, through the episodic back-arc spreading process, younger basins developed eastward leaving the Kyushu-Palau Ridge and West Mariana Ridge as remnant arcs. This scenario, apart from exact dates for each event, has generally been supported by later results, including those of the DSDP and related investigations. One important revision to this scenario came from subduction models depicted earlier (Ben-Avraham and Uyeda, 1982). The eastward stepwise advance of the proto-Mariana arc in the above scenario requires episodicity of the "roll-back vector" as envisaged by Kanamori (1977). However, an alternative possibility (Fig. 2) is that the proto-Mariana arc has stayed at essentially the same position as today by the anchoring effect of the slab. If that was the case, the growth of the back-arc basins has progressed through the westward retreats of the over-riding plate just as the present situation of the Philippine Sea appears to be. This model (Ben-Avraham and Uyeda, 1982) is more consistent with the geologic argument (Matsuda, 1978) that the Izu-Mariana Ridge, except its northward migration, has stayed at the same position relative to the Japanese islands since the late Mesozoic time. One important point not to be overlooked is the possibility that the northward motion of the entire Philippine Sea region continued to the present as postulated from a DSDP paleomagnetic study (Louden, 1976; Kinoshita, 1980). If that was the case, some of the events depicted in Fig. 2 could have taken place at lower latitude than in Fig. 2 relative to, say, the Japanese islands. At any rate, back-arc spreading was active during the 30–17 mybp and 17–6 mybp periods. If the revised model (Fig. 2) is accepted, the episodic growth of the basins implies the episodic westward motion of the over-riding plate. Then the subduction along the western rim of the Philippine Sea must have been episodic also. This hypothesis should be testable from the investigation of the geologic history of the Philippine Archipelago.

3. Possible Role of Collision/accretion on the Episodicity of Subduction and Back-
    arc Spreading

For simplicity, let us consider a hypothetical two dimensional case as in Fig. 3. In Fig. 3a, subduction of the oceanic plate, $\bar{O}$, is in progress at IA (island arc). Landward of IA is another oceanic plate, P, which is not subducting at CM (continental margin). If we assume that the continent, C, is stationary, so is the oceanic plate, P. In fact, C and P form a single over-riding plate of which velocity vector $V_U = 0$. If we further assume that the roll-back velocity of the trench line V_T is small, subduction at IA is neutral, There is no back-arc spreading. If V_U has an eastward component greater than that of V_T, the subduction would be the Chilean (or Peruvian)-type. In this type of subduction,

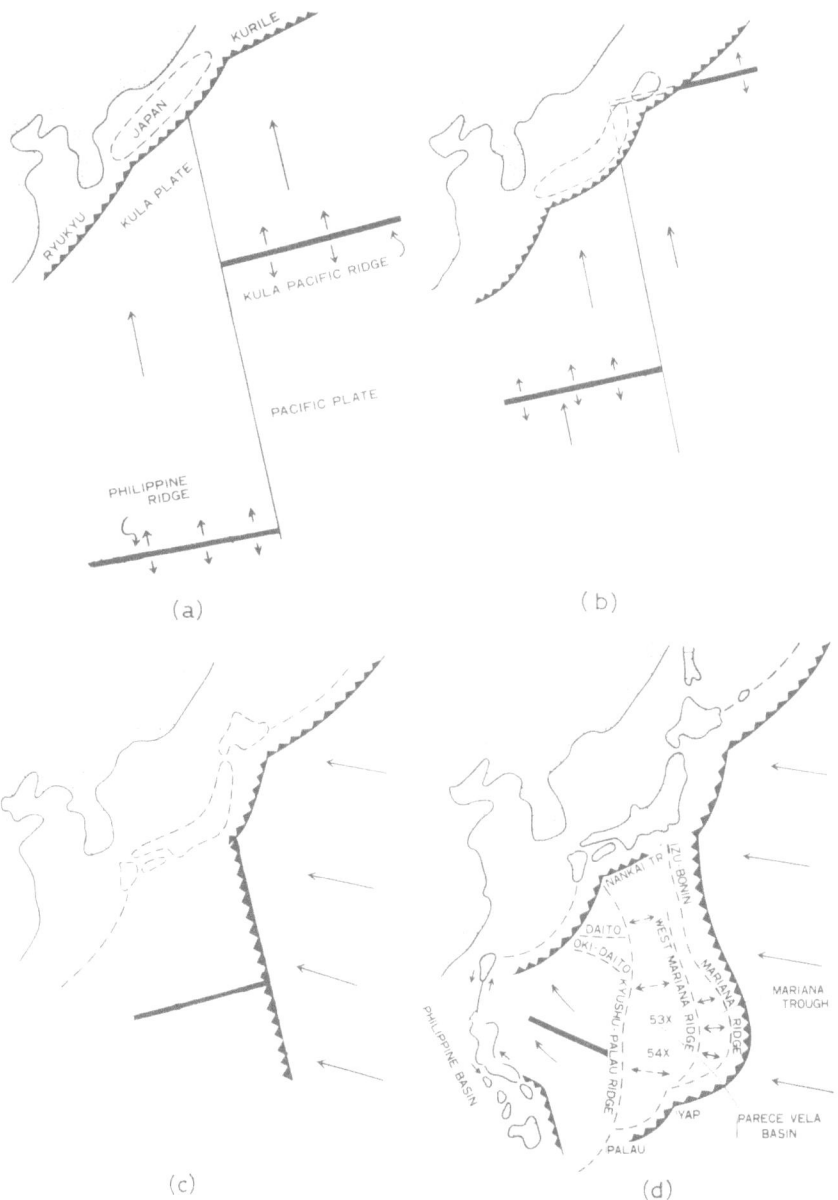

Fig. 2. Revised model of the evolution of the Philippine Sea. Note that the original position of the transform fault is set at the present position of the Izu-Bonin Mariana Ridge.

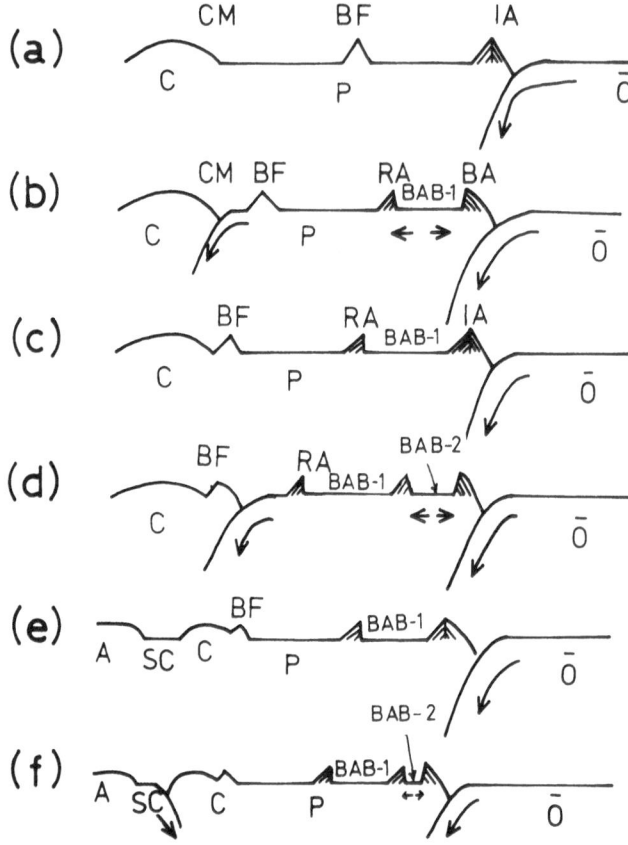

Fig. 3. Simplified two dimensional scenario for episodic back-arc spreading due to collision. Ō: oceanic plate (e.g. Pacific plate), P: back-arc plate (e.g. West Philippine Sea plate), C: continental plate (e.g. Philippines), IA: island arc, BF: buoyant feature, CM: continental margin, RA: remnant arc, BAB: back-arc basin, A: Asiatic plate, SC: South China Sea plate.

no spreading but strong thrust faulting and folding or crustal thickening would take place in the back-arc region.

Now, if subduction starts at CM (Fig. 3b), C and P are not parts of a single over-riding plate anymore. V_U at IA has a westward component so that the island arc (IA) is put under extensional tectonics. IA would eventually be split at its weakest zone, that is the line of the active volcanoes. A new back-arc basin (BAB-1) would be formed between IA and RA (remnant arc). Along with the westward motion of the over-riding plate, P, a buoyant feature (BF) on it would approach the trench at CM and would collide with it. If BF is large enough, subduction of P would be suspended and so would be the back-arc spreading (Fig. 3c). After a while, subduction may resume by stepping out of the subduction zone to the east (Fig. 3d), then back-arc spreading will also resume and form another new back-arc basin (BAB-2). It must be noted that, from this stage on, each remnant arc generated by back-arc spreading will act as the buoyant feature. In this way, the present model has a built-in mechanism for episodicity.

In the case of the Philippines, C is again separated from the main Asia, A, by another marginal sea, SC (South China Sea) as shown in Fig. 3e. Therefore, resumption of subduction may involve a reversal of polarity as in Fig. 3f. Such subduction would result in the westward V_U of plate, P, now combined with C, since plate A is stationary. Namely a course (a) → (b) → (e) → (f) may take place instead of (a) → (b) → (c) → (d).

The simple scheme, outlined above, may be a viable one to explain the episodicity of back-arc spreading and changes in the mode of subduction that are too local and too rapid to be explained by Dewey's model which would call for changes in the roll-back vectors due to variation in seafloor ages. In the present model, initiation or resumption of subduction presents a problem. It is not easy to deal with this problem by the simple two dimensional case. In a three dimensional case, which of course is more realistic, it would be quite possible that the slab pull outside the collision zone would keep pulling the plate, P, so that after a while subduction would resume.

In the following, we will look into the possibility of application of this type of model to the case of the Philippine Sea, since we know from the literature that there is ample evidence for arc-arc collisions, involving polarity changes of subduction, in the Philippine Archipelago (MURPHY, 1973; BEN-AVRAHAM, 1978; KARIG, 1973; BOWIN *et al.*, 1978). Figure 4 is the summary of collisional history of the Philippines as outlined in CCOP-IOC SEATAR volume, 1980, p. 112. To this summary, we have added the timing of events related to back-arc spreading in the Philippine Sea. The timings of events proposed by various authors show considerable variations and are neither compatible among themselves or with the chronology we propose. In the next section, we will present our version of collisional history.

4. Collision History of the Philippine Islands

The Philippine Archipelago has a very complicated physiographic structure (Fig. 5). Its tectonic evolution was classified into four stages (BALCE *et al.*, 1981). The demarkation of these stages is based on major unconformities and the occurrence of similar stratigraphic sequences. These four stages are: (1) Pre-Jurassic or Basement Stage, (2) Middle Jurassic to Oligocene Stage, (3) Late Oligocene to Late Miocene Stage, (4) Late Miocene to Recent Stage.

In this paper, we suggest that the last three stages of Balce *et al.* may be modified slightly as follows: (2) Middle Jurassic to Early Oligocene Stage, (3) Middle Oligocene to Late Miocene Stage, (4) Late Miocene to Recent Stage.

Moreover, both TAYLOR and HAYES (1980) and McCABE *et al.* (1982) have suggested that there was a collision in the central Philippines around the early to middle Miocene. Although the timing and mechanism of these collisions are not yet fully understood, there appears to be a lull in volcanism and intense folding throughout the Central Philippines in the middle Miocene. These facts suggest that the Middle Oligocene-Late Miocene Stage is divisible into two substages; a Middle Oligocene-Early Miocene substage and a Middle Miocene to Late Miocene substage.

This paper will primarily deal with events that occurred since the Eocene. In relation to the above discussed problem of the episodicity in the Philippine Sea region,

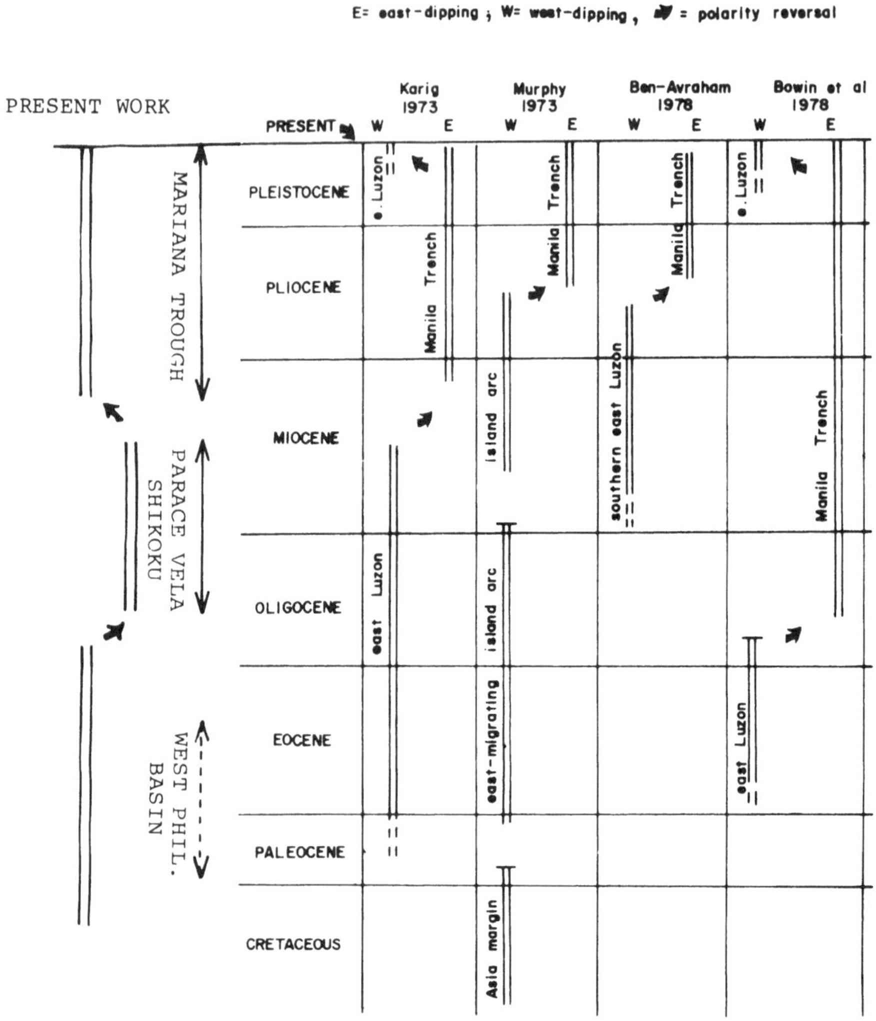

E= east-dipping ; W= west-dipping , ➶ = polarity reversal

Fig. 4.   Comparison of published subduction polarity histories of Luzon (CCOP-IOC SEATAR, 1980) and the polarity history based on the present study (left column).

the authors will attempt to give evidence which allows one to speculate on the tectonic evolution of the Philippines since Eocene time. From these pieces of evidence, we will be able to predict the sense of motion of the Philippine Sea plate during this period.

4.1   Late Jurassic to Oligocene stage

The eastern Philippines (Eastern Physiographic Province, in Fig. 5) is characterized by Cretaceous to Oligocene arc related sequences of volcanic and plutonic rocks which are in thrust contact with ultramafic rocks. The Eastern Sierra Madre Mountain Range are Eocene in age with a marked break in the ages at about the Oligocene time. Field observations of the rocks from the southern portion of this range show

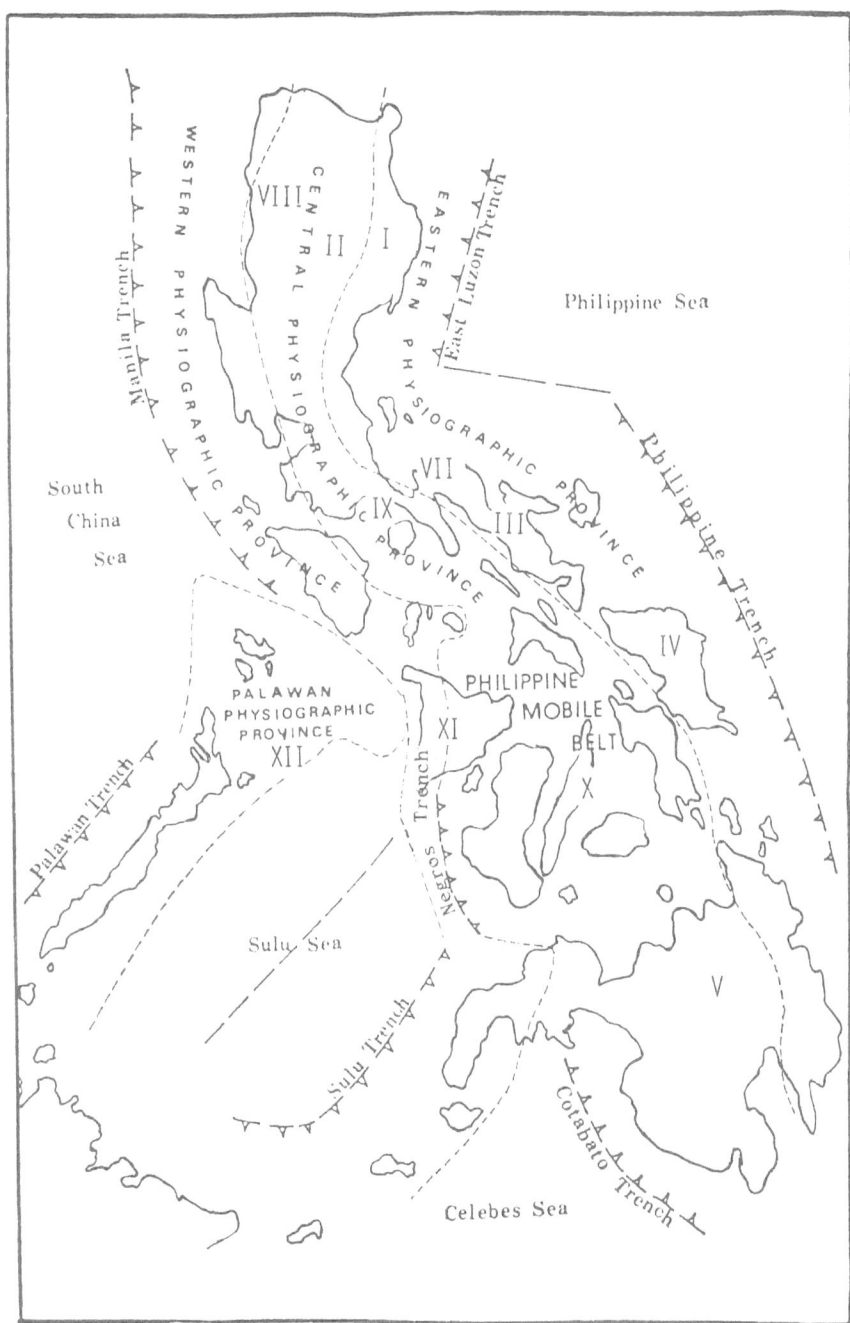

Fig. 5. Major physiographic elements in the Philippines (BALCE *et al.*, 1981). I : Sierra Madre, II : Cagayan Valley, III : Bicol, IV : Samar, V : Mindanao, VI : Polillo Island, VII : Camarines Norte, VIII : Central Cordillera of Luzon, IX : Marinduque, X : Cebu; XI : Panay, XII : Palawan Block.

predominantly dioritic plutonic rocks which have been intruded into a thick sequence of volcanic flows and related sedimentary rocks. East of the Sierra Madre Range is a belt of upthrusted mafic and ultramafic rocks, pelagic sediments, cherts, and interbedded basaltic volcanics (HASHIMOTO *et al.*, 1978). These sequences are unconformably overlain by the Oligocene to middle Miocene clastic sediments (DURKEE and PEDERSON, 1961) of the Cagayan Valley.

Similar sequences to those discussed above are also observed in the Bicol Region of Luzon, Samar Island, and Central Mindanao. This long continuous belt of arc related sediments, volcanics and plutonic rocks strongly supports the suggestion that during the early Tertiary, the Philippines was situated over a westward dipping subduction zone. The lack of post Eocene aged plutonic and volcanic rocks in the eastern Philippines suggests that during this period subduction ceased in the eastern Philippines. This suggestion is also supported by the fact that in the Sierra Madre Range, Polillo Island, Camarines Norte, Samar and Northern Mindanao, we find westward thrusted ophiolites that are overlain by late Oligocene to middle Eocene aged clastic sediments and limestones (BALCE *et al.*, 1981).

Although the reasons why this subduction halted is unknown, we suggest that during the Eocene/Oligocene period the westward subducting arc collided with an island arc complex that was located on the downgoing plate. Evidence for this collision is seen both in the Bicol Region and in Central Mindanao (Fig. 6).

The Bicol Region (Fig. 6a) is composed of two parallel belts of greenschist facies metamorphic rocks which are separated by a median valley. In both of these sequences, probably Paleocene aged felsic plutonic rocks are intruded into the sections. Situated between these two high temperature metamorphic terranes are the Bicol Basin and small slivers of fault contacted mafic and ultramafic rocks. All these older sequences are overlain by folded late Oligocene to early Miocene sediments (BALCE *et al.*, 1981). We suggest that this region is an old collision zone between the eastern and western belts. This collision is supported by the zone of ophiolitic rocks situated between these two zones and by the fact that both of these terranes are probably Paleogene in age. The Oligocene sediments cover both these belts showing that these terranes were sutured together prior to this date.

Figure 6b shows a portion of Mindanao. The geology of this region is similar to the geology that we observed in Bicol. In the northern portion of the map, Cretaceous and Paleogene greenschist rocks are in thrust contact with ophiolitic rocks. Both these juxtaposed different terranes are covered by late Oligocene clastic sediments which put a minimum age of pre-late Oligocene on this thrusting. Situated in south-central portion of Mindanao is the Daguma Range. This range consist of Cretaceous to Eocene aged volcanic terrane and related plutonic rocks. North of the Daguma Range is a small belt of ophiolitic rocks. Although the contact between the Daguma Range and the ophiolitic rocks has not been described, the map shows that both of these units are overlain by Oligocene to recent clastic sediments and volcanics, indicating that they have been in structural contact since this time.

4.2 *Middle Oligocene to Late Miocene Stage*

The Middle Oligocene-Late Miocene stage is characterized by large amounts of

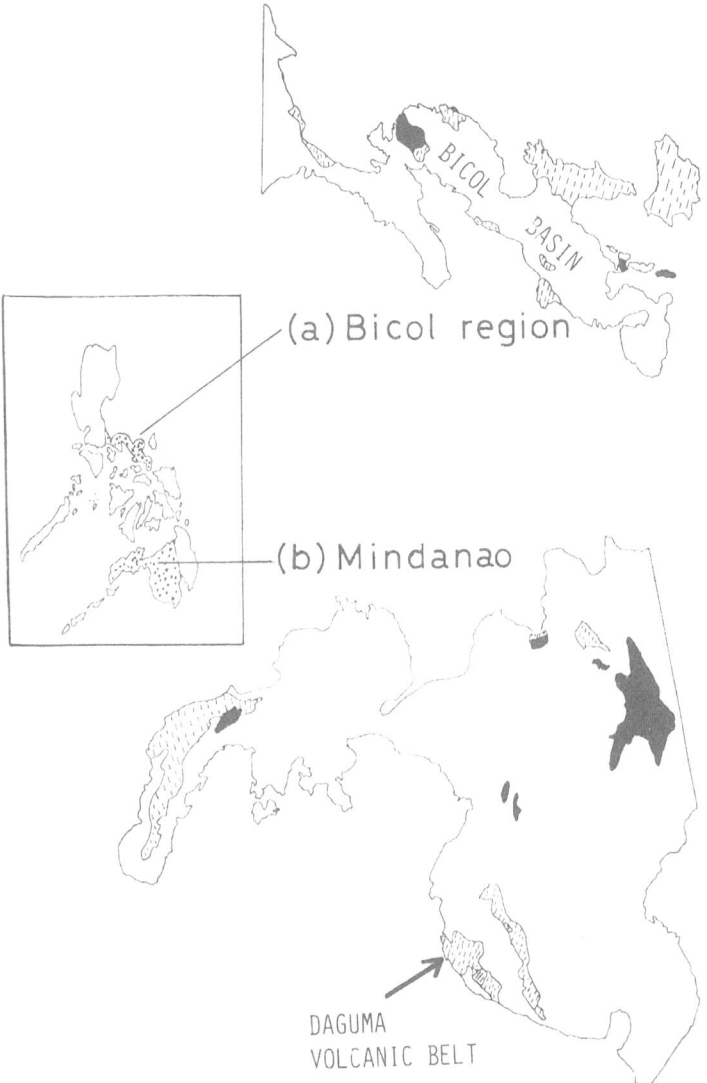

Fig. 6.   Simplified geologic maps of (a) Bicol Region and (b) Mindanao. Black area: ultramafics. Stippled area: arc related greenschist facies metamorphics.

volcanic, pyroclastic rocks with related felsic plutonic rocks throughout the central cordillera of Luzon and the island of Marinduque to the south. During this period, plutonic activity switched from the eastern side of Luzon to the western side, and the area east of this western belt of plutonic activity became a basin in which relatively shallow water sediments accumulated (DURKEE and PEDERSON, 1961). Similar shallow water sediments were also deposited in Cebu and Panay. These facts suggest that prior to the middle Oligocene, subduction flipped from the eastern to the western side of the archipelago.

As earlier stated the authors are in favor of a division of this stage into two substages, which correspond to before and after the Miocene collision which occurred in the central Philippines. HAMILTON (1979) suggested that western Panay is an area of possible collision based on the juxtaposition of two different terranes in the northwestern section of Panay. This collision is also supported by CARDWELL *et al.* (1980), who show a break in the subduction zones between the Manila Trench and the Negros Trench. Recently, McCABE *et al.*, (1982) have examined the island of Panay and have found both geologic and paleomagnetic evidence which suggests that the collision occurred after the early Miocene. This data is further supported by lack of volcanic activity in the central Philippines after the middle Miocene.

In addition, the late Miocene is a period of intense folding throughout the central Philippines (BALCE *et al.*, 1981). The direction of this folding is normal to the proposed collision direction.

4.3   Late Miocene to Recent Stage

The occurrence of middle Miocene aged volcanic and plutonic rocks in Leyte and Eastern Mindanao (BALCE *et al.*, 1981) and late Miocene aged volcanic rocks in the eastern Bicol (DUMAPIT, 1973), suggest that during the late Miocene, subduction was resumed along a long linear belt from Mindanao to Luzon. Although the reason for the change of the polarity of the arc is unknown, we suggest that collision of the central Philippines with the Palawan block initiated this polarity change. This polarity change may have been partial along the whole Philippine Archipelago in the sense that the east-dipping subduction at Manila and Negros trenches are also still active. Existence of igneous activity up to the present in the western Philippines and seismicity (CALDWELL *et al.*, 1980) indicates that east-dipping subduction is still occurring.

The paleomagnetic studies offer some supporting evidence for this scenario. Northern drift of the Philippines which was consistent with the northern drift of the Philippine Sea plate (Fig. 7), was halted during the Miocene and the counter-clockwise rotation of the northern Philippines stopped in the late Miocene (HSU, 1972; FULLER *et al.*, 1982). One possible explanation for these changes in paleomagnetic motions would be the flipping of the subduction zone which would change the plate affinity of the Philippine islands from the Philippine Sea plate to the South China Sea plate. This change in affinity would not require the Philippine microplate to experience parallel motion to the Philippine Sea plate and could be responsible for the cancellation of the tectonic rotation.

5.   Conclusions

To summarize, the Cenozoic evolution of the Philippines can be separated into several different stages. Figure 8 and Table 1 show these stages.

From the Cretaceous to the late Eocene/early Oligocene, a westward dipping subduction zone existed. This subduction formed the eastern belt of plutonic and volcanic rocks in the Philippine islands. Sometimes in the late Eocene/early Oligocene time, this proto-Philippine arc collided with another arc which was on the downgoing plate. The result of this collision halted the subduction process. It should also be noted

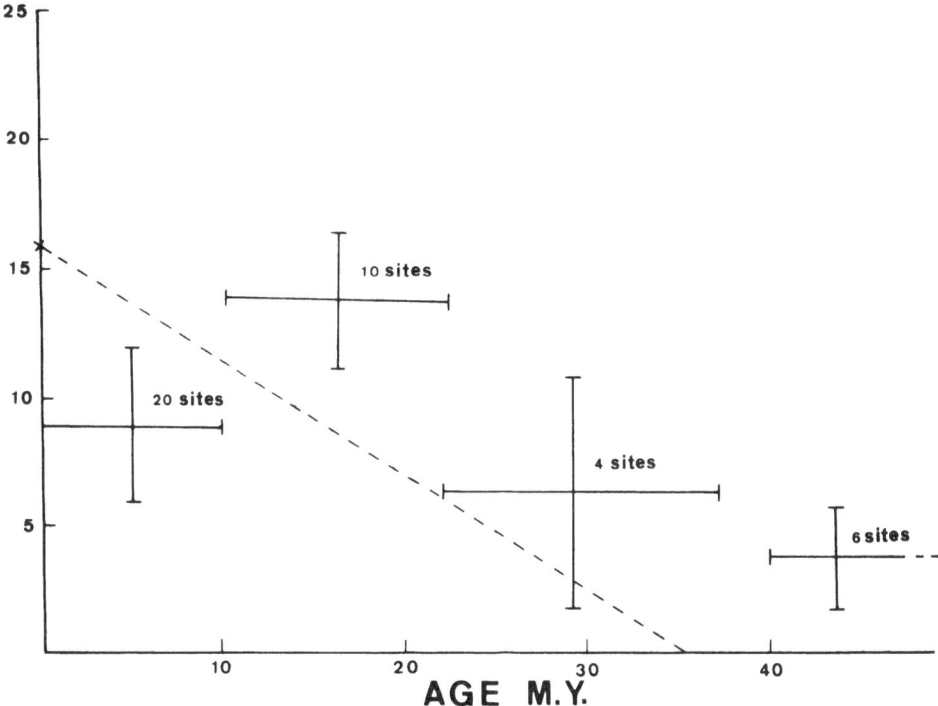

Fig. 7. Paleomagnetic latitude in °N vs age of the Philippines (FULLER *et al.*, 1982). The dotted line is the same for the Philippine Sea plate by LOUDEN, 1977.

however, that the Pacific plate changed its direction of motion at about the same time. There is a possibility that the subduction was halted as a result of this change in the direction of Pacific plate motion. Namely, it is quite possible, as mentioned earlier, that this change in the plate motion generated the Kyushu-Palau arc as a result of stepping out of the subduction zone to this more easterly boundary, which was presumably a long transform fault. This shifting in the location of subduction might have halted the subduction in the eastern Philippines at least temporarily. However, at this stage, we prefer the former explanation for inhibition of subduction which is based on geologic reconstruction. We now consider it likely that the collision mentioned above took place at almost the same time as the change in the Pacific plate motion. This collision may be responsible for the initiation of subduction along the Kyushu-Palau transform fault.

In the Oligocene, subduction started in the west Philippines. This subduction brought the Philippines westward relative to the Eurasian plate. The Parece Vela and Shikoku Basins, that were located on the eastern side of the Philippine Sea plate were opened during this period. This would eventually bring the Philippines in contact with the Palawan microcontinent. Sometime during the Middle Miocene, the Philippines located above the eastward dipping subduction zone collided with the Palawan block. This collision resulted in the slowing or halting of subducting along the western boundary of the Philippine Sea plate. MROZOWSKI and HAYES (1979) have shown that

CRETACEOUS – EOCENE   LATE EOCENE

OLIGOCENE

E – MID MIOCENE   RECENT

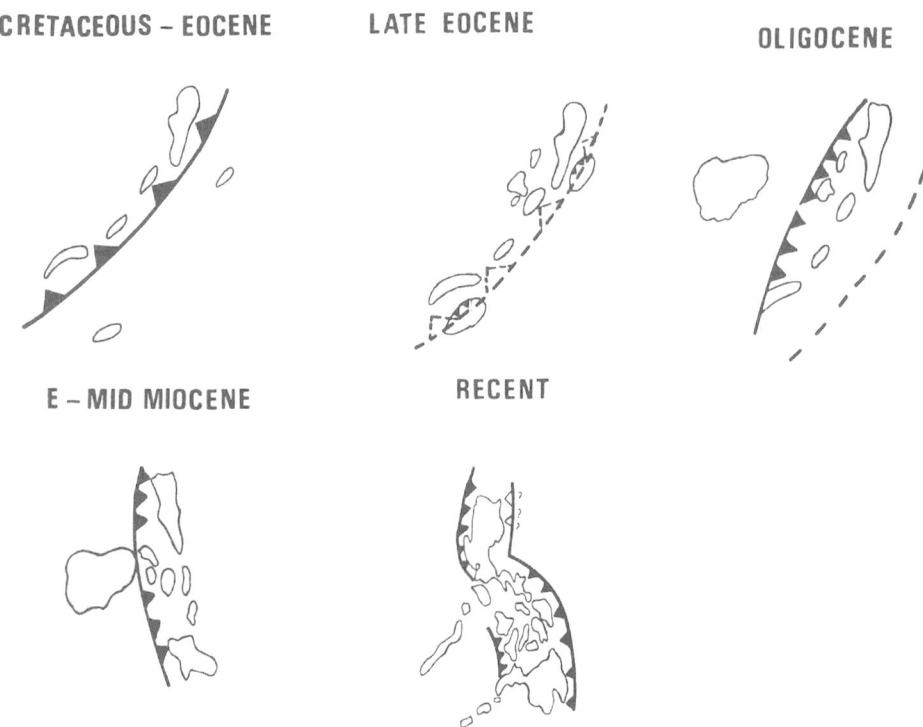

Fig. 8.   Possible stages in the Cenozoic evolution of the Philippines.

Table 1.   Possible stages in the Cenozoic evolution of the Philippines and the events in the Philippine Sea region.

Age	Western Philippine Sea	Eastern Philippine Sea
Late Miocene–Recent	West dipping subduction	Opening of Mariana Trough
Early Miocene–Late Miocene	Collision in central Philippines	No back-age spreading
Oligocene–Early Miocene	East dipping subduction	Opening of Parace Vela and Shikoku Basins
Late Eocene (?)–Early/mid Oligocene	Collision in east Philippines	No back-arc spreading
Late Eocene	Change in Pacific plate motion West dipping subduction	

spreading in the Parece Vela Basin halted during this period. After a while, a flip of subduction took place and the subduction zone was brought back to the eastern side of the Philippines. The late Miocene volcanic rocks in the eastern Bicol suggest that subduction was re-activated along the eastern boundary during this period. This subduction also would result in a westward migration of the Philippine Sea plate. This motion corresponds to the opening of the Mariana Trough.

The present model explains the episodic spreading as a result of episodic retreating motion of the over-riding plate. DSDP leg 58 results (KINOSHITA, 1980), however, indicated that the northward motion of the Philippine Sea plate has been almost steady since 40 m.y. ago (see also Fig. 7). This apparent inconsistency may be resolved if we consider that the collisions along the Philippines hindered the westward motion of the Philippine Sea plate only.

REFERENCES

BALCE, G. R., O. A. CRISPIN, S. M. SAMANIEGO, and F. E. MIRANDA, Metallogenesis in the Philippines, Geol. Surv. Japan Rep., 261, 125–148, 1981.

BEN-AVRAHAM, Z., The evolution of marginal basins and adjacent shelves in east and southeast Asia, Tectonophys., 45, 269–288, 1978.

BEN-AVRAHAM, Z. and S. UYEDA, Entrapment origin of marginal seas, GSA-AGU Geodyn. Ser., in press, 1982.

BOWIN, C. O., R. S. LU, C. C. LEE and H. SCHOUTEN, Plate convergence and accretion in Taiwan-Luzon region, Am. Assoc. Petrol. Geol. Bull., 62, 1645–1672, 1978.

CARDWELL, R. K., B. L. ISACKS, and D. E. KARIG, The spatial distribution of earthquakes, focal mechanism solutions, and subducted lithosphere in the Philippines and northeastern Indonesian Islands, in The Tectonic and Geologic Evolution of Southeast Asian Seas and Islands, edited by D. E. Hayes, Geophys. Monogr. Am. Geophys. Union, 23, 1–35, 1980.

CCOP-IOC SEATAR, Studies in the Asian Tectonics and Resources, 257 pp., CCOP Project Office, 1980.

CHASE, C., Extension behind island arcs and motions related to hot-spots, J. Geophys. Res., 83, 5385–5387, 1978.

DEWEY, J. F., Episodicity, sequence and style at convergent plate boundaries, in The Continental Crust and Its Mineral Resources, edited by D. STRANGWAY, Geol. Assoc. Canada, Spec. Paper, 20, 553–574, Waterloo, Canada. 1980.

DUMAPIT, P. T., Groundwater geology of Bicol Peninsula, J. Geol. Soc. Phil., 27(2), 24–44, 1973.

DURKEE, E. F. and S. L. PEDERSON, Geology of northern-Luzon, Philippines, Am. Assoc. Petrol. Geol. Bull., 45, 137–416, 1961.

FULLER, M. D., I. WILLIAMS, R. MCCABE, J. ALMASCO, R. ENCINA, and J. WOLFE, Paleomagnetism of Luzon, in Geophys. Monogr. Ser., edited by D. E., HAYES, in press, 1982.

HAMILTON, W., Tectonics of the Indonesian Region, in Geol. Surv. Prof. Paper 1078, 345 pp., U.S. Government Printing Office, Washington, D. C., 1979.

HASHIMOTO, W. et al., Numulites from the Lubingon crystalline schist of Bongabon, Nueva Ecija and their significance on the geologic development of the Philippines, Proc. Japan Acad., 54, Ser. B., 1–4, 1978.

HSU, I-chi, Magnetic properties of igneous reocks in the northern Philippines, Ph. D Thesis, Washington Univ., St. Louis, Missouri, 164 pp., 1972.

HUSSONG, D. M., and S. UYEDA, Tectonics in the Mariana Arc: results of recent studies, including DSDP Leg 60, Oceanologica Acta, Suppl., 4, 203–212, 1981.

HYNDMAN, R., Plate motions relative to the deep mantle and the development of subduction zones, Nature, 238, 263–265, 1972.

KANAMORI, H., Seismic and aseismic slip along subduction zones and their tectonic implications, in Island Arcs, Deep Sea Trenches, and Back-Arc Basins, edited by M. Talwani and W. C. Pitman III, pp. 163–174, Am. Geophys. Un., 1977.

KARIG, D. E., Structural history of the Mariana Island Arc System, *Geol. Soc. Am. Bull.*, **82**, 323–344, 1971.

KARIG, D. E., Plate convergence between the Philippines and the Ryukyu Islands, *Marine Geol.*, **14**, 153–168, 1973.

KINOSHITA, H., Paleomagnetism of sediment cores from DSDP Leg 58, Philippine Sea, in *Initial Report of the Deep Sea Drilling Project, Vol. 58*, pp. 765–768, U.S. Government Printing Office, Washington, D.C., 1980.

KOBAYASHI, K. and N. ISEZAKI, Magnetic anomalies in Japan Sea and Shikoku Basin and their possible tectonic implications, in *The Geophysics of the Pacific Ocean Basin and Its Margins*, edited by G. Sutton M. H. Manghnami and R. Moberly, *Geophys. Monogr. Am. Geophys. Union*, **19**, 235–251, 1976.

LOUDEN, K., Magnetic anomalies in the western Philippine Basin, in *The Geophysics of the Pacific Basin and its Margins*, edited by G. Sutton, M. H. Manghnami and R. Moberly, *Geophys. Monogr. Am. Geophys. Union*, **19**, 253–267, 1976.

MATSUDA, T., Collision of the Izu-Bonin Arc with Central Honshu: Cenozoic tectonics of the Fossa Magna, Japan, *J. Phys. Earth*, **26**, Suppl., S409–S421, 1978.

McCABE, R., J. ALMASCO, and W. DIEGOR, Geologic and paleomagnetic evidence for possible Miocene collision in Western Philippines, Central Philippines, *Geology*, in press, 1982.

MOLNAR, P. and T. ATWATER, Interarc spreading and cordilleran tectonics as alternates related to the age of subducted oceanic lithosphere, *Earth Planet. Sci. Lett.*, **41**, 330–340, 1978.

MROZOWSKI, C. L. and D. E. HAYES, The evolution of the Parece Vela Basin, eastern Philippine Sea, *Earth Planet. Sci. Lett.*, **46**, 49–67, 1979.

MURPHY, R. W., The Manila Trench-West Taiwan Foldbelt: A flipped subduction zone, *Bull. Geol. Soc. Malaysia*, **6**, 27–42, 1973.

SHIH, T. C., Marine magnetic anomalies from the western Philippine Sea: Implications for the evolution of marginal basins, in *The Tectonic and Geologic Evolution of Southeastern Asian Seas and Islands*, edited by D. E. Hayes *Am. Geophys. Union, Monogr.*, 49–75, 1980.

TAYLOR, B., and D. E. HAYES, The tectonic evolution of the South China Basin, in *The Tectonic and Geologic Evolution of Southeast Asian Seas and Islands*, edited by D. E. Hayes, *Geophys. Monogr. Am. Geophys. Union*, **23**, 89–105, 1980.

TULLIS, T. E., Evidence that lithosphere slabs act as anchors (abstract), *EOS, Trans. Am. Geophys Un.*, **53**, 522, 1972.

UYEDA, S., Subduction zones: facts, ideas and speculations, *Oceanus*, **22**, 52–62, 1979.

UYEDA, S., Subduction zones and back arc basins—A review, *Geol. Rundsch.*, **70**, 552–569, 1981.

UYEDA, S., Subduction zones (Introduction to Comparative Subductology), *Tectonophys.*, **81**, 133–159, 1982.

UYEDA, S. and Z. BEN-AVRAHAM, Origin and development of the Philippine Sea, *Nature*, **240**, 176–178, 1972.

UYEDA, S. and H. KANAMORI, Back-arc opening and the mode of subduction, *J. Geophys. Res.*, **84**, 1049–1061, 1979.

Accretion Tectonics in the Circum-Pacific Regions, edited by M. Hashimoto and S. Uyeda, 307–318.
Copyright © 1983 by Terra Scientific Publishing Company (TERRAPUB), Tokyo.

Tertiary Accretion of Ophiolite Seamounts, North Island, New Zealand

R. N. BROTHERS

Geology Department, Auckland University, New Zealand

Tertiary accretionary tectonics caused Early Miocene obduction of ophiolite massifs which were emplaced by southwestward gravity sliding across an auto-chthonous substrate of basement Permian-Jurassic metagreywackes and compacted Cretaceous-Oligocene sediments. Ophiolite paleontological and K-Ar ages are Cretaceous to Early Miocene, and diversity in rock types is from alkaline or tholeiitic to calc-alkaline. On-land distribution of the massifs indicates a source near the NE edge of the North Island, along a transform plate interface (Vening Meinesz Fracture Zone). Within this fracture zone, old and sheared slices of MORB ocean crust were intruded and added to by alkaline igneous activity, to produce compound ophiolite associations of apparent seamount character. The accretionary obductive event determined the setting for the initiation and orientation of volcanic arcs in a succeeding period of regional calc-alkaline eruptions.

1. Geological Setting

In the Northland peninsula (Fig. 1) the apparent basement is formed by Permian-Jurassic metagreywackes which lie along the northeastern coastline and are overlain westwards by Early Cretaceous-Oligocene sediments. These post-Jurassic strata outcrop over an area of approximately 10,000 km^2 and they are folded, fractured, sheared and sometimes tectonically transposed to give complicated reversals in age sequences; from microfossils the total stratigraphic thickness is about 7,000 m and from geophysics a maximum depth to the metagreywacke basement is in the order of 4–5 km. Syn-sedimentational deformation (HAY, 1960, 1975) produced notable unconformities and interrupted sedimentation patterns in the Paleocene (coinciding with the end of Tasman Sea spreading), in the Early and Late Eocene, and in the Oligocene thus compacting and lithifying a suitable substrate for westwards gravity-sliding of ophiolite masses which were obducted in the lowermost Miocene. Some 25 separate ophiolite massifs (Tangihua Volcanics and Wangakea Volcanics) stand up to 600 m above the general land surface and are seated on the Cretaceous-Oligocene sequence; the massifs are fault-bounded against the surrounding sediments and internally they sometimes carry extensive shear systems. In surface area these igneous piles cover about 1,500 km^2 in Northland, with the largest individual mass 40 km in diameter, and in volume they approach 300 km^3 which is probably a minimum figure since drillholes show that some have concealed depressed bases. Geophysical evidence indicates that these massifs are shallow, rootless bodies and areas of deep erosion reveal

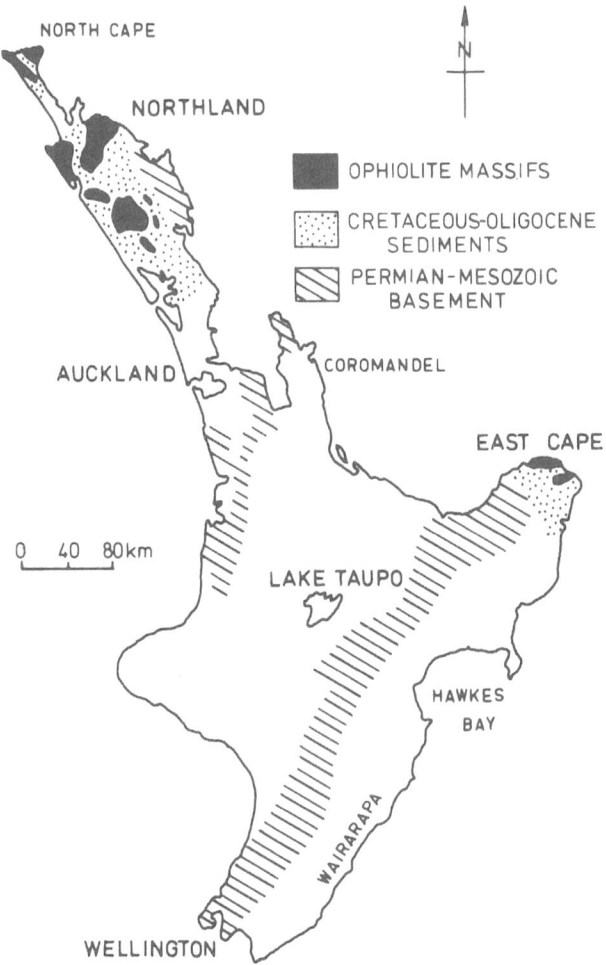

Fig. 1. Outline map of the distribution of ophiolite massifs in North Island, New Zealand.

windows of the sedimentary substrate (Brothers, 1974).

At East Cape, two massifs (Matakaoa Volcanics) cover amost 200 km² and reach an elevation of 1,000 m; they are separated by an east-west depression filled with Late Tertiary and Quaternary sediments, and the southern mass is faulted against Cretaceous formations.

2.   Massif Lithology and Structure

Within individual massifs (e.g. Brothers, 1965; Briggs and Searle, 1975; Bennett, 1976; Pirajno, 1980) it is possible to map an ophiolitic association, sometimes in continuous sequence, downwards from lavas to sheeted dolerite dikes and into noncumulate or cumulate plutonic phases. The uppermost extrusive rocks contain pillowed and massive flows with interspersed pillow-selvedge and lithic breccias; intercalated sedimentary units, usually only metres in dimension, are discontinuous lenses of deeply-coloured, oxidised, siliceous argillites and micrites. Within these sediments exhalative precipitation from submarine hot springs produced minor deposits of heavy metals as Mn oxides and massive sulphides (pyrite-chalcopyrite-sphalerite-marcasite-galena) with traces of gold and silver. The sheeted dolerites form series of parallel to sub-parallel dikes, including multiple intrusive sequences, that often can be distinguished only by chill-surface relationships between adjacent coarsely-textured hypabyssal units. This level also contains some small differentiated bodies of pegmatite, granophyre and diorite. Below the dolerites, wispy-textured static gabbros are the common plutonic phase with rare apophyses that penetrate upwards to the level of the pillow lavas; at North Cape and East Cape cumulate layered ultramafics and gabbros with coarse grain-size (up to 15 cm) complete the downwards succession of lithologies. Depleted components of possible mantle origin, such as tectonised and serpentinised harzburgite or lherzolite, have not been recognised within the North Island ophiolites.

Transition between the main lithologies (flows, dikes, plutonics) is most apparent from regional mapping since the boundaries are irregular in detail; on a broad scale these units show only gentle fold flexures, but they are cut by an abundance of faults that juxtapose contrasted rock types and make thickness estimates difficult. Measured thicknesses for each of the main units are usually in the order of 1–2 km. On an outcrop scale the rocks are sheared, sometimes in highly localised zones of dislocation that are thoroughly chloritised and zeolitised.

3. Ophiolite Petrology and Petrochemistry

The igneous suite is highly variable and ranges in character from quartz-normative tholeiitic to undersaturated alkaline even within one massif (e.g. BRIGGS and SEARLE, 1975) and rock types include serpentinite, cumulate harzburgite and lher-zolite, olivine clinopyroxenite and wehrlitic dikes, and hornblendite; teschenite, picrite, eucrite, and olivine, hypersthene and hornblende gabbros; diorite, tron-dhjemite, alkali syenite, and granophyre; augite, hypersthene, oligoclase, and kaersutite-titanaugite basalts and dolerites; variolites and camptonitic basalts. Lowgrade metamorphism is widespread, with replacement of igneous phases by zeolite associations (analcime-stilbite-heulandite-laumontite-natrolite-thomsonite-chabazite-apophyllite) and prehnite-pumpellyite, pumpellyite-actinolite, and quartz-albite-chlorite-epidote-actinolite-sphene greenschist assemblages. For the North Cape area, LEITCH (1978) identified an early phase of greenschist metamorphism in a water-saturated environment at a minimum temperature of 320°–350°C, and a late phase of zeolitic alteration below 165°–180°C.

Petrochemical classification of the ophiolite suite gives ambiguous results (Fig. 2). The alkali-silica ratios accurately reflect the petrographic range from tholeiite to alkali basalt, but in contrast the trace element discriminants for basalts and dolerites in a single massif at East Cape variously give conflicting identifications. The Ti-Cr plot

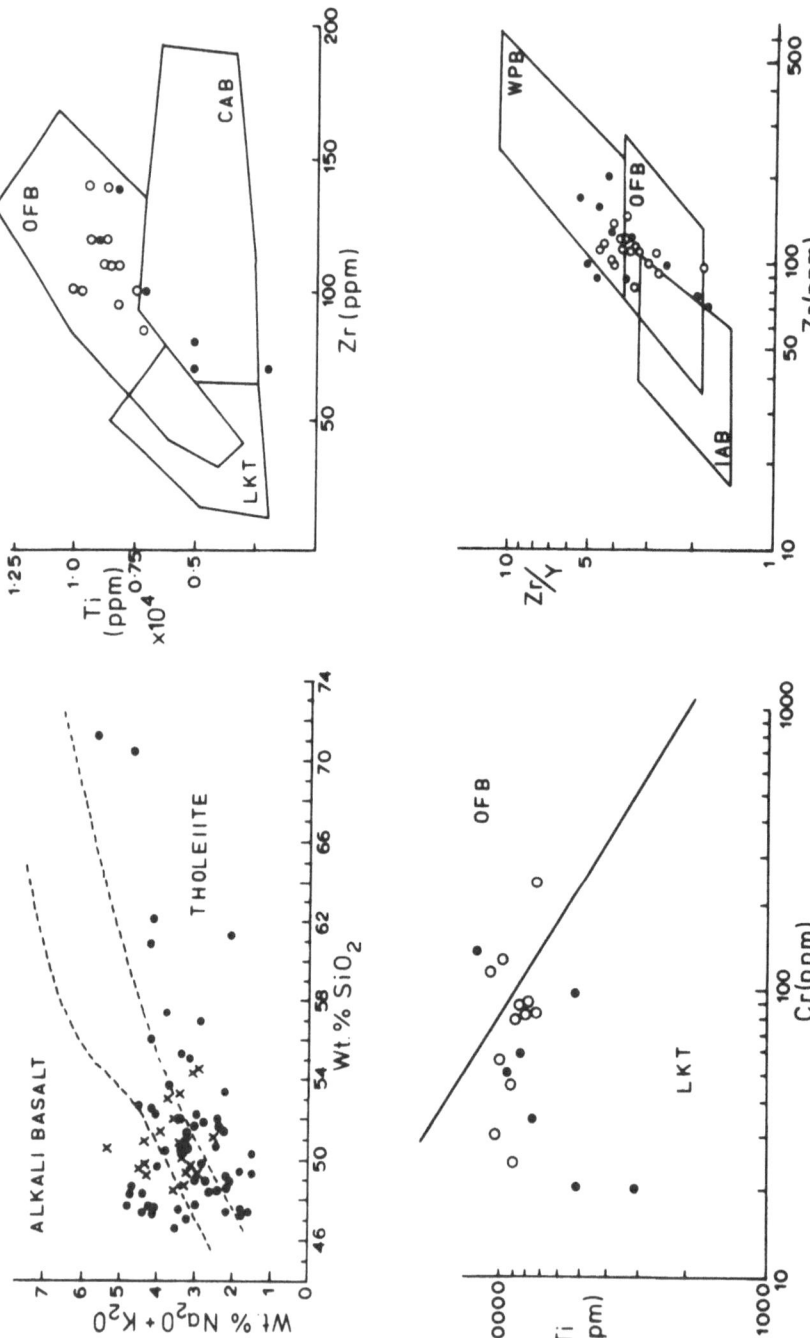

Fig. 2. Geochemical characterisation of basic igneous rocks from North Island ophiolite associations. The alkalis-silica diagram summarises data for volcanic rocks and dolerites from Northland (solid circles) and East Cape (crosses). Trace element plots refer only to the East Cape massif for basalts (open circles) and dolerites (solid circles) based on data from PIRAJNO (1980).

places most of the samples as low-K tholeiites, Ti-Zr indicates a dominance of ocean-floor basalts, while the Zr/Y-Zr diagram notably has a vacant field for island arc basalts and a cluster of data points for ocean floor basalts and within-plate basalts. The ambiguity of the trace element data for discrimination of basic rock types is similar to that described from Macquarie Island (GRIFFIN and VARNE, 1980).

4.   Age of the Ophiolites

An age range from Early Cretaceous to Early Miocene is indicated by paleon-tological and K-Ar evidence (Fig. 3) from widely-separated localities.

The fine-grained sediments interbedded with the ophiolite volcanics occasionally yield foraminifera, and more rarely macrofossils, which give ages to individual massifs of late Early Cretaceous (Albian) to Middle Eocene (Lutetian) at both East Cape and Northland (HORNIBROOK and HAY, 1978; STRONG, 1980). In addition, andesitic volcaniclastics associated with some massifs (LEITCH, 1970; KATZ, 1976) contain microfossils of Early Miocene (Aquitanian) age, thus implying not only longevity of ophiolite accummulation but also a notable change in igneous rock type during the last eruptive phase.

K-Ar dates listed by BROTHERS and DELALOYE (1983) cover the same time-span as the paleontological indicators, but with a marked clustering of ages towards the younger end of the scale (Fig. 3). The data was determined on plagioclase feldspar, hornblende and biotite separates from volcanic, hypabyssal and plutonic rocks from thirteen massifs, and the extended age range can be found within one massif as well as across the whole North Island ophiolite belt. For example, from some Northland massifs, at Opouteke basalt (hornblende 97 Ma) is intruded by gabbro (hornblende 42 Ma) and at Tangihua gabbro (hornblende 100 Ma) has a granophyre offshoot (biotite 60 Ma). The initial ratio for $^{40}Ar/^{36}Ar$ at 300.8 $\pm$ 6.5 is sufficiently close to the ratio for atmospheric argon as to indicate a lack of any argon overpressures that might account for the older K-Ar dates. For the same reason it would appear that the ophiolite series has not been significantly de-gassed in bulk by a combination of deep burial (with elevated temperatures) and a vigorous physical event (tectonic shock by obduction). An isochron age of 33.6 $\pm$ 2.1 Ma (Oligocene), given by BROTHERS and DELALOYE (1983), corresponds to the onset of sea-floor spreading that developed the South Fiji Basin adjacent to the North Island, a process which most likely developed the tectonic strain capable of inducing variable homogenisation of argon throughout the ophiolite suite located along the plate boundary.

5.   Longevity and Style of Ophiolite Genesis

Four lines of evidence suggest that the ophiolite massifs had a long history of igneous accretion, and originally as units within seamounts of major dimensions. *The paleontological evidence* gives an age range from Early Cretaceous to Eocene and possible lowest Miocene for East Cape, and Early to uppermost Cretaceous and Early Miocene for Northland. *The K-Ar ages* are from Early Cretaceous to Early Miocene for a variety of lithologies throughout the North Island ophiolites. *Age variation within*

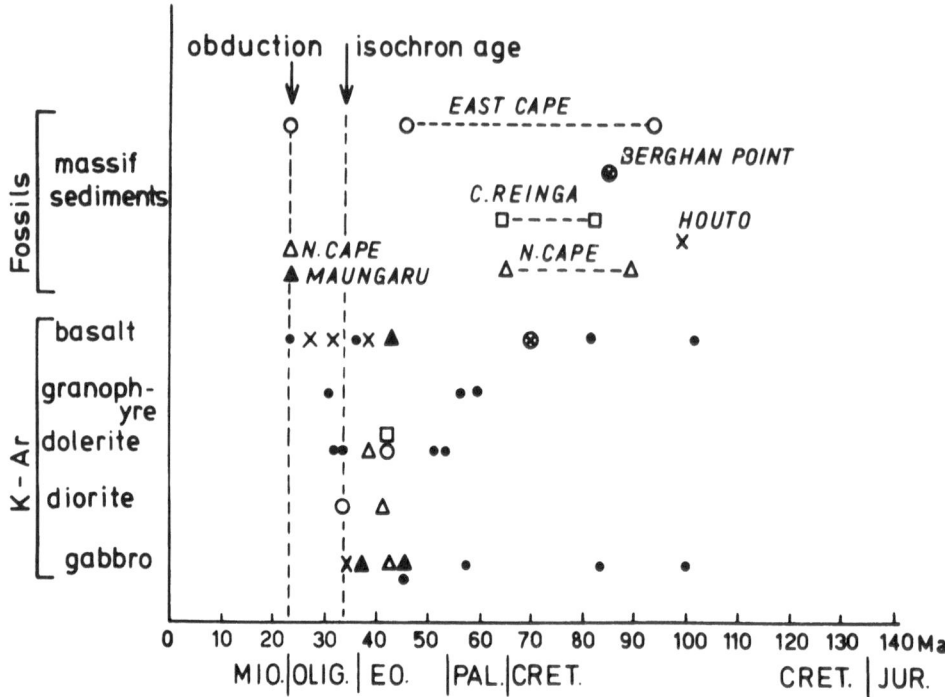

Fig. 3. Paleontologic and K-Ar ages for ophiolite massifs in the North Island. Solid circles represent ages for various Northland massifs for which K-Ar dates are available. Other symbols identify specific massifs, at East Cape and in Northland, from which both K-Ar and paleontologic ages have been determined (From Brothers and Delaloye, 1983).

individual massifs is indicated by mapping, fossils and K-Ar data in Northland: at Whangaroa an older plutonic-hypabyssal complex, schistose and metamorphosed to low-grade greenschist, is penetrated by less-deformed and unaltered dikes of basalt and dolerite; at a low level in the Houto massif, shattered and sheared olivine basalt pillows enclose sediments with Early Cretaceous macrofossils, but higher in the sequence undeformed and fresh camptonitic basalt pillows give Late Eocene to Early Miocene K-Ar ages; at Opouteke the older basalts (97–66 Ma) are intruded by 58–30 Ma gabbros and granophyres. *Lithologic diversity* is a characteristic of the whole North Island ophiolite suite, with some pronounced petrological contrasts between and within individual massifs: at Cape Reinga primitive tholeiites and two-pyroxene dolerites are intruded by dikes of alkali basalt and dolerite, but the extrusive phases of the Doubtless Bay mass are magnesian tholeiites-ferrobasalts-icelandites; at East Cape the igneous sequence includes quartz-normative basalt-gabbro-diorite, alkaline teschenite-picrite-dolerite, and undersaturated gabbro-syenite associations; and late andesitic phases occur at Northland and East Cape.

The evidence indicates continuity of igneous activity at highly localised sites of sporadic eruption and intrusion, and over a long period of time during which earlier rocks suffered deformation and low-grade metamorphism. The petrological complexities of the igneous assemblage, from saturated tholeiitic to undersaturated alkaline and calc-alkaline are not typical of oceanic crust. Nor are they indicative of a simple arc association, since the chemically-contrasted members of the suite do not show any normal consanguinity by way of lithologic compatibility in time and space. Similarly, the absence of subaerial volcanic debris precludes any concept of a system of oceanic islands related to a volcanic zone. Because obduction of the massifs was directed towards the southwest, ophiolite accumulation apparently took place in a near-shore position along the northeastern continental borderland of the North Island. The two clusters of submarine loci, parallel to Northland and close to East Cape, suggest an oceanic tectonic lineament of semi-vertical transform character joining two widely-separated sites of long-continued basic magma generation. In such a setting, the age range of the ophiolites (Cretaceous-Miocene) does not necessarily imply continuous magmatic injection over that interval at those sites, because old and sheared oceanic crust (with a tholeiitic MORB origin) could have been tectonically fractured into large blocks which were intruded and added to by fracture zone alkaline igneous activity, thus giving the final compound ophiolite association a multi-stage seamount character.

Alternatively, these two localities may represent the termination of ancient spreading ridges against the North Island continental mass. In the adjacent Pacific marginal basin system (Fig. 5), the oldest magnetic lineation is Anomaly 12 at 32.5 Ma (MALAHOFF et al., 1982) so that direct evidence is lacking for pre-Oligocene sea-floor spreading systems. Nevertheless, it is clear that a subductive interface did not pass under the North Island during the Cretaceous-Oligocene interval; calc-alkaline igneous rocks of that age are generally absent throughout the North Island where regional epi-continental volcanism commenced only after the ophiolite obduction event (BROTHERS and DELALOYE, 1983).

6.   Accretion Tectonics

The distribution and lithologic content of the northern Northland massifs (Fig. 4) conform to a regular pattern for disruption and decoupling of contrasted rock types, of differing yield strengths, within the ophiolite sequence. The deeper cumulate and gabbro phases are located along the east coast, gabbros and sheeted dolerites are common through the middle reaches, and pillow basalts are dominant along the western side. Relatively simple dislocation at site, and westwards transposition of ophiolite segments, produced dispersal from northeast to southwest with the topmost pillow units travelling furthest from source. Each ophiolite slice is a thin igneous unit which, as part of a high-density regional slide, was capable of free epidermal gliding for distances of up to 80 km across Northland on the westerly-tilted surface of the compacted in situ Cretaceous-Oligocene sedimentary substrate.

The date of the obductive event is estimated to be 23 Ma (BROTHERS, 1974), at about the Oligocene-Miocene boundary, since the youngest sediments in the Northland substrate are uppermost Oligocene and ophiolite debris first appears in Early Miocene conglomerates of Northland and East Cape. Dismemberment of the ophiolite source and obduction of the massifs must have been a rapid process, in the order of only one or two million years, in response to a brief but profound tectonic

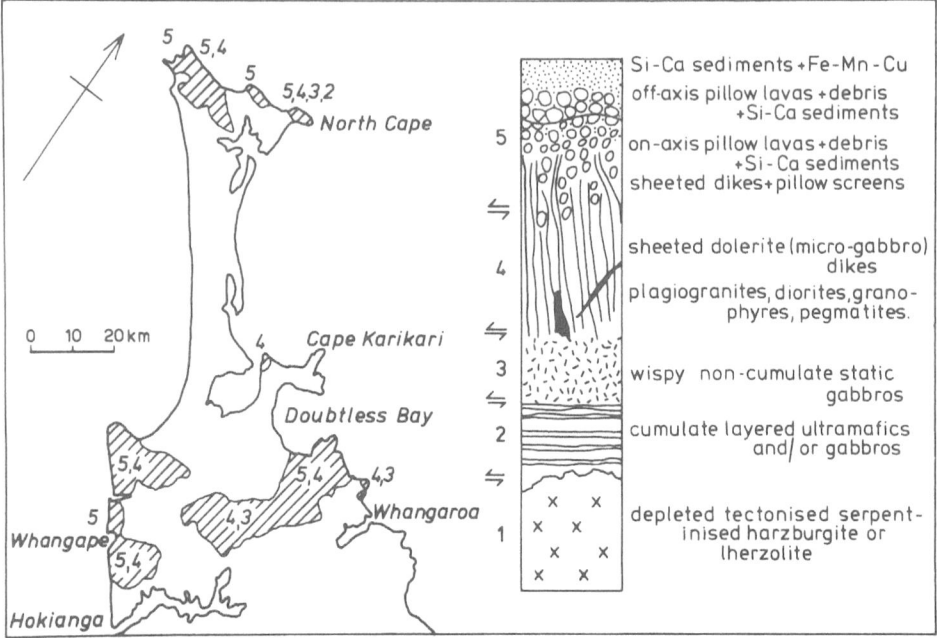

Fig. 4.   Identification of the lithologies of ophiolite massifs in northernmost Northland, in terms of a decoupled column of oceanic crust. The direction of movement during obduction and gravity-sliding was from northeast to southwest, with the dismembered uppermost lithologic units travelling furthest downslope as individual massifs (From BROTHERS and DELALOYE, 1983).

adjustment along a margin between North Island and the Pacific Plate.

The Upper Tertiary tectonic phases occurred in a rapid sequence (Table 1) within which the ophiolite accretion event was a critical factor in determining the initiation, and arc orientation, of the succeeding Miocene-Holocene regional volcanism.

(1)   35.5–25.5 Ma: Oligocene sea-floor spreading in the marginal South Fiji Basin produced magnetic lineaments oriented NE-SW (Fig. 5). Differential adjustment between North Island and the mobile ocean crust was taken up by right lateral strike-slip movement along the Vening Meinesz line (BROTHERS and DELALOYE, 1976; MALAHOFF et al., 1982), a transform fracture zone which was the most likely site for the ophiolite seamounts (BROTHERS and DELALOYE, 1983). Within Northland, nepheline-bearing lavas were erupted during the Oligocene along a narrow NE-SW belt (Zone 3, Fig. 5) which appears to have been a direction of important, and highly localised, continental rifting parallel to the oceanic magnetic lineaments.

(2)   25.5–23 Ma: The Oligocene phase of spreading in South Fiji Basin ceased at Anomaly 8 (25.5 Ma). Collision impact along the plate-plate interface (Vening Meinesz Fracture Zone) then caused compressive uplift and obductive accretion of the seamounts, at about 23 Ma, onto the North Island sialic crust (from Zone 1, Fig. 5). Tectonic impingement which contributed to thrusting of the hot high-standing oceanic crust and ophiolite seamounts, over the cold lowlying continental edge, may have been

Table 1.   Comparison of events in Northland with the Tasman and southwest Pacific Basins (data from MALAHOFF *et al.*, 1982).

Northland	Ma			S.W. Pacific and Tasman
	0			
Development of NE-SW Taupo Volcanic Zone and NW-SE Hauraki and East Coromandel Rifts with associated volcanics				Renewal of marginal basin activity with NE-SW magnetic lineaments aligned with Taupo Volcanic Zone
	8			
Calc-alkaline volcanism established in NE-SW Zone 4	16	rotation		Marginal basin
Calc-alkaline eruptions commenced in NE-SW Zone 3	21			quiescence
			fast	
Obduction of North Island ophiolites from NW-SE Zone 1	23	of	slow	
	25			
Post-Teurian pre-Miocene sedimentation		rate	South Fiji Basin developed; NE-SW magnetic lineaments aligned with Zone 3 in Northland	
	35			
Nephelinic basic intrusive arc initiated in NE-SW Zone 3	40	Plate	slow	
			fast	
—Major unconformity— Teurian deformation, change of sediment type and provenance	60	Pacific	Opening of Tasman Basin	Separation of New Zealand from Antarctica
	80			
Cretaceous Paleocene (Teurian) sedimentation with acid plutonic and metamorphic provenance				
	100			

Vertical rotated labels: "Ophiolite seamount growth along Indian-Pacific plate junction"; "Plate rotation rate of ... fast / slow"; "Pacific"

caused by acceleration in Pacific Plate rotation (from <0.5° per million years to 1.3° per million years) that took place in the interval 25–20 Ma (CLAGUE and JARRARD, 1973).

(3)   21 Ma: As a result of this accretionary event, the Indian and Pacific Plates became locked along the line of the Vening Meinesz Fracture Zone, bringing about a radical change in plate response to the prevailing E-W regional compression vector (WALCOTT, 1978). A new subduction system developed parallel to the Hikurangi Trench, with a convergence zone that passed northwest at a shallow angle under the North Island to produce the first of the regional calc-alkaline volcanic arcs (21 Ma)

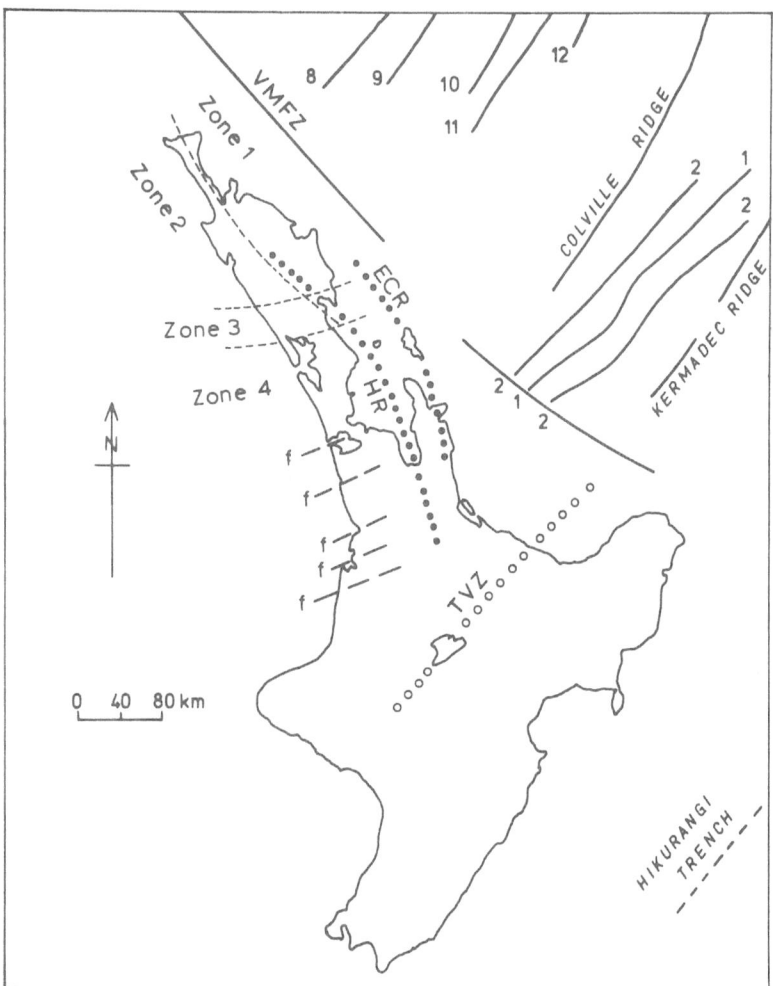

Fig. 5. Cenozoic tectonic and volcanic lineaments in North Island and in the offshore region (from Malahoff *et al.*, 1982). Oligocene-Early Miocene components are: Zone 1 as the source of obducted ophiolites in Northland and East Cape. Zones 3 and 4 containing NE-SW volcanic arcs, Vening Meinesz Fracture Zone (VMFZ), and magnetic anomalies 12–8 (Oligocene, 32.5–26.8 Ma). Late Miocene-Quaternary lineaments are: Taupo Volcanic Zone (TVZ), Hauraki Rift (HR), East Coromandel Rift (ECR), and magnetic anomalies 2–1 (1.8 Ma to the present). Broken lines labelled *f* are the trends of some major Late Cenozoic faults normal to the rift system: other faults with similar trend, or parallel to the rifts, have been omitted because of map scale.

parallel to the NE-SW Oligocene rift (Zone 3, Fig. 5). From Miocene to Holocene the convergence angle gradually steepened below the North Island, so that the maximum ages for the calc-alkaline volcanics become younger southeastward (through Zone 4, Fig. 5) across a NE-SW arc pattern, towards the Taupo Volcanic Zone (Challis, 1978).

(4)   10 Ma (or less): In the back-arc region to the north of the Taupo Volcanic Zone, long tensional rifts (Hauraki and East Coromandel Rifts, Fig. 5) developed in a north to northwest direction as a result of Late Cenozoic upwarping above a linear upper mantle swell (HOCHSTEIN, 1978; HOCHSTEIN and NIXON, 1979). Mantle plumes parallel to the rifts produced young alkaline basalts (HEMING, 1980); NE-SW compressional faults are oriented transverse to the rifts.

The on-land geology contains a sequence of early rifting (Oligocene), accretionary ophiolite obduction (Early Miocene), subduction and calc-alkaline volcanism (Early Miocene-Quaternary) and late rifting (Pliocene-Quaternary) that may be a part of the continuing disruption and northwards rafting of Gondwanaland segments, a fragmentation that appears to have started in the Permian and to be still in process.

An abridged version of this paper, describing the ophiolite massifs, will appear in "Ophiolites of Continents and Comparable Oceanic Floor Rocks," the report of I.G.C.P. Project No. 39. A more detailed statement on the topic, dealing particularly with K-Ar data for the ophiolites and with their autochthonous substrate, is given by BROTHERS and DELALOYE (1983).

REFERENCES

BENNETT, M. C., The ultramafic-mafic complex at North Cape, northernmost New Zealand, *Geol. Mag.,* **113**, 61–76, 1976.

BRIGGS, R. M. and E. J. SEARLE, Tangihua Volcanics in the Opouteke-Pakotai area, Northland, New Zealand, *N.Z. J. Geol. Geophys,* **18**, 327–341, 1975.

BROTHERS, R. N., Tangiteroria District-Tangihua Volcanics, in *New Zealand Volcanology: Northland-Coromandel-Auckland,* edited by B.N. Thompson and L.O. Kermode, *Information Series,* **49**, pp. 46–50, N.Z.D.S.I.R., Wellington, 1965.

BROTHERS, R. N., Kaikoura orogeny in Northland, New Zealand, *N.Z. J. Geol. Geophys.,* **17**, 1–18, 1974.

BROTHERS, R. N. and M. DELALOYE, On-land ophiolites, North Auckland, New Zealand, *25th Int. Geol. Cong., Abstr.,* **1**, 45–46, 1976.

BROTHERS, R. N. and M. DELALOYE, Obducted ophiolites of North Island, New Zealand: age, emplacement and tectonic implications for Tertiary and Quaternary volcanicity, in press, *N.Z. J. Geol. Geophys.,* **25**(3), 1983.

CHALLIS, G. A., Volcanism and volcanic trends in the geological history of New Zealand, in *The Geology of New Zealand,* edited by R. P. Suggate, pp. 664–667, Government Printer, Wellington, 1978.

CLAGUE, D. A. and R. D. JARRARD, Tertiary Pacific plate motion deduced from the Hawaiian-Emperor chain, *Bull. Geol. Soc. Am.,* **84**, 1135–1154, 1973.

GRIFFIN, B. J. and R. VARNE, The Macquarie Island ophiolite complex: mid-Tertiary oceanic lithosphere from a major ocean basin, *Chem. Geol.,* **30**, 285–308, 1980.

HAY, R. F., The geology of the Mangakahia Subdivision, *N.Z. Geol. Surv. Bull.,* **61**, 109, 1960.

HAY, R. F., Sheet N7 Doubtless Bay (1st Edition), *Geological Map of New Zealand 1:63360,* D.S.I.R., Wellington, 1975.

HEMING, R. F., Patterns of Quaternary basaltic volcanism in the northern North Island, New Zealand, *N.Z. J. Geol. Geophys.,* **23**, 335–344, 1980.

HOCHSTEIN, M. P., Geothermal systems in the Hauraki Rift Zone (New Zealand)—an example for geothermal systems over an inferred upper mantle swell, in *Alternative energy sources: an international compendium vol. 6 geothermal energy and hydro power,* edited by T. N. Veziroglu, pp. 2599–2610, Hemisphere Pub. Corp., Washington, D.C., 1978.

HOCHSTEIN, M. P. and I. M. NIXON, Geophysical study of the Hauraki Depression, New Zealand, *N.Z. J. Geol. Geophys.,* **22**, 1–19, 1979.

HORNIBROOK, N. de B. and R. F. HAY, Late Cretaceous agglutinated foraminifera from sediments interbedded with the Tangihua Volcanics, Northland, New Zealand. "The Crespin Volume: Essays in honour of Irene Crespin." *Aust. Bur. Min. Res. Bull.*, **192**, 67–72, 1978.

KATZ, H. R., Cretaceous foraminifera from the Matakaoa Volcanic Group—comment, *N.Z. J. Geol. Geophys.*, **19**, 943–945, 1976.

LEITCH, E. C., Contributions to the geology of northernmost New Zealand: II—The stratigraphy of the North Cape district, *Trans. Roy. Soc. N.Z. (Earth Sci.)*, **8**, 45–68, 1970.

LEITCH, E. C., Hydrothermal metamorphism of the Whangakea Basalt, New Zealand, *N.Z. J. Geol. Geophys.*, **21**, 287–292, 1978.

MALAHOFF, A., R. H. FEDEN, and H. S. FLEMING, Magnetic anomalies and tectonic fabric of marginal basins north of New Zealand, *J. Geophys. Res.*, **87**, 4109–4125, 1982.

PIRAJNO, F., Subseafloor mineralisation in rocks of the Matakaoa Volcanics around Lottin Point, East Cape, New Zealand, *N.Z. J. Geol. Geophys.*, **23**, 313–334, 1980.

STRONG, C. P., Early Paleogene foraminifera from Matakaoa Volcanic Group (Note), *N.Z. J. Geol. Geophys.*, **23**, 267–272, 1980.

WALCOTT, R. I., Present tectonics and late Cenozoic evolution of New Zealand, *Geophys. J. Roy. Astron. Soc.*, **52**, 137–164, 1978.

Accretion Tectonics in the Circum-Pacific Regions, edited by M. Hashimoto and S. Uyeda, 319–331.

Roles of Seamount, Rise, and Ridge in Lithospheric Subduction

Yoshibumi TOMODA and Hiromi FUJIMOTO

Ocean Research Institute, University of Tokyo, Tokyo 164, Japan

Interference between trench and seamounts or rises is discussed from the viewpoint of the sub-Moho structures of the latter. Seamounts or rises can be classified into three types; (1) seamounts which can easily subduct (as represented by Kashima No. 1 Seamount), (2) sizable seamounts which will take a long time to subduct, (3) rises which will never subduct.

It is shown by gravity anomaly data that the deep structures of the different types of seamounts or rises are different from one another. These various structures of seamounts or rises are correlatable to their sizes and would result in a complicated accretion process at trench axes. A rise with a thick crust may accrete to the landward plate at the trench axis, and if the subduction forces are large enough, a new subduction zone would be formed seaward of the old trench, a jump of the plate boundary would take place.

1. Introduction: Residual Gravity Anomaly and Estimation of the Thickness of the Lithosphere

In this paper, interference between a trench and a seamount or a seamount chain will be discussed from the viewpoint of their deep structures below the Moho, by use of both gravity anomalies and velocity structures derived from the explosion seismology.

The "residual gravity anomaly (R.G.A.)" introduced by YOSHII (1973) is defined as the difference between observed gravity anomaly and the gravity anomaly expected from the velocity structure. In other words, R.G.A. is a kind of Bouguer gravity anomaly reduced to the depth of the Moho, showing the structure below the Moho. The actual calculation of R.G.A. is made by the following formula,

$$\text{R.G.A.} = \Delta G' + 2\pi k^2 \sum_i H_i (\rho_L - \rho_i)$$

where $\Delta G'$: free air gravity anomaly, H_i: thickness of the i-th layer, ρ_L: 3.3 g/cm^3 corresponds to the layer of $V_p = 8.1$ km/sec, ρ_i: density of the i-th layer, and k^2: universal constant of attraction.

Thickness of the lithosphere is calculated from R.G.A., assuming that the density (ρ_L) of the lithosphere below the Moho and that of the asthenosphere (ρ_A) are uniform, and that the density contrast between the two is known.

According to the seismic studies, P_n velocities in the upper mantle are in the range between 7.7 and 8.9 km/sec. These values have little correlation with the age of the ocean floor. When these scattered values are used, no significant results would be obtained concerning the thickness of the lithosphere. Therefore, in order to consider the deeper structure we will hereafter rely upon one important assumption that the density of the lithosphere below the Moho is uniform and is larger than that of the asthenosphere i.e. $\rho_L > \rho_A$ (Yoshii, 1973).

This gravimetrical assumption implies that the asthenosphere corresponds to zone of partial melt from a petrological point of view, and corresponds to the low velocity layer from a seismological point of view. In the present paper, it is assumed that $\rho_L - \rho_A = 0.1$ g/cm^3.

2. General Features of the Residual Gravity Anomalies (R.G.A.) in the Pacific

Figure 1 shows R.G.A. of the Pacific region based on velocity structures at 475 observation points. The map also defines the thickness of the lithosphere. R.G.A. of 80 mgals approximately corresponds to a deviation of the thickness of lithosphere by 20 km when the density contrast between the lithosphere and the asthenosphere is assumed to be 0.1 g/cm^3. The maximum difference in R.G.A. in the region amounts to 500 mgals which is equivalent to a variation of thickness of the lithosphere by about 120 km.

The subduction zones such as the Japan, Izu-Ogasawara and Mariana trenches correspond to the area at which R.G.A. is larger than about 700 mgals and the lithosphere is thicker than about 100 km. This belt of high R.G.A. extends south to the Ontong-Java Plateau. The same area is also characterized by a strongly magnetized body lying at the depth of 50–100 km as deduced from the marine magnetic anomaly of intermediate wave length (Nomura, 1979).

According to Yoshii's model, the lithosphere becomes thick in proportion to the square root of the age of the ocean floor, and the map in Fig. 1 generally agrees with this "thickening model."

R.G.A. on the Hawaiian Ridge is smaller than 600 mgals. This shows that the lithosphere may be made thinner by the heat supply from the plume in the asthenosphere, as pointed out by Ogawara and Kono (1981).

3. Thickness of the Lithosphere at Trenches, Seamounts, Rises and Ridges

3.1 Outer gravity high seaward of trenches and the thickness of the lithosphere

Figure 2 shows the outer gravity high and velocity structure seaward of the Japan Trench as well as the residual gravity anomaly (R.G.A.). The profiles show that the outer gravity high seaward of the trench cannot be explained by the structure shallower than the Moho.

According to Ogawara and Kono (1981), the thickness of the lithosphere should attain its asymptotic value at the age of 90–100 m.y., because the heat coming from the asthenosphere and the heat lost by conduction through the lithosphere will balance. Rapid thickening of the lithosphere seaward of the trench given from R.G.A. shown in Fig. 2 seems to suggest the existence of extra cooling near the trench. Gigantic faults recognized near the trench may be responsible for the cooling process. An alternative explanation may be the existence of high density material beneath the outer gravity

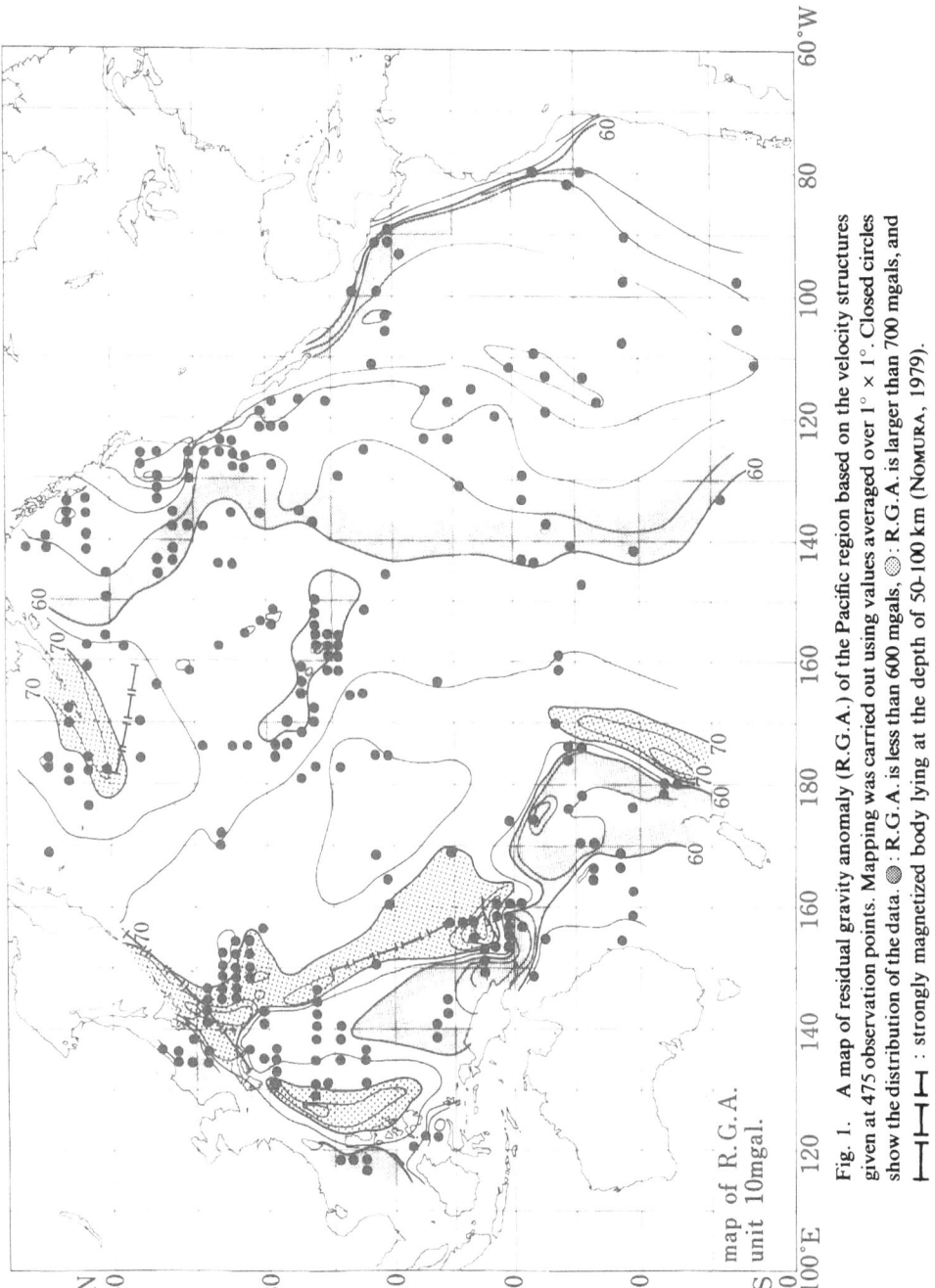

map of R.G.A.
unit 10mgal.

Fig. 1.   A map of residual gravity anomaly (R.G.A.) of the Pacific region based on the velocity structures given at 475 observation points. Mapping was carried out using values averaged over 1° × 1°. Closed circles show the distribution of the data. ⊙ : R.G.A. is less than 600 mgals, ◎ : R.G.A. is larger than 700 mgals, and ⊢—⊢—⊣ : strongly magnetized body lying at the depth of 50-100 km (Nomura, 1979).

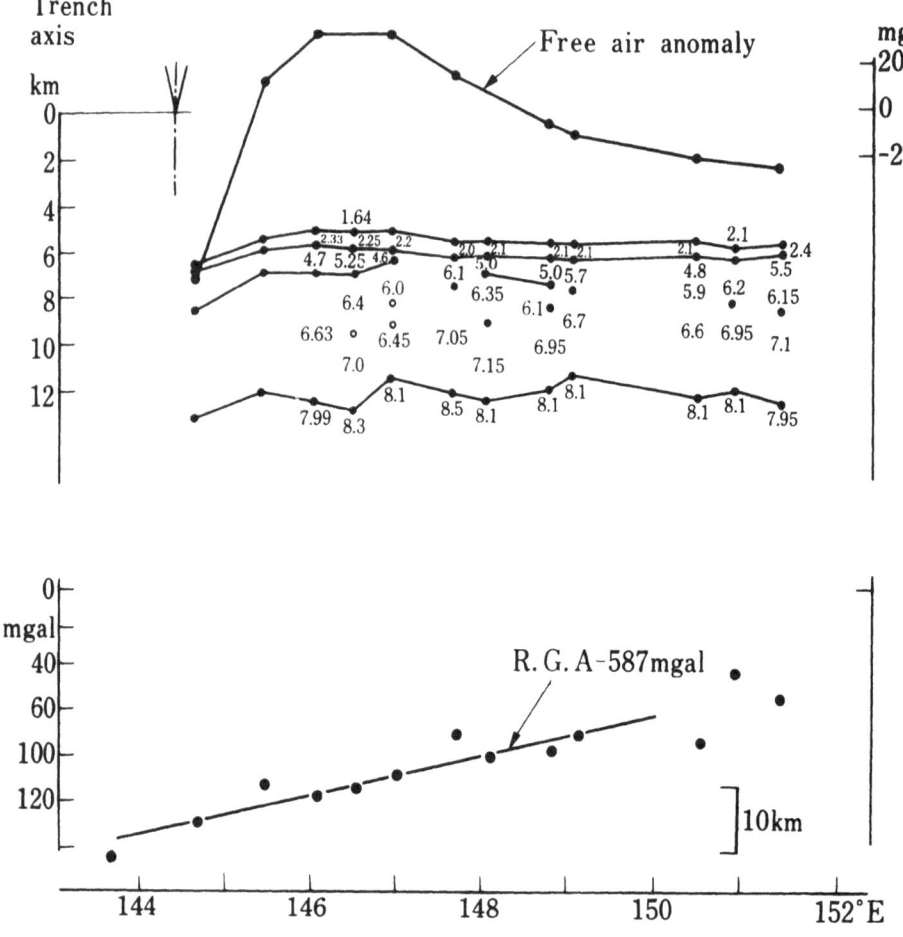

Fig. 2.   The outer gravity high, velocity structure, and R.G.A. seaward of the Japan Trench.

high if Pratt's model of isostasy is assumed.

Positive free air anomaly extending over a wide area means that the isostatic equilibrium is not achieved, and it may be caused by rapid thickening of the lithosphere as mentioned above. If this is the case, it may contribute to the enhanced driving force of subduction, if the force is mainly due to "negative buoyancy" of the slab (ELSASSER, 1971).

In order to explain the positive free air anomaly seaward of the trench, two models have so far been proposed. A model shown in Fig. 3(a) explains the positive anomaly as the gravity effect of a high density material which is situated in the deeper part beneath the trench (MORGAN, 1965). The other model shown in Fig. 3(b) explains the anomaly as caused by the mantle material swollen by the bending of an elastic plate (WATTS and TALWANI, 1974).

Results shown in Fig. 2 suggest that the positive free air anomaly seaward of the

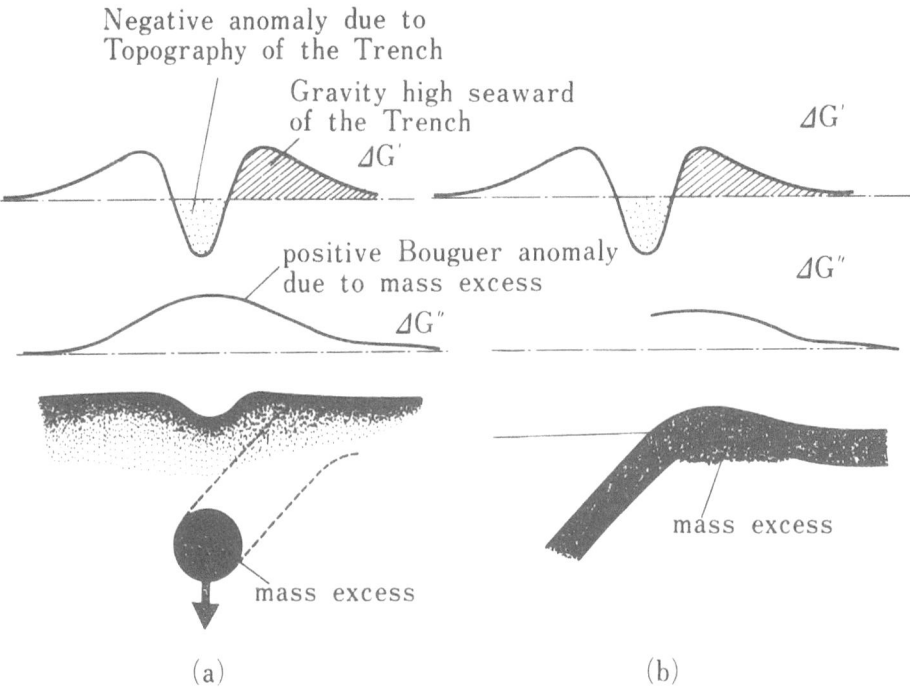

Fig. 3. In order to explain the positive free air anomaly seaward of the trench, two models have so far been proposed: (a) high density material in deeper part beneath the trench (MORGAN, 1965), (b) the mantle material swollen by bending of the elastic plate (WATTS and TALWANI, 1974).

trench is not caused by bending of the crust but caused by mass excess in the sub-Moho. Therefore, the model 3(a) is preferable for our results if we are to choose one of the two models.

3.2   Thickness of the lithosphere beneath a seamount chain

The velocity structure beneath the Hawaiian Ridge is well known (FURUMOTO *et al.*, 1968). Therefore, R.G.A. or the thickness of its lithosphere can be investigated as follows. Figure 4 shows the velocity structure, free air anomaly, and R.G.A. of the southwest to northeast section across the Hawaiian Ridge. Negative R.G.A. exists beneath the ridge. Its horizontal extent and the amplitude are 500–700 km and −60 − 70 mgals, respectively.

The anomaly in the thickness of the lithosphere can easily be calculated using R.G.A. It is shown that the lithosphere under the ridge is about 20 km thinner than in the surrounding area. As pointed out by OGAWARA and KONO (1981), the thinning of the lithosphere may be caused by the heat supply from the asthenosphere due to the hot spot. It is suggested that the weight of the Hawaiian Ridge is not supported by the density contrast between the crust and the mantle, but balanced by the buoyancy of the asthenosphere.

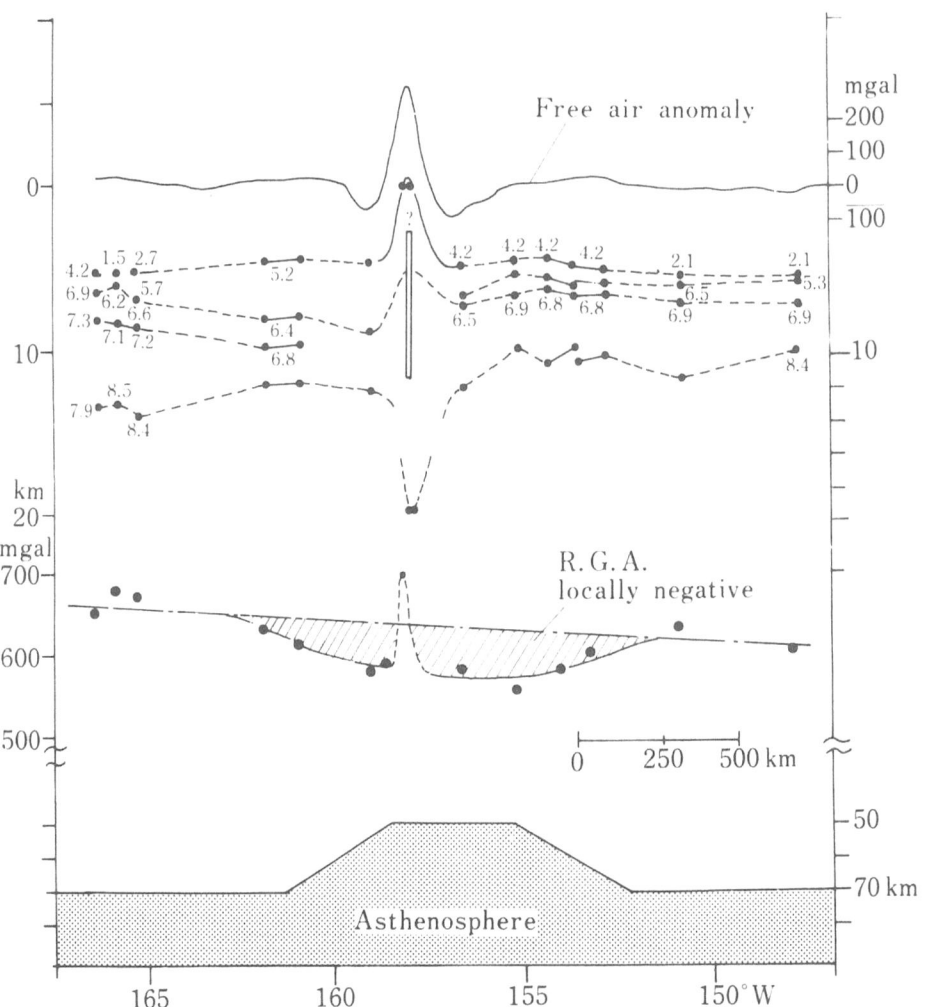

Fig. 4.   Velocity structure, free air anomaly, and R.G.A. of the southwest to northeast section across Oahu Island.

Many seamounts exsist in the western and central Pacific, but their velocity structures are little known. Therefore, their deep structures are derived here from gravity anomalies. The density of the seamount body is first estimated from free air anomaly and topography, when good correlation between the two exsists. Then the gravitational effect of the seamount body is removed, and a modified type of Bouguer anomaly is obtained. To explain the Bouguer anomaly by the structure above the Moho, three kinds of subterranean structures can be considered, as shown in (a) ∼ (c) of Fig. 5(1). Among these three, model (a) which satisfies Airy-Heiskanen isostasy does not agree with observed gravity anomaly (A in Fig. 5(2)). Model (b) is an isostatic model taking into account the elasticity of the crust. MCKENZIE and BOWIN (1976) derived a transfer function between topography and free air anomaly for the model, and the method is applied for the determination of the thickness of the elastic plate using the deviation from the Airy-Heiskanen isostasy. Although the topographic moat recognized in the northeast of the Hawaiian Ridge seems to be consistent with this model, the thickness of the sediment around the seamounts in the western and central Pacific, surveyed by acoustic reflection, shows that the sediment becomes monotonously thicker with distance from the crest of seamounts (HEEZEN *et al.*, 1973). In addition to such seismic results, the crustal bending model does not agree with the observed results (B in Fig. 5(2)).

In the case of model (c), the crust is thicker in the area three to five times larger than the horizontal scale of the seamount. However, seismic results indicating such a crustal bulge have not been obtained so far. As isostatic equilibrium is not achieved in this model, the topography would not be kept for a geological time.

Considering such conditions, the model (d) in Fig. 5(1) assuming that the weight of the seamount is supported by the buoyancy of the asthenosphere is proposed here, as was also shown in the case of the Hawaiian Ridge.

3.3   Rise or ridge supported by the buoyancy of the crust
There exist regions of large crustal thickness such as the Shatsky Rise or the Ontong-Java Plateau in the ocean. Figure 6 shows the velocity structure and R.G.A. across the Shatsky Rise along 32°N latitude. The eastern half of the velocity structure of the Shatsky Rise is not used because the seismic profiles were not reversed.

A remarkable characteristic of R.G.A. over the Shatsky Rise is that the variation in R.G.A. is only several mgals, in spite of the fact that the thickness of the crust varies by nearly 10 km. The variation in the thickness of the lithosphere expected from R.G.A. is only 2 km. It can be said that the expected variation is almost negligible compared to the whole thickness amounting to several tens of kilometers.

The free air anomaly there shows that isostatic equilibrium is attained, and nearly constant R.G.A. indicates that it is achieved almost entirely by the buoyancy of the crust.

The main difference between these rises and seamounts, described in the previous section, lies in the stability of the root supporting the topography. The thick crust as seen beneath the Shatsky Rise can probably persist stably, but the thin lithosphere beneath a seamount would disappear by the cooling process.

4.   Interference Between the Seamount and the Trench

Free air anomalies seaward of trenches (Fig. 2) indicate that the deviation from isostatic equilibrium is explained by a rapid increase in the thickness of the lithosphere near the trench. It results in the enhanced source of the driving force of the lithospheric subduction.

Therefore, if seamounts or a seamount chain having thin lithosphere exists near the trench, "positive gravity anomaly seaward of the trench" is interfered and the driving force of subduction will decrease. Simplified map of free air anomaly (Fig. 7) shows that each of the Japan, Izu-Ogasawara, and Mariana Trenches has its own

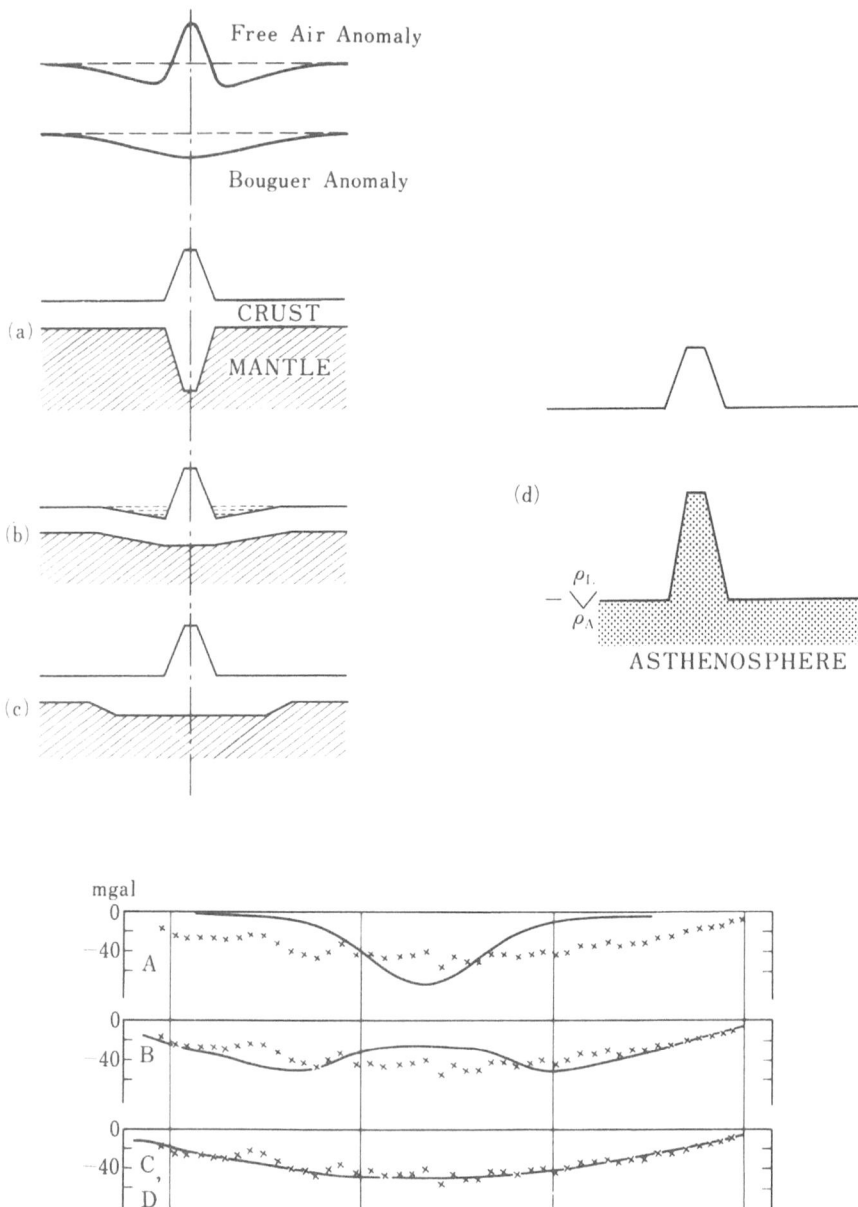

Fig. 5. (1) Four models of subterranean structure of a seamount. (2) Distribution of a modified type of Bouguer anomalies around the seamount. Observed anomalies were obtained after gravity effect of the seamount (Shunsetsu Seamount: 23°50′N, 148°45′E) body was removed and shown with the mark "x". Solid lines in A, B, C, and D show calculated anomalies corresponding to the models (a), (b), (c), and (d).

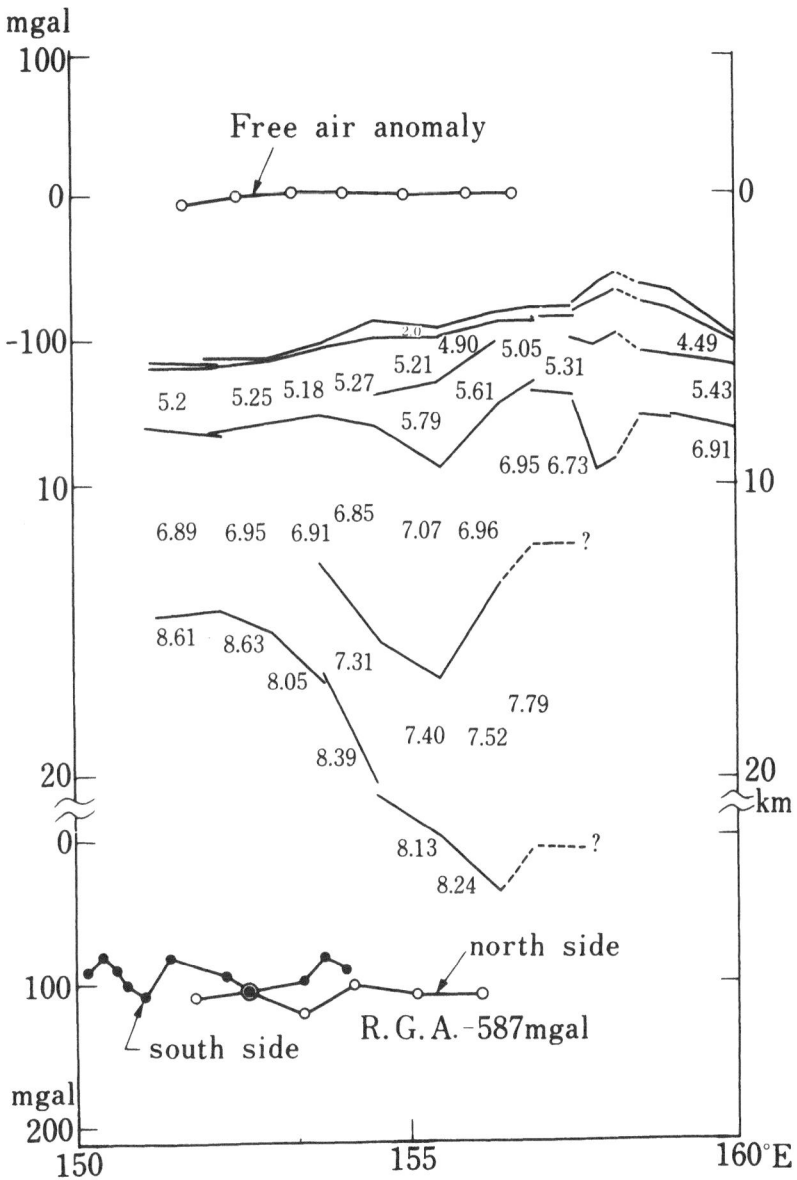

Fig. 6.   Velocity structure, free air anomaly, and R.G.A. across the Shatsky Rise.

"outer gravity high." The "outer gravity high" is cut off by the seamounts or seamount chain. Where a seamount chain is impinging on the region between two trenches corresponds to the region of low seismic activity except shallow focus earthquakes (KATSUMATA and SYKES, 1969; BARAZANGI and DORMAN, 1969).

If subduction is obstructed by the thin lithosphere accompanied by a seamount, and if the driving force of subduction seaward of the seamount becomes large by thickening of the lithosphere, new subduction would take place seaward of the

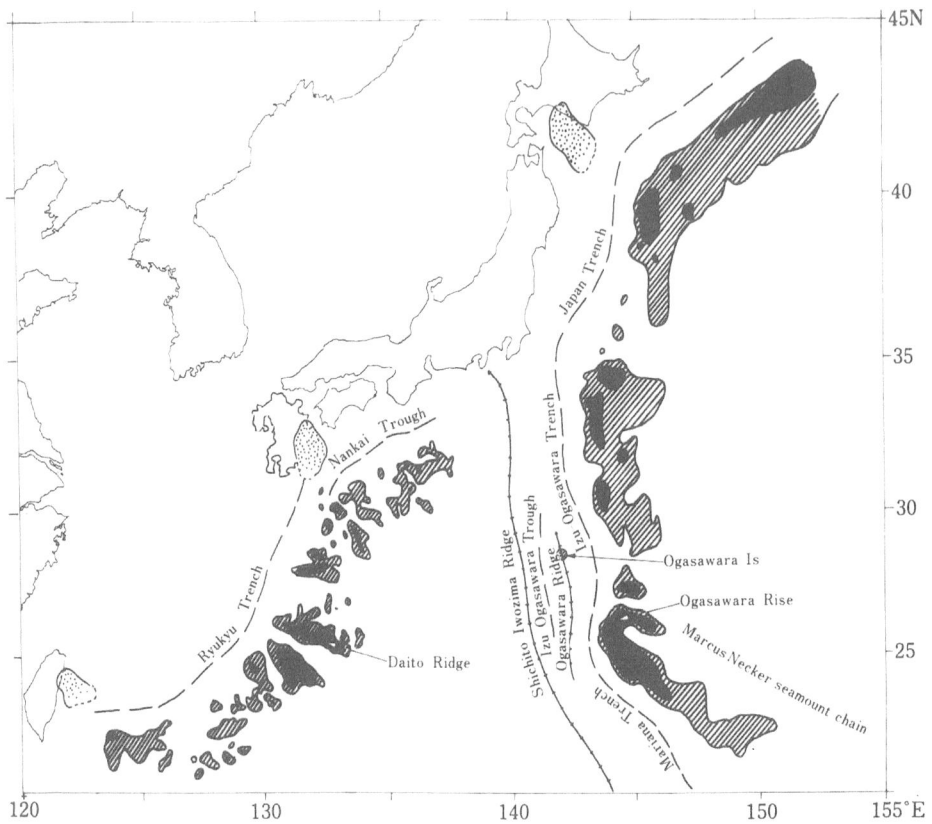

Fig. 7.   Simplified map of free air anomaly in the western Pacific ● : outer gravity high seaward of trenches (larger than + 40 mgals). ● : outer gravity high seaward of trenches (larger than + 20 mgals). ⊙: gravity lows landward of trench junctions (less than − 40 mgals).

seamount. In such a case, delayed subduction would take place landward of the trench as the root of the seamount preventing the subduction slowly disappears due to a cooling process (sizable A type in Fig. 8). Gravity lows landward of trench junctions (off Urakawa in Hokkaido, off Miyazaki in Kyushu, and east off Taiwan in Fig. 7) may be explained to be the delayed subductions.

The Kashima No. 1 seamount was broken down and seems to be subducted at the axis of the Japan trench (MOGI and NISHIZAWA, 1980). According to the precise gravimetry about the seamount, the Bouguer anomaly simply shows the regional trend of the trench margin and nothing which suggests the existence of the root. The originally thin lithosphere beneath the old Kashima No. 1 seamount would have become thicker and resulted in the breakdown and subduction (type A in Fig. 8).

In the case of the Shatsky Rise, which is supported by the buoyancy of the crust (type B in Fig. 8), it would be difficult to subduct. A new subduction would take place seaward of such a rise, ridge, or plateau when it arrives at the trench in a similar fashion as in the case of the early stage of sizable A at the trench.

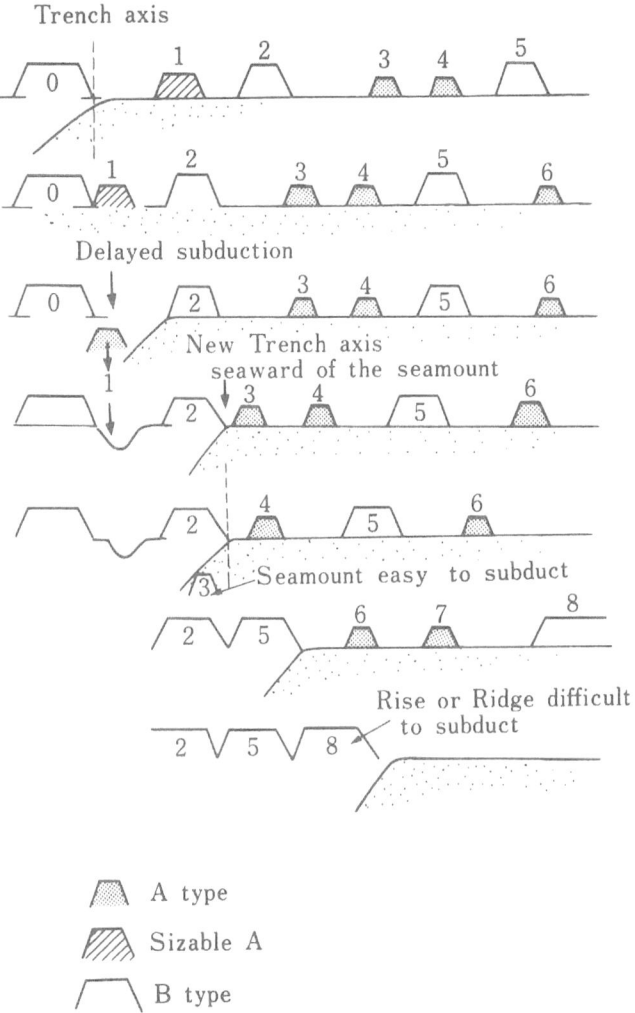

Fig. 8.   Seamounts or rises can be classified into three types; (1) A type: seamounts supported by buoyancy of the asthenosphere which is easy to subduct. (2) Sizable A type: large seamounts supported by buoyancy of the asthenosphere which take a long time to subduct. (3) B type: rises supported by buoyancy of the crust, which is difficult to subduct. These various structures of seamounts or rises, correlatable to their sizes, would result in complicated accretion processes at trench axes. A rise with a thick crust may accrete at the trench axis and if the subduction forces are large enough, a new subduction zone would be formed seaward of the original trench so that a jump of the plate boundary would take place.

In the Ogasawara region, where the Izu-Ogasawara Trench and the Ogasawara Rise are supposed to collide, an active island arc represented by the Iwo-jima Islands and an inactive island arc represented by the Ogasawara Islands run parallel to each other. Ogasawara Trough is situated between the two. The Ogasawara Islands and the Daito Ridge are similar from a geochronological and paleontological point of view

(KANEOKA et al., 1970; OZIMA et al., 1980). If the Ogasawara Islands and the Daito Ridge were separated by the opening of the northern part of the Philippine Sea, it seems difficult to explain why large inactive island arc (the Ogasawara Islands) exist seaward of the active island arc (the Iwo-jima Islands). However, the situation would be explained if we assume the process of the formation of new subduction seaward of the Ogasawara Islands. The Ogasawara Islands which belonged to the Marcus Necker Seamount chain and moved to the old Ogasawara Trench—now called Ogasawara Trough—by the plate motion and a new subduction took place seaward of Ogasawara Islands.

Formation of a new subduction zone explained above causes a jump of the plate boundary and results in the "transfer of the plate" for the seamount. That is to say, the Ogasawara Islands which had belonged to the Pacific Plate have been transferred to the Philippine Sea Plate.

Unlike the Mariana Trough, heat flow in the Ogasawara Trough is lower than the average value (Results of heat flow measurement in the Ryohumaru cruise 1974 77— private communication from Prof. H. Kinoshita, Chiba Univ.—, and KH 78–2 cruise). This fact also seems to support the view that the Ogasawara Trough is a relic of trench.

We would like to express sincere thanks to the organizers of the Oji International Seminar for giving us a chance to present the paper.

We wish to thank Prof. K. Kobayashi, S. Uyeda, and R. Carlson for reading the manuscript and for giving many valuable advices.

We also wish to thank the attendants at the Oji Seminar for valuable discussions.

REFERENCES

BARAZANGI, M. and J. DORMAN, World seismicity maps compiled from ESSA, Coast and Geodetic Survey, epicenter data, 1961–1967, *Bull. Seism. Soc. Am.*, **59**, 369–380, 1969.

ELSASSER, W. M., Sea-floor spreading as thermal convection, *J. Geophys. Res.*, **76**, 1101–1112, 1971.

FURUMOTO, A. S., G. P. WOOLLARD, J. F. CAMPBELL, and D. M. HUSSONG, Variation in the thickness of the crust in the Hawaiian Archipelago, in *The Crust and Upper Mantle of the Pacific Area*, edited by L. Knopoff, C. Drake, and P. Hart, *AGU Monogr.*, **12**, 94–111, 1968.

HEEZEN, B. C., J. L. MATTHEWS, T. CATALANO, J. NATLAND, A. COOGAN, M. THARTP, and M. PAWSON, Western Pacific guyots, in *Initial Reports of the Deep Sea Drilling Project*, edited by B. C. HEEZEN and I. D. MACGREGOR, *Vol. 20*, pp. 653–723, U. S. Government Printing Office, Washington, D.C., 1973.

KANEOKA, I., N. ISSHIKI, and S. ZASHU, K-Ar ages of the Izu-Bonin Islands, *Geochem. J.*, **4**, 53–60, 1970.

KATSUMATA, M. and L. R. SYKES, Seismicity and tectonics of the Western Pacific: Izu-Mariana-Caroline and Ryukyu-Taiwan Regions, *J. Geophys. Res.*, **74**, 5923–5948, 1969.

McKENZIE, D. P. and C. BOWIN, The relationship between bathymetry and gravity in the Atlantic Ocean, *J. Geophys. Res.*, **81**, 1903–1915, 1976.

MOGI, A. and K. NISHIZAWA, Breakdown of a seamount on the slope of the Japan Trench, *Proc. Japan Acad.*, *Ser. B*, **59**, 257–259, 1980.

MORGAN, W. J., Gravity anomalies and convection currents, *J. Geophys. Res.*, **70**, 6175–6204, 1965.

NOMURA, M, Marine geomagnetic anomalies with intermediate wavelengths in the Western Pacific Region, *Bull. Ocean Res. Inst., Univ. Tokyo*, **11**, 1–42, 1979.

OGAWARA, H. and Y. KONO, Thinning of the oceanic lithosphere—evolution of the Hawaiian-Emperor Chain, *J. Seism. Soc. Japan*, **34**, 385–400, 1981(in Japanese).

OZIMA, M., Y. TAKIGAMI, and I. KANEOKA, ^{40}Ar–^{39}Ar geochronological studies on rocks of Deep Sea Drilling Project Sites 443, 445, and 446, in *Initial Report of the Deep Sea Drilling Project*, edited by G. de VRIES KLEIN, K. KOBAYASHI *et al.*, *Vol. 58*, pp. 917–920, U. S. Government Printing Office, Washington, D.C., 1980.

WATTS, A. B. and M. TALWANI, Gravity anomalies of deep sea trenches and their tectonic implications, *Geophys. J. R. Astr. Soc.*, **36**, 57–90, 1974.

YOSHII, T., Upper mantle structure beneath the North Pacific and the marginal seas, *J. Phys. Earth*, **21**, 313–328, 1973.

Geophysics and Metal Logenesis

Accretion Tectonics in the Circum-Pacific Regions, edited by M. Hashimoto and S. Uyeda, 335–344.

Forces Acting on the Subducted Lithosphere Revealed by the Geometry of Deep-Seismic Zones

Yoshiaki IDA

Ocean Research Institute, University of Tokyo,
Nakano-ku, Tokyo 164, Japan

Elastic bending of the descending lithosphere was calculated and compared with the observed geometry of Benioff-Wadati zones. The lithosphere was assumed to be a thin elastic plate with an effective elastic thickness of 20 km. Negative buoyancy and other probable forces acting on the subducted slab are not strong enough by themselves to produce such a steep dip of the slab as observed in the Mariana arc. On the other hand, the dip angle of a slab can be made even close to the vertical by an interplate pull force acting at shallow depths, if a dominant compression in the oceanic lithosphere coexists. The observed variation of dip angles over a wide range is explained by the change of the interplate interaction. In particular, a steep inclination occurs when two plates tend to be pulled from each other. The force system to cause a steep dip is realized if excess material, such as remelted oceanic crust is supplied to the mantle wedge above the slab.

1. Introduction

The geometry of subducted lithosphere, determined from the distribution of deep earthquake foci (e.g., ISACKS and BARAZANGI, 1977), reveals a wide variety of dip angles from as low as $10°$ in the Peru-Chile arc to almost vertical in the Mariana arc. What kind of forces act on the subducted slab, and govern their dip angles? The motivation to this paper is to answer this question, and to help us understand mechanical structure of convergent plate boundaries, and the dynamics of accretion tectonics.

Dynamic and thermal considerations of subduction zones predict that the slab should be subject to several forces (DAVIES, 1980). There may be a buoyancy force due to the density difference between the subducted slab and surrounding mantle. The motion of the slab will be accompanied by viscous drag. If the pressure is different for the upper and lower surfaces of the slab, such a pressure difference will tend to bend or stretch the slab. In the shallower part of the slab, some force that represents the mechanical interaction of the two plates would also act. In this paper, we try to evaluate the effects of these possible forces on the geometry of the descending slab, by a simple calculation of elastic bending.

The analysis of bending requires information on the rheological property of the lithosphere, which has been determined in several ways (FORSYTH, 1979). Both field observations of lithospheric flexure (BODINE *et al.*, 1981) and laboratory experiments of rock deformation (GOETZE and EVANS, 1979) have revealed that the lithosphere should involve some inelastic deformation, especially in its deeper part. Here we employ a simple model in which some portion of the lithospheric thickness is purely elastic, while the rest is able to flow and relax stress quickly enough. This model is quite tractable mathematically, because the model is equivalent to a thin elastic plate whose thickness is effectively reduced by the presence of the ductile part.

It has been argued, however, that such a simplification is not always adequate to study the deformation of the lithosphere. The elastic plate sometimes involves bending stresses too large to be supported by the slab. To avoid this difficulty, several calculation that have taken visco-elasticity, plasticity or fractured layering of the lithosphere into account have been made (BODINE and WATTS, 1979; CAREY and DUBOIS, 1981; LAGO and CAZENAVE, 1981). The rheological properties of the slab have also been discussed following the recognition of double seismic zones (FUJITA and KANAMORI, 1981). For instance, the slab was assumed to be elastico-plastic by ENGDAHL and SCHOLZ (1977), and visco-plastic by SLEEP (1979), in interpreting the earthquake mechanisms in the upper and lower layers of double seismic zones.

Nevertheless we here treat the simple elastic model, because inelastic property always contains some ambiguous parameters that have not yet been well established. The present simple model gives a qualitative answer to the question of what forces are effective in producing the ovserved geometry of slabs. At least, the study would be helpful as a preliminary step to more sophisticated models.

In this paper, the lithosphere is regarded as a thin elastic plate. Similar mathematical techniques have been employed by many authors (e.g., WATTS and TALWANI, 1974; BODINE and WATTS, 1979). In those treatments, it is ordinarily assumed that bending involves only infinitesimal displacements so that the deformed plate is almost held along the horizontal line. The present treatment is free from this restriction, and it is able to represent any deformation with finite angle to the horizontal. Such a generalization is essential for the study of the slab geometry.

2. Method of Calculation

We calculate the bending of the lithosphere, assuming that the lithosphere behaves like a two-dimensional thin elastic plate whose elastic property is specified by single parameter D called flexural rigidity:

$$D = H^3 E/12(1 - v^2) \tag{1}$$

where E and v are the Young's modulus and Poisson's ratio, respectively, and H is the effective thickness, all of the elastic plate. It is not reasonable to choose as H the entire thickness of plate down to the low-velocity zone. The rheological property of mantle rocks at high temperatures suggests that only upper 20 to 30 km of the lithosphere could be elastic without dominant flow (GOETZE and EVANS, 1979). More direct observations on lithospheric flexure give values of D ranging from 10^{23} to 10^{24} N.m for old lithosphere in the vicinity of most trenches (BODINE et al., 1981).

For the purpose of demonstration, we tentatively adopt a value of $D = 10^{23}$ N.m, corresponding to an effective elastic thickness of 20 km. It is noted here that this value represents the lower limit of D and thus a more or less too weak lithosphere. We can convert the result of our calculations quite easily for other choice of D, because all the equations used here can be rewritten in terms of dimensionless variables by use of the units in Table 1. For a new choice of D, revised values of variables are determined by multiplying the scale factors associated with the change in the units.

In the approximation for a one-dimensional thin elastic plate, elastic bending is described by the following equation that relates the curvature β to the bending force T_n at each point on the plate;

$$D(d\beta/ds) = -T_n. \tag{2}$$

This equation is obtained from the condition of moment balance coupled with the definition of flexural rigidity, i.e., the ratio of the bending moment to the curvature. Here ds is a line segment along the plate; namely s is the length along the plate measured from a reference point. The bending force T_n corresponds to the shear stress supported on a hypothetical cross section perpendicular to the plate, and thus it represents the normal (i.e., perpendicular to the plate) component of the total force that acts across the cross section. T_s is the tangential (i.e., parallel to the plate) component of the same force, corresponding to the extensional ($T_s > 0$) or compressive ($T_s < 0$) stress sustained by the lithosphere.

Let (x, y) represent the position of an arbitrary point on the plate. We then have the following geometrical relationships that are equivalent to the definition of curvature:

$$(d^2x/ds^2) = -\beta(dy/ds) \tag{3}$$
$$(d^2y/ds^2) = \beta(dx/ds). \tag{4}$$

Furthermore the condition of force balance yields

$$(dT_s/ds) = \beta T_n - f_s \tag{5}$$
$$(dT_n/ds) = -\beta T_s - f_n \tag{6}$$

where f_s and f_n are, respectively, the tangential and normal components of the net force that is externally applied to the unit length of the plate.

Equations (2) to (6) constitute a set of ordinary differential equations for five unknown variables β, x, y, T_s, and T_n. These variables are thus determined as a function

Table 1.   Units of the variables used in the numerical calculation.

Quantity	Variable	Unit	Tentative scale
Length	x, y, s, H	L	100 km
Curvature	β	$1/L$	0.01/km
Stress	f_s, f_n	D/L^3	1 kbar
Traction	T_s, T_n	D/L^2	100 kbar km
Density	ρ	D/gL^4	0.1 g/cm^3

The units consist of the flexural rigidity D and kinematic thickness L of the lithosphere, and the gravitational acceleration g. The tentative scales, used in deriving Figs. 1 to 3, correspond to $D = 10^{23}$ N.m, $L = 100$ km and $g = 10^3$ cm/s^2.

of s, if the force (f_s, f_n) is given at each point. For the distribution of f_s and f_n, we may take any form that is prescribed by a thermal or mechanical consideration of subduction dynamics. Actual calculations were made, using dimentionless variables based on the units in Table 1. The dimensionless variables scaled by those units satisfy the same equations (2) to (6) with $D = 1$.

The calculation also requires boundary conditions. Equations (2) to (6) completely determine the unknown variables as a function of s, if the values of β, T_s, T_n, x, y, dx/ds and dy/ds are specified at an end point, say at $s = 0$. Among them, the values of x and y at $s = 0$ simply fix the origin of coordinate system, and have no physical meaning. For the other variables, reasonable values at $s = 0$ were selected so as to reproduce the topography of outer rise. Namely we first calculated the bending of the oceanic plate, starting from a suitable point. In this calculation, the buoyancy force due to the density difference between sea-water and the mantle was substituted into f_s and f_n. We selected initial conditions that reproduced reasonable height and location of outer rise (WATTS and TALWANI, 1974), and we continued the calculation toward the earth's interior only for these cases.

This criterion could determine only a permissible range of the boundary conditions, but the ambiguity turned out not to seriously influence the geometry of the slab in the mantle. Therefore throughout the present paper we use as representative boundary conditions $\beta = 10^{-4}/\mathrm{km}$, $dx/ds = 1$, $dy/ds = 0$, and $T_n = 0$ at distance 250 km seaward of the deep-sea trench axis. Here we take the x-axis horizontal, and the y-axis vertical upward. The boundary value of T_s will be regarded as a variable parameter, because this quantity, which represents a tectonic stress in the oceanic plate, has a significant influence on the bending of the slab.

The problem of bending is often solved, using other boundary conditions that specify, for instance, x, y, dx/ds and dy/ds at the both ends of the plate. In the present treatment boundary conditions are concerned with only one end at $s = 0$, which is subject to twice requirements involving the quantities up to higher derivatives. We employ this type of boundary conditions, because we do not know the state at the other end, which is situated in the mantle. Fortunately the result of calculation again shows that the bending geometry is rather insensitive to conditions at the deeper end.

3.   Results

The geometry of the slab was first calculated for the simple case where no force acts on the slab after it enters the earth's interior. In this case, the slab simply continues to sink, without changing the initial low dip angle significantly. In fact, the elastic beam slightly waves at relatively shallow depths, but such a behavior does not affect the gross geometry of the slab. It is thus assured that the variety of dip angle reflects the presence of some forces acting on the slab. We here demonstrate the effects of some typical forces on slab bending.

Figure 1 displays the effect of negative buoyancy due to the density difference between the slab and the surrounding mantle. Let us suppose that a block of material of thickness L is identified with the slab and subject to negative buoyancy. We here distinguish the kinematic thickness L of the slab from the elastic thickness H. For

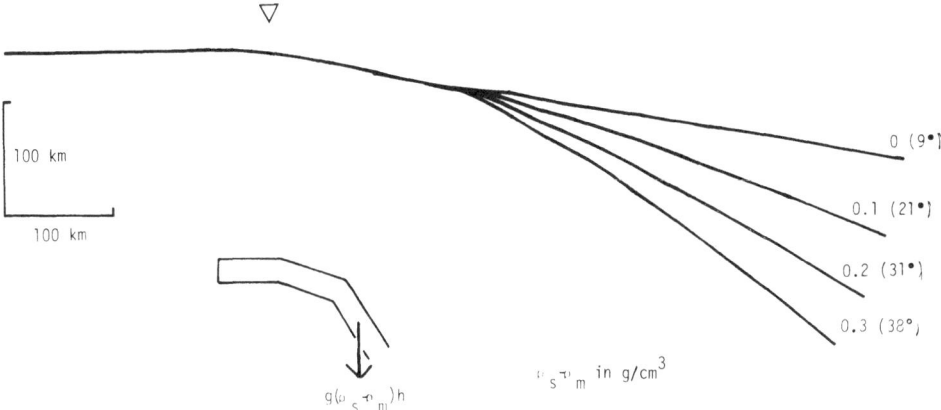

Fig. 1.   The effect of negative buoyancy on the bending of a slab. The mean density difference between the slab and the surrounding mantle is given in g/cm³ on each curve. The final dip angle is shown in parenthesis.

simplicity, it is assumed that the density difference is distributed uniformly within the thickness L, and that the negative bouyancy becomes effective only after the slab has entered the mantle and moved over a certain distance, which is also put to be L. We are assuming $L = 100$ km in Fig. 1.

In Fig. 1, numerical values assigned to the curves show density difference in g/cm³. The figures in parenthesis give resulting dip angles. The triangle marks the position of the deep-sea trench.

In Fig. 1, we find that negative buoyancy has an appreciable effect on the dip angle. It should be noted, however, that the density difference represents a certain average over the entire thickness of slab, and thus is probably less than 0.1 g/cm³ (DAVIES, 1980). Therefore the effect of the negative buoyancy is quantitatively too small to explain the observed wide distribution of dip angles, especially very high dip angles close to the vertical.

We next evaluate how much the bending geometry could be influenced by the force that pushes the slab obliquely downward. This force results from a non-isostatic pressure difference between the upper and lower surfaces of the slab at the same depth in the mantle. Such a pressure difference may be created by non-uniform distribution of crustal weight, stress field, or mantle flow, even if the effect is not often taken into account in the ordinary analysis of subduction dynamics (DAVIES, 1980). Later we shall consider an increased pressure in the mantle wedge above the slab due to the accumulation of excess material.

The effect of the pressure difference across the slab is given in Fig. 2. Here it is assumed that the prescribed pressure difference, which is denoted in kbar by the number attached to each curve, acts on each slab segment only after it has moved over distance L from the trench. This figure shows that the effect is significant, only if the pressure difference has a magnitude greater than about 1 kbar. We note that the pressure difference of 1 kbar corresponds to a vertical movement of the crust of about 4 km. Therefore it is unlikely that the pressure difference that can be maintained in the

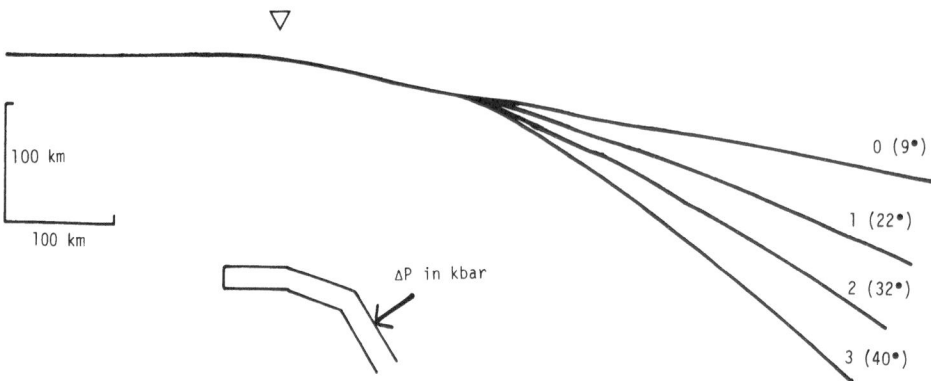

Fig. 2.   The effect of over-slab pressure on the bending of the slab. The pressure is given in kbar on each curve.

mantle is much greater than 1 kbar, and produces very high dip angles.

In addition to the normal force due to the pressure difference, there can be a tangential force due to a viscous drag resisting the motion of slab. Our calculations show, however, that the viscous force has much less effect on the slab geometry than the pressure difference. The slab also may be subject to some other type of forces in its deeper part, as has been proposed by DAVIES (1980). However the sagging of slab in the mesosphere cannot contribute to steeper dip by the nature of the force. The density change associated with the olivine-spinel phase transition nor makes the curvature significantly greater.

It is thus derived from our calculation that observed high dip angles can not be explained as the effect of likely forces expected in the earth's interior. This is true even if a simple superposition of those effects is considered. In the later section, this conclusion will be examined more carefully with the consideration of uncertain rheological property of the slab. Here we continue the study of bending, still holding the already adopted assumption on the rheology.

In seeking solutions with a high dip angle, some calculations were made for more complicated force systems that additionally included forces acting at shallower depths. In the observed Benioff-Wadati zones that have high dips (e.g., ISACKS and BARAZANGI, 1977), the steep dips start rather abruptly, after the slab has sunk through a few hundred kilometers at relatively shallow dip. This seems to suggest that some force, working at shallower depth, may play an important role in increasing the dip angle. The effect of such interplate interaction was examined by our numerical calculation.

First the tangential component that is regarded as a frictional resistance in the plate boundary does not make an appreciable contribution to the slab bending. The normal component that pushes the slab is also not significant. However, a normal force that pulls the slab sideways has a marked effect, when combined with a compressive stress in the oceanic lithosphere. Figure 3 shows how effectively the dip angle is increased by rather small changes in normal pull force. Here the pull force is assumed to

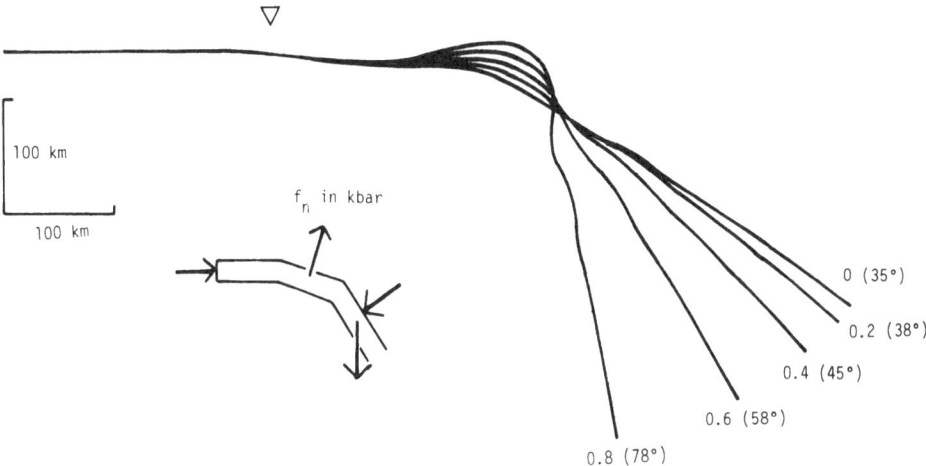

Fig. 3.   The effect of an interplate pull force on the bending of the slab. In addition to the pull force, given in kbar on each curve, a negative buoyancy of 0.1 g/cm^3 and a pressure difference of 1 kbar are acting on the deeper part of the slab. A compressive stress of 1 kbar is also sustained in the oceanic lithosphere.

be uniformly distributed in the upper part of the slab, over the length L. In this figure, the normal pull is given in kbar on the curves. Our result suggests that even very steep dip angles can be obtained by adjusting the magnitude of an interplate pull force. The pull force is a consept similar to the "suction" that was introduced in the treatment of the driving forces of plate motion (FORSYTH and UYEDA, 1975).

It is emphasized that the high dip angles obtained in Fig. 3 are not caused by the pull at the plate boundary alone. A compressive stress in the oceanic lithosphere and some force, such as negative buoyancy and pressure difference, in the mantle are also necessary, in addition to the interplate force. Such combination of forces could successfully interpret the observed wide distribution of slab dip angles. In the next section, we consider possible origin and geological meaning of such a force system.

4. Discussion

It has been shown in the previous section that many of the probable forces acting on the slab are not strong enough to cause the high dip angles of some descending slabs. The results in Figs. 1 and 2 show that, to cause high dips, the forces due to either negative buoyancy or pressure difference must be about ten times stronger than reasonable estimates. In fact, this conclusion depends on the assumed bending strength of the slab. According to the units given in Table 1, the effect of the forces (f_s, f_n) is inversely proportional to the flexural rigidity D. Therefore we have another alternative that the value of D should be an order of magnitude smaller, i.e., $D = 10^{22}$ N.m.

However such a small flexural rigidity is not acceptable, because the value of D employed here is close to the lower limit for old trenches (BODINE et al., 1981). It is reminded that the Pacific plate approaching to the Mariana trench is older than 130 m.y., and thus the Mariana trench should have greater flexural rigidity than we have assumed. The flexural rigidity could be reduced by temperature elevation in the mantle. The dip is, however, almost determined by the flexure in the shallower part of the slab, which is not significantly heated (e.g., ANDERSON et al., 1978). Therefore we can not expect a very marked temperature effect.

We have already shown that some forces other than the negative buoyancy and the pressure difference are less effective on the bending. After all, we have to recognize that high dip angles are not realized without such a combined effect as in Fig. 3.

This conclusion may not be justified if an inelastic rheology is dominant in the deformation. If the slab literally contains a purely elastic layer of thickness H, the layer should be subjected to fairly large elastic stresses. In the case where the density difference is equal to $0.1 \ g/cm^3$ (Fig. 1), or in the case where the pressure difference is equal to 1 kbar (Fig. 2), a maximum curvature of about $2 \times 10^{-3}/km$ occurs in the 200 km interval past the trench. This value of curvature should be accompanied by stresses, about 20 kbar on the top and bottom of the elastic layer, if $H = 20$ km. Such a large stress cannot be sustained entirely by elastic strength, and some inelastic flow must be involved in the deformation. Therefore the study including the inelastic effect is very important.

It does not seem to be true, however, that the above-estimated magnitude of stress is large enough to completely eliminate the elastic response of lithosphere. The inelastic effect may require some modifications of results obtained here. Nevertheless it would be still valid that no single force can explain the wide variety of the dip of slabs. In this context, the combined effect in the force system in Fig. 3 is worthy of further consideration.

In this force system, steep bending is caused by the combination of the following three forces; (1) upward pull at the plate boundary, (2) compressive stress in the oceanic lithosphere, and (3) some force in the mantle that acts to bend the slab downward. Here we employ the pressure difference as the last force for the reason mentioned below. The force system is schematically displayed in Fig. 4.

Three such forces can be simultaneously produced in nature by a single tectonic event, which has been proposed (IDA, 1978, 1981) to explain various processes systematically in subduction zones. The desired force system can be produced if excess material is injected into the wedge shaped mantle volume above the slab (See Fig. 4). Namely such excess mass pushes the slab, and causes the pressure to bend the slab downward. To balance the pressure that pushes the oceanic plate away seaward from the trench, a compressive stress should appear in the oceanic lithosphere. The volumetric addition to the wedge stretches the overlying landward plate, so that the two plates tend to separate, giving rise to a pull force between the two plates in the plate boundary.

Positive gravity anomaly is commonly observed in island-arcs and back-arc basins (WATTS and TALWANI, 1974). This gravity anomaly has been interpreted as reflecting the presence of such excess material (IDA, 1978).

A likely source for the excess material is oceanic crust that has been subducted, remelted and stored due to its buoyancy in the mantle above the slab. The descending lithosphere can continuously supplies remelted crust to the mantle and gradually accumulate this excess material in the volume above the slab.

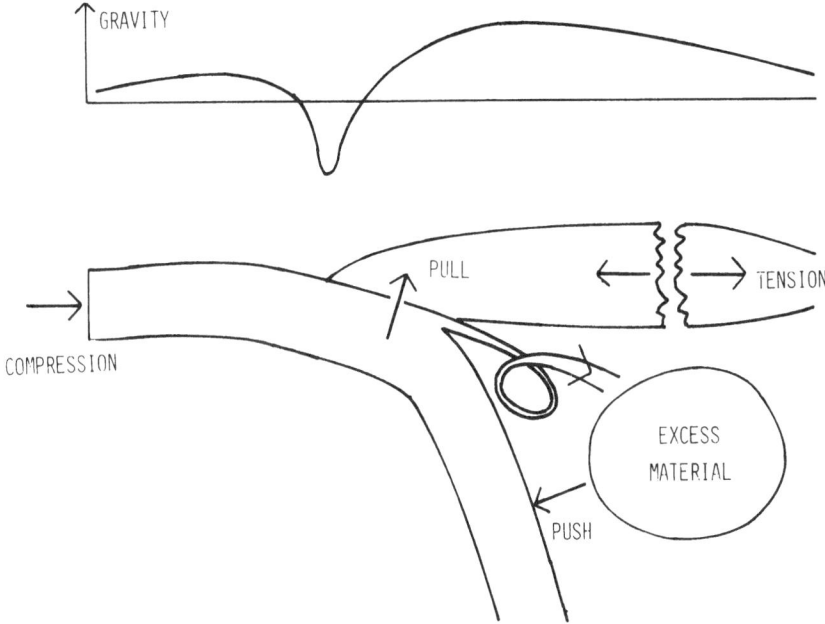

Fig. 4.   A force system that causes a high slab dip angle. The system consists of the pull in a plate boundary, a compressive stress in the oceanic plate, and a pressure pushing the slab downward. Those forces are simultaneously produced when excess material, such as remelted oceanic crust, is injected into the mantle wedge above the slab.

There are other possibilities of the source for the excess material. UYEDA and KANAMORI (1979) and FUJITA and KANAMORI (1981) referred to eastward flow of the upper mantle relative to the plate. Such a mantle flow may accumulate material in the end of the mantle separated by the slab. ROOTS (1982) also proposed a convection that involves material coming from the center to the edge of the continents. At present, we do not have sufficient evidence to determine which of these possibilities is most likely. In the following, we here examine the first mechanism due to melting of subducted oceanic crust in more detail.

The melting of subducted crust could be achieved by the following process (IDA, 1981). Since the subduction of cool lithosphere causes a horizontal temperature gradient in the mantle, violent convection of melt should develop, through narrow paths between still solid rock. Such convection could carry the heat necessary for successive melting of the subducted crust, resulting in a well-developed asthenosphere with abundant melt, an accepted feature landward of the trench.

It is known that high dip angles, as found in the Mariana region etc., occur in conjunction with tensional stress landward of the trench, and often associated with back-arc spreading (UYEDA and KANAMORI, 1979). The tensional stress landward of the trench has already been explained by the proposed force system in Fig. 4. It is also noted that the presence of compressive stress seaward of the trench, as pointed out for the Izu-Bonin and southern Mariana trenches by BODINE and WATTS (1979), is consistent with the present model. Back-arc spreading is understood in this model to be started by the accumulation of excess material. This increases tension in the arc more and more, and finally tears that part of lithosphere to allow a volcanic extrusion corresponding to the formation of a new mid-ocean ridge system (IDA, 1978, 1981).

The author thanks Prof. S. Uyeda of University of Tokyo, Dr. D. Roots of MacQuarie University and Dr. K. Fujita of Michigan State University for the critical reading of the manuscript and many valuable comments.

REFERENCES

ANDERSON, R. N., S. E. Delong, and W. M. SCHWARZ, Thermal model for subduction with dehydration in the downgoing slab, *J. Geol.*, **86**, 731–739, 1978.

BODINE, J. H. and A. B. WATTS, On lithospheric flexure seaward of the Bonin and Mariana trenches, *Earth Planet. Sci. Lett.*, **43**, 132–148, 1979.

BODINE, J. H. M. S. STECKLER, and A. B. WATTS, Observations of flexure and the rheology of the oceanic lithosphere, *J. Geophys. Res.*, **86**, 3695–3707, 1981.

CAREY, E. and J. DUBOIS, Behaviour of the oceanic lithosphere at subduction zones; plastic yield strength from a finite-element method, *Tectonophysics*, **74**, 99–110, 1981.

DAVIES, G. F., Mechanics of subducted lithosphere, *J. Geophys. Res.*, **85**, 6304–6318, 1980.

ENGDAHL, E. R. and C. H. SCHOLZ, A double Benioff zone beneath the central Aleutians: An unbending of the lithosphere, *Geophys. Res. Lett.*, **4**, 473–476, 1977.

FORSYTH, D. W., Lithospheric flexure, *Rev. Geophys. Space Phys.*, **17**, 1109–1114, 1979.

FORSYTH, D. W. and S. UYEDA, On the relative importance of the driving forces of plate motion, *Geophys. J. R. Astron. Soc.*, **43**, 163–200, 1975.

FUJITA, K. and H. KANAMORI, Double seismic zones and stresses of intermediate depth earthquakes, *Geophys. J. R. Astron. Soc.*, **66**, 131–156, 1981.

GOETZE, C. and B. EVANS, Stress and temperature in the bending lithosphere as constrained by experimental rock mechanics, *Geophys. J. R. Astron. Soc.*, **59**, 463–478, 1979.

IDA, Y., Oceanic crust in the dynamics of plate motion and back-arc spreading, *J. Phys. Earth*, **26**, S55–S67, 1978.

IDA, Y., Thermal circulation of partial melt and volcanism behind trenches, *Oceanol. Acta*, *v.N°SP*, 241–244, 1981.

ISACKS, B. L. and M. BARAZANGI, Geometry of Benioff zones: Lateral segmentation and downward bending of the subducted lithosphere: in *Island Arcs, Deep See Trenches and Back-Arc Basins*, pp. 99–114, edited by M. Talwani and W. C. Pitman III, Am. Geophys. Union, Washington, D.C., 1977.

LAGO, B. and A. CAZENAVE, State of stress in the oceanic lithosphere in response to loading, *Geophys. J. R. Astron. Soc.*, **64**, 785–799, 1981.

ROOTS, R. D., Convection hierarchy tectonics—A speculative synthesis of continental and oceanic geology (preprint), 1972.

SLEEP, N. H., The double seismic zone in downgoing slabs and the viscosity of the mesosphere, *J. Geophys. Res.*, **84**, 4565–4571, 1979.

UYEDA, S. and H. KANAMORI, Back-arc opening and the mode of subduction, *J. Geophys. Res.*, **84**, 1049–1061, 1979.

WATTS, A. B. and M. TALWANI, Gravity anomalies seaward of deep-sea trenches and their tectonic implications, *Geophys. J. R. Astron. Soc.*, **36**, 57–90, 1974.

Accretion Tectonics in the Circum-Pacific Regions, edited by M. Hashimoto and S. Uyeda, 345–348.

Overthrust and Underthrust as a Fundamental Process of Continental Accretion

Wenyou ZHANG and Jiayou ZHONG

Institute of Geology, Academia Sinica, Peking, China

Underthrust and overthrust are originally relative displacement of two walls of a thrust fault. In the light of geomechanics, it is a tectonic expression of shear process. But now, many structural geologists only emphasize underthrust, and name the large-scale underthrust zone subduction or Benioff zone. They consider the island arc trench system along the margin of northwest Pacific Ocean and the mountain range arc-trench system on the margin of southeast Pacific Ocean are actual examples. Moreover, they deduce the older tectonic structures of Palaeozoic and pre-palaeozoic eras from these younger ones took place since Mesozoic and Cenozoic eras. On the other hand, they sometimes call the large-scale overthrust-line tectonic structure obduction zone, or as some people call it the flap, which is usually considered to be far less important than the subduction zone. Further they use this concept to interpret the formation of tectonic-lithofacies and metallogenic zones and the many phenomena of geophysics and geochemistry.

What are the factors which control the pattern of a thrust fault zone, the overthrust or underthrust? Not much attention has been paid to this problem yet. In relation to the field observations, model experiments were done by the senior author in the 1950's. It has been proved that the factor which affects the pattern of a thrust fault is not the source of force, but the boundary conditions. Recently we repeated another series of model experiments, considering the differences in density and strength between the crust and the upper mantle. Two kinds of mixtures in different proportion of vaseline and paraffin were used. Their density and strength are 1.03, 0.95 g/cm^3 and 144, 118 g/cm^2, respectively. We cast the materials into layers of various thickness, then put them into squeeze trough on a horizontal level side by side. A piston compressed the layers edgewise horizontally. The result showed that the lighter side is easily thrust over the denser side. It also indicates that the side of higher elevation, i.e. the thicker side, is easily thrust over the lower side, and the side of lower strength is easily thrust over the higher strength side. In other words, geological bodies with higher density, lower elevation and higher strength are liable to form the footwall of a thrust fault, i.e. the underthrust side. Besides, there is another important phenomenon especially to be noticed in our experiments. That is the inclination of the contact plane of two blocks with different density or strength. The variation of the dipping usually plays a very important role controlling underthrust or overthrrust. Generally the contact plane inclines towards the hangingwall of a thrust, even if the density or strength of the two sides are exchanged. This problem needs further study.

Then why do underthrust or subduction action, but not overthrust or obduction?

What they consider are nothing but that the oceanic crust is cooler than the continental crust, with higher Q-value, rigidity and density. In addition, Benioff seismic zone is dipping towards the continent, and the high pressure low temperature zone of a paired metamorphic belt lies on the ocean side and low pressure high temperature zone exists on the continental side. Furthermore, the zoning of magma shows a tendency to increase the acidity in ocean and alkalinity in continent, and the basic composition increases towards the ocean side. From the geomechanical point of view, all these phenomena can be explained in terms of seeing a continental crust thrusted over an oceanic crust. In the 1960's, we tried to interpret the cause of the formation of the andesite zone. We assumed that it was the mix-melted products of collision of continental and oceanic crusts with thrust slips and displacements influencing the mantle. Therefore the upwelling of mantle materials was probable. The above-mentioned explanation is as acceptable as the theory of underthrust. We also noticed that the zoning of magma of island arc-trench-backarc basin system on the northwest Pacific margin is relatively wide and clear, but the zoning of magma of mountain arc-trench system on the southeast Pacific margin, such as the Andes Mountains, is comparatively narrow and not so clear. These appearances may be due to the difference of dipping angles of thrust fault. The former is gentle and the latter is steep. This problem needs further study. With regard to the concept of "structure level," to this problem should be paid more attention. Because the present existing island arc-trench system is only the shallow level tectonic pattern, the deep level tectonic pattern may not be conordant with the shallow one. This point has been proved in our model experiments. So it is not proper to indiscriminately imitate the idea of plate tectonics and island arc-trench tectonic patterns in dealing with the problem of continents, especially the problem of deep structure level of Precambrian.

Both in Prof. J. Logan's rock mechanics experiments (A and M University, Texas, U.S.A.) and our model experiments it has been observed that the development of a fault zone usually varies with depth. In general, the fault zone in the upper position is more complicated that in the lower. Thus the present existed island arc-trench system might be corresponding to a deep geosuture on a continent. The ophiolitic suite on a continent, the representation of old oceanic crust, in the depth might become a line of beaded ultrabasic intrusive bodies.

Now we come back to the problem of overthrust or underthrust of a thrust fault zone. Obviously those who excessively stress underthrust or subduction zone and indiscriminately apply the present island arc-trench structure to Precambrian, have not properly thought over the problems of geomechanics and structure level mentioned above. Judging from the point of density and strength, it is hard to think that the oceanic crust with P-wave velocities of about 6.5 km/sec might underthrust into the mantle-crust mix which has P-wave velocities between 7.4 and 7.7 km/sec, and sometimes the underthrust might penetrate into such a depth as 700 km. It is just like a boat rowing on a river. It is obvious that the higher the head of the boat, the easier the resistance to be overcome, and the lower the head of the boat, the more difficult the boat to row. the continental crust is more liable to overthrust than the oceanic crust is to underthrust. Furthermore, the former is more easily pushed into the lighter layer such as atmosphere than the latter is driven into denser substratum such as the mantlecrust mix. Thus further study on overthrust and structure level is of great significance.

From geomechanical point of view, overthrust and underthrust are two components of same tectonic shearing displacement.

Field observations show that overthrust more frequently occurs than underthrust, because rock layers under compression are usually more easily squeezed upward than downward in case of buoyancy overcoming load pressure.

According to model analysis under copression, lighter layers usually thrust on denser ones, layers of lower strength on those of higher strength, and the original dip of the dividing plane between two different parts usually towards the overthrusting side.

With the above facts in view, it is obvious that under compression continental crust is more likely to thrust over the oceanic, rather than oceanic crust thrusts under continental one, because the former is higher and softer than the latter, besides the original contact plane usually dips towards continent.

Both in oceanic and continental crust under compression usually occur folding, thrusting, colliding together and further leading to continental accretion, while under tension, faulting, rifting, pulling apart and further leading to oceanic spreading with mantle rising. Thus, the former process may be named as "continentization" and the latter, "oceanization." In considering the two worldwide tectonic processes, it is probable that the occurrence of the so called passive continental margin with normal faulting and graben type depressing along the present Atlantic Ocean are mainly due to tension, whereas the so called active continental margin with thrust faulting and arc-trench type structures mainly due to compression. Thus, we may tentatively suppose that in future geological times Atlantic Ocean tends to further opening with middle oceanic spreading while the Pacific Ocean is subjected to further closing with marginal continental accretion.

The process of continental accretion by thrusting may be preliminarily classified into two types:

The former usually makes older continental blocks to fuse together as a united younger continent with older continental relics as marginal thrust zones and older oceanic crustal relics as suture lines such as the Qinghai-Xizang (Tibet) plateau and Indian block collided together during Cenozoic time with the Himalaya and Yalu-zagbo ophiolite zone as continental marginal zones and oceanic suture lines respectively. The successive continental collision between the Tarim-North China block and the Xizang-South China blocks from Paleozoic to Mesozoic times left the Kunlun-Qilian-Qinling ranges as continental marginal zones and oceanic suture lines.

The successive marginal growth of South China continental block (The Yangtze platform) due to Caledonian and Hercynian movements toward the East and South China Seas may be tentatively assumed as an example of adhesive process of continental accretion. Another example is the successive marginal growth of Siberia continental block with adhesion of the Jining, the Caledonian and the Hercynian geosynclinal folding zones from north to south and finally united with Tarim-North China block. The recent tectonic framework around the Pacific Ocean including the trench-island arc basin system in the northwestern Pacific and the trench-mountain-arc system in the eastern Pacific shows that the adhesion process of continental accretion is still going on.

Thus our tentative concluding remarks may reached. In courses of continental accretion developments the collision process may be considered as the final stage of adhesion process. Under compression the thrusting usually begins along the contact zone of continental and oceanic margins and leads to the formation of arc-trench system sometimes with arc-back basins due to local tension. As compression continuing, the continents approach to each other with ocean contracting inbetween. Finally, they collide together. However it is only a crude assumption. Further geological and geophysical detail studies are still needed.

Accretion Tectonics in the Circum-Pacific Regions, edited by M. Hashimoto and S. Uyeda, 349–355.
Copyright © 1983 by Terra Scientific Publishing Company (TERRAPUB), Tokyo.

Accretion Tectonics and Metallogenesis

Chikao NISHIWAKI* and Seiya UYEDA**

*Institute Intern. Min. Devel., Kamiide, Fujinomiya 418–02, Japan
**Earthquake Research Institute, Univ. of Tokyo,
Tokyo 113, Japan

An attempt is made to discover what the genetic implications to collision/accretion tectonics might be for metallogenesis. The distribution of all the major porphyry copper deposits in the southwest Pacific region seems to be confined in the arcs that were under the collision tectonics at the time of ore emplacement. The central mountain range of New Guinean subcontinent is selected as an area of typical collision/accretion terrane where many porphyry-type copper concentrations have been discovered and explored recently. The collision took place in the very late Cenozoic period and the zone is characterised by a broad terrane of imblication structure with numerous thrust faults. The zone also have many contemporaneous intrusive porphyry stocks, majority of which are mineralized in copper including several very large porphyry copper deposits. Apparent affinity of collision tectonics for certain subsurface metallogenesis of shallow to medium depth, such as porphyry-type disseminated deposit, is also well recognized here owing to its very young (3 to 1.5 MY) geology. High compressional horizontal stress environment during the period of collision and accretion is suspected to be one of the major factors which control the porphyry-copper metallogenesis.

UYEDA and KANAMORI (1979) and UYEDA (1981; 1982), pointed out the importance of the difference in the mode of subduction between the Mariana type and the Chilean type subduction zones. The Mariana type subduction is characterised by extensional stress conditions and is often associated with Kuroko or similar volcanogenic massive sulphide deposits, while the Chilean type subduction is under compressional stress environments and is frequently associated with porphyry copper deposits, e.g. those along the cordillera of the Pacific North and South Americas. Recently developed porphyry copper deposits in the Southwestern Pacific island arc region might have also been related to the compressional stress environments which prevailed at the time of their Cenozoic emplacement, and the origin of the compressional stress might have been related to collisions (UYEDA and NISHIWAKI, 1980) (see Fig. 1).

The tectonic environments at the time of ore emplacement of major 23 porphyry copper deposits in the Southwest Pacific region were investigated by NISHIWAKI (1981). He demonstrated that the distribution of porphyry copper deposits is confined in the zone of collision tectonics in the broad sense, and further found that 19 among 23 deposits were under compressional horizontal stress with its maximum direction roughly parallel to the estimated direction of plate convergence or the collision. Other

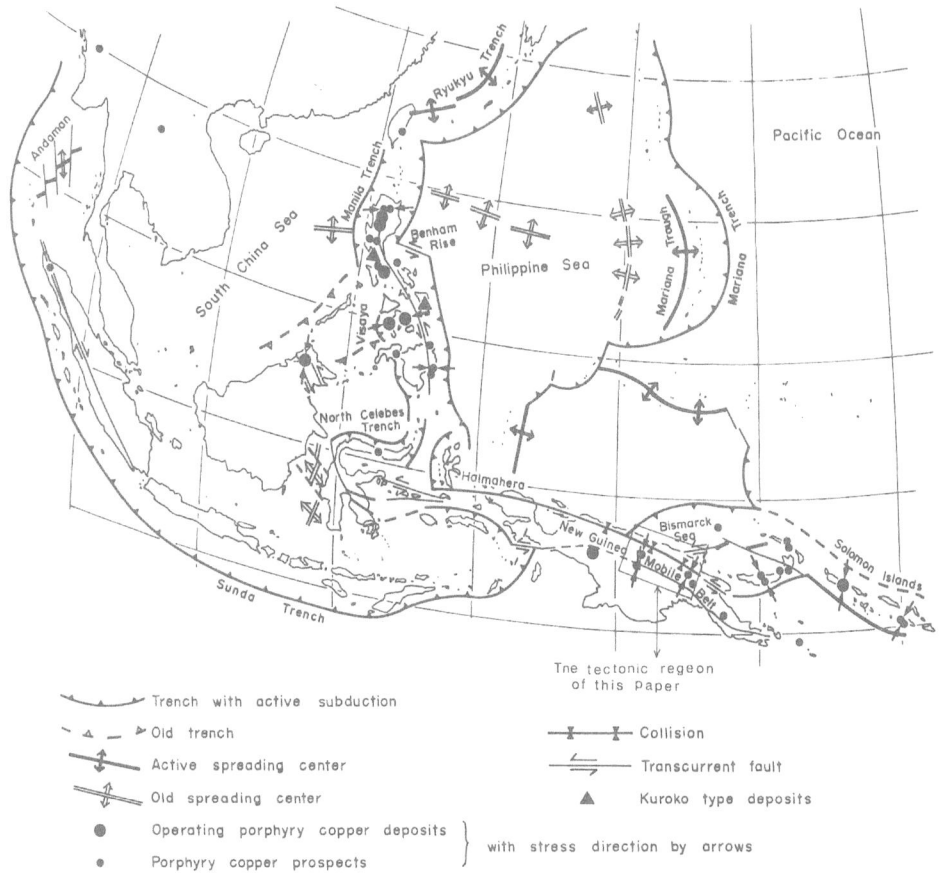

Fig. 1.   Distribution of porphyry copper deposits in the southwestern Pacific.

four deposits were developed along conspicuous transcurrent faults or their branches under the shear stress associated with their large strike-slip movement. This shear stress ultimately results in the compressional stress more or less parallel to the direction of the plate movement which often is roughly orthogonal to the general arc direction (NISHIWAKI and UYEDA, 1980, NISHIWAKI, 1981).

One of the most typical areas among these collision terranes is the central mountain reange of Papua-New Guinea at about 136°E, 3° to 4°S to 145°E, 6° to 7°S (Fig. 2). The range has altitude of 3,000 to 5,000 m, and is situated at the extreme nothern border of the northward moving continental Australian plate. The area has extensive upper Mesozoic shelf sediments on the upper Paleozoic base. These are covered again by well developed Tertiary limestone and thick clastic trough sediments with a few very pronounced ophiolite and melange zones. General strike of the entire system is roughly E to ESE (HAMILTON, 1979; Dow, 1977). Hamilton further recognized the collision between the northward moving Australian plate and a

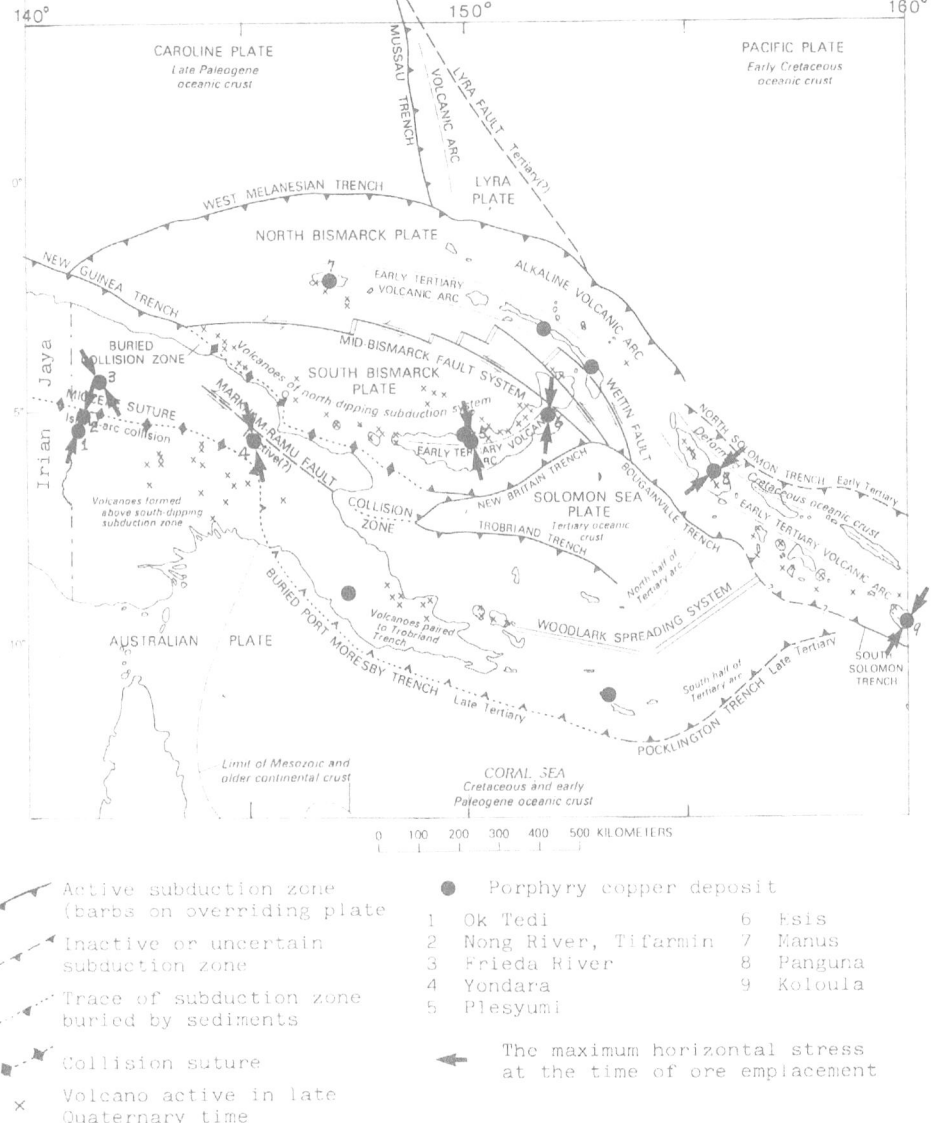

Fig. 2. Major tectonic elements and porphyry copper mines and prospects in Papua New Guinea and the Solomon archipelago. After Fig. 146 of HAMILTON (1980), the porphyry coppers superimposed by the author.

southward moving island arc at the southern edge of the Caroline plate, followed by a polarity reversal of subduction from northward to southward under the Australian plate (Fig. 3). This piggybacking of an island arc might have caused the high elevation of the mountain range.

The latest collision event took place in the Quaternary period. A broad zone of

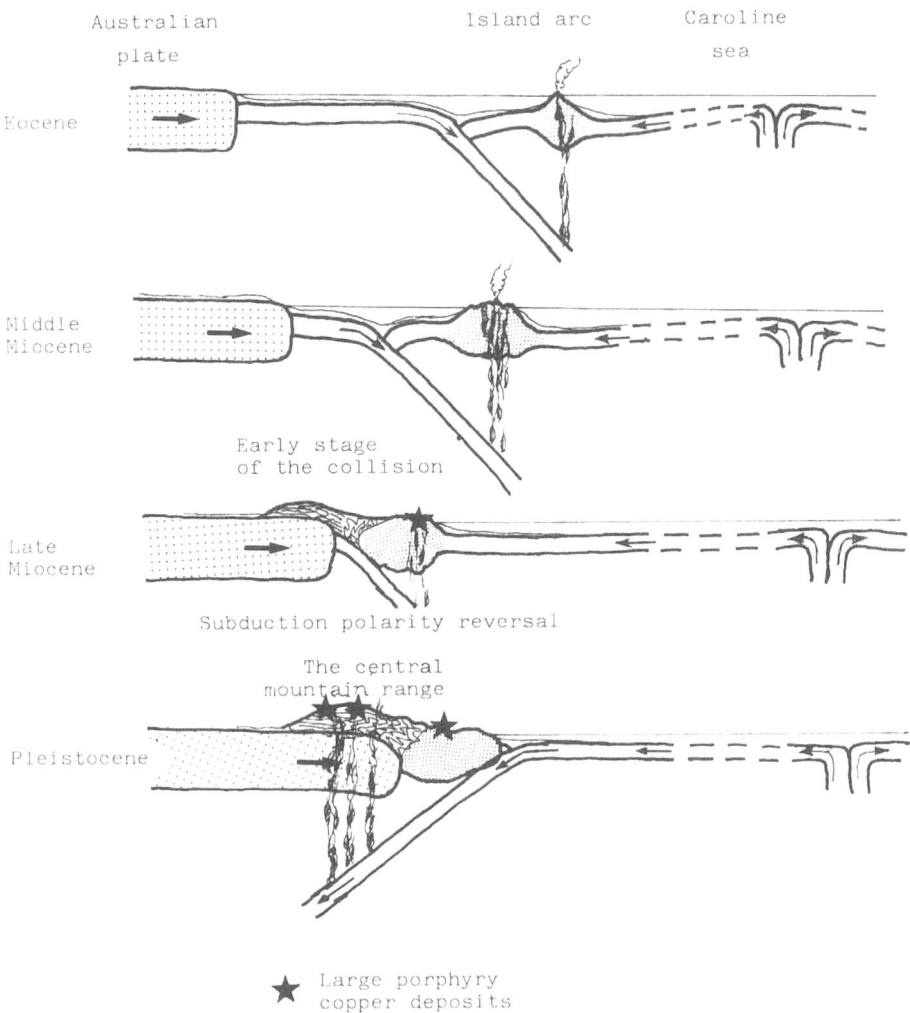

Fig. 3.   Schematic model of collision of the Australian plate with a Tertiary island arc and its relation to porphyry copper deposits.

numerous nappés and imblication structures of flat to low angle thrust faults is the most characteristic and common feature of surfacial or shallow manifestation of the collision. The direction of the maximum horizontal stress at that time can be inferred by the strike and dip of faults and folds, the direction of alignment, elongation or protrusion of intrusive stocks or ore bodies, and the orientation of dykes and veins. It is inferred here as N10°–20°E, nearly orthogonal to the general structural trend, the direction of the mountain range and the supposed arc.

Around Tfarmin (Fig. 4), Miocene limestone beds are frequently found to be thrusted over a not well lithified Quaternary gravel bed. There are many porphyry

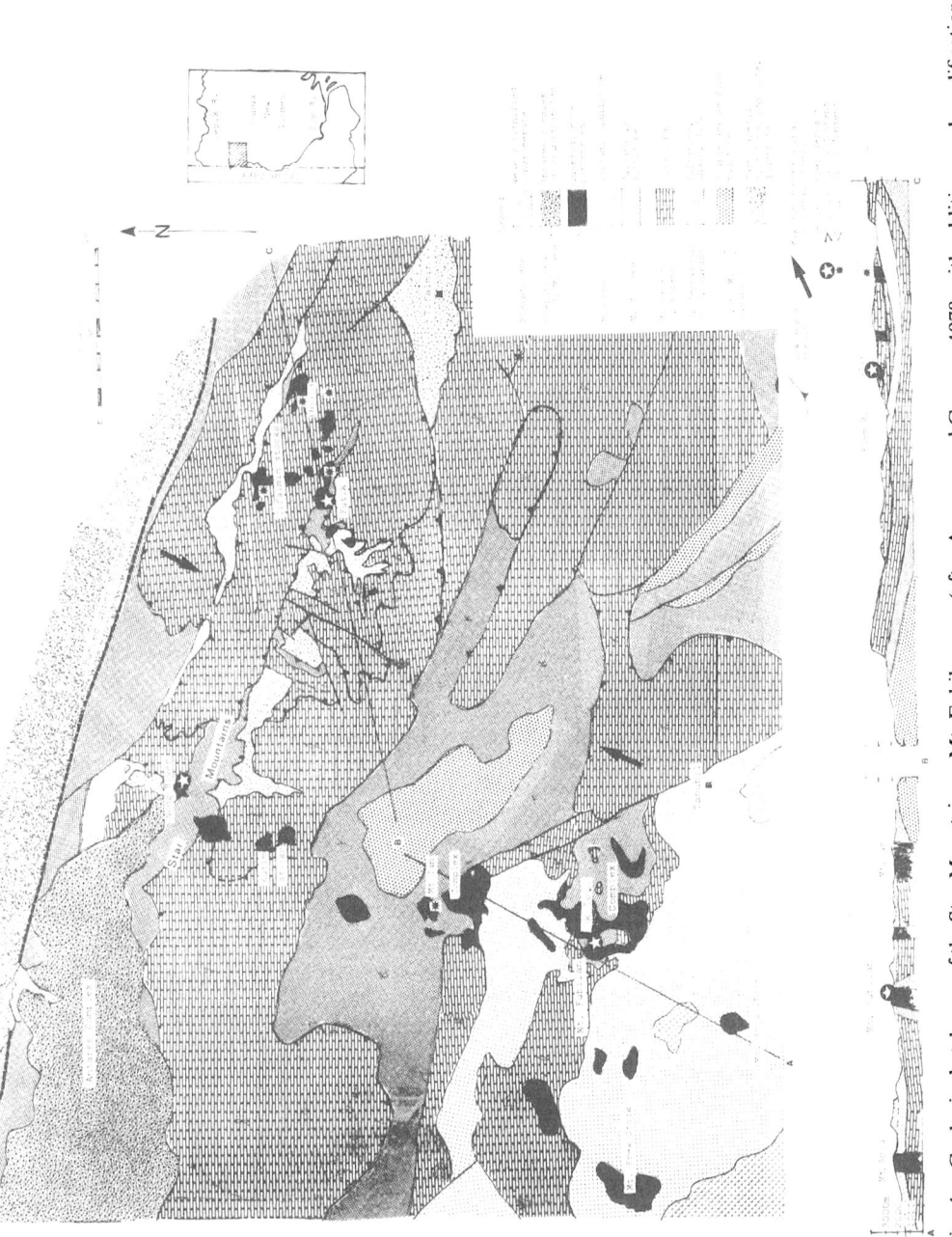

Fig. 4. Geologic sketch map of the Star Mountains—Mt. Fubilan area (after ARNOLD and GRIFFIN, 1978, with additions and modifications).

intrusive stocks in the overlying thrusted blocks. Some of them are associated with the typical porphyry-type copper mineralization ornamented by copper and/or iron skarn ore replaceing the Miocene limestone of the block. Age of these porphyries in the Tfarmin, Star Mountain to OK Tedi area is determined to be in the range of 3 to 1.5 m.y. (Page, 1975). Geology and mineral deposits of the area are well documented by McGee (1976), Arnold and Griffin (1978), Davies (1978), Howell et al. (1978), Asami and Britten (1980) and many otheres in addition to the authors cited before.

The occurrence of porphyry coppers in this collision/accretion terrane is not only frequent but also includes very large copper concentrations such as OK Tedi, Frieda River or Erzberg which may rank in the world's largest blacket of copper deposits. It can also be speculated that when erosion goes deeper the above-stated very broad zone of numerous faults may turn into a simpler, narrower and probably steeper boundary of accretion terranes. At the same time, the mode of copper concentration in the deep may be different from the large porphyry type and the copper content may be less.

It has been shown that the collision tectonics seems to have a remarkable affinity for certain subsurface metallognesis at shallow to medium depth, such as porphyry-type copper concentration. High compressional horizontal stress environment during the collision/accretion period is suspected to be one of the major factors controling the genetical implications. One of the authors further investigated the implications of the tectonic stress to the physico-chemical conditions in the case of shallow igneous stock intrusion, its solidification and related mineral concentration (Nishiwaki, 1982).

REFERENCES

Arnold, G. O. and T. J. Griffin, Intrusions and porphyry copper prospects of the Star Mountains, Papua New Guinea, *Econ. Geol.*, **73**, 785–795, 1978.

Asami, N. and R. M. Britten, The porphyry copper deposits at the Frieda River prospect, Papua New Guinea, *Mining Geol.*, *Spec. Issue*, **8**, 117–139, 1980.

Davies, H. L., History of the Ok Tedi porphyry copper prospect, Papua New Guinea; Part 1. The year 1966 to 1976, *Econ. Geol.*, **73**, 796–802, 1978.

Dow, D. B., A geological synthesis of Papua New Guinea, *Bur. Min. Res. Geol. Geoph. Australia, Bull.*, 201, 1977.

Hamilton, W., Tectonics of Indonesian region, *U.S. Geol. Surv., Prof. Paper,* 1078, 1979.

Howell, W. J. S., R. S. H. Fardon, R. J. Carter, and E. D. Bumstead, History of the Ok Tedi prospect, Papua New Guinea, Part 2. The year 1975 to 1978, *Econ. Geol.*, **73**, 802–809, 1978.

McGee, W. A., Summary of mineral exploration in the Star Mountains, *Geol. Surv. Papua New Guinea, Report*, 76/10, 1976.

Nishiwaki, C., Tectonic control of porphyry copper genesis in the southwestern Pacific island arc region, *Mining Geol.*, **31**, 131–145, 1981.

Nishiwaki, C. and S. Uyeda, Tectonic control of porphyry copper genesis; 4th. Joint Meeting, *Min. Met. Inst. Japan and Am. Inst. Min. Eng. (AIME)* Pre-print, Tech. Session A.1. p. 17, 1980.

Nishiwaki, C., Tectonic stress and metallogenesis, primarily with reference to porphyry copper genesis, *Mining Geol.*, **32**, 291–304, 1982.

Page, R. W., Geochronology of late Tertiary Quaternary mineralized intrusive porphyriesin the Star Mountains of Papua New Guinea and Irian Jaya, *Econ. Geol.*, **70**, 928–936, 1975.

Uyeda, S., Subduction zones and back-arc basins—A review, *Geol. Rundsch.*, **70**, 552–569, 1981.

Uyeda, S., Subduction zones: An introduction to comparative subductology, *Tectonophys*, **81**, 133–159, 1982.

UYEDA, S. and H. KANAMORI, Back-ark opening and the mode of subduction, *J. Geophys. Res.*, March 1979, 1049–1061, 1979.

UYEDA, S. and C. NISHIWAKI, Stress fierd, metallogenesis and mode of subduction, in *The continental crust and its mineral deposits*, edited by D. W. Strangway, *Geol. Assoc. Canada, Spec. Paper*, **20**, 323–339, 1980.

Index

Abashiri Tectonic Line, 92, 97, 124, 130
Abyssal tholeiite, 65, 223, 264, 272
Accretion, 3, 10, 27, 30, 38, 176, 193, 229, 314, 335, 347, 354
Accretionary complex, 214
Accretionary prism, 116, 204, 216, 226
Accretionary wedge, 29, 60, 136
Alkali basalt, 52, 71, 80, 97, 110, 196, 223, 250, 264, 272, 309, 317
Alkali igneous rocks, 71, 313
Allochthonous terrane, 3, 6, 11, 14, 31, 37, 146
Amalgamation, 22, 27, 30, 176
Arc-trench system, 98, 107, 116, 140, 150, 292, 345
Asthenosphere, 319, 323

Back-arc basin, 52, 54, 60, 69, 246, 294, 317, 342, 346
Batholith, 64
Butsuzo Tectonic Line, 210, 219, 232

Calk-alkali igneous rocks, 62, 64, 159, 216, 313
    collision type, 69, 73
    continental margin type, 69, 73
    intracontinent type, 69, 73
    island-arc type, 45, 69, 73
Chichibu belt, 211, 232,
Circum Hida belt, 181, 195
Collision, 6, 10, 13, 14, 27, 38, 52, 63, 100, 131, 133, 136, 176, 193, 246, 284, 292, 300, 314, 346, 349, 354
Conrad discontinuity, 213
Convergent plate margin, 47, 52, 60, 117, 175, 335

Earthquake foci, 8, 213, 335

Flysch, 23, 49, 62, 97, 98, 109, 140, 209, 222, 231
Fore-arc basin, 60, 117, 150, 246

Gondwana, 12, 13, 14, 317
Gravity anomalies, 97, 272, 319, 325, 328, 342

Hida (metamorphic) terrane (belt), 71, 171, 172, 181, 195

Hidaka (metamorphic) belt, 96, 110, 130, 142, 153
Hidaka Orogeny, 91, 94, 100, 107, 144, 149
High $P(P/T)$ metamorphic rocks, 80, 96, 109, 195, 216

Ishikari (-Teshio) belt, 94, 127, 140
Island arc, 13, 47, 60, 116, 136, 214, 247, 250, 294, 329, 342, 346, 351

Kamuikotan (metamorphic) belt, 96, 109, 142, 150, 262
Koryak Highlands, 49, 63
Kula plate, 52, 241, 257, 294
Kuril arc, 97, 100, 118, 123, 150
Kurosegawa Tectonic Zone, 209, 211, 219, 262

Laramide Orogeny, 31
Lithosphere, 13, 80, 139, 319, 323, 335

Marginal sea, 6, 214
M-discontinuity (Moho), 213, 319, 324
Median Tectonic Line, 196, 207, 232
Melange, 49, 109, 127, 209, 214, 219, 221, 233, 245, 256, 262, 350
Mineoka ophiolite, 246, 249, 254, 263
Mino terrane (belt), 171, 172, 181
Mongolian-Okhotsk system (suture), 63, 66, 74, 80, 84

Nemuro belt, 97, 114, 142

Obduction, 13, 116, 250, 257, 261, 313, 345
Ocean floor basalt, 64, 223, 311
Oceanic plateau, 3, 11, 13, 29, 52
Oceanic reef complex, 196, 204
Oceanic ridge, 29, 52, 82, 136, 254, 257, 344
Oceanic rise, 13
Okhotsk Paleoland, 99, 114, 117, 144
Okhotsk-Chukotka volcanic (magmatic) belt, 45, 49, 62, 66, 75
Olistostrome, 49, 100, 172, 181, 232, 245
Onton-Java Plateau, 4, 5, 13, 320, 325
Ophiolite, 12, 45, 54, 60, 96, 109, 140, 153, 223, 245, 261, 272, 300, 309, 314, 346, 350

Oshima (-Rebun) belt, 107

Pacific plate, 31, 145, 250, 257, 294, 303, 314, 330, 341
Pacifica, 10, 12, 145
Paired metamorphic belts, 91, 149, 346
Paleocurrent, 98, 99, 113, 114, 176, 249
Paleolatitude, 31, 172, 192, 238, 270
Paleomagnetism, 6, 31, 172, 181, 223, 233, 266, 272,
    274, 277, 278, 302
Panthalassa, 32
Pelagic sediments (deposits), 28, 81, 144, 181, 199,
    204, 214, 221, 233, 245, 249, 300
Philippine Sea plate, 226, 246, 257, 285, 302, 305, 330
Pluton, 27, 130, 136, 181, 215

Sambagawa metamorphic belt, 82, 209, 232
Seamount, 13, 29, 38, 52, 196, 204, 223, 263, 311, 314,
    319, 324
Shatsky Rise, 4, 6, 9, 325, 328
Shimanto belt, 181, 207, 217, 221, 226, 232
Sorachi-Yezo belt, 109
South-Anyui (Anyuy) system (suture), 47, 65, 71
Strike-slip fault (mobile zone), 5, 24, 31, 38, 80, 97,
    125, 132, 142, 217, 229, 246, 284, 314, 350
S-type igneous rocks, 75, 157
Subduction, 13, 60, 82, 135, 217, 229, 284, 294, 339,
    349

Subduction complex, 23, 225, 231
Subduction zone, 5, 13, 43, 80, 131, 139, 285, 291, 320,
    330, 335, 345
  Chilean type, 291, 349
  Mariana type, 291, 349
Suture, 10, 27, 38, 43, 96, 131, 209, 215, 219, 347

Terrane, 6, 22, 24, 45, 51
  composite, 23, 45
  disrupted, 23
  metamorphic, 23
  stratified, 22
  tectonostratigraphic, 37, 45, 169
Terrane analysis, 21, 32, 175
Tholeiite, 110, 155, 159, 223, 250, 309
Tokoro belt, 96, 142, 150
Transform fault, 245, 254, 294, 314
Trench, 6, 99, 117, 135, 139, 196, 320, 325, 346
Tripple junction, 246, 257
Turbidite, 94, 98, 100, 112, 114, 222, 232, 249

Uda-Murgal volcanic (magmatic) belt, 62, 71, 80

Volcanic arc, 23, 133, 135, 150

Wrangellia, 6, 23, 24, 25, 27, 32